DYE LASER PRINCIPLES

With Applications

QUANTUM ELECTRONICS–PRINCIPLES AND APPLICATIONS

EDITED BY

PAUL F. LIAO
Bell Communications Research, Inc.
Red Bank, New Jersey

PAUL L. KELLEY
Lincoln Laboratory
Massachusetts Institute of Technology
Lexington, Massachusetts

A complete list of titles in this series appears at the end of this volume.

DYE LASER PRINCIPLES

With Applications

Edited by

F.J. Duarte

Eastman Kodak Company
Research Laboratories
Rochester, New York

Lloyd W. Hillman

Department of Physics
The University of Alabama in Huntsville
Hunstville, Alabama

ACADEMIC PRESS, INC.

Harcourt Brace Jovanovich, Publishers

Boston San Diego New York
London Sydney Tokyo Toronto

This book is printed on acid-free paper. ♾

ACADEMIC PRESS, INC.
1250 Sixth Avenue, San Diego, CA 92101

United Kingdom Edition published by
ACADEMIC PRESS LIMITED
24–28 Oval Road, London NW1 7DX

Library of Congress Cataloging-in-Publication Data

Dye laser principles / edited by Frank J. Duarte, L.W. Hillman.
p. cm. — (Quantum electronics—principles and applications)
Includes bibliographical references.
ISBN 0-12-222700-X (alk. paper)
1. Dye lasers. 2. Quantum electronics. I. Duarte, Frank J.
II. Hillman, Lloyd William, date. III. Series.
QC688.D94 1990
621.36′64—dc20 89-17912
CIP

Printed in the United States of America
90 91 92 93 9 8 7 6 5 4 3 2 1

Contents

Contributors

The numbers in parentheses indicate the pages on which the authors' contributions begin.

M.A. Akerman (413), *Oak Ridge National Laboratory, Oak Ridge, Tennessee 37831*

J.-C. Diels (41), *Department of Physics and Astronomy, University of New Mexico, 800 Yale Boulevard, North East, Albuquerque, New Mexico 87131*

F.J. Duarte (1, 133, 239), *Photographic Research Laboratories, Eastman Kodak Company, Kodak Park, B-59, Rochester, New York 14650-1744*

Leon Goldman (419), *U.S. Naval Hospital, Balboa Park, San Diego, California 92134*

Lloyd W. Hillman (1, 17), *Department of Physics, The University of Alabama in Huntsville, Huntsville, Alabama 35899*

Leo Hollberg (185), *National Institute of Standards and Technology, Boulder, Colorado 80303*

Guilford Jones II (287), *Department of Chemistry, Boston University, Boston, Massachusetts 02215*

David Klick (345), *Massachusetts Institute of Technology, Lincoln Laboratory, Lexington, Massachusetts 02173-0073*

Preface

The dye laser is a remarkable and highly versatile source of tunable coherent radiation. Its success and importance are demonstrated by its wide applicability. From basic science, such as physics and spectroscopy, to medicine and industry, the dye laser has been shown to be an extremely flexible and useful tool. In this book, a number of topics representative of dye laser research and development are considered and discussed. The treatment starts with basic principles of coherence and propagation and then considers the physics and technology of ultrashort pulse generation. The design and theory of dispersive narrow-linewidth pulsed oscillators is then followed by a detailed description of continuous wave dye lasers and a discussion on the technology of pulsed dye lasers. Next, attention is focused on the molecular structure and photophysics of dyes. The book concludes with three additional chapters on industrial applications, laser isotope separation, and medical applications.

It is important to indicate that many of the topics considered here transcend dye lasers and are applicable to lasers in general and other areas of quantum electronics.

The style of presentation is determined by our attempt to write at a level appropriate to seniors and graduate students. However, extensive references given in most of the chapters should also make this book attractive to current researchers.

During the gestation period of this volume, it has been our pleasure to work and interact with a fine group of contributors. Certainly, they should be given credit for all the good features of this book. As editors, we assume responsibility for any shortcomings.

Finally, we would like to thank J. Donaldson for compiling and verifying the references of some chapters. Also, one of us (F. J. D.) is particularly grateful to the U.S. Army Missile Command (Redstone Arsenal, Alabama) for funding some of the work discussed in this book.

F.J. Duarte
Lloyd W. Hillman

Chapter 1

INTRODUCTION*

F.J. Duarte

Photographic Research Laboratories
Photographic Products Group
Eastman Kodak Company
Rochester, New York

and

Lloyd W. Hillman

Department of Physics
The University of Alabama in Huntsville
Huntsville, Alabama

1. INTRODUCTION

Dye lasers are perhaps the most versatile and one of the most successful laser sources known today. Indeed, at the time of the discovery of this class of lasers by Sorokin and Lankard (1966), few could have anticipated their spectacular diversification and their significant contribution to basic physics, chemistry, biology, and additional fields.

* The brief historical section was compiled with the help and cooperation of J.-C. Diels, G. Jones, and L. Hollberg.

ISBN 0-12-222700-X

Dye lasers offer to researchers both pulsed and continuous wave (cw) operation that is tunable from the near-UV to the near-IR. Such versatility emerges from the large choice of molecular dye species available (see Appendix) coupled with the wide variety of excitation sources (see, for example, Table 6.1 in Chapter 6). A closer examination of dye-laser characteristics highlights the availability of the following output modes:

(i) Tunable cw oscillation from ~400 to ~1000 nm.
(ii) Very stable cw single-mode operation at linewidths less than 1 kHz (see, for example, Hough *et al.* (1984)).
(iii) Tunable pulsed oscillation from ~320 to ~1200 nm.
(iv) Conversion efficiencies exceeding 50% for some pulsed laser-pumped dye lasers (see, for example, Morey (1980), Bos (1981)).
(v) High pulse energies. A single coaxial flashlamp-pumped dye laser has been reported to yield 400 J in a 10-μs pulse (Baltakov *et al.*, 1974), and an excimer laser-pumped dye laser has been reported to produce 800 J in a 500-ns pulse (Tang *et al.*, 1987).
(vi) Ultrashort pulses: 27 fs (Valdmanis *et al.*, 1985).
(vii) Very long pulses (well into the μs regime under flashlamp pumping).
(viii) High pulse repetition frequency (prf): in excess of 10 kHz under copper vapor laser excitation (see, for example, Duarte and Piper (1984)).
(ix) High average powers. The ability of dye lasers to operate at high average powers is partly due to its liquid state, which facilitates the important process of heat removal. Single flashlamp dye lasers have been reported to provide up to 200 W of average power (Mazzinghi *et al.*, 1981), and a flashlamp-pumped system has demonstrated 1.4 J per pulse at 850 Hz (~1.2 kW average) (Morton and Draggoo, 1981). The copper-laser-pumped dye-laser system at the Lawrence Livermore National Laboratory has demonstrated output powers in the 600–800-W range (Paisner, 1989).
(x) Broad bandwidths.
(xi) Narrow-linewidth pulsed operation including single-longitudinal-mode oscillation at low and high prf (see, for example, Littman (1978), Saikan (1978), Bos (1981), Bernhardt and Rasmussen (1981), Duarte and Piper (1984)).

This unique flexibility furnishes many economic and engineering design advantages. Emission at any particular wavelength is achieved by simply changing the dye, which offers significant savings in capital and operating cost. Also, numerous alternatives are available in system integration.

Further testimony of the success of the dye laser can be provided by a broad listing of its applications:

(i) industrial applications,
(ii) medical applications,
(iii) military applications,
(iv) large-scale laser isotope separation (LIS),
(v) study of fundamental physics,
(vi) numerous types of spectroscopy techniques,
(vii) laser radar as well as light detection and ranging (lidar).

The large number and wide range of dye-laser applications certainly provide proof of their flexibility and versatility.

In this introduction, we limit the discussion to some aspects of broad interest in dye-laser physics and technology.

2. BRIEF HISTORICAL SURVEY

The field of dye lasers offers a rich and extensive literature. In this regard, we note that Magyar (1974) in his classified bibliography of journal dye-laser publications for the early period of 1966–1972 included 454 references! This number represents only publications in the infancy of dye lasers. It is our opinion that a fair historical perspective requires a thorough consideration of a large number of publications. Such proper chronological review is beyond the scope of this introduction.

In this section, we provide a limited and brief historical description of the field of dye lasers as illustrated by the following subfields: pulsed dye lasers, cw dye lasers, ultrafast dye lasers, and laser dyes.

2.1. Pulsed Dye Lasers

The first dye laser was the ruby laser-pumped dye laser introduced by Sorokin and Lankard (1966). This report was quickly followed by the papers of Schäfer *et al.* (1966) and Spaeth and Bortfeld (1966). Shortly afterward, dye-laser excitation at shorter wavelengths utilizing second harmonic emission from solid-state lasers, such as ruby and neodymium, was reported (see, for example, Sorokin *et al.* (1967), Soffer and McFarland (1967), Schäfer *et al.* (1967), Kotsubanov *et al.* (1968, 1969), Wallace (1971)).

The use of the nitrogen laser as a direct UV excitation source for dye lasers was reported by several authors within a short period (Lankard and von Gutfeld, 1969; Lidholt, 1970; Capelle and Phillips, 1970; Myer *et al.*, 1970; Broida and Haydon, 1970). The excimer laser was introduced as a dye-laser pump a few years later via the KrF laser (see, for example, Sutton and Capelle (1976) and Godard and de Witte (1976)) and the XeCl laser (see, for example, Uchino *et al.* (1979)).

Laser-pumped oscillator-amplifier configurations were introduced by Hänsch *et al.* (1971) and Itzkan and Cunningham (1972).

High-prf operation of dye lasers was demonstrated using a copper-vapor laser as an excitation source by Hargrove and Kan (1977) and Pease and Pearson (1977). These authors reported dye-laser oscillation at a prf of 6 kHz.

The first flashlamp-pumped dye laser was reported by Sorokin and Lankard (1967) and Schmidt and Schäfer (1967). Flashlamp-pumped dye lasers were operated in an oscillator-amplifier configuration by Huth (1970) and Flamant and Meyer (1971). A master-oscillator forced-oscillator arrangement was described by Magyar and Schneider-Muntau (1972).

Frequency doubling of tunable dye-laser radiation using an ADP crystal was introduced by Bradley *et al.* (1971) and Dunning *et al.* (1972). Raman shifting of dye-laser emission utilizing compressed hydrogen was announced by Schmidt and Appt (1972).

Dye lasers in the solid state were introduced by Soffer and McFarland (1967) and Peterson and Snavely (1968). Lasing of dyes in the vapor phase has also been reported (see, for example, Steyer and Schäfer (1974)).

In the area of frequency selectivity, the following developments may be considered important:

(i) mirror-grating resonator (Soffer and McFarland, 1967),
(ii) mirror-grating resonator in conjunction with an intracavity etalon (Bradley *et al.*, 1968),
(iii) multiple-prism tuning (Strome and Webb, 1971; Schäfer and Müller, 1971),
(iv) telescopic narrow-linewidth resonator (Hänsch, 1972).
(v) single-prism expander (Myers, 1971; Stokes *et al.*, 1972; Hanna *et al.*, 1975),
(vi) pressure tuning of telescopic dye-laser resonator (Wallenstein and Hänsch, 1974),
(vii) grazing-incidence narrow-linewidth resonator (Shoshan *et al.*, 1977; Littman and Metcalf, 1978),
(viii) multiple-prism laser-beam expansion (Novikov and Tertyshnik, 1975; Klauminzer, 1978; Wyatt, 1978; Duarte and Piper, 1980),
(ix) synchronous tuning of grazing-incidence dye laser (Liu and Littman, 1981),
(x) prism preexpanded grazing-incidence oscillator (Duarte and Piper, 1981),

(xi) dispersion theory of multiple-prism laser-beam expansion (Duarte and Piper, 1982),

(xii) analytical and numerical design of achromatic multiple-prism laser-beam expanders (Barr, 1984; Duarte, 1985; Trebino, 1985),

2.2. Continuous-Wave Dye Lasers

The first cw dye laser was demonstrated by Peterson *et al.* (1970). These authors utilized rhodamine 6G dye excited by the $4p^4D_{5/2} \rightarrow 4s^2P_{3/2}$ transition (at 514.5 nm) of an Ar ion laser.

Important developments in cw dye-laser technology include the demonstration of a three-mirror folded cavity system by Kohn *et al.* (1971), the introduction of the dye jet by Runge and Rosenberg (1972), and the report on the use of birefringent filters (Bloom, 1974).

One of the unique features of the cw dye laser is its ability to provide tunable single-frequency oscillation (see, for example, Hercher and Pike (1971) and Barger *et al.* (1973)). A further important landmark has been the demonstration of very stable single–longitudinal-mode operation at subkilohertz linewidths (Drever *et al.*, 1983; Hough *et al.*, 1984).

Efficient high-power operation of single-frequency cw dye lasers has been reported by Schröder *et al.* (1977) and Jarrett and Young (1979). The latter authors reported single-frequency oscillation at an output power of 0.9 W using rhodamine 6G and an Ar ion laser pump providing 4 W at 514.5 nm. Johnston *et al.* (1982) reported on a stabilized single-frequency ring dye-laser system yielding 5.6 W using rhodamine 6G dye and a 24 W Ar ion laser pump. This dye laser was designed to provide continuous wavelength coverage throughout the visible spectrum using several dyes.

The wavelength range of cw dye lasers has been extended to the UV spectral region using harmonic and sum frequency generation (see, for example, Blit *et al.* (1978)). Pine (1974) utilized mixing in $LiNbO_3$ to obtain tunable radiation in the infrared.

2.3. Ultrashort Pulse Dye Lasers

Passive mode locking in pulsed solid-state lasers, using liquid-saturable absorbers, was first reported by Mocker and Collins (1965) in a ruby laser and by DeMaria *et al.* (1966) in a Nd^{3+} glass laser. Self-mode locking in a rhodamine 6G flashlamp-pumped dye laser, using the saturable absorber 3,3′-diethyloxadicarbocyanine iodide, was reported by Schmidt and Schäfer (1968).

Passive mode locking in a cw dye laser was described by Ippen *et al.* (1972). The colliding pulse concept was introduced by Ruddock and Bradley (1976), and it was applied to a ring dye laser by Fork *et al.* (1981) who reported a pulse length of 90 fs. The first ring dye laser utilizing intracavity pulse compression (dispersion compensation) was introduced by Dietel *et al.* (1983) who reported pulse lengths of 53 fs.

Using intracavity compression with negative dispersion, Valdmanis *et al.* (1985) reported 27-fs pulses and Fork *et al.* (1987) reported pulses as short as 6 fs utilizing amplification and extracavity compression.

2.4. Laser Dyes

The first organic dye reported to lase was chloro-aluminum phthalocyanine, utilized by Sorokin and Lankard (1966) in their transverse ruby-laser-pumped dye laser. As of 1990, perhaps the most successful and widely used dye is rhodamine 6G, which was first employed by Sorokin *et al.* (1967) under transverse excitation from a frequency-doubled ruby laser, and Sorokin and Lankard (1967) using flashlamp pumping. The solvent used by these authors was ethyl alcohol. Another very important class of dyes is the coumarins, which were introduced in flashlamp-pumped dye-laser systems by Snavely *et al.* (1967) and Snavely and Peterson (1968).

Efficient and photochemically stable coumarin dyes were introduced by Reynolds and Drexhage (1975). The importance of rigidized molecular structures to the decrease of laser-output variations as a function of temperature was highlighted by Drexhage (1977).

For comprehensive listings of dyes and corresponding references, readers should consult Drexhage (1977) and Maeda (1984).

The use of oxygen as a triplet quencher, an important effect in flashlamp-pumped dye lasers, is discussed and evaluated by Snavely and Schäfer (1969), Marling *et al.* (1970a,b), and Schäfer and Ringwelski (1973).

Among the advances related to basic photochemistry and photophysics of organic dyes is the evolution of dye structures. New dyes have emerged in which offending, photochemically reactive positions have been removed, substituted, or otherwise protected. New classes of dyes such as the pyridinium oxazoles (Fletcher *et al.*, 1987; Kauffman and Bentley, 1988) and the bimanes (Pavlopoulos *et al.*, 1986) appear promising for lasing in the blue–green spectral region. A variety of tethered dyes have also evolved (Schäfer, 1983) in which a basic chromophore is modified with groups that act as energy-transfer agents (for down-shifting) or triplet quenchers.

The influences of solvent medium on laser-dye photophysics are now better understood. Several studies have documented the solvent tuning of fluorescence properties which include not only the wavelength and Stokes-shift of emission, but also, in some cases, the fluorescence yield and lifetime (Jones *et al.* 1985; Grenci *et al.*, 1986).

Considerable progress has also been made in the elucidation of mechanisms of photochemical degradation (see Chapter 7, Section 3).

Dyes are also used as nonlinear media in the vapor phase (Aleksandrov *et al.*, 1988), liquid phase (see, for example, Hoffman *et al.*, 1989), and solid-state phase (see, for example, Knabke *et al.*, 1989).

3. AIMS OF THIS BOOK

One of the principal aims of this book is to provide the means for a pedagogical introduction to the field of dye lasers for seniors and graduate students. An additional goal is to provide an updated description of the field of dye lasers for workers in the field who are using dye lasers as research and industrial tools. In this regard, this book may be cataloged as a monograph and textbook.

These aims may appear as broad and excessively ambitious. However, we believe that the diversity and scope of the field justifies such optimism. For instance, we as physicists have little or limited knowledge on the chemistry of dyes. Similarly, someone specializing in the synthesis of laser dyes may be quite unfamiliar with some of the engineering and technological issues involved in dye lasers. In addition, those involved in the technology can always find some refreshing idea in the more fundamental aspects of the physics of dye lasers.

At this stage we note that many of the topics discussed in this book transcend the context of the dye laser and are applicable to lasers in general. Here, we refer with some specificity to the areas of excitation mechanisms, coherence and propagation, frequency control in cw lasers, spectral diagnostics, theory and cavity calculations in ultrafast lasers, ultrashort pulse measurements, techniques of linewidth narrowing in pulsed lasers, and dispersion theory. In this regard, this text will provide a broad base of knowledge to the student. Hopefully, the attempt to present the subject matter in a pedagogical order in conjunction with problems provided at the end of each chapter will help to accomplish this goal.

3.1. Book Organization

This book contains several topics important to present dye-laser research and development. In addition, as mentioned earlier, many of the topics considered here are also applicable to lasers in general. Ideally, each of the topics treated can be approached independently, and as such, there is no recommended order of study. In this regard, the only chapter that is considered of a fundamental nature, and that may be recommended to have priority, is Chapter 2. This chapter provides a review of basic excitation mechanisms, coherence, propagation, and other fundamental issues of physics associated with lasers.

Chapter 3 gives a thorough review and perspective on ultrafast lasers and techniques of pulse compression. This chapter introduces the reader to passive mode locking and the theory associated with ultrafast lasers including the density matrix, the colliding pulse effect, cavity equations, dispersion, coherence effects, and solitons. Other sections discuss diagnostic techniques and hybrid mode locking.

Narrow-linewidth pulsed-dye-laser oscillators are considered in Chapter 4. This chapter introduces basic concepts of linewidth narrowing and describes the performance of different oscillator designs. The intracavity dispersion of oscillators incorporating multiple-prism grating assemblies is considered in detail, including the design of zero-dispersion multiple-prism laser-beam expanders. This chapter then considers the application of the generalized dispersion theory to pulse compression in ultrafast lasers. The use of grazing-incidence and multiple-prism grating techniques in a variety of gas lasers is also discussed.

The physics and technology of continuous-wave dye lasers is considered in Chapter 5. This chapter discusses issues associated with the design of cw dye lasers, including constraints such as triplet population, lifetime, and heat removal. Resonator design and elements, such as cells, dye jets, and pump sources, are also described. Frequency-control and frequency-stabilization methods are considered in detail. In this area, the chapter expands on many approaches, including atomic and molecular locks, polarization, rf heterodyne, pzt, galvo, and electro-optical methods. The section on laser diagnostics includes descriptions of several elements, such as spectrum analyzers, wavelength meters, and Fizeau interferometers.

Issues of technology associated with pulsed dye lasers are discussed in Chapter 6. The material in this chapter may be considered complementary to Chapter 4. Topics considered here include pump sources, excitation geometries, dye cells, amplified spontaneous emission (ASE), polarization effects, thermal effects, and tuning methods. In the description of laser-excitation sources, information on the spectral characteristics of these sources is provided in addition to typical efficiencies and prf figures. Sources considered are excimer lasers, nitrogen lasers, HgBr lasers, recombination lasers, metal-vapor lasers, and solid-state lasers. Emission characteristics of different flashlamp-pumped dye-laser configurations are also provided. Narrow-linewidth laser-pumped oscillator-amplifier systems and multistage flashlamp-pumped systems are also described.

Chapter 7 deals with issues of dye molecular structure and stability. In this regard, the author considers the fundamentals of photophysics including topics such as solvent effects, triplets and triplet energy transfer, and singlet energy transfer between dyes. Topics on dye photodegradation include mechanisms of photoinduced electron transfer, photochemistry of rhodamine and coumarin dyes, and photodegradation under lasing conditions. The properties of dye in unusual media such as detergents, cyclodextrins, and polymers and plastics are also discussed.

The last three chapters are meant as an overview toward important applications of dye lasers and as such may be considered an independent set of topics. Chapter 8 describes the use of excimer laser-pumped dye lasers in industry. This contribution deals with the technological requirements for industrial dye lasers and the application of these dye lasers in material processing and chemical production. Chapter 9 provides a literature survey on the use of dye lasers in the field of laser isotope separation (LIS). Chapter 10 deals with medical applications of dye lasers. This contribution describes the class of dye lasers of interest to medicine, and it includes particular applications such as cancer phototherapy, lithotripsy, removal of birth marks, and angioplasty.

The appendix identifies commonly used dyes listed in order of lasing wavelength. The molecular structure is included to help identify the dye compound without ambiguity. The information given in the table has been reproduced from *KODAK Laser Dyes* (Birge, 1987) courtesy of Eastman Kodak Company. Also in this appendix, we include a list of abbreviations used for mode locking dyes in Chapter 3.

3.2. Additional Topics

The field of dye lasers and related subjects has progressed at a considerable rate in the past few years, especially in the area of applications. In this regard, it is virtually impossible to provide a comprehensive and extensive review of the subject in a limited space. Thus, our consideration of applications has been limited only to certain areas that we felt had not been covered elsewhere. This limitation is particularly relevant in the case of spectroscopic applications and techniques.

In the area of dye lasers we devote only one subsection (2.6 in Chapter 6) to the coverage of flashlamp-pumped dye lasers, which deserve at least a complete chapter. Similarly, the review of dye-laser Raman shifting is rather limited, and little is said about dye lasers in the solid-state and gaseous phases. In addition, the subject of distributed feedback dye lasers is mentioned only in a peripheral sense.

4. CHALLENGES TO DYE LASERS

As of 1990, dye lasers face several definite challenges. First, we have genuine questions of fundamental interest to the field of physics of dye lasers. Then we address more practical issues such as the problem of photodegradation. A further question is the ability of the dye laser to survive in certain regions of the spectrum where other active media, such as tunable solid-state lasers, may provide an attractive alternative.

First, we should state that a better understanding of physics issues, such as intensity and linewidth instabilities, not only may help us to engineer and design better dye lasers but could also provide valuable contributions to the field of lasers and quantum electronics in general.

The question of photodegradation in dye lasers brings several very interesting issues to the surface. First, there is no doubt that from an academic perspective one desires developments at the molecular level that could eventually lead to a new generation of very stable dye species. To a certain degree, some progress has been achieved in this area with new, more stable, more efficient, and more soluble dyes (Pavlopoulos *et al.*, 1989; Chen *et al.*, 1988).

In addition to the perspective just advocated, one may approach the task by designing more reliable systems. In this approach, the logic has two alternatives; the most important element is to ask the right question.

As an example of the first alternative, we turn to the excitation of rhodamine 6G by green lasers such as the copper and frequency-doubled Nd:YAG laser. In both cases the first singlet electronic level (S_1) of the dye molecules is excited via the $S_0 \rightarrow S_1$ transition. For both excitation sources, dye-laser efficiencies as high as 55% have been reported (Morey, 1980; Bos 1981). In addition, rhodamine 6G under high-prf copper-vapor-laser excitation has been found to be very stable (see, for example, Duarte and Piper (1984); Paisner (1989)). In other words, excitation of higher electronic singlet states such as S_2 and S_3 leads to lower efficiencies and higher photodegradation.

An example of the second alternative is the case of flashlamp-pumped excitation whereby efficiencies and operational reliability are improved. This approach involves the successful use of converter dyes (Everett *et al.*, 1986) and the utilization of carbon filters in the flow system of the active medium (Everett and Zollars, 1988). Without going into further detail, let us reiterate that it is not very difficult to design efficient laser-pumped or flashlamp-pumped systems where the dye-lifetime issue can be neutralized.

The third topic concerns the perceived challenge to dye lasers in the near infrared by tunable solid-state lasers. It is our opinion that the introduction of this type of solid-state laser enhances the options available to users and thus represents a positive development. In this regard, it is important to consider output characteristic requirements in conjunction with issues of engineering and cost. For instance, it is clear that for certain applications, such as spectroscopy in the near IR, tunable solid-state lasers such as alexandrite and titanium sapphire are quite attractive.

In the case of pulsed emission in the near IR, it may be useful to remember that this spectral region can be reached fairly efficiently by utilizing Raman shifting of tunable emission in the visible, where dye lasers have been shown to provide very efficient single-longitudinal-mode oscillation (see, for example, Bos (1981)). The option of Raman-shifting visible-dye-laser emission should also be considered if a particular application requires tunable radiation, somewhere in the near IR, at a few hundred Joules per pulse or at a few hundred watts of average power. Furthermore, we should not ignore the fact that dyes provide an active medium that is relatively inexpensive and easy to replace.

REFERENCES

Aleksandrov, K.S., Aleksandrovsky, A.S., Karpov, S.V., Lukinykh, V.F., Myslivets, S.A., Popov, A.K., and Slabko, V.V. (1988). Dye vapors—new nonlinear optical material for VIS, UV and VUV generation?. In *Proceedings of the International Conference on Lasers '87* (Duarte, F.J., ed.). STS Press, McLean, Va., pp. 378–380.

Baltakov, F.N., Barikhin, B.A., and Sukhanov, L.V. (1974). 400-J pulsed laser using a solution of rhodamine 6 G in ethanol. *JETP Lett.* **19,** 174–175.

Barger, R.L., Sorem, M.S., and Hall, J.L. (1973). Frequency stabilization of a cw dye laser. *Appl. Phys. Lett.* **22,** 573–575.

Barr, J.R.M. (1984). Achromatic prism beam expanders. *Opt. Commun.* **51,** 41–46.

Bernhardt, A.F., and Rasmussen, P. (1981). Design criteria and operating characteristics of a single-mode pulsed dye laser. *Appl. Phys. B* **26,** 141–146.

Birge, R.R. (1987). *KODAK Laser Dyes*, KODAK publication JJ-169, Eastman Kodak, Rochester, New York.

Blit, S., Weaver, E.G., Rabson, T.A., and Tittel, F.K. (1978). Continuous wave uv radiation tunable from 285 to 400 nm by harmonic and sum frequency generation. *Appl. Opt.* **17,** 721–723.

Bloom, A.L. (1974). Modes of a laser resonator containing tilted birefringent plates. *J. Opt. Soc. Am.* **64,** 447–452.

Bos, F. (1981). Versatile high-power single–longitudinal-mode pulsed dye laser. *Appl. Opt.* **20,** 1886–1890.

Bradley, D.J., Gale, G.M., Moore, M., and Smith, P.D. (1968). Longitudinally pumped, narrow-band continuously tunable dye laser. *Phys. Lett.* **26A,** 378–379.

Bradley, D.J., Nicholas, J.V., and Shaw, J.R.D. (1971). Megawatt tunable second harmonic and sum frequency generation at 280 nm from a dye laser. *Appl. Phys. Lett.* **19,** 172–173.

Broida, H.P., and Haydon, S.C. (1970). Ultraviolet laser emission of organic liquid scintillators using a pulsed nitrogen laser. *Appl. Phys. Lett.* **16,** 142–144.

Capelle, G., and Phillips, D. (1970). Pumping organic dyes with a nitrogen laser. *Appl. Opt.* **9,** 517–518.

Chen, C.H., Fox, J.L., Duarte, F.J., and Ehrlich, J.J. (1988). Lasing characteristics of new coumarin-analog dyes: broadband and narrow-linewidth performance. *Appl. Opt.* **27,** 443–445.

DeMaria, A.J., Stetser, D.A., and Heynau, H. (1966). Self mode-locking of lasers with saturable absorbers. *Appl. Phys. Lett.* **8,** 174–176.

Dietel, W., Fontaine, J.J., and Diels, J.-C. (1983). Intracavity pulse compression with glass: a new method of generating pulses shorter than 60 fs. *Opt. Lett.* **8,** 4–6.

Drever, R.W.P., Hall, J.L., Kowalski, F.V., Hough, J., Ford, G.M., Munley, A.J., and Ward, H. (1983). Laser phase and frequency stabilization using an optical resonator. *Appl. Phys. B* **31,** 97–105.

Drexhage, K.H. (1977). Structure and properties of laser dyes. In *Dye Lasers* (Schafer, F.P., ed.). Springer-Verlag, New York, pp. 144–193.

Duarte, F.J. (1985). Note on achromatic multiple-prism beam expanders. *Opt. Commun.* **53,** 259–262.

Duarte, F.J., and Piper, J.A. (1980). A double-prism beam expander for pulsed dye lasers. *Opt. Commun.* **35,** 100–104.

Duarte, F.J., and Piper, J.A. (1981). Prism preexpanded grazing-incidence grating cavity for pulsed dye lasers. *Appl. Opt.* **20,** 2113–2116.

Duarte, F.J., and Piper, J.A. (1982). Dispersion theory of multiple-prism beam expanders for pulsed dye lasers. *Opt. Commun.* **43,** 303–307.

Duarte, F.J., and Piper, J.A. (1984). Narrow linewidth, high prf copper laser-pumped dye-laser oscillators. *Appl. Opt.* **23,** 1391–1394.

Dunning, F.B., Stokes, E.D., and Stebbings, R.F. (1972). The efficient generation of coherent radiation continuously tunable from 2500 Å to 3250 Å. *Opt. Commun.* **6,** 63–66.

Everett, P.N., Aldag, H.R., Ehrlich, J.J., Janes, G.S., Klimek, D.E., Landers, F.M., and Pacheco, D.P. (1986). Efficient 7-J flashlamp-pumped dye laser at 500 nm wavelength. *Appl. Opt.* **25,** 2142–2147.

Everett, P.N., and Zollars, B.G. (1988). Engineering of a multi-beam 300-watt flashlamp-pumped dye-laser system. In *Proceedings of the International Conference on Lasers '87* (Duarte, F.J., ed.). STS Press, McLean, Va., pp. 291–296.

Flamant, P., and Meyer, Y.H. (1971). Absolute gain measurements in a multistage dye amplifier. *Appl. Phys. Lett.* **19,** 491–493.

Fletcher, A.N., Henry, R.A., Pietrak, M.E., Bliss, D.E., and Hall, J.H. (1987). Laser dye stability, part 12. The pyridinium salts. *Appl. Phys. B.* **43,** 155–160.

Fork, R.L., Greene, B.I., and Shank, C.V. (1981). Generation of optical pulses shorter than 0.1 psec by colliding pulse mode locking. *Appl. Phys. Lett.* **38,** 671–672.

Fork, R.L., Brito Cruz, C.H., Becker, P.C., and Shank, C.V. (1987). Compression of optical pulses to six femtoseconds by using cubic phase compensation. *Opt. Lett.* **12,** 483–485.

Godard, B., and de Witte, O. (1976). Efficient laser emission in paraterphenyl tunable between 323 and 364 nm. *Opt. Commun.* **19,** 325–328.

Grenci, S.M., Bird, G.R., Keelan, B.W., and Zewail, A.H. (1986). Practical broad-band tuning of dye lasers by solvent shifting. *Laser Chem.* **6,** 361–371.

Hanna, D.C., Karkkainen, P.A., and Wyatt, R. (1975). A simple beam expander for frequency narrowing of dye lasers. *Opt. Quantum Electron.* **7,** 115–119.

Hänsch, T.W. (1972). Repetitively pulsed tunable dye laser for high resolution spectroscopy. *Appl. Opt.* **11,** 895–898.

Hänsch, T.W., Varsanyi, F., and Schawlow, A.L. (1971). Image amplification by dye lasers. *Appl. Phys. Lett.* **18,** 108–110.

Hargrove, R.S., and Kan, T. (1977). Efficient, high average power dye amplifiers pumped by copper vapor lasers. *IEEE J. Quantum Electron.* **QE-13,** 28D.

Hercher, M., and Pike, H.A. (1971). Single mode operation of a continuous tunable dye laser. *Opt. Commun.* **3,** 346–348.

Hoffman, R.C., Stetyick, K.A., Potember, R.S., and McLean, D.G. (1989). Reverse saturable absorbers: indanthrone and its derivatives. *J. Opt. Soc. Am. B.* **6,** 772–777.

Hough, J., Hils, D., Rayman, M.D., Ma, L.S., Hollberg, L., and Hall, J.L. (1984). Dye-laser frequency stabilization using optical resonators. *Appl. Phys. B.* **33,** 179–185.

Huth, B.G. (1970). Direct gain measurements of an organic dye amplifier. *Appl. Phys. Lett.* **16,** 185–188.

Ippen, E.P., Shank, C.V., and Dienes, A. (1972). Passive mode locking of the cw dye laser. *Appl. Phys. Lett.* **21,** 348–350.

Itzkan, I., and Cunningham, F.W. (1972). Oscillator-amplifier dye-laser system using N_2 laser pumping. *IEEE J. Quantum Electron.* **QE-8,** 101–105.

Jarrett, S.M., and Young, J.F. (1979). High-efficiency single-frequency cw ring dye laser. *Opt. Lett.* **4,** 176–178.

Johnston, T.F., Brady, R.H., and Proffitt, W. (1982). Powerful single-frequency ring dye laser spanning the visible spectrum. *Appl. Opt.* **21,** 2307–2316.

Jones, G.II, Jackson, W.R., Choi, C.Y., and Bergmark, W.R. (1985). Solvent effects on emission yield and lifetime for coumarin laser dyes. Requirements for a rotary decay mechanism. *J. Phys. Chem.* **89,** 294–300.

Kauffman, J.M. and Bentley, J.H. (1988). Effects of various anions and zwitterions on the lasing properties of a photostable cationic laser dye. *Laser Chem.* **8,** 49–59.

Klauminzer, G.K. (1978). Optical beam expander for dye laser. U.S. Patent no. 4, 127, 828.

Knabke, G., Franke, H., and Frank, W.F.X. (1989). Electro-optical properties of a nonlinear dye in poly(methyl methacrylate) and poly(α-methyl styrene). *J. Opt. Soc. Am. B* **6,** 761–765.

Kohn, R.L., Shank, C.V., Ippen, E.P., and Dienes, A. (1971). An intracavity-pumped cw dye laser. *Opt. Commun.* **3,** 177–178.

Kotsubanov, V.D., Naboikin, Y.V., Ogurtsova, L.A., Podgornyi, A.P., and Pokrovskaya, F.S. (1968). Laser action in solutions of organic luminophors in the 400–650 nm range. *Opt. Spectrosk* **25,** 406–410.

Kotsubanov, V.D., Naboikin, Y.V., Ogurtsova, L.A., Podgornyi, A.P., and Pokrovskaya, F.S. (1969). Xanthene dye series laser excited by second-harmonic radiation from a neodymium laser. *Sov. Phys.—Tech. Phys.* **13,** 923–924.

Lankard, J.R., and von Gutfeld, R.J. (1969). Organic lasers excited by a pulsed N_2 laser. *IEEE J. Quantum Electron.* **QE-5,** 625.

Lidholt, L.R. (1970). Dye laser pumped by an ultraviolet nitrogen laser. *IEEE J. Quantum Electron.* **QE-6,** 162.

Littman, M.G. (1978). Single-mode operation of grazing-incidence pulsed dye laser. *Opt. Lett.* **3,** 138–140.

Littman, M.G., and Metcalf, H.J. (1978). Spectrally narrow pulsed dye laser without beam expander. *Appl. Opt.* **17,** 2224–2227.

Liu, K., and Littman, M.G. (1981). Novel geometry for single-mode scanning of tunable lasers. *Opt. Lett.* **6,** 117–118.

Maeda, M. (1984). *Laser Dyes.* Academic Press, New York.

Magyar, G. (1974). Dye lasers—a classified bibliography 1966–1972. *Appl. Opt.* **13,** 25–45.

Magyar, G., and Schneider-Muntau, H.J. (1972). Dye laser forced oscillator. *Appl. Phys. Lett.* **20,** 406–408.

Marling, J.B., Gregg. D.W., and Thomas, S.J. (1970a). Effect of oxygen on flashlamp-pumped organic-dye lasers. *IEEE J. Quantum Electron.* **QE-6,** 570–572.

Marling, J.B., Gregg, D.W., and Wood, L. (1970b). Chemical quenching of the triplet state in flashlamp-excited liquid organic lasers. *Appl. Phys. Lett.* **17,** 527–530.

Mazzinghi, P., Burlamacchi, P., Matera, M., Ranea-Sandoval, H.F., Salimbeni, R., and Vanni, U. (1981). A 200 W average power, narrow bandwidth, tunable waveguide dye laser. *IEEE J. Quantum Electron.* **QE-17**, 2245–2249.

Mocker, H.W., and Collins, R.J. (1965). Mode competition and self-locking effects in a Q-switched ruby laser. *Appl. Phys. Lett.* **7**, 270–273.

Morey, W.W. (1980). Copper vapor laser pumped dye laser. In *Proceedings of the International Conference on Lasers '79* (Corcoran, V.J., ed.). STS Press, McLean, Va., pp. 365–373.

Morton, R.G., and Draggoo, V.G. (1981). Reliable high average power high pulse energy dye laser *IEEE J. Quantum Electron.* **QE-17** (12), 222.

Myer, J.A., Johnson, C.L., Kierstead, E., Sharma, R.D., and Itzkan, I (1970). Dye laser stimulation with a pulsed N_2 laser line at 3371 Å. *Appl. Phys. Lett.* **16,** 3–5.

Myers, S.A. (1971). An improved line narrowing technique for a dye laser excited by a nitrogen laser. *Opt. Commun.* **4,** 187–189.

Novikov, M.A., and Tertyshnik, A.D. (1975). Tunable dye laser with a narrow emission spectrum. *Sov. J. Quantum Electron.* **5,** 848–849.

Paisner, J.A. (1989). High power dye laser technology (panel discussion). In *Proceedings of the International Conference on Lasers '88* (Sze, R.C., and Duarte, F.J., eds.). STS Press, McLean, Va., pp. 773–790.

Pavlopoulos, T.G., Boyer, J.H., Politzer, I.R., and Lau, C.M. (1986). Laser action from syn-(methyl, methyl)bimane. *J. Appl. Phys.* **60,** 4028–4030.

Pavlopoulos, T.G., Shah, M., and Boyer, J.H. (1989). Efficient laser action from 1,3,5,7,8-pentamethylpyrromethene-BF_2 complex and its disodium 2,6-disulfonate derivative. *Opt. Commun.* **70,** 425–427.

Pease, A.A., and Pearson, W.M. (1977). Axial mode structure of a copper vapor pumped dye laser. *Appl. Opt.* **16,** 57–60.

Peterson, O.G., and Snavely, B.B. (1968). Stimulated emission from flashlamp-excited organic dyes in polymethyl methacrylate. *Appl. Phys. Lett.* **12,** 238–240.

Peterson, O.G., Tuccio, S.A., and Snavely, B.B. (1970). CW operation of an organic dye solution laser. *Appl. Phys. Lett.* **17,** 245–247.

Pine, A.S. (1974). Doppler-limited molecular spectroscopy by difference-frequency mixing. *J. Opt. Soc. Am.* **64,** 1683–1690.

Reynolds, G.A., and Drexhage, K.H. (1975). New coumarin dyes with rigidized structure for flashlamp-pumped dye lasers. *Opt. Commun.* **13,** 222–225.

Ruddock, I.S., and Dradley, D.J. (1976). Bandwidth-limited subpicosecond pulse generation in mode-locked cw dye lasers. *Appl. Phys. Lett.* **29,** 296–297.

Runge, P.K., and Rosenberg, R. (1972). Unconfined flowing-dye films for cw dye lasers. *IEEE J. Quantum Electron.* **QE-8,** 910–911.

Saikan, S. (1978). Nitrogen-laser-pumped single-mode dye laser. *Appl. Phys.* **17,** 41–44.

Schäfer, F.P. (1983). New developments in laser dyes. *Laser Chem.* **3,** 265–278.

Schäfer, F.P., and Müller, H. (1971). Tunable dye ring-laser. *Opt. Commun.* **2,** 407–409.

Schäfer, F.P., and Ringwelski, L. (1973). Triplet quenching by oxygen in a rhodamine 6 G laser. *Z. Naturforsch.* **28a,** 792–793.

Schäfer, F.P., Schmidt, W., and Volze, J. (1966). Organic dye solution laser. *Appl. Phys. Lett.* **9,** 306–309.

Schäfer, F.P., Schmidt, W., and Marth, K. (1967). New dye lasers covering the visible spectrum. *Phys. Lett.* **24A,** 280–281.

Schmidt, W., and Appt. W. (1972). Tunable stimulated Raman emission generated by a dye laser. *Z. Naturforsch.* **27a,** 1373–1375.

Schmidt, W., and Schäfer, F.P. (1967). Blitzlampengepumpte farbstofflaser. *Z. Naturforschg.* **22a**, 1563–1566.
Schmidt, W., and Schäfer, F.P. (1968). Self-mode-locking of dye-lasers with saturable absorbers. *Phys. Lett.* **26A**, 558–559.
Schröder, H.W., Stein, L., Frölich, D., Fugger, B., and Welling, H. (1977). A high-power single-mode cw dye ring laser. *Appl. Phys.* **14**, 377–380.
Shoshan, I., Danon, N.N., and Oppenheim, U.P. (1977). Narrowband operation of a pulsed dye laser without intracavity beam expansion. *J. Appl. Phys.* **48**, 4495–4497.
Snavely, B.B., Peterson, O.G., and Reithel, R.F. (1967). Blue laser emission from a flashlamp-excited organic dye solution. *Appl. Phys. Lett.* **11**, 275–276.
Snavely, B.B., and Peterson, O.G. (1968). A-2-experimental measurement of the critical population inversion for the dye solution laser. *IEEE J. Quantum Electron.* **QE-4**, 540–545.
Snavely, B.B., and Schäfer, F.P. (1969). Feasibility of cw operation of dye-lasers. *Phys. Lett.* **28A**, 728–729.
Soffer, B.H., and McFarland, B.B. (1967). Continuously tunable, narrow-band organic dye lasers. *Appl. Phys. Lett.* **10**, 266–267.
Sorokin, P.P., and Lankard, J.R. (1966). Stimulated emission observed from an organic dye, chloro-aluminum phthalocyanine. *IBM J. Res. Develop.* **10**, 162–163.
Sorokin, P.P., and Lankard, J.R. (1967). Flashlamp excitation of organic dye lasers: a short communication. *IBM J. Res. Develop.* **11**, 148.
Sorokin, P.P., Lankard, J.R., Hammond, E.C., and Moruzzi, V.L. (1967). Laser-pumped stimulated emission from organic dyes: experimental studies and analytical comparisons. *IBM J. Res. Develop.* **11**, 130–147.
Spaeth, M.L., and Bortfeld, D.P. (1966). Stimulated emission from polymethine dyes. *Appl. Phys. Lett.* **9**, 179–181.
Steyer, B., and Schäfer, F.P. (1974). A vapor-phase dye laser. *Opt. Commun.* **10**, 219–220.
Stokes, E.D., Dunning, F.B., Stebbings, R.F., Walters, G.K., and Rundel, R.D. (1972). A high efficiency dye laser tunable from the UV to the IR. *Opt. Commun.* **5**, 267–270.
Strome, F.C., and Webb, J.P. (1971). Flashtube-pumped dye laser with multiple-prism tuning. *Appl. Opt.* **10**, 1348–1353.
Sutton, D.G., and Capelle, G.A. (1976). KrF-laser-pumped tunable dye laser in the ultraviolet. *Appl. Phys. Lett.* **29**, 563–564.
Tang, K.Y., O'Keefe, T., Treacy, B., Rottler, L., and White, C. (1987). Kilojoule output XeCl dye laser: optimization and analysis. In *Proceedings: Dye Laser/Laser Dye Technical Exchange Meeting, 1987* (Bentley, J.H., ed.). U.S. Army Missile Command, Redstone Arsenal, Al., pp. 490–502.
Trebino, R. (1985). Achromatic N-prism beam expanders: optimal configurations. *Appl. Opt.* **24**, 1130–1138.
Uchino, O., Mizunami, T., Maeda, M., and Miyazoe, Y. (1979). Efficient dye lasers pumped by a XeCl excimer laser. *Appl. Phys.* **19**, 35–37.
Valdmanis, J.A., Fork, R.L., and Gordon, J.P. (1985). Generation of optical pulses as short as 27 femtoseconds directly from a laser balancing self-phase modulation, group-velocity dispersion, saturable absorption, and saturable gain. *Opt. Lett.* **10**, 131–133.
Wallace, R.W. (1971). Generation of UV from 2610 to 3150 Å. *Opt. Commun.* **4**, 316–318.
Wallenstein, R., and Hänsch, T.W. (1974). Linear pressure tuning of a multielement dye laser spectrometer. *Appl. Opt.* **13**, 1625–1628.
Wyatt, R. (1978). Narrow linewidth, short pulse operation of a nitrogen-laser-pumped dye laser. *Opt. Commun.* **26**, 429–431.

Chapter 2

LASER DYNAMICS

Lloyd W. Hillman

Department of Physics
The University of Alabama in Huntsville
Huntsville, Alabama

1. INTRODUCTION TO LASER DYNAMICS

Dynamics means motion. Dynamics addresses the forces at work in a system, specifically, how they actively interplay and change. At the root of all dynamics is a flow or exchange of energy. For example, in mechanical systems, we encounter forces that change the momentum and energy of particles in the system.

Consider the dynamics of a simple mass connected to an elastic spring. If we displace the mass from equilibrium and then release it, the mass oscillates back and forth. This is an example of a mechanical harmonic oscillator. We associate with the motion a transfer of energy between two states. These are the kinetic energy of the moving mass and the potential energy stored in the spring. When we experimentally build such a mechanical oscillator, we discover another dynamic property. Once set in motion, the oscillation of the mass does not continue forever. Rather the

ISBN 0-12-222700-X

overall motion becomes smaller and smaller. After a while, the mass returns to its equilibrium position; we see no further movement. Physically, this occurs because the oscillating mass interacts with the surrounding air, transferring its kinetic energy into acoustic energy. In addition, the spring absorbs a small amount of the system's mechanical energy, which raises the spring's temperature.

We can counteract the decay of the mechanical motion by supplying additional energy to the system. If we supply this energy at the same rate at which the system loses it, we can achieve a stationary oscillation. We therefore can view this driven harmonic oscillator as a type of transducer. It converts the energy that we supply into both acoustic energy and thermal energy. Although this system is simple, it illustrates the basic ideas of energy conversion from one form into another. A more complicated system is a car. A car's engine converts the chemical binding energy in gasoline into the kinetic energy that propels us around town. The dynamics of a system describes the physics of this energy conversion.

A laser is a special optical transducer. A laser converts energy supplied by a pump source into coherent or ordered light. When we design lasers, we seek both to maximize this energy conversion and to control the coherence properties of the light. By studying and modeling the dynamics of a laser, we can determine the absolute limits that we can achieve. In this chapter, we discuss the physical principles that govern a laser's dynamics.

All lasers consist of three main parts:

- pump or energy source,
- gain or laser medium,
- optical cavity or resonator.

The lasers we build may include other parts as well, such as prisms, gratings, and saturable absorbers, to name a few. We add such elements to tailor and control the spectral and time-dependent properties of the laser's output. You will find the details of such engineering in other chapters of this book. In this chapter, we limit our discussion to the physical dynamics that are intrinsic to the three main parts found in all lasers: the pump, the gain medium, and the optical cavity.

All lasers operate under similar principles; however, each laser has unique characteristics that govern and limit our ability to control it. We shall address both general properties and specific issues associated with dye lasers. In a dye laser, we supply the pump energy in the form of light, from either flashlamps or another laser. The gain medium consists of organic dye molecules dissolved in a fluid solvent, such as methanol, ethanol, dimethylformamide, or just water. In addition, there are reports of doping organic dyes into plastics, glasses, and jello and achieving laser action. Depending upon the flow rates, pumping, and cavity configuration, the dye concentration in the solvent typically ranges from 10^{-5} to 10^{-3} molar. The solvent serves several useful purposes. One of its chief attributes is thermal cooling. In addition, interactions between the dye, solvent, and other additives can effect the wavelength and efficiency of the optical gain. Again the reader can find the details of such interactions in other chapters of this book.

The dye molecules absorb the pump light. They then release this energy by three processes. Some of the energy goes into thermal heating of the dye and solvent. The dye molecules radiate the rest of the energy in the form of light. There are two ways that the dye molecule may emit the light. The first process is *spontaneous emission*. In the spontaneous-emission process, the dye molecule lowers its energy by spontaneously emitting a photon or quantum of light energy. The direction and phase of the emitted photon is random or incoherent. Its energy is lost from the system. The second optical process is *stimulated emission*. Stimulated emission occurs when there are other photons present that interact with the dye molecule. Under stimulated emission, the dye molecule again loses its energy by emitting a photon. This photon, however, has the same phase and direction as the other photons present. We say that this photon is *coherent* since its energy coherently adds to the energy of the other photons. Stimulated emission is the gain process that increases or amplifies the optical energy coherently.

Where do the photons that cause stimulated emission come from? Here the optical cavity or resonator plays a key role. The cavity consists of a system of mirrors, prisms, and other optical elements. The elements of the cavity define a self-repeating path that rays of light can follow. There are two ways that we can arrange the geometry of the cavity. We can align the mirrors so that the rays of light travel back and forth along the same path, called a standing-wave cavity. Or we arrange the cavity so that the light travels around a closed loop, called a ring cavity. Along the path we may place prisms and gratings, so that only light of a single wavelength follows a closed path. We add such dispersive elements to tune or select the wavelength or frequency of the laser.

In constructing a dye-laser cavity, we use broad-band, high-reflecting mirrors and top-quality optical elements. Our objective is to minimize the loss of power from the light as it travels around the cavity. We, however, choose one mirror to be partially transmissive. This mirror is the output coupler, or port, where we extract the energy from the laser in the form of coherent light.

2. ELEMENTARY RATE-EQUATION ANALYSIS

2.1. Elements of Laser Dynamics

The level of detail needed in a model for the dynamics of a laser depends on both the laser and its application. In most cases, simplified working models provide adequate information for making informed engineering decisions. With this in mind, we begin with a simple, or first-order, model for a dye laser called the rate-equation model. The key elements in this model involve the coupled dynamics of three processes:

- rate of excitation of the dye molecules by the pump,
- rate of stimulated emission by excited dye molecules,
- rate of energy loss from the cavity.

The utility of this simple model is that it provides physical insight about the generic dynamics of all lasers. We can use the rate-equation model to find the threshold pumping rate and the energy-conversion efficiency of a laser. Furthermore, under many conditions, the rate equations accurately describe the evolution of single-mode, pulsed lasers.

To obtain the rate equations, we begin with a simplified quantum model for a dye molecule. We assume that we can describe the quantum-mechanical state of the dye molecule with four energy levels. We label the states by their energy eigenvalues: $|E_0\rangle$, $|E_1\rangle$, $|E_2\rangle$, and $|E_3\rangle$. We illustrate the four-level model in Fig. 2.1.

From a heuristic viewpoint, we say a dye molecule is in one of these four states. The interaction between a dye molecule and the light in the laser

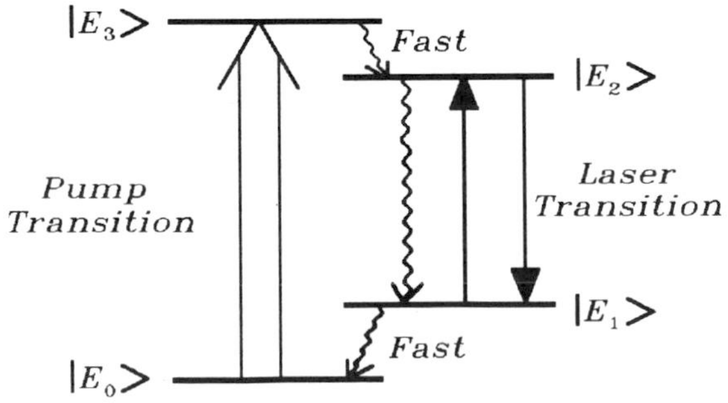

Fig. 2.1 Energy-level diagram for four-level laser. The wiggly lines represent incoherent relaxation processes such as spontaneous emission. The double arrow connecting levels $|E_0\rangle$ and $|E_3\rangle$ is the pumping transition. The two solid arrows connecting levels $|E_1\rangle$ and $|E_2\rangle$ are the coherent optical transitions of the laser. The upward arrow represents stimulated absorption; the downward arrow represents stimulated emission.

cavity causes the molecule to undergo transitions from one quantum state to another. The gain material consists of numerous dye molecules within some volume inside the laser's cavity. At any time there is a given number of dye molecules in each of the four levels. We assume that the distribution of dye molecules in the solvent is uniform. It is therefore common practice to speak about the *population density* of each state, which we define:

$$\frac{|E_j\rangle}{\text{population density of the state}}: \qquad \rho_j \equiv \frac{|E_j\rangle}{\text{number of molecules in volume of gain material}} = \frac{N_j}{V_{\text{gain}}} \tag{2.1}$$

In addition, we define the total number of dye molecules per unit volume as the number density (or #-density),

$$\text{\#-density:} \qquad \mathcal{N} \equiv \frac{\text{total number of molecules}}{\text{volume of gain material}} = \rho_0 + \rho_1 + \rho_2 + \rho_3. \tag{2.2}$$

Equation (2.2) also expresses the population conservation condition that must hold for a closed four-level model.

2.2. The Pumping Process and Excitation Rate

At thermal equilibrium, all the dye molecules are in the lowest or ground state, $|E_0\rangle$. This sets the initial condition of the system to the

$$\text{thermal equalibrium:} \qquad \rho_0 = \mathcal{N}. \tag{2.3}$$

Using flashlamps or another laser, we pump or excite the dye molecules from the ground state, $|E_0\rangle$, into the highest state, $|E_3\rangle$. In other words, the dye molecules gain energy by absorbing the optical energy supplied by the pump. The rate at which we excite the dye molecules is proportional to three quantities. First, the rate is proportional to the flux of power[1] (power per unit area) from the pump into the gain material. This flux of power is equal to the energy density of the light from the pump in the gain material, U_P, times the speed of light, c. Second, the pump rate depends on the number of dye molecules in the ground state. Finally, there is factor that characterizes the absorptive properties of a dye molecule in the ground state. We express this constant in terms of the absorption cross section, (ω_P) or $\sigma_P(\omega_P)$,[2] divided by the energy of a pump photon, $\hbar\omega_P$. Physically,

[1] Commonly, we call this the pump intensity.

[2] When dealing with discrete levels, we use the notation $\sigma_{30}(\omega)$ to mean the cross section for the optical transition from state $|E_0\rangle$ to $|E_3\rangle$. Since the absorption band of dye molecules consists of many levels, we use the simplified expression $\sigma_P(\omega)$ meaning the cross section of the pump's light.

the absorption cross section represents the effective size of the dye molecule that intercepts and scatters the flux of energy from the pump. For rhodamine 6G, the absorption cross section at $\lambda_P = 530$ nm (the peak) is 3.8×10^{-16} cm^2. For organic dyes, the upper level $|E_3\rangle$ is a manifold or band of closely spaced levels. The absorption cross section depends on the frequency of the pump's light. Since the pump is often broadband or nonmonochromatic (e.g., flashlamp), we must integrate over the energy-density spectrum of the pump, $\tilde{U}_P(\omega)$, to obtain the total pump rate,[3]

$$\underset{|E_0\rangle \to |E_3\rangle}{\text{total pump rate}}: \quad \mathcal{R}_{\text{pump}}(0 \to 3) = V_{\text{gain}}\,\rho_0 \int d\omega\, \frac{\sigma_p(\omega) c \tilde{U}_p(\omega)}{\hbar\omega} = V_{\text{gain}}\,\rho_0 R_{\text{pump}}, \tag{2.4}$$

where V_{gain} is the volume of the gain medium that spatially overlaps the lasing mode of the cavity. Furthermore, we assume that the quantities $\tilde{U}_P(\omega)$ and ρ_0 are spatially uniform in the gain volume. Note that $V_{\text{gain}}\,\rho_0 = N_0$ is the total number of dye molecules in the ground state $|E_0\rangle$ within the active region of the laser cavity.

Most of the dye molecules excited to the upper state $|E_3\rangle$ quickly relax to the excited state $|E_2\rangle$. The energy the dye molecules lose from this process raises the temperature of the solvent. This decay happens on the time scale of a few femtoseconds (i.e., 10^{-15} seconds!). Effectively, we find that the excitation rate into the state $|E_2\rangle$ is equal to the pump rate from $|E_0\rangle$ to $|E_3\rangle$,

$$\underset{|E_0\rangle \to |E_2\rangle}{\text{excitation rate}} \cong \underset{|E_0\rangle \to |E_3\rangle}{\text{total pump rate}}. \tag{2.5}$$

Under this approximation, we set $\rho_3 = 0$.

2.3. Spontaneous Emission and Stimulated Emission

It is the density of dye molecules in the excited state $|E_2\rangle$ that leads to optical gain. We note however that there are two competing processes by which a dye molecule can transfer its energy during its transition from state $|E_2\rangle$ to $|E_1\rangle$. The first process is *spontaneous emission*. During spontaneous emission, the dye molecule emits an incoherent photon. The phase and direction of this photon are random or incoherent. The laser loses the energy of this photon. The rate at which spontaneous emission occurs depends on only two quantities: the number density in state $|E_2\rangle$, ρ_2, and

[3] Throughout, we use script $\mathcal{R}$ to mean the total rate and italic R to mean the rate per dye molecule.

the Einstein-A coefficient between levels $|E_2\rangle$ and $|E_1\rangle$, A_{12},

spontaneous emission rate $|E_2\rangle \rightarrow |E_1\rangle$:

$$\mathcal{R}_{\text{spon. emission}}(2 \rightarrow 1) = A_{12} V_{\text{gain}} \rho_2 = \frac{1}{T_{12}} V_{\text{gain}} \rho_2 . \tag{2.6}$$

We call the quantity $T_{12} = (A_{12})^{-1}$ the lifetime of the upper level $|E_2\rangle$ or the gain lifetime. For rhodamine 6G, this lifetime is $T_{12} = 3.7$ nsec.

The second process is *stimulated emission*, which is the process that yields coherent optical gain. The rate of stimulated emission is proportional to three quantities: the number density in state $|E_2\rangle$, ρ_2; the flux of optical power circulating in a mode of the cavity that resonantly couples to the laser transition, $c\tilde{U}_L(\omega_{12})$; and the emission cross section, $\sigma_{12}(\omega_{12})$ or $\sigma_L(\omega_{12})$, divided by $\hbar\omega_{12}$:

stimulated emission rate $|E_2\rangle \rightarrow |E_1\rangle$:

$$\mathcal{R}_{\text{stim. emission}}(2 \rightarrow 1) = \frac{\sigma_L(\omega_{12}) c U_L}{\hbar\omega_{12}} V_{\text{Gain}} \rho_2 . \tag{2.7}$$

In Eq. (2.7), we set $U_L = \tilde{U}_L(\omega_{12})\,\delta\omega$, where is the laser's bandwidth. Since $\tilde{U}_L(\omega_{12})$ is the energy-density spectrum, U_L is the energy density of the lasing mode. For rhodamine 6G the emission cross section is $1.2 \times 10^{-16}\ \text{cm}^2$ at $\lambda_L = 580$ nm. During stimulated emission, the dye molecule emits a coherent photon. This photon coherently adds its energy to the cavity mode that contains the photons that stimulated it. All photons in a single cavity mode are indistinguishable; the stimulated photon's direction, polarization, and frequency are identical to the other photons in the mode.

2.4. Stimulated Absorption and Relaxation to Ground State

Dye molecules in the lower state $|E_1\rangle$ are also subject to two competing processes. First, a dye molecule in state $|E_1\rangle$ may absorb a photon from the lasing mode. This raises its energy back to the upper state $|E_2\rangle$ and lowers the optical energy in the laser cavity. This process is *stimulated absorption*. The rate of stimulated absorption is like stimulated emission except that it is proportional to the number of dye molecules in the state $|E_1\rangle$ (i.e., $N_1 = V_{\text{gain}}\rho_1$),

stimulated absorption rate $|E_1\rangle \rightarrow |E_2\rangle$:

$$\mathcal{R}_{\text{stim. absorption}}(1 \rightarrow 2) = \frac{\sigma_L(\omega_{12}) c U_L}{\hbar\omega_{12}} V_{\text{gain}} \rho_1 . \tag{2.8}$$

The second alternative is for the dye molecule to incoherently lose its energy and return to the ground state $|E_0\rangle$. The ground-state recovery rate is proportional to an effective Einstein-A coefficient between the levels $|E_1\rangle$ and $|E_0\rangle$,

$$\text{ground state recovery rate}\ |E_1\rangle \rightarrow |E_0\rangle: \quad \mathcal{R}_{1\rightarrow 0} = A_{01} V_{\text{gain}}\, \rho_1 = \frac{1}{T_{01}} V_{\text{gain}}\, \rho_1 . \tag{2.9}$$

For rhodamine 6G the lifetime T_{01} of the lower state $|E_1\rangle$ is a few picoseconds (i.e., 10^{-12} seconds).[4] Because this lifetime is so short, we often approximate that $\rho_1 \approx 0$.

2.5. Cavity Modes and Dynamics

We stated that stimulated emission creates a coherent photon, while stimulated absorption annihilates a photon. The dye molecules emit these photons into a mode of the laser cavity. The term *mode* needs definition. We construct a laser cavity by arranging mirrors such that rays of light travel back and forth through the gain medium. Technically, the modes of the cavity are stationary solutions to Maxwell's equations (Fox and Li, 1961; Kogelnik and Li, 1966). We find the modes by assuming that the mirrors are perfect reflectors and the cavity is empty. Like the pattern of an oscillating piano wire, the modes of the cavity define the pattern or distribution of the electromagnetic field in the cavity. Each mode has a definite spatial dependence and a resonant frequency.

Mathematically, the modal functions form a complete, orthogonal set of spatial functions. We can decompose or expand any field distribution within the cavity into its projected modal parts. Physically, the mirrors define a direction or axis of the cavity. The cavity mirrors confine the electromagnetic field so that spatially the optical energy lies close to the cavity's axis. Each mode has both *transverse* i.e., in the directions perpendicular to the cavity axis) and *longitudinal* (i.e., in the direction of the cavity axis) spatial variations. Often, we treat lasers by assuming only a single mode. Furthermore, we ignore the transverse and longitudinal dependence and treat the energy density of the lasing mode, U_L, as being effectively uniform within a volume defined by the mode, V_{mode}. We therefore express the total number of photons in the mode of the laser as,

$$\text{number of photons in cavity mode}: \quad Q = V_{\text{mode}} \frac{U_L}{\hbar \omega_{\text{L}}} . \tag{2.10}$$

[4] Although we often set the lower state's lifetime to a few picoseconds, the fact that the lower state is a band affects the dynamics. The real lifetime of the levels may be much longer. For further discussion, see Fu and Haken (1987a).

Equivalently, the flux of optical energy or intensity (power/unit area) of the mode is

intensity of cavity mode: $$\mathscr{I}_{\text{cav}} = cU_L. \tag{2.11}$$

As the light circulates around the cavity, the mirror and other elements of the cavity scatter the light. This scattering is a source of loss, decreasing the power in the cavity. Because we construct the optical cavity using high-quality components, the dominant loss is through the output coupler. If the cavity round-trip length is L and the reflectivity of the output coupler is $\mathbf{R}$, then the rate at which the cavity loses photons due to the output coupler is

photon loss rate from output coupler: $$\mathscr{R}_{\text{cavity loss}} = -\left(\frac{c}{L}\ln(R)\right)Q = \gamma_c Q = \frac{1}{T_C}Q, \tag{2.12}$$

where γ_c is the cavity-decay constant and $T_c = (\gamma_c)^{-1}$ is the cavity-decay time.

Finally we ask the question, how much optical power do we extract from the laser? If the geometry of the laser's cavity is a unidirectional ring and the transmissivity of the output coupler is T, then the power we extract from the laser is

output power from laser (ring cavity): $$\mathscr{P}_{\text{out}} = \mathbf{T}A_{\text{mode}}\mathscr{I}_{\text{cav}}, \tag{2.13}$$

where A_{mode} is the transverse cross-sectional area of the mode or laser beam. For a standing-wave cavity, we obtain one-half this value, since the photons in the cavity travel both forward and backward. We note that the total density of photons, which determines the rate of stimulated emission, is the sum of the two.

2.6. Population-Rate Equations

We obtain the elementary population-rate equations by combining the principles of pumping, relaxation, stimulated emission, stimulated absorption, and cavity loss. We note that the total optical gain is proportional to the difference between the population density in states $|E_2\rangle$ and $|E_1\rangle$, or $\rho_2 - \rho_1$. We therefore define the inversion of the gain, w, to be this difference normalized to the number density:

inversion: $$w = \frac{\rho_2 - \rho_1}{\mathscr{N}}. \tag{2.14}$$

By this definition, we see that, $-1 \leq w \leq +1$. The inversion represents the relative amount of excitation. Since the decay from both $|E_3\rangle$ to $|E_2\rangle$ and $|E_1\rangle$ to $|E_0\rangle$ is very rapid for organic dyes, we further approximate that $\rho_3 \approx \rho_1 \approx 0$. Because of this approximation, the dye molecules are in either state $|E_2\rangle$ or $|E_1\rangle$. By specifying only the inversion, which now satisfies the inequality $0 \leq w + 1$, we therefore describe all population densities of the gain medium. Because the #-density of the gain material is large, we can treat the inversion as a differentially continuous function of time. We apply the principles outlined in Eqs. (2.1)–(2.9) and obtain two coupled differential equations that describe the time dynamics of the inversion and the cavity intensity:

Inversion:

$$\dot{w}(t) = R_{\text{pump}}[1 - w(t)] - \frac{\sigma_L}{\hbar\omega_L} \mathcal{I}_{\text{cav}}(t)\, w(t) - \frac{1}{T_{12}} w(t) \tag{2.15}$$

Cavity Intensity:

$$\dot{\mathcal{I}}_{\text{cav}}(t) = \left(\frac{\mathcal{N} V_{\text{gain}} c \sigma_L}{V_{\text{mode}}}\right) \mathcal{I}_{\text{cav}}(t)\, w(t) - \frac{1}{T_C} \mathcal{I}_{cav}(t). \tag{2.16}$$

By the definitions given in Eqs. (2.10) and (2.11), we can rewrite the two population-rate equations in terms of the number of photons in the cavity.

The rate equations given in Eqs. (2.10) and (2.11) are the starting point for all laser analysis. We can use them to find the threshold pump rate and the efficiency of the power conversion by the laser. We however note that they lack one important parameter. If initially the cavity intensity is zero, Eqs. (2.10) and (2.11) predict that the intensity remains zero no matter how hard you pump the system. This is because we did not include spontaneously emitted photons that the dye molecules emit into the cavity mode. This process is random. We cannot model it as a deterministic process. Often we do account for these photons by adding "extra photons" to the cavity mode. For accurate modeling, we must randomly add photons to the cavity at a rate that is proportional to the inversion.

2.7 Extension of Rate Equations for Organic Dyes

Equations (2.14) and (2.15) model the basic dynamics of dye lasers based upon the simplified four-level model given in Fig. 2.1. The real level structure of organic dyes is considerably more complex. The energy levels of a dye are not simple isolated states. Rather, the states of the dye molecules form bands, as illustrated in Fig. 2.2. The band structure results from the vibrational and rotational degrees of freedom associated with the

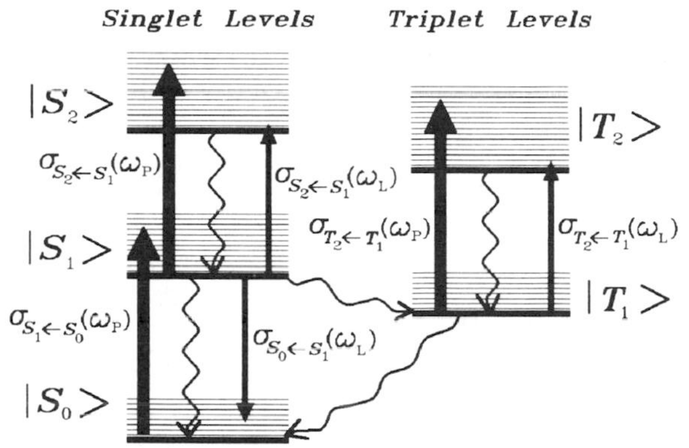

Fig. 2.2 Band-structure and transition paths of a dye molecule. There are two manifolds of bands: the singlet levels and the triplet levels. The heavy arrows represent pump absorption processes, the lighter arrows represent laser field–induced transitions. The wiggly lines represent incoherent interband decay routes. For clarity, the rapid intraband decay lines are not drawn.

binding of the atoms that make up the dye molecule. We designate the lowest band $|S_0\rangle$ and the lowest level in this band $|S_{0,G}\rangle$. The configuration of the electrons in the $|S_0\rangle$ band is such that their total quantum-mechanical spin adds up to zero. We call such a state a singlet.

When the dye molecule absorbs a pump photon, it raises the energy of the dye molecule by exciting the outermost electron into a new orbit. This redistributes the electronic charge that binds the atoms of the dye molecule together. Under this new charge distribution, the binding forces acting on the atoms are not in equilibrium. We call this a *Frank-Condon state*. The atoms quickly redistribute their positions to minimize the total binding potential. The dipolar interaction of the dye with the solvent assists in the dissipation of this excess energy. A few femtoseconds after the absorption of the photon, the dye molecule relaxes to the lowest level or equilibrium state of the $|S_1\rangle$ band or $|S_{1,G}\rangle$.

There are now five competing processes by which the dye molecule may leave the $|S_{1,G}\rangle$ state. We discussed the two main paths in Section 2.3. These are by either spontaneous emission or stimulated emission. Under these processes, the dye molecule emits a photon and relaxes to a Frank-Condon state in the $|S_0\rangle$ band. A few picoseconds after this emission, the dye molecules relaxes to the ground state, $|S_{0,G}\rangle$.

In addition, a dye molecule in the $|S_{1,G}\rangle$ state may absorb either another pump photon or a laser photon. This raise the energy of the molecule to a Frank-Condon state in the $|S_2\rangle$ band. We refer to this process as *excited-state absorption*. During this process, the absorbed photon excites the

Table 2.1

Cross Sections for Rhodamine 6G ($\lambda_P = 530$ nm, $\lambda_L = 580$ nm)

Ground-state pump's absorption cross section:	$\sigma_{S_1 \leftarrow S0}(\omega_P) = 3.8 \times 10^{-16}$ cm^2
Ground-state laser's absorption cross section :	$\sigma_{S_1 \leftarrow S_0}(\omega_L) = 1 \times 10^{-19}$ cm^2
Excited-state laser's emission cross section :	$\sigma_{S_0 \leftarrow S_1}(\omega_L) = 1.2 \times 10^{-16}$ cm^2
Excited-state pump's absorption cross section :	$\sigma_{S_2 \leftarrow S_1}(\omega_P) = 4 \times 10^{-17}$ cm^2
Excited-state laser's absorption cross section :	$\sigma_{S_2 \leftarrow S_1}(\omega_L) = 1 \times 10^{-17}$ cm^2
Triplet-states pump's absorption cross section:	$\sigma_{T_2 \leftarrow T_1}(\omega_P) = 1 \times 10^{-17}$ cm^2
Triplet-states laser's absorption cross section :	$\sigma_{T_2 \leftarrow T_1}(\omega_L) = 1 \times 10^{-17}$ cm^2

outer electron to yet another orbit. Again, the position of the atoms rapidly adjusts to the equilibrium state $|S_{2,G}\rangle$. Dye molecules in the $|S_{2,G}\rangle$ level decay back to the $|S_1\rangle$ band in a few picoseconds. The rate of excited-state absorption depends on the number of molecules in the state $|S_{1,G}\rangle$ or the inversion, the energy density of the pump and laser's field, and their respective excited-state absorption cross sections. In Table 2.1, we list the various cross sections for rhodamine 6G. Since the excited-state relaxation rate back to the state $|S_{1,G}\rangle$ is so rapid, effectively excited-state absorption has no effect on the population in the $|S_{1,G}\rangle$ level. It does however lead to heating of the gain media and to an additional loss of energy from the laser's field.

The fifth process by which a dye molecule may leave the state $|S_{1,G}\rangle$ is through decay to the triplet band $|T_1\rangle$. The configuration of the electrons in the triplet bands is such that their total quantum mechanical spin adds up to one. In the decay process from $|S_{1,G}\rangle$ to $|T_{1,G}\rangle$, the spin of the excited electron flips. For rhodamine 6G the associated decay lifetime of the state $|S_{1,G}\rangle$ into the state $|T_{1,G}\rangle$ is $T_{TS} = 5 \times 10^{-8}$ sec. The state $|T_{1,G}\rangle$ is metastable. Dye molecules in the triplet state $|T_{1,G}\rangle$ decay back to the singlet state $|S_{0},_{G}\rangle$ with a long lifetime $T_{TS} = 5 \times 10^{-8}$ sec. In Table 2.2 we list the intraband decay lifetimes for rhodamine 6G.

Dye molecules that decay to the triplet levels lead to two detrimental losses in the laser system. First, they lower the overall populations in the

TABLE 2.2

Intraband Decay Lifetimes for Rhodamine 6G

$\lvert S_1\rangle \rightarrow \lvert S_0\rangle$:	$T_{S_0 \leftarrow S_1} = T_1 = 3.7 \times 10^{-9}$ sec
$\lvert S_2\rangle \rightarrow \lvert S_1\rangle$:	$T_{S_1 \leftarrow S_2} \approx 0$ (Fast!)
$\lvert S_1\rangle \rightarrow \lvert T_1\rangle$:	$T_{T \leftarrow S} = 5.0 \times 10^{-8}$ sec
$\lvert T_2\rangle \rightarrow \lvert T_1\rangle$:	$T_{T_2 \leftarrow T_1} \approx 0$ (Fast!)
$\lvert T_1\rangle \rightarrow \lvert S_0\rangle$:	$T_{S \leftarrow T} = 5.0 \times 10^{-8}$ sec

singlet levels. Second, a dye molecule in the triplet state $|T_{1,G}\rangle$ may absorb either a pump photon or a photon from the laser's field. We refer to this as *triplet-state absorption*. As with excited-stated absorption, triplet-state absorption leads to excess heating of the gain medium and to loss of energy from the laser field. There are two methods employed to eliminate the effects of the triplet states. The first is flowing of the dye. New dye replaces the heated dye in the cell or jet at a specific flow rate R_{flow} (i.e., the inverse of the time a dye molecule is in the active gain region). This removes the heated gain material where it can then be cooled. In addition, flowing the dye removes the dye molecules trapped in the triplet level. In fact because of triplet-state absorption, cw operation of a dye laser requires a rapid flowing of the dye through a nozzle. In addition, by adding agents to the solvent, such as oxygen (see Chapter 6, Section 2.6), we can quench the effects of the triplet-state absorption. Such additives can inhibit the decay to the triplet levels and increase the rate of decay out of the triplet level.

To model the effects of excited-state absorption, triplet-state absorption, and dye flow, we need to supplement the fourth-level model given by Eqs. (2.15) and (2.16). We need to add one more dynamic variable to account for the population in the triplet level $|T_1\rangle$. We define a new parameter, τ, triplet excitation, which relates to the triplet population density.

$$\tau = \frac{\rho_{T_1}}{\mathcal{N}} = \frac{\text{density in } |T_1\rangle}{\#\text{-density}}. \tag{2.17}$$

We therefore model the dynamics of a dye laser with the three coupled population-rate equations:

$$\begin{aligned} \dot{w}(t) = R_{\text{pump}}[1 - w(t) - \tau(t)] - \frac{\sigma_L}{\hbar\omega_L}\mathcal{I}_{\text{cav}}(t)\,w(t) \\ - \left(\frac{1}{T_1} + \frac{1}{T_{TS}} + R_{\text{flow}}\right)w(t) \end{aligned} \tag{2.18}$$

$$\dot{\tau}(t) = \frac{1}{T_{TS}}w(t) - \left(\frac{1}{T_{ST}} + R_{\text{flow}}\right)\tau(t) \tag{2.19}$$

$$\dot{\mathcal{I}}_{\text{cav}}(t) = \left(\frac{\mathcal{N}V_{\text{gain}}c}{V_{\text{mode}}}\right)\{[\sigma_L - \sigma_E]\,w(t) - \sigma_T\tau(t)\}\mathcal{I}_{\text{cav}}(t) - \frac{1}{T_C}\mathcal{I}_{\text{cav}}(t) \tag{2.20}$$

In Eqs. (2.18)–(2.20) we simplified the notation of terms without sacrificing clarity of their meaning. All cross sections refer to the laser's field, namely, σ_L, σ_E, and σ_T are the cross section for laser emission, excited-state absorption, and triplet-state absorption cross sections. T_1 is the lifetime of the inversion associated with decay from the upper singlet band $|S_1\rangle$ to the lower singlet band $|S_0\rangle$, while T_{TS} and T_{ST} are the decay times associated with decay from the upper singlet band to the triplet band and from the triplet band to the lower singlet band, respectively.

Equation (2.28) differs from Eq. (2.14) in two ways. As population enters the triplet levels, τ takes on a positive value. This decreases the population that can reenter the ground state. Also there are additional decay terms due to decay into the triplet manifold and due to flowing of the dye. The addition of dye flow, triplet-state absorption, and excited-state absorption has no effect on the rate of stimulated emission. Equation (2.20) differs from Eq. (2.15) by the addition of the two loss terms due to excited-state absorption and triplet-state absorption.

3. COHERENT DYNAMICS

Lasers are special because the light they emit is highly coherent—both spectrally and spatially. It is these attributes that make laser light so useful. Often it is said that laser light is monochromatic. This statement is incorrect. There is always a finite linewidth. But what determines it? This takes a deeper understanding of the laser's dynamics. The closer we scrutinize the properties of laser light, the more complex our model becomes. In addition to the intrinsic dynamics of the laser, we must account for the laser's reaction to environmental noise that is always present. This task is beyond the scope of this chapter. We limit our discussion to the development of semiclassical laser theory.

3.1. Semiclassical Radiation Theory

The first extension beyond the rate-equation model is to include coherence in the model for a laser (Lamb, 1964). Laser light is a form of electromagnetic radiation, which obeys Maxwell's equations. Maxwell's equations describe how electromagnetic waves propagate. For a nonmagnetic material, we can reduce Maxwell's equations to a single equation for the electric field,

$$\nabla^2\mathbf{E}(\mathbf{r},t) - \frac{1}{c^2}\ddot{\mathbf{E}}(r,t) = \frac{4\pi}{c^2}\ddot{\mathbf{P}}(r,t). \tag{2.21}$$

The quantity $\mathbf{P}(\mathbf{r},t)$ is the polarization density. It arises from the gain material. The quantum properties of the dye molecules determine the relationship between $\mathbf{P}(\mathbf{r}, t)$ and $\mathbf{E}(\mathbf{r}, t)$. With no material present, the value of $\mathbf{P}(\mathbf{r}, t)$ is zero. For such a case, we recognize that Eq. (2.21) reduces to the wave equation.

For a planewave traveling in the $\hat{z}$-direction, we can express the electric field as:

$$\mathbf{E}(\mathbf{r}, t) = 2\,Re(\mathscr{E}(\mathbf{z}, t)\exp\{i[\mathrm{k}\mathbf{z} - \omega_L t - \phi(\mathbf{z}, t)]\}), \tag{2.22}$$

where ω_L is the laser's frequency and $k = \frac{\omega_L}{c}$. If no material is present, that is $\mathbf{P}(\mathbf{r}, t) = 0$, then the amplitude, $\mathscr{E}(\mathbf{z}, t)$, and phase factor, $\phi(\mathbf{z}, t)$ for the planewave will remain constant as the electromagnetic field propagates. In the gain material $\mathbf{P}(\mathbf{r}, t) \neq 0$, we find that the amplitude and phase of the field change as it propagates. Variation in the amplitude equates to either loss or gain; the phase change is dispersion. In the non-coherent rate-equation model only amplitude changes enter the dynamics through the emission and absorption of photons. Coherent theory dynamically includes both amplitude and phase changes of the laser light.

Our objective is to understand how the light interacts with the gain medium to produce a polarization density which is the source term of Eq. (2.21). This is the basis of coherent laser theory. We rely on a simple model for the gain medium. We assume that the gain medium consists of a homogeneous collection of noninteracting dye molecules. This is the same model as we used for the rate equations. We however approach the quantum mechanics for the dye molecules differently. Although it is the basis for much of our intuition, the photon picture of absorption and emission conceals the important coherent or wave aspects of the interaction.

Recall that quantum mechanics yields the stationary states for the dye molecule. Associated with each state is a wavefunction for the dye molecule's outer electron. We interpret the squared modulus of this wavefunction as the probability density of the electron's position. We find these wavefunctions by solving the time-independent Schröndinger equation for the outer electron bound to the dye molecule. Loosely speaking, the electron distributes itself around the molecular structure. To illustrate the concept, we can picture a single electron bound to a positively charged nucleus. Figure 2.3 illustrates two wavefunctions for an electron in either state $|E_1\rangle$ and $|E_2\rangle$. In both cases the charge distribution is symmetric about the nucleus. The quantum mixing of the two states by the radiation field results in a charge distribution with a net dipole moment. This dipole oscillates at the transition frequency and radiates a coherent field. The total atomic charge distribution has no dipole moment when the electron is in one of its stationary states.

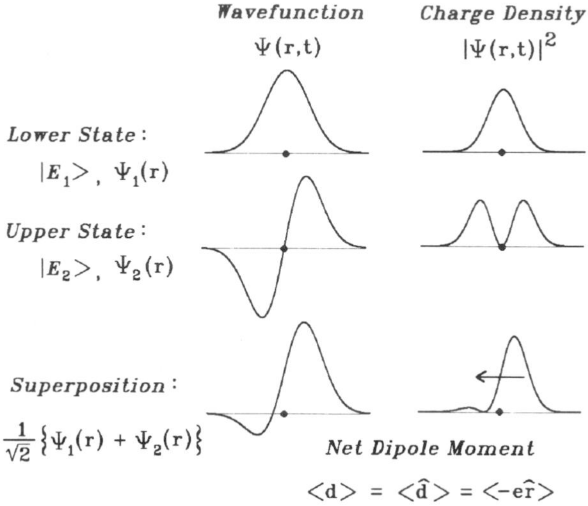

Fig. 2.3 Quantum picture of the coherent dipole. We illustrate two wavefunctions for an electron in either state $|E_1\rangle$ and $|E_2\rangle$. In both cases the charge distribution is symmetric about the nucleus. The quantum mixing of the two states by the radiation field results in a charge distribution with a net dipole moment. This dipole oscillates at the transition frequency and radiates a coherent field.

When a radiation field interacts with the dye molecule, the electron's wavefunction changes according to the time-dependent Schrödinger equation. The electron is no longer exactly in one quantum level or another. Rather, its wavefunction is a quantum-state mixture or superposition. An example of a mixed-state wavefunction appears in Fig. 2.3. In this superposition state, the distribution of the electronic charge is asymmetric about the atom's nucleus. This charge separation corresponds to a classical microscopic dipole.

At this point, we concentrate only on the coherent dynamics for the two states resonant to the laser's field, $|E_1\rangle$ and $|E_2\rangle$ in Fig. 2.1. For this two-level system, the dynamic evolution of the polarization and inversion follow the equations

$$\ddot{\mathbf{P}}(\mathbf{r},t) + \frac{2}{T_2}\dot{\mathbf{P}}(\mathbf{r},t) + \omega_a^2\mathbf{P}(\mathbf{r},t) = -\frac{2\mathcal{N}\omega_a\mathbf{d}^2}{\hbar}\, w(\mathbf{r},t)\mathbf{E}(\mathbf{r},t) \tag{2.23}$$

and

$$\dot{w}(\mathbf{r},t) = \frac{2}{\mathcal{N}\hbar\omega_a}\dot{\mathbf{P}}(\mathbf{r},t)\cdot\mathbf{E}(\mathbf{r},t) - \frac{1}{T_1}[w(\mathbf{r},t) - w_{\mathrm{eq}}]. \tag{2.24}$$

In Eqs. (2.23) and (2.24), w_{eq} is the stationary inversion caused by the pump, ω_a is the transition frequency, and **d** is the dipole moment between the states $|E_1\rangle$ and $|E_2\rangle$. T_1 and T_2 are the longitudinal lifetime, or the lifetime of the inversion, and the transverse lifetime, or the lifetime of the induced coherent polarization. Note that Eq. (2.24) is equivalent to Eq. (2.15) except that the driving term is different. Classically, the term $\dot{\mathbf{P}} \cdot \mathbf{E}$ is the amount of work done on a dipole by an electric field. The work done on the dipole raises the potential energy of the molecule, which corresponds to increasing its inversion. Furthermore, we note that Eq. (2.23) is of the form of a driven harmonic oscillator. Once excited, the polarization oscillates at its resonance frequency. This oscillating polarization radiates an electromagnetic field which coherently adds to the laser field.

3.2. Maxwell-Bloch Equations

Equations (2.21), (2.23), and (2.24) are the basic semiclassical equations that describe the interaction between light and matter. Instead of using them in their present form, we more often make an approximation that is valid in the optical region of the electromagnetic spectrum. The rapid oscillation of the electric field and the polarization is considerably faster than the relaxation rates of their amplitudes. We therefore make a slowly varying envelope approximation. Under this approximation, we note that the amplitude and phase in Eq. (2.22) satisfy the inequalities

$$\begin{aligned} \left|\frac{\partial \mathscr{E}}{\partial z}\right| &\ll |k\mathscr{E}|, \\ \left|\frac{\partial \mathscr{E}}{\partial t}\right| &\ll |\omega\mathscr{E}|, \\ \left|\frac{\partial \phi}{\partial z}\right| &\ll |k\phi|, \text{ and} \\ \left|\frac{\partial \phi}{\partial t}\right| &\ll |\omega\phi|. \end{aligned} \tag{2.25}$$

By Eq. (2.22), we assume the electric field propagates in the $\hat{z}$-direction. We therefore can write the polarization in a similiar form,

$$\mathbf{P}(r,t) = \mathscr{N}\mathbf{d}Re\{[u(z,t) - iv(z,t)]\exp\{i[kz - \omega_L t - \phi(z,t)]\}\}, \tag{2.26}$$

where $u(z,t)$ and $v(z,t)$ are slowly varying functions as before,

$$\left|\frac{\partial u}{\partial z}\right| \ll |ku|,$$

$$\left|\frac{\partial u}{\partial t}\right| \ll |\omega u|,$$

$$\left|\frac{\partial v}{\partial z}\right| \ll |kv|, \text{ and} \tag{2.27}$$

$$\left|\frac{\partial v}{\partial t}\right| \ll |\omega v|.$$

Using the slowly varying quantities, we can rewrite the basic semiclassical equations keeping only the dominant terms. We call this new set of equations the Maxwell-Bloch equations:

$$\left(\frac{\partial}{\partial z} + \frac{1}{c}\frac{\partial}{\partial t}\right)\mathscr{E}(z,t) = \frac{\mathcal{N}2\pi\omega_a d}{c} v(z,t) \tag{2.28}$$

$$\mathscr{E}(z,t)\left(\frac{\partial}{\partial z} + \frac{1}{c}\frac{\partial}{\partial t}\right)\phi(z,t) = -\frac{\mathcal{N}2\pi\omega_a d}{c} u(z,t) \tag{2.29}$$

$$\dot{u}(z,t) = -[\Delta - \dot{\phi}(z,t)]v(z,t) - \frac{1}{T_2}u(z,t) \tag{2.30}$$

$$\dot{v}(z,t) = [\Delta - \dot{\phi}(z,t)]u(z,t) + \kappa\mathscr{E}(z,t)w(z,t) - \frac{1}{T_2}v(z,t) \tag{2.31}$$

$$\dot{w}(z,t) = -\omega\mathscr{E}(z,t)v(z,t) - \frac{1}{T_1}[w(z,t) - w_{\text{eq}}], \tag{2.32}$$

where the coupling coefficient κ and the detuning parameter Δ are

$$\kappa = \frac{2d}{\hbar} \quad \text{and} \quad \Delta = \omega_a - \omega_{\text{L}}. \tag{2.33}$$

The Maxwell-Bloch equations are used extensively in quantum optics (Allen and Eberly, 1975). Chapter 3 discusses their use in describing short pulse formation in mode-locked lasers.

3.3. Laser Operation and Instabilities

A laser is an optical feedback oscillator. Its operation principle is like any electronic feedback oscillator: amplification plus positive feedback equals oscillation. To illustrate this, we consider a ring cavity as shown in

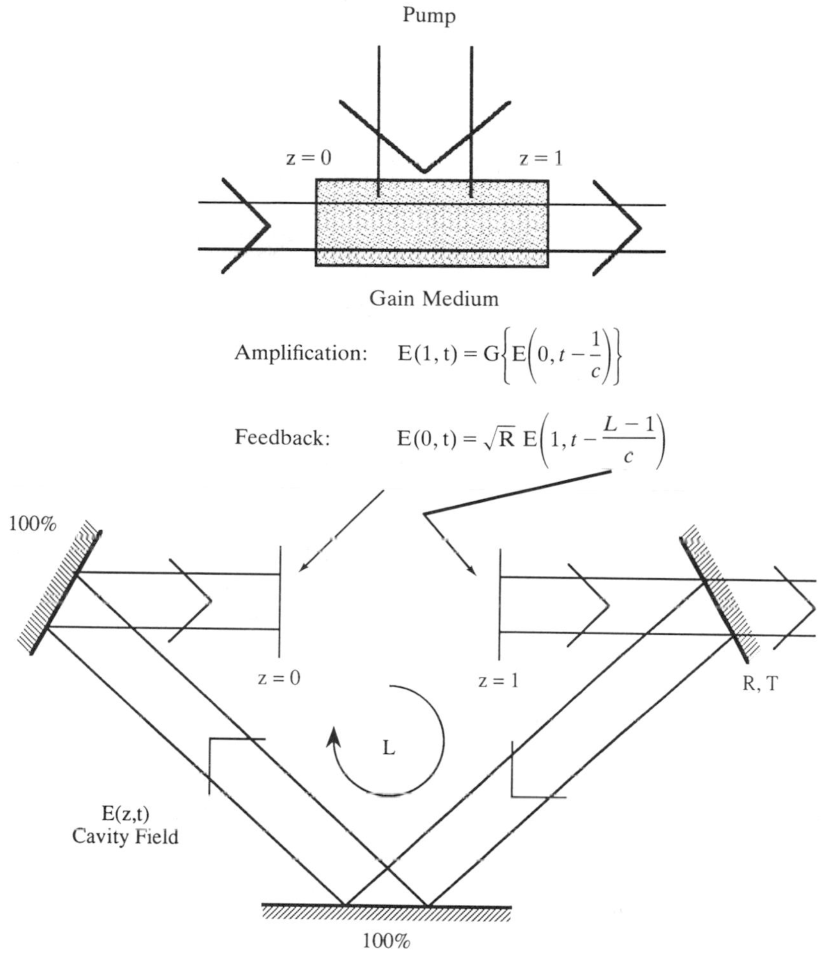

Fig. 2.4 The parts of a simple ring laser. The gain medium amplifies the field; the cavity furnishes feedback. The laser is therefore an optical feedback oscillator (from Hillman, 1988).

Fig. 2.4. The gain medium amplifies the light; the cavity mirrors supply the feedback. The length of the cavity and the material dispersion determine the positive feedback frequencies. (The laser is therefore an optical feedback oscillator.) We call this set of frequencies the laser's modes.

The basic operation of a laser is simple. A pumping process excites the gain medium and creates a population inversion. This sets a value for w_{eq} in Eq. (2.32) and establishes the gain. The gain medium amplifies optical fields with frequencies near the transition resonance. The inverse of the polarization decay time, $(1/T_2)$, sets the bandwidth of optical gain. Coherent optical oscillations grow when the gain at a modal frequency is greater than the total round-trip loss. This is the well-known threshold condition for a laser.

In all feedback oscillators, nonlinearities of the amplifier eventually limit the growth and amplitude of the oscillations. The amplification process in a laser is self-limiting because the light actively modifies the response of the gain material. As the intensity of the optical field increases, the gain decreases. We refer to this change as *saturation*. Gain saturation results from quantum transitions between the upper and lower levels due to stimulated emission. This lowers the population inversion and hence the gain. When the saturated gain is equal to the round-trip loss, no further growth in optical field is possible. Gain saturation therefore stabilizes the amplitude of the light in the cavity. An increase in the pump or excitation rate leads to a proportional growth in the light intensity.

Just above threshold, a laser operates at a single frequency. Does the single-frequency oscillation remain dynamically stable at higher pumping rates? According to many textbooks, an ideal homogeneously broadened laser will only oscillate at a single frequency. This conclusion is wrong! It is this prediction about the dynamic stability where the coherent semiclassical model differs from the noncoherent rate-equation models (Risken and Nummedal, 1968).

What determines the stability of a single-frequency laser oscillator? The theory of feedback oscillators provides the answer. The time dynamics consist of two parts: the stationary oscillation and the decay of transients. The excitation of the transients by stochastic noise, such as spontaneous emission, leads the finite linewidth of the laser (Haken, 1966). Furthermore, a single-mode laser is only stable if the gain for all modes, except one, is less than the round-trip loss. This one mode oscillates; the others decay. To answer the question of stability, we therefore must determine the gain for the nonlasing modes.

In the stability analysis of a single-frequency laser, we treat the nonlasing modes as probe fields. This problem is the same as finding the gain of a probe field in a pump-probe experiment (Hillman *et al.*, 1982). The field of the laser is the pump; the fields for the other modes are the probes. We refer the reader to other papers for the specific details on this calculation.[5] As the intracavity field gets intense, it dynamically alters the response of the gain medium. By parametric coupling, energy can be transferred to the other cavity modes. In fact, this parametric gain can be so great that

[5] Two entire special issues of the *Journal of the Optical Society of America* are devoted to this topic: *J. Opt. Soc. Am. B*, Vol. 2 No. 1 (Jan 1985) and Vol. 5, no. 5 (May 1988). In addition, see Boyd *et al.* (1986).

they begin to lase. When this happens, the single-frequency oscillation is unstable!

The instability found in the coherent theory is intrinsic to all lasers' dynamics. Only the coherent nonlinear interactions between a two-level resonance and the optical field are necessary. Unfortunately, the instability predicted by the simple two-level resonance does not fit observed instabilities in dye lasers (Hillman *et al.*, 1984). This raises the question on what role the band structure of a dye molecule plays in bringing about dynamic instabilities. Work by H. Fu and H. Haken (1987b, 1988) shows the structure of resonant bands does affect a dye laser's stability. In addition there are other factors that may affect a dye laser's stability, such as the transverse mode structure (Narducci and Abraham, 1987) and turbulence in the dye flow (Duarte *et al.*, 1988). These are research questions for the 1990s.

PROBLEMS

1. The laser emission cross section for a Nd: glass laser is $\sim 3 \times 10^{-20}$ cm^2, which is typical of solid state lasers. Compare this number with the emission cross section for rhodamine 6G. List and discuss the many advantages that a dye laser system has over a solid state laser.
2. Dye lasers have revolutionized spectroscopy and photochemistry by giving tunable sources of light that selectively excite resonances of molecules. Using the appendix, select a dye laser system for exciting (a) sodium D lines at 589 and 589.6 nm, (b) the potassium D lines at 766.5 and 769.9 nm, and (c) rubidium D lines at 780 and 794.8 nm.
3. Thermal cooling of the dye is very important for cw operation of a dye laser. Assume that we have a 100% quantum efficient dye laser, that is, for every photon absorbed, one dye laser photon is created. You pump the dye laser with a 2 Watt argon ion laser $\lambda_P = 514$ nm); the dye absorbs 80% of the pump photons. The laser oscillates at $\lambda_L = 590$ nm. What is the maximum energy conversion efficiency for this laser? The dye is dissolved in water and flows through a jet nozzle whose thickness is 1 mm. Assume that all the excess pump energy heats the dye solution. At the jet, the spot size of the argon ion laser beam is 10 μm in diameter. What is the volume of the gain that is excited? What must the flow velocity of the jet be for a maximum raise of 20°C in the solvent temperature that passes through the gain region. [Hint: 1 calorie = 4.18 J, 1 calorie raises the temperature of 1 gram of water 1°C, and 1 gram of water equals 1 cm^3.]
4. One advantage of dye lasers is that you can control the optical gain by changing the concentration of the dye. Suppose you build a cw ring dye laser using an output coupler with a reflectivity R and a jet with thickness *t*.

For this laser, show that the number density of the dye in the solvent must satisfy the inequality

$$\mathcal{N} > -\frac{\ln(\mathrm{R})}{\sigma_L \mathrm{t}}.$$

What assumptions did you make in arriving at this result? The atomic weight of rhodamine thetrafluoroborate is 530. How many grams of this dye must be added to 3.785 liters of solvent to achieve a dye concentration of 10^{-3} M?

(5) For modeling, we often simplify Eqs. (2.15) and (2.16) by combining the physical constants and defining two new variables,

$$I(t) = \frac{\sigma_L}{\hbar\omega_L}\mathcal{I}_{\mathrm{Cav}}(t) \qquad \text{and} \qquad G = \left(\frac{\mathcal{N} V_{\mathrm{Gain}}\, c\sigma_L}{V_{\mathrm{mode}}}\right).$$

Verify that Eqs. (2.15) and (2.16) can be rewritten as

Inversion $\quad \dot{w}(t) = R_{\mathrm{Pump}}[1 - w(t)] - (1/T_1)[1 + I(t)]w(t),$

Cavity Intensity $\quad \dot{I}(t) = [Gw(t) - (1/T_C)]I(t).$

For cw pumping, R_{Pump} is a constant. The inversion and cavity intensity reach stationary values, which means both $\dot{w} = 0$ and $\dot{I} = 0$. Verify that at steady state the inversion, w_{ss}, and intensity, I_{ss}, are

$$w_{ss} = [1/(GT_C)] \qquad \text{and} \qquad I_{ss} = R_{\mathrm{Pump}} T_1 [GT_c - 1] - 1$$

Discuss the physical meaning of these results.

(6) Numerical investigation of the rate equations. Investigate the time dynamics of the rate equations given in problem 5 by programing them on a personal computer and numerically integrating them in time. Assume at $t = 0$ the $w(0) = 0$ and that R_{Pump} is a constant for $t > 0$. Why must you set the initial value of the intensity $I(0)$ equal to some small but finite value? Plot the dependence of the inversion and intensity it as a function of time. Verify that inversion and intensity approach the stationary values given in Problem 5. Discuss how the dynamics differ for the two cases: (i) $T_C \gg T_1$ and (ii) $T_1 \gg T_C$.

(7) Repeat Problems 5 and 6 but start with Eqs. (2.18)–(2.20), which include the effects of triplet-state absorption, excited-state absorption, and dye flow. Investigate how each process affects the steady state and the dynamics of the laser.

REFERENCES

Allen, L., and Eberly, J.H. (1975). *Optical Resonance and Two-Level Atoms*. Wiley, New York.

Boyd, R.W., Raymer. M.G., and Narducci, L.M., eds. (1986). *Optical Instabilities*. Proceedings of the International Meeting on Instabilities and Dynamics of Lasers and Nonlinear Optical Systems, University of Rochester (June, 1984), Cambridge University Press, Cambridge, England.

Duarte, F.J., Ehrlich, J.J., Patterson, S.P., Russell, S.D, and Adams, J.E. (1988). Linewidth instabilities in narrow-linewidth flashlamp-pumped dye laser oscillators. *Appl. Opt.* **27,** 843–846.

Fox, A.G., and Li, T. (1961). Resonant modes in a maser interferometer. *Bell Sys. Tech. J.* **40,** 489–508.

Fu, H., and Haken, H. (1987a). Semiclassical dye-laser equations and the unidirectional single-frequency operation. *Phys. Rev. A.* **36,** 4802–4816.

Fu, H., and Haken, H. (1987b). A band-model for dye lasers and the low threshold of the second instability. *Opt. Commun.* **64,** 454–456.

Fu, H., and Haken, H. (1988). Semiclassical theory of dye lasers: the single-frequency and multifrequency steady state of operation. *J. Opt. Soc. Am. B* **5,** 899–907.

Haken, H. (1966). Theory of intensity and phase fluctuations of a homogeneously broadened laser. *Z. Phys.* **190,** 327–356.

Hillman, L.W. (1988). Coherent dynamics of lasers: a source of instabilities and noise. In *Proceedings of the International Conference on Lasers '87* (Duarte, F.J., ed.). STS Press, McLean, Va., pp. 4–10.

Hillman, L. W., Boyd, R. W., and Stroud, C.R., Jr. (1982). Natural modes for the analysis of optical bistability and laser instability. *Opt. Lett.* **7,** 426–428.

Hillman, L. W., Krasinski, J., Boyd, R. W., and Stroud, C.R., Jr. (1984). Observations of higher-order dynamical states of a homogeneously broadened laser. *Phys. Rev. Lett.* **52,** 1605–1608.

Kogelnik, H., and Li, T. (1966). Laser beams and resonators. *Appl. Opt.* **5,** 1550–1567.

Lamb, W. E. (1964). Theory of an optical maser. *Phys. Rev.* **134,** A1429–A1450.

Narducci, L. M., and Abraham, N.B. (1987). *Lectures on Laser Physics and Laser Instabilities*. Taylor and Francis, New York.

Risken, H., and Nummedal, K. (1968). Self-pulsing in lasers. *J. Appl. Phys.* **39,** 4662–4672.

Chapter 3

FEMTOSECOND DYE LASERS

J.-C. Diels

Department of Physics and Astronomy
University of New Mexico
Albuquerque, New Mexico

ISBN 0-12-222700-X

1. INTRODUCTION

The very large gain bandwidth of dye lasers make them attractive candidates for ultrashort pulse generation. For instance, an argon-laser-pumped Rh6G laser can be tuned from 560 to 635 nm, or, in frequencies, from .53 to .47 10^{15} s^{-1}. The bandwidth is thus, in principle, capable of producing pulses as short as 16 fs. Bandwidth considerations are neither a necessary nor sufficient condition for ultrashort pulse generation. The large homogeneously broadened bandwidth of dye lasers implies that *linear* amplification of ultrashort pulses is possible. Computer modeling has shown that pulses can evolve through an amplifier, to reach a steady-state shape, with a duration shorter than the inverse linewidth of the gain medium (Arecchi and Bonifacio, 1965; Diels and Hahn, 1976). For the intense steady-state pulses, the rate of population transfers between the levels involved in the gain transition is comparable to the inverse of the pulse duration. It can be said therefore that, in this condition of coherent transient amplification, the relatively narrow gain line is always sufficiently power-broadened as to sustain the bandwidth of the ultrashort steady-state pulse. Whether coherent transient interaction effects play a role in the ultrashort pulse formation is still a matter of debate. As we shall show later, other nonlinear elements and effects play an important role in the fs laser operation.

Next to the gain medium, the most essential component of the fs laser is the saturable absorber, which also consists of an organic dye solution. Liquid dyes provide unique switches for fs signals not only because of bandwidth considerations, but also because of the large average power densities at which these nonlinear media can operate. The nature of the broadening of the absorber is essential. To say that the line is "homogeneously broadened" implies that the full bandwidth will be saturated by an intense signal at any frequency within the line. Such a global saturation implies that the correlation time between components of the line has to be faster than the excitation time. Measurements of phase-relaxation times imply that the absorbing lines can be considered to be homogeneously broadened up to at least 100 fs. For times in the tens of fs, the exact mechanism of intraband relaxation is still under investigation, and it may play a role in the ultrashort-pulse–formation process. Dispersive effects associated with the dye saturation play another important role in the pulse formation. At the high-energy densities reached in the focal points in the cavity, other nonlinear effects (Kerr effect, multiphoton absorption) complicate further the theory and understanding of the passively mode-locked dye laser. Historically, the mode-locked dye-laser development has been marked by successive realizations of the role played by any of the aforementioned mechanisms in the laser operation.

2. PASSIVELY MODE-LOCKED LASERS

2.1. Historical Introduction

The evolution of the mode-locked dye laser has followed a semi-empirical development in the 1970s and 1980s, each technical modification reflecting a step toward a better understanding of the ultrashort pulse formation. Today's passively mode-locked dye laser is the result of successive evolutionary mutations of the basic linear design of Ippen *et al.* (1972), which included all the basic building blocks of the present lasers. Gain is provided by an argon ion pump laser focused in a jet of Rh 6G. The beam inside the cavity has another focal spot in an absorber section, in which saturation of 3,3′-diethyloxadicarbocyanine iodide (DODCI) is used as the nonlinearity to phase lock the cavity modes. Ruddock and Bradley (1976) improved this basic linear design by introducing what is now called colliding pulse mode locking, reaching a pulse duration of 0.3 ps, a twofold improvement over the previous design. The basic idea is to use a standing-wave configuration (for instance by positioning the dye against the output mirror [Ruddock and Bradley, 1976]), to saturate the absorber. Because the mutual saturation coefficient is twice the self-saturation coefficient for a homogeneously broadened line, three times less energy density is required in the "colliding-pulse" configuration as compared to the traveling-wave configuration, to obtain the same level of saturation. The emphasis was next in ensuring that the cavity was broadband, by eliminating all intracavity transmissive elements and mixing the absorbing and amplifying jets in a single jet. A pulse duration of 120 fs was obtained with such a cavity (Diels *et al.*, 1978a; Diels *et al.*, 1980), where "edge mirrors" were used to restrict the wavelengths over which the laser can oscillate, to the 600–630-nm range. With emphasis still on a broadband cavity (edge mirrors and no intracavity prism), the colliding-pulse condition was reenacted through the ring configuration of Fork and Shank (1981). The "100-fs" mark was passed, with a reported pulse duration of 90 fs, which was subsequently upgraded to 65 fs (Shank, 1982), after Mourou and Sizer (1982) reported 70 fs in a linear hybrid mode-locked laser (all dyes mixed in a single jet, as in Diels *et al.*, (1980)).

It was realized next that prisms did not necessarily prevent a cavity from being broadband. The cavity optics could be such as to geometrically demagnify the angular dispersion effect (Dietel *et al.*, 1983). The prism was mounted on a translation stage, in order to provide an adjustable amount of intracavity dispersion. The ultrashort pulses that were being down-chirped in the saturable absorber excited below resonance were subsequently compressed by propagation through normally dispersive glass

(Dietel *et al.*, 1983; Fontaine *et al.*, 1983). As will be shown in the subsequent chapters, there is an optimum focusing in the absorber resulting in a maximum downchirp. Subsequent control of intracavity dispersion leads to the production of either linearly downchirped pulses or bandwidth-limited pulses of 55-fs duration (Dietel *et al.*, 1983; Diels *et al.*, 1985b). With very tight focusing in the absorber jet, however, the Kerr-effect–induced modulation can be made to dominate. With a negatively dispersive four-prism sequence (Gordon and Fork, 1984), the upchirped pulses can be compressed inside the cavity, resulting in a minimum pulse duration of less than 30 fs (Valdamanis *et al.*, 1985; Wang *et al.*, 1986; Wilhelmi *et al.*, 1986; Kubota *et al.*, 1988).

2.2. Ring Dye Lasers

Passively mode-locked ring dye lasers have become the most popular source of fs pulses. A general description of most common cavity configurations will be given in this section. The various parameters affecting the laser operation will be discussed in detail in Section 3 ("Theory of Operation").

We start with a cavity configuration including only one or two intracavity prisms (Dietel *et al.*, 1982, 1983; Fontaine *et al.*, 1983), as sketched in Fig. 3.1. A simple topological representation is used in Fig. 3.1b, where

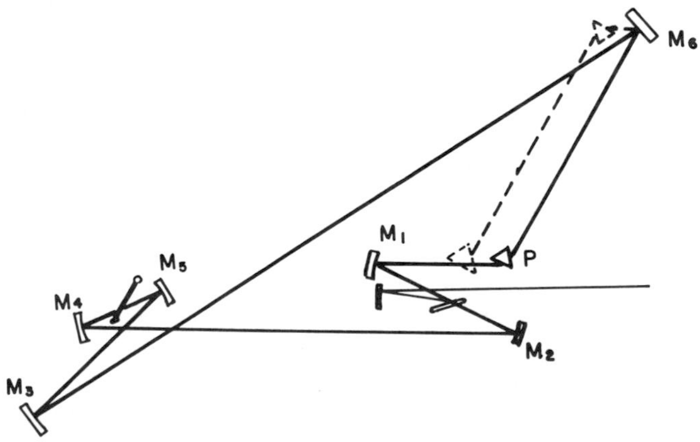

Fig. 3.1a Ring laser cavity with one (solid line) or two (dashed line) intracavity prisms. The focusing mirrors in the absorber (M_4 and M_5) and in the amplifier (M_1 and M_2) typically have radii of curvature of 30 and 50 mm, respectively. M_6 is a flat output mirror. The position and curvature of M_3 has a dominant influence on the spot sizes in the absorber and amplifier section. (See Section 5.3 of this chapter.) With a single prism in the cavity, the curvature of M_3 (typically 1 m) plays a crucial role in determining the cavity bandwidth.

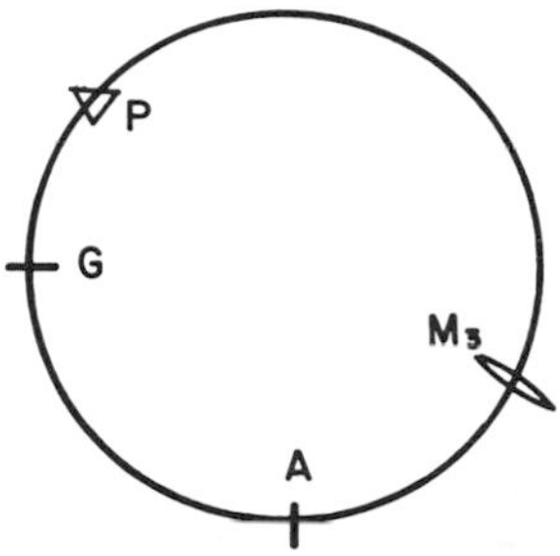

Fig. 3.1b Topology of the cavity sketched in Fig. 3.1a. The shortest distance between the spots in the gain (G) and absorber (A) jets is 1/4 of the cavity perimeter.

flat mirrors are ignored, curved mirrors are represented by lenses, and light beams are represented by arcs of circle. In the cavity including only one prism, broad bandwidth operation is ensured by the proximity of the dispersive prism and the mirror M_3 (for the clockwise propagating pulses) as well as by the curved mirror M_1 (for the counterclockwise propagating pulses). A self-compensating arrangement of two prisms is also often used with this cavity (dashed line) (Diels *et al.*, 1985b). There are two basic conditions to fulfill both in the layout and alignment of the cavity. In both arrangements sketched in Fig. 3.1, the cavity alignment has to be such as to provide an uneven number of beam waists in the cavity. Indeed, with an even number of foci in a resonator round-trip, a nonuniform transverse spectral distribution would result. With three waists in the cavity, the effect of the angular dispersion of the prism is inverted at each round-trip. The second condition is a topological requirement for all fs ring lasers. The dimensioning of the cavity should be such that the optical path between the dye jets is one quarter of the cavity perimeter. Indeed, only if in Fig. 3.1b arc AG = 1/3 of arc APG will the counter-propagating pulse enter the gain medium at equal time intervals. This condition is needed to have the two outputs of the laser balanced, since the gain recovery time is then equal between successive passages of each counter-propagating pulse. The exact tolerance on this requirement varies from cavity to cavity and also with the particular focusing condition. A mismatch of a few cm is generally acceptable for most lasers.

The energy densities in the absorber and amplifier jets are important parameters in passively mode-locked lasers that determine the initial phase of the pulse evolution, before the pulses are short enough for dispersive effects to dominate. It is for that reason that the particular cavity configuration and alignment are so important, since, as shown in Section 3.5 of this chapter, virtually any ratio of spot size in the amplifier and absorber media can be obtained by adjusting various focusing parameters (angles and distances). The role of the saturable absorber is the same as for solid-state mode-locked or Q-switched lasers in the initial stage of pulse evolution. The pulse leading edge will be reduced by absorption, which affects less the latter part of the pulse. Saturable absorption alone cannot lead to ultrashort pulse generation (i.e., pulses shorter than the absorber lifetime). If, as is generally the case, the energy-relaxation time of the absorber is much longer than the pulse duration, the saturable "switch" remains open, with a net gain for any perturbation at the end of the pulse tail to grow. There has to be a mechanism that limits development of the trailing edge of the pulse. This mechanism is depletion of the gain medium, which leaves a net negative gain in the cavity after passage of the pulse. The recovery time of the gain medium will determine the maximum pulse-repetition rate in a cavity. The interplay between gain and absorption saturation, as it is affected by the respective focusing parameters, will have a determining influences on the shape of the pulse developing in the passively mode-locked laser. The standing-wave saturation that takes place in the saturable absorber in the ring laser relaxes roughly by a factor of 3 the focusing constraints. In other words, for the same focal spot size in the absorber and amplifier, the beam focused in the absorber medium has its intensity "magnified" by a factor of 3 with respect to the saturation intensity. This is simply due to the fact that, in a purely homogeneously broadened medium, the mutual saturation coefficient equals twice the self-saturation coefficient.

Other pulse-compression mechanisms leading to the ultimate output of a fs laser and detailed in Section 3.6 are normal or anomalous dispersion on down- or upchirped pulses, and coherent propagation effects. The ultimate steady-state pulses result from a balance between pulse compression and broadening mechanisms that limit the minimum achievable pulse duration. Some of the broadening mechanisms contribute a stabilizing influence on the laser output. Pulsewidth-limitation factors include cavity bandwidth, dispersion, four-wave mixing type coupling and two-photon absorption in the dye jets.

As will be detailed in the following theoretical Section (3.5), the function of the glass prisms in the cavity sketched in Fig. 3.1 is to compensate a negative frequency modulation (downchirp). Such a downchirp results from near resonant saturation of the amplifier or absorber dyes, provided the focal spots in the dye jets remain larger than 10 microns. For tighter focusing in the dye jets, the contribution of the saturation to the phase modulation becomes negligible, while the Kerr effect from the solvent introduces a significant upchirp.

The relative importance of the Kerr effect in various laser configurations can be calculated easily. A typical laser will have an average output power in each beam of 20 mW, which, at the cavity rate of 100 MHz, corresponds to an energy per pulse of 0.2 nJ. The peak power is 3 kW for 50-fs pulses, which translates into an intracavity peak power of 100 kW. Typical operation for the laser of Fig. 3.1 requires a beam waist of 15 microns in the absorber jet (cf. Section 3.5.3). Assuming a Gaussian beam profile in the absorber, we thus have a peak power density of

$$I_p = 10^5 \mathrm{W}/\{\pi \cdot w_0^2/2\} = 2.5 \times 10^{10}\ \mathrm{W/cm^2}. \tag{3.1}$$

Two contributions of the Kerr effect have to be added. For the portion of the jet where the pulses overlap, the Kerr effect has to be spatially averaged over the standing wave, giving a contribution twice as large as that for a single pulse:

$$\Delta\phi_1 = 2n_2(I_p\tau \cdot c) \cdot [2\pi/\lambda]. \tag{3.2}$$

That contribution to the phase shift in the middle of the pulse has to be added to the contribution from each single pulse, integrated over the remaining portion of the dye jet:

$$\Delta\phi_2 = n_2 I_p d \cdot 2\pi/\lambda. \tag{3.3}$$

For the particular example being considered, with a jet thickness of 80 microns, the total phase shift in the middle of the pulse, due to the Kerr effect, is $\Delta\phi_1 + \Delta\phi_2 = 0.5^0 + 1^0 = 1.5^0$, where we have used for the Kerr coefficient a value of 1.10^{-15} cm^2/W. Such a small phase shift cannot make a noticeable contribution to the pulse spectrum. By contrast, in the antiresonant ring laser described shortly, the beam waist in the absorber can be as small as 1.1 micron. Therefore, even with the assumption that the pulses do not overlap in the absorber, we find a phase shift of nearly 2π at the center of the pulse, hence a positive frequency modulation that will significantly broaden the pulse spectrum and dominate the resonant contribution due to saturation of the absorber.

For these tight focusing configurations, negative dispersion is required to compensate the upchirp caused by the Kerr effect. Fork *et al.* (1984) have shown that such a negative dispersion can be obtained with a sequence of prisms. The sequence of four prisms introduced in the cavity and sketched in Fig. 3.2, has the property of being broadband (negligible effect of angular dispersion of the prisms), while having a second-order dispersion that can be adjusted from a negative value to a positive one. Very large compression factors are possible if very tight focusing is made in the absorber jet (large Kerr effect).

A negative dispersion, as that provided by a four-prism sequence, is required whenever the phase modulation induced by intracavity nonlinear effects is positive (upchirp). A four-prism sequence, such as that shown in

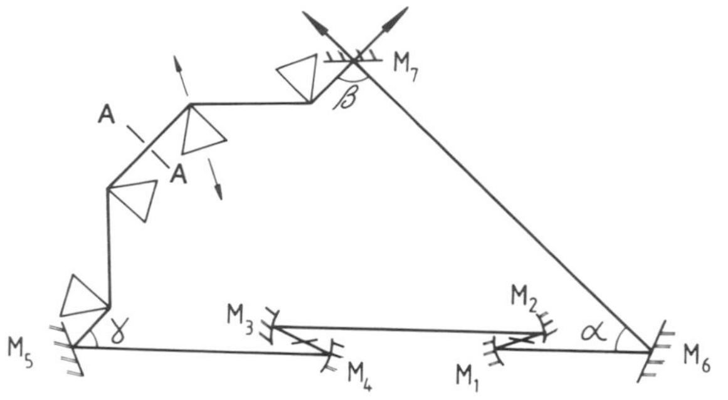

Fig. 3.2 Sketch of the four-prism cavity configurations used by Avramopoulos *et al.* (1988) and French and Taylor (1987). The values used for the angles α, β and γ were 70, 89, and 27°, respectively. The large value of β was used to shift the spectral region of high reflectivity of the mirror toward the yellow (French and Taylor, 1987). Folding angles of 7° were used. It is possible to control the cavity bandwidth with an aperture placed between the two central prisms (where the laser spectrum is spatially distributed) [from French *et al.* 1987b]. © 1988 IEEE.

Fig. 3.2, is not the only way to achieve the negative group velocity dispersion required for the compression of upchirped pulses. Reflective coatings can have negative and positive dispersion, as pointed out by De Silvestri *et al.* (1984a,b). However, it was demonstrated experimentally and numerically that the dispersion of standard reflection coatings (used near the center of their reflection band in order to have the uniform broad bandwidth needed for mode locking) is negligible compared to the negative or positive group velocity dispersion needed to compensate the frequency modulation induced in the jets (Diels *et al.*, 1984b; Dietel *et al.*, 1984a). The use of Gires-Tournois interferometers has been proposed by Dugay (1969) for compression of mode-locked pulses from solid-state lasers. Heppner and Kühl (1985) have implemented this concept for the femtosecond ring dye laser. If the dispersion function of the Gires-Tournois or of the four-prism sequence (of Fig. 3.2) is expanded in a power series of frequencies, the two types of dispersers differ at the third order. There is a need for a well-predetermined dispersion function up to higher orders than two, in order to compensate for the nonlinear chirp induced by the Kerr effect. The spectra of the shortest pulses obtained with the cavity of Fig. 3.2 are highly asymmetric, indicating that the pulse cannot be described by a real envelop function, but still have an important residual chirp. Fork *et al.* (1987) have demonstrated that, in carefully designing a negative group velocity dispersion delay line, it is possible to achieve nearly perfect extracavity pulse compression. Bandwidth-limited pulses as short as 6 fs were obtained by a pulse compressor that included prisms and gratings. A noise analysis by Kühlke *et al.* (1986a,b) demonstrates another need for a careful shaping of the *intracavity* dispersion function. An improved stability of the mode-locked operation was observed with a combination of four-prism sequence and an intracavity Gires-Tournois.

The question is still open whether there is a real need for multiplying the number of intracavity elements, thereby increasing the linear losses and the complexity of the laser alignment. The flexibility of design of a four-prism sequence has not yet been fully exploited, as will become clear in Section 3.6 dealing with this particular problem. Could pulses of less than 10 fs be obtained directly from the laser oscillator, without extracavity pulse compression? Attempts are being made by some groups to shape the reflection curve of mirrors to obtain a broader net bandwidth of the cavity as well as the desired dispersion function.

Passively mode-locked lasers are famed to be uncontrollable in wavelength, a rumor widely publicized by manufacturers of synchronously mode-locked lasers. Any strong positive control on the frequency of any femtosecond laser will affect its bandwidth and, hence, its pulse duration. However, for a given amplifier–absorber combination, a tuning range of 20 nm can be covered through adjustment of the saturable absorber concentration (Fontaine *et al.*, 1983; Vanherzeele *et al.*, 1984a). Other common tuning techniques involve the use of masks at the symmetry axis of the prism sequence (A--A in Fig. 3.2), where there is a spatial distribution of the laser spectrum, or tuning the angle of incidence on broadband mirrors to shift the reflectivity edge, or a combination of both techniques (French and Taylor, 1987). In addition, combinations of dyes have been found in 1986–89 to cover most of the visible spectrum with sources of femtosecond pulses, as will be detailed in Section 6.

The mode of operation of any femtosecond laser is a particularly sensitive function of the alignment procedure. This sensitivity is due to the fact that the competing nonlinear processes involved in the pulse formation, as discussed in Section 3, depend on the particular spot sizes. Systematic alignment procedures require a careful analysis of the beam parameters in the cavity and are reported in Section 3.5.

2.3. Linear Dye Lasers

The ring-laser cavity is not generally viewed as a "user-friendly" instrument. Considering for instance the most commonly used ring laser with four intracavity prisms (Fig. 3.2), one counts 19 optical surfaces in the resonator (including the jet surfaces), which represents a combined loss of 10% if only 0.5% is lost at each reflection–transmission. The laser is

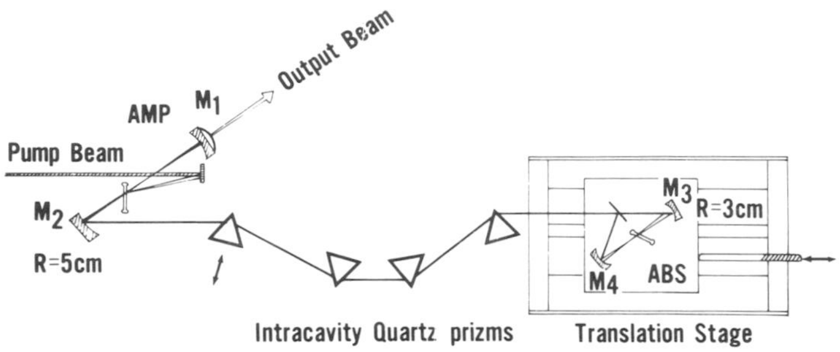

Fig. 3.3 Sketch of the antiresonant ring dye laser cavity. Mirror M_1 is a recollimating output mirror. For cavity lengths of 1.5 m or larger, the curvature of the mirrors should be at least 75 mm for M_1 and M_2, 50 mm for M_3 and M_4. In Table 3.1 (Section 3.5.2), θ is the angle of incidence on mirror M_2 (or half the folding angle). The angle at M_3 and M_4 is 16° [from Jamasbi *et al.*, 1988].

therefore unforgiving for poor alignment procedure and lossy optics. Another unfriendly feature of this laser is the sensitivity of the output beam alignment to any mirror adjustment. In an effort to incorporate the advantages of the ring fs laser in a simpler cavity, a linear laser terminated by an antiresonant ring has been proposed (Siegman, 1981) and demonstrated for both passive operation (Diels *et al.*, 1984c, 1986a,b) and hybrid operation (Diels *et al.*, 1984c, 1986b; Jamasbi *et al.*, 1988; Vanherzeele *et al.*, 1984b; Norris *et al.*, 1985). The antiresonant ring terminating the cavity (Fig. 3.3) acts as a broadband reflector (Siegman, 1973), and provides a standing wave in the absorber. The saturation enhancement however is not identical to that of the ring laser. Let I be the intensity of the pulse circulating in the cavity and I_s the saturation intensity of the absorber. The intensity-dependence of the absorption is $dI/dz = -aI/\{1 + bI/Is\}$ where $b = 3$ for the ring laser and 1.5 for the antiresonant ring laser. The reason for the different coefficients b is that, for the antiresonant ring-dye laser, unlike the situation in the ring laser, the beam is split in two parts before being sent to the jet. It is therefore essential in the case of the linear laser to design a cavity geometry that ensures tight focusing in the absorber jet. Methods to determine cavity configurations that satisfy the focusing requirements are discussed in Section 3.5. Details of the cavity configurations and procedures of alignment have been published (Jamasbi *et al.*, 1988).

With ajustable linear dispersion inserted in the cavity, the output characteristics of the passively mode-locked antiresonant ring-dye laser are similar to those of the ring laser. The pulse duration is typically 60 fs, for an average output power of 25 mW. An unusual feature of this laser is its ability to produce multiple pulse train (more than one pulse/cavity roundtrip) without degradation of the pulse duration, and with equal pulse spacing, with an accuracy better than 100 fs (Jamasbi *et al.*, 1988). Since the pulse-repetition rate is continuously adjustable, this laser can easily be pumped synchronously (Diels *et al.*, 1986a). The "hybrid mode-locking" characteristics are nearly identical to those of the passively mode-locked laser. Output pulses of 60 fs, for an average power of 100 mW, were produced.

In passive operation, and in contrast to the ring laser, it has not been possible to optimize the intracavity dispersion to obtain pulses free of phase modulation. The spectrum is not symmetric, and the interferometric autocorrelation indicates nonlinear chirp (Jamasbi *et al.*, 1988). This is not a fundamental limitation of the laser, but an indication that the dispersion function that has been chosen (sequence of four fused silica prisms) does not match the nonlinear frequency modulation induced in the dye jets. As will be discussed in Section 3.6, there are a large number of tools still available to optimize the intracavity dispersion to second and higher order.

3. THEORY OF OPERATION

3.1. Numerical versus Analytical Approaches

A global theory of operation of the mode-locked dye laser does not exist as of 1989. Various models have been studied, each of them putting emphasis on a particular aspect.

Analytical approximation and numerical approaches have been used to study pulse evolution and steady-state pulses. The analytical approach has the advantage of giving a better identification of the respective role of the various physical processes involved in the pulse formation.

Analytical models have to transform the physical system into a continuous mathematical model, where the function of the various cavity elements is continuously distributed in the cavity. In addition, one has to assume that the pulse evolution during any round-trip is infinitesimally small. What is lost in the model of a continuous medium filling the cavity is the eventual role of the order of the elements and oscillatory solutions specific to difference equations (cf. Section 3.3.1). The order of operations becomes important for very short pulses and/or large individual components. In the case of a large Kerr-effect modulation, requiring a large amount of quadratic negative dispersion for compensation, the same compression will not be obtained if the pulse is sent first through the dispersive elements. Indeed, because of pulse broadening by dispersion, not enough intensity is left for the nonlinear process. As an analytical model is expanded to include more physical phenomena, it will generally be solved by numerical approximation methods. It still remains basically an analytical approach, as opposed to the numerical one discussed shortly.

Numerical models can handle large intracavity absorptions, gain, losses, and dispersion in any round-trip. This flexibility can also turn to a weakness for the numerical simulation. To achieve convergence in a reasonable number of round-trips, investigators are sometimes forced to use artificially (i.e., non–physically realistic) large nonlinearities.

First we shall derive some general equations for the mode-locked dye laser. In the following subsections, each individual aspect of the laser operation will be treated separately.

3.2. Light Interaction with a Two-Level System

3.2.1. Density Matrix Equations

Because of the very fast relaxation rates in dye molecules, the absorption and/or gain lines can generally be treated as simple homogeneously broadened two-level systems. We shall start from the semiclassical equations for the interaction of near resonant radiation with a two-level system. These interaction equations are valid for either single traveling waves or counterpropagating fields, and they can be extended to the treatment of two-photon resonant interaction that also appears to play a role in the mode-locked operation. Some approximations can be made because of the very fast phase relaxation time of dyes, ranging from 12.5 fs for direct dipole transitions in Malachite green (Shank *et al.*, 1988) to 50 fs for two-photon transitions in DODCI (Diels and McMichael, 1986). Ultimately assuming this phase relaxation time to be infinitesimal leads to the rate-equation approximation.

Let us designate by $|a\rangle$ the ground state, $|b\rangle$ the excited state of a two-level system excited by a field $E = (\tilde{\mathscr{E}}/2)e^{i\omega t}$, at frequency ω, of complex amplitude $\tilde{\mathscr{E}}$. The density matrix equation for this two-level system is:

$$\dot{\rho} = \frac{1}{ih}[H_0 - pE, \rho], \tag{3.4}$$

where H_0 is the unperturbed Hamiltonian, and p the dipole moment. Equation (3.4) leads to the following differential equations for the diagonal and off-diagonal matrix elements:

$$\dot{\rho}_{bb} - \dot{\rho}_{aa} = \frac{2p}{h}[i\rho_{ab}E^* - i\rho_{ba}E] \text{ and} \tag{3.5}$$

$$\dot{\rho}_{ab} = i\omega_0\rho_{ab} + \frac{ipE}{\hbar}[\rho_{bb} - \rho_{aa}]. \tag{3.6}$$

It is generally convenient to define a complex "pseudo polarization" amplitude $\tilde{Q}(i\rho_{ab}p = (\tilde{Q}/2)\, e^{i\omega t})$, of which the real part will describe the attenuation (amplification) of the electromagnetic field. The complete system of interaction and propagation equations can be written as

$$\dot{\tilde{Q}} = i(\omega_0 - \omega)\tilde{Q} - \kappa\tilde{\mathscr{E}}\, w - \tilde{Q}/T_2 \tag{3.7}$$

$$\dot{w} = \frac{\kappa}{2}[\tilde{Q}^*\tilde{\mathscr{E}} + \tilde{Q}\tilde{\mathscr{E}}^*] - \frac{w - w_0}{T_1} \tag{3.8}$$

$$\frac{\partial\tilde{\mathscr{E}}}{\partial z} = -\frac{\mu\omega c}{2}\int \tilde{Q}(\omega_0)g(\omega_0)d\omega_0\,, \tag{3.9}$$

where $w = p(\rho_{bb} - \rho_{aa})$, $\kappa\mathscr{E}$ is the Rabi frequency, with $\kappa = p/\hbar$, ω_0 is the resonant frequency of the two-level system, and $g(\omega_0)$ describes the inhomogeneous line profile. T_1 and T_2 are, respectively, the energy and phase-relaxation times. Most of the energy-conserving interactions are generally lumped in the phase-relaxation time T_2. The phase of the various molecules excited simultaneously will average to zero after a time that is long compared to T_2, because of elastic collisions. As a dye molecule moves around in a solvent environment, it experiences changes in local field and hence, changes in resonant frequency because of time-dependent Stark shifts. Therefore the resonant frequency ω_0 becomes $\omega_0 + \delta$, where $\delta(t)$ is a random variable. This "cross-relaxation" is generally represented by a relaxation time T_3. Experiments of degenerate four-wave mixing (DFWM) have shown that the fast relaxation is dominated by processes best described by a relaxation T_2 (Diels and McMichael, 1986). As in most current models of the passively or actively mode-locked dye laser, we shall therefore ignore the effect of cross-relaxation. Maxwell's equation for propagation (Eq. (3.9)) is written in the slowly varying envelope approximation, and in the retarded time-frame of reference.

Another common set of notations uses only real quantities, such as the in-phase (u) and out-of phase (v) components of the polarization, and, for the electric field, its (real) amplitude $\mathscr{E}$ and its phase φ. Defining

$$\tilde{Q} = (iu + v)e^{i\varphi} \tag{3.10}$$

and substituting in the preceding system of equations leads to the usual form of the Maxwell-Bloch system of equations,

$$\dot{u} = (\omega_0 - \omega - \dot{\varphi})v - u/T_2 \tag{3.11}$$

$$\dot{v} = -(\omega_0 - \omega - \dot{\varphi})u - \kappa\mathscr{E}w - v/T_2 \tag{3.12}$$

$$\dot{w} = \kappa\mathscr{E}v - \frac{w - w_0}{T_1}, \tag{3.13}$$

with the initial value for w at $t=0$: $w_0 = p(N_2 - N_1)$,

$$\frac{\partial \mathscr{E}}{\partial z} = -\frac{\mu\omega c}{2}\int_{-\infty}^{\infty} v(\omega_0) g(\omega_0) d\omega_0 \tag{3.14}$$

$$\frac{\partial \varphi}{\partial z} = -\frac{\mu\omega c}{2}\int_{-\infty}^{\infty} \frac{u(\omega_0)}{\mathscr{E}} g(\omega_0) d\omega_0 \tag{3.15}$$

The vector representation of Feynman, Hellwarth, and Vernon (Feynman *et al.*, 1956) for the interaction equations is particularly useful in the description of coherent phenomena. For the system of Eqs. (3.11)–(3.13), the interaction can be described as the rotation of a pseudopolarization vector $\mathbf{P} = P(u, v, w)$ around a pseudoelectric vector $\mathbf{E} = E(\kappa\mathscr{E}, 0, -\Delta\omega)$ (Fig. 3.4a). For the system of Eqs. (3.7) and (3.8), the complex electric field is represented in the complex plane Q_i, Q_r (Fig. 3.4b).

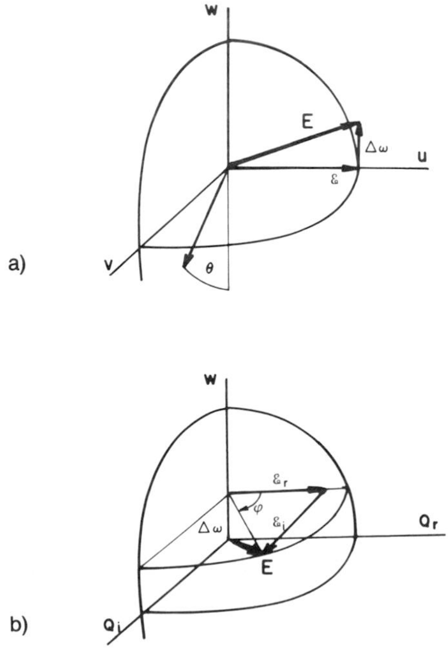

Fig. 3.4 Vector diagram for Bloch's equations. The motion of the pseudopolarization vector **P** (initially pointing downward along the w axis) is a rotation around the pseudoelectric vector **E** with an angular velocity proportional to the amplitude of that vector $|\mathbf{E}|$ (upper figure). In the complex amplitude representation (lower figure), the phase of the electric field determines the particular vertical plane containing the pseudoelectric field vector.

3.2.2. Conservation Equation

The duration of the ultrashort pulses being considered here is much shorter than the energy relaxation time T_1. The pulse energy is defined as

$$\mathcal{T} = \int_{-\infty}^{\infty} \frac{\mathcal{E}^2\, dt}{2\sqrt{\frac{\mu}{\epsilon}}}. \tag{3.16}$$

We shall use the notation T for energy densities (T_s for saturation energy densities). The pulse energy will be designated by $\mathbf{w} = ST$, where S is the beam cross section (not to be confused with the lower case s parameter of the ring laser cavity defined in Section 3.3). A simple energy-conservation law can be derived by integrating Eq. (3.14) over time:

$$\begin{aligned} \frac{d\mathcal{T}}{dz} &= \sqrt{\frac{\epsilon}{\mu}} \int dt \left\{ \mathcal{E} \frac{\partial \mathcal{E}}{\partial z} \right\} \\ &= -\frac{\mu\omega c}{2} \sqrt{\frac{\epsilon}{\mu}} \int v\mathcal{E} g\, dt\, d\omega_0 \\ &= \frac{h\omega}{2} \int \frac{\Delta w}{p} g\, d\omega_0, \\ w_0 &= -pN_0. \end{aligned} \tag{3.17}$$

3.2.3. Rate Equations

In most situations, T_2 is sufficiently short compared with the pulse duration, so that the interactions Eqs. (3.7), (3.11), and (3.12) can be considered in steady state. We can therefore solve for the polarization as a function of the quantity w, to find the rate equation:

$$\dot{w} = -\frac{\mathcal{E}^2(\kappa^2 T_1 T_2)}{1 + \Delta\omega^2 T_2^2} \frac{w}{T_1} - \frac{w - w_0}{T_1} \tag{3.18}$$

with $\Delta\omega = \omega_0 - \omega$.

The quantity $1/(1 + \Delta\omega^2 T_2^2)$ is the Lorentz profile of the homogeneously broadened line. A saturation field E_{s0} at resonance can be defined from Eq. (3.18) as

$$E_{s0} = \frac{1}{\kappa\sqrt{T_1 T_2}}. \tag{3.19}$$

Equation (3.18) can be written in terms of the saturation field, E_{s0}, and the population difference ΔN:

$$\begin{aligned}\Delta\dot{N} &= -\left[\frac{\mathscr{E}}{E_{s0}}\right]^2 \frac{1}{1+\Delta\omega^2 T_2^2}\frac{\Delta N}{T_1} - \frac{\Delta N - \Delta N_0}{T_1} - \frac{R}{2}\left[\Delta N + \Delta N_0\right] \\ &= -\left[\frac{\mathscr{E}}{E_{s0}}\right]^2 \frac{\Delta N}{T_1} - \frac{\Delta N - \Delta N_0}{T_1} - \frac{R}{2}\left[\Delta N + \Delta N_0\right] \\ &= -\frac{I}{I_s}\frac{\Delta N}{T_1} - \frac{\Delta N - \Delta N_0}{T_1} - \frac{R}{2}\left[\Delta N + \Delta N_0\right],\end{aligned} \tag{3.20}$$

where E_s and I_s are, respectively, the off-resonant saturation fields and intensity, and eventual contribution from other levels has been incorporated as a "pumping rate" $R(t)$. The coefficient of the pumping rate R ensures that the total number of participating molecules is conserved. The substitutions

$$\frac{1}{T'_1} = \frac{1}{T_1} + \frac{R}{2}, \; R' = -R\,\Delta N_0 \text{ and } I'_s = [1 - RT_1]I_s \tag{3.21}$$

lead to a more common form of the rate equation:

$$\frac{\partial\,\Delta N}{\partial t} = -\frac{\Delta N\, I}{I'_s T'_1} - \frac{\Delta N - \Delta N_0}{T'_1} + R'. \tag{3.22}$$

A high pumping rate leads thus to an effective shortening of the lifetime of the transition, an effect that should not be overlooked in mathematical simulations of the synchronously pumped mode-locked lasers.

Using Maxwell's propagation equations and a retarded time frame of reference lead to:

$$\frac{\partial\mathscr{E}}{\partial z} = \frac{\mu\omega c}{2}\kappa T_2 w\mathscr{E}. \tag{3.23}$$

Since in most practical situations with mode-locked dye lasers, the pulse duration is much shorter than the energy relaxation time T_1, the inverse of the rate pumping, and the pump pulse duration τ_p, the second term of the right-hand side of Eqs. (3.18) and (3.20) can be neglected, leading to the simple solution for the population difference:

$$\Delta N(t) = \Delta N_e \exp\left\{-\int_{-\infty}^{t} I\,dt/I_s T_1\right\} = \Delta N_e \exp\left\{-\int_{-\infty}^{t} I\,dt/\mathscr{T}_s\right\}. \tag{3.24}$$

In Eq. (3.24), ΔN_e is the population difference at equilibrium, positive for the amplifying medium, negative for the absorber. The evolution of the population of the two-level systems is solely determined by a saturation parameter

$$j(t) = -\int_{-\infty}^{t} I\,dt/\mathcal{T}_s = -W(t)/S\mathcal{T}_s, \qquad (3.25)$$

which is the energy density accumulated by the pulse at time t, normalized to the saturation energy density $\mathcal{T}_s = I_s T_1$. Since both the real and imaginary parts of the polarization are proportional to the density of participating atoms given by a $\exp\{-j_g\}$ and $\exp\{-j_a\}$ for the amplifying and absorbing media, respectively, the parameters j will be determinant in shaping the pulse both in amplitude and phase.

3.2.4. *Two-Photon Resonant Interaction*

In most passively or hybridly mode-locked dye lasers, the radiation is below resonance with the center of the absorbing transition. Most dyes, however, such as DODCI, have an excited-state absorption for higher-energy photons. Therefore, there will often be a (dipole-forbidden) two-photon transition near resonance with the laser radiation. The two-photon resonance enhances four-wave mixing processes (Diels and McMichael, 1986) which have been shown to contribute to satellite formation and play an important role in pulse stabilization in ring lasers (Diels *et al.*, 1982a) (see also Section 3.4). Two-photon absorption is also likely to play a role in pulse shaping and stabilization in a mode-locked laser, although this effect has not yet been investigated as of 1989.

As long as the intermediate level is more than one Rabi cycle off-resonance (Besnainou *et al.*, 1984), a similar derivation as in Section 3.2.1. can be made to yield the two-photon analog of Eqs. (3.7), (3.8), and (3.9):

$$\dot{Q} = i(\omega_0 - 2\omega + \Delta_2)\tilde{Q} - \frac{E_{01}E_{12}}{\Delta\omega_1} w - \tilde{Q}/T_2 \qquad (3.26)$$

$$\dot{w} = 4\,\mathrm{Re}\left\{\frac{E_{01}E_{12}}{\Delta\omega_1}\tilde{Q}^*\right\} \qquad (3.27)$$

$$\frac{\partial\tilde{\mathcal{E}}}{\partial z} = -\alpha_2\tilde{Q}\tilde{\mathcal{E}}^*. \qquad (3.28)$$

In these equations, $\Delta\omega_1$ is the detuning of the intermediate levels (the single-photon transition frequency as compared to the frequency of the two-photon resonant photon), while $E_{01} = \kappa_{01}\mathcal{E}$ and $E_{12} = \kappa_{12}\mathcal{E}$ are the Rabi frequencies corresponding to the transitions from the ground and from the excited states, respectively. $\Delta_2 = \{|E_{01}^2 - E_{12}^2|/\Delta\omega_1\}$ is the Stark shift induced by the electromagnetic pulse. The same rate-equation approximations that have been applied to the single-photon transition can be used in the system of Eqs. (3.26)–(3.27). One should be cautious however that the basic rate-equation approximation will be violated sooner for the two-photon than for the single-photon transition. Indeed, the shortest phase-relaxation time measured for a two-photon transition in dyes is 50 fs (Diels and McMichael, 1986), as opposed to 13 fs for a single-photon transition (Shank *et al.*, 1988).

3.3. The Basic Cavity Equations

3.3.1. *Growth of the Pulse Energy*

We assume that the pulses are shorter—and remain shorter—than the energy-relaxation time, and longer than the phase-relaxation time. The condition under which this is the case will be investigated in the next subsection. Within the basic approximation, very simple integral expressions can be found for the growth (or decay) of energy in such a mode-locked cavity.

In the following we shall designate by the indices $i = a$ and $i = g$ the absorbing and amplifying media, respectively. The pulse energy is the value of the function $W(t) = S \int I\, dt$ at $t = \infty$:

$$\mathrm{W}(t = \infty) = \mathcal{T}_i S_i \tag{3.29}$$

where S_i are the cross sections in the respective media. The growth (decay) of energy in any saturable element of the laser is found by combining Eqs. (3.17) and (3.24):

$$\frac{d\mathrm{W}}{dz} = S_i \alpha_{i0} \mathcal{T}_{si}[1 - e^{-\mathrm{W}/S_i\mathcal{T}_{si}}], \tag{3.30}$$

where the constant α_{i0}—positive for the amplifying medium, negative for the absorber—is the linear gain (absorption) coefficient of the unsaturated medium. Equation (3.30) can readily be integrated over the thickness d_i to yield:

$$1 - e^{+\mathrm{W}(d_i)/S_i\mathcal{T}_{si}} = e^{\alpha_i 0 d_i}[1 - e^{+\mathrm{W}(0)/S_i\mathcal{T}_{si}}]. \tag{3.31}$$

Beer's law is indeed the first-order approximation of Eq. (3.31). In order to find a simple expression for the cavity round-trip, we consider the simple sequence sketched in Fig. 3.5. The linear losses of energy are represented by a mirror reflection coefficient r. The pulse of energy $r\mathrm{W}_1$ enters the absorber, to exit with an energy W_2, which is taken as initial condition for the gain medium. Let us use the following reduced variables: $A = -\alpha_{a0}d_a$ and $G = \alpha_{g0}d_g$ are the optical thicknesses of the absorbing and amplifying jets, respectively. $y = \mathrm{W}_3/\{S_g T_g\}$ and $x = \mathrm{W}_1/\{S_g T_g\}$ are normalized input and output energies of the cavity for each cycle increment. We introduce a

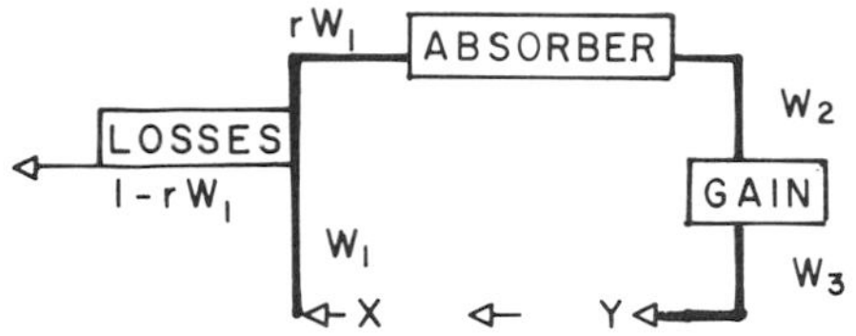

Fig. 3.5 Diagram of the energy gain and losses of a pulse circulating in one round-trip of the cavity.

parameter $s = kS_g\mathcal{T}_g/\{S_a\mathcal{T}_a\}$ to compare the saturation in the absorber relative to that in the amplifier. The number k is used to account for a reduction in saturation energy density in the case of standing-wave excitation ($k = 3$ for standing-wave excitation in the homogeneously broadened absorber, as discussed in the next section). With these notations, the energy growth in this cavity can be represented by

$$\{[1 - \exp(-G)] + \exp(y - G)\}^s$$
$$= [1 - \exp(-A)] + \exp(-A + rsx). \tag{3.32}$$

If we choose the sequence (mirror)—(gain)—(absorber), the cavity equation takes the form

$$\{1 - \exp(A) + \exp(A + sy)\} = \{1 - \exp(G) + \exp(G + rx)\}^s. \tag{3.33}$$

With the sign convention chosen here, both absorption and gain coefficients A and G are positive numbers. In the small signal approximation, we find that we have indeed

$$y = e^{(G-A)}rx. \tag{3.34}$$

The threshold condition for cw operation is

$$G > A + L \tag{3.35}$$

where $L = -\ln r$. The passively mode-locked laser however need only be above threshold for the saturated absorption only:

$$L < G < A + L. \tag{3.36}$$

The gain coefficient could be lower than that corresponding to cw operation. In that case, there is a marked "hysteresis" in the laser operation, which does not immediately reset itself after the intracavity laser beam has been interrupted.

Even though A and G are the *unsaturated* absorption and gain, Eq. (3.33) takes full account of the saturation in each medium. No assumption is made on the energy increment per round-trip, which can be large or

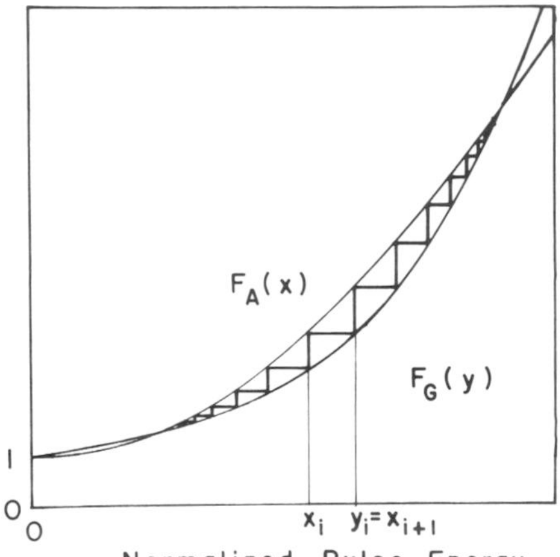

Fig. 3.6 Geometric representation of the energy of a pulse circulating in the cavity with a saturable absorber and a saturable gain. $F_G(Y)$ and $F_A(X)$ are the left- and right-hand sides of Eq. (3.33), respectively. The stability factor is defined as the ratio of the slopes of F_G and F_A at their stable crossing point.

small. The only assumption is that the losses or gain remain independent of the pulse shape and duration. Equations (3.32) and (3.33) are thus valid in the range of operation for which the pulse duration is shorter than the energy relaxation time of the dye, and longer than the inverse of the cavity bandwidth.

It should be noted that each element in this nonlinear equation is not necessarily small and that, therefore, linearization and small difference calculation are not always legitimate operations. For instance, saturated gain factors in excess of 2 and 6 have been measured in cw and synchronously pumped dye jets (McMichael, 1984; Diels *et al.*, 1984a).

The pulse evolution is calculated by introducing an initial value for the normalized input energy x, solving for the output energy y after the first round-trip, which is reintroduced as input energy x. This procedure is recycled, until steady state ($y = x$) is reached. Equation (3.33) can be solved graphically (Fig. 3.6), using exponentials of which the ordinate and abscissa have been translated (except that the right-hand side has been put to the power s). Each horizontal "step" in Fig. 3.6 is the energy increment at a particular round trip. In addition to the trivial solution $x = 0$, the two curves of Fig. 3.6 can have two crossing points, corresponding to an unstable ($x = x_1$) and a stable ($x = x_2$) solution of Eq. (3.36). The physical meaning of the intersection at x_1 (which exists in the case that the unsaturated gain is less than the unsaturated losses) is that there is a minimum fluctuation energy required to saturate the absorber and bring the laser above threshold. When that condition exists, the laser exhibits a marked hysteresis (Yoshizawa and Kobayashi, 1984). The operating energy of the laser is given by the second intersection at x_2. Equation (3.33) can be solved on a small personal computer to provide an insight into some main trends of the mode-locked laser:

threshold conditions,

minimum values of the parameter s,

output energy and its dependence on output coupling,

stability.

In the case where the unsaturated gain is less than the unsaturated losses, the minimum intracavity energy x_1 required to start mode-locked oscillation should be small (of the order of the gain or loss fluctuations). Within the approximation $sx \ll 1$ and $rx \ll 1$, Eq. (3.33) can be solved to yield

$$x_1 = \frac{2\{e^{A-G} - r\}}{s(s-1)r^2e^G + r^2 - se^{A-G}}. \tag{3.37}$$

A low threshold x_1 can still be found, even if the laser is operating below the cw threshold condition $r = \exp[A - G]$, provided a large value of s is chosen (tight focusing in the absorber). The second solution x_2 is the intracavity pulse energy. An approximate solution of (3.33) is found by assuming $x_2 \gg 1$, and neglecting 1 compared to the exponentials:

$$x_2 = \frac{\left[G - \dfrac{A}{s}\right]}{(1-r)}. \tag{3.38}$$

Since $x_2(1-r)$ is the output (assuming no other linear losses), the maximum output pulse energy normalized to the saturation energy in the gain medium, is simply G. Given a perturbation from steady-state Δx_2, the return to equilibrium will be fast if the ratio of the derivatives of the functions on both sides of the equality (Eq. 3.33) is largest. Consistent with the approximation $x_2 \gg 1$ of Eq. (3.38), we find for the ratio of these derivatives:

$$F'_A/F'_G = 1/r. \tag{3.39}$$

Numerical calculations show indeed that, in the case of large s factor and large gain, $x_2 > 1$, the maximum stability is indeed obtained for large output coupling. The increased stability is however at the expense of the output energy, as shown in Fig. 3.7, where the stability factor

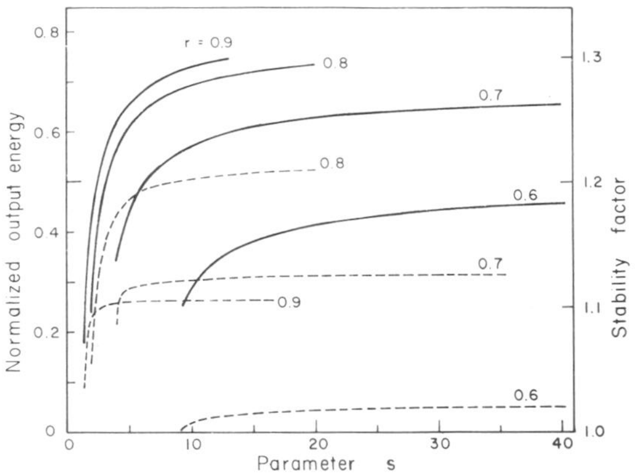

Fig. 3.7 Output energy, $x(1 - r)$, normalized to the saturation energy in the gain medium (solid line) and stability factor (dashed line) as a function of the parameter s, for various values of the reflectivity of the output mirror. The linear gain G and the linear absorption A are 0.25 and 0.22, respectively.

$[F'_A/F'_G]_{x=x_2}$ and the output energy $(1 - r)x_2$ are plotted as a function of s factor and output coupling, for relatively high gain cw pumped cavity. As compared to the cw lasers, optimum output energy is achieved for a smaller transmission factor of the output mirror. The approximation (Eq. 3.39) gives indeed the asymptotic value of the stability parameter. However, this limit value is approached at much smaller s values for small output couplings. Therefore, for optimum output power and stability of the cw pumped mode-locked laser, the linear losses, including the output coupling, should be kept below 20%.

Equation (3.33) applies also to the hybrid mode-locked laser, in a mode of operation dominated by saturable absorption. To model such a laser we can choose a gain factor G as large as 1.1 (Diels *et al.*, 1984a). Using a saturable absorption coefficient of $A = 0.8$, solutions of Eq. (3.33) consistent with the model are found for values of the feedback coefficient r between .6 and .8. Again, as for the passively mode-locked laser, optimum output power and stability are found for relatively low values of the output mirror transmission, as illustrated in Fig. 3.8.

An interesting sophistication of this simple model can be made by introducing losses that are proportional to the energy in the *previous* round-trip. Such losses do occur in most absorbing dyes, since the absorbed energy has generally a radiationless transition with a rate of a few ns to another group of levels, which can absorb the radiation at the next round-trip (triplet-triplet absorption). Losses proportional to the energy in the previous round-trip result in a reduction of the feedback (r) of the

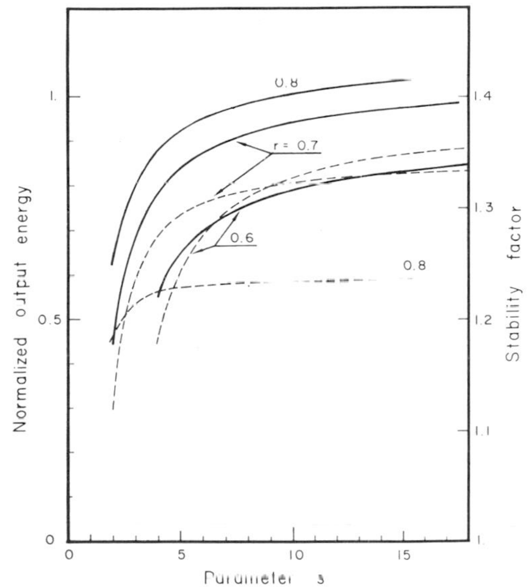

Fig. 3.8 Output energy $x(1 - r)$, normalized to the saturation energy in the gain medium (solid line) and stability factor (dashed line) as a function of the parameter s, for various values of the reflectivity of the output mirror. The linear gain G and the linear absorption A are 0.8 and 0.7, respectively.

cavity and are described by replacing the constant r by $r = r_0\,(1 - asx_1)$ in Eq. (3.33). The effect can be quite dramatic, even in this very simple model. Instead of reaching a steady state, the pulse energy can fluctuate in the cavity in a periodic fashion, with one or more characteristic frequencies, as illustrated in Fig. 3.9. This is another example of a mathematical system that can lead to deterministic chaos. The particular case selected for Fig. 3.9 has very large gain and s factor. Periodic solution can however be found in less extreme conditions. This model offers one possible explanation to the oscillation routinely observed in 1980s (see, for instance, Diels and Sallaba, 1980; Diels *et al.*, 1980) for the oscillations observed in the output of mode-locked lasers.

3.3.2. Emergence of a Pulse

After developing a simple model that describes the growth of energy of a pulse inside a cavity, the next step is to show that the energy will remain concentrated in a short pulse. As for the solid-state lasers, pulses will emerge from the random noise of the laser amplifier medium (see, for instance, Chekalin *et al.*, 1974). In this section the evolution of an arbitrary pulse toward a steady-state shape, as influenced by gain and absorption

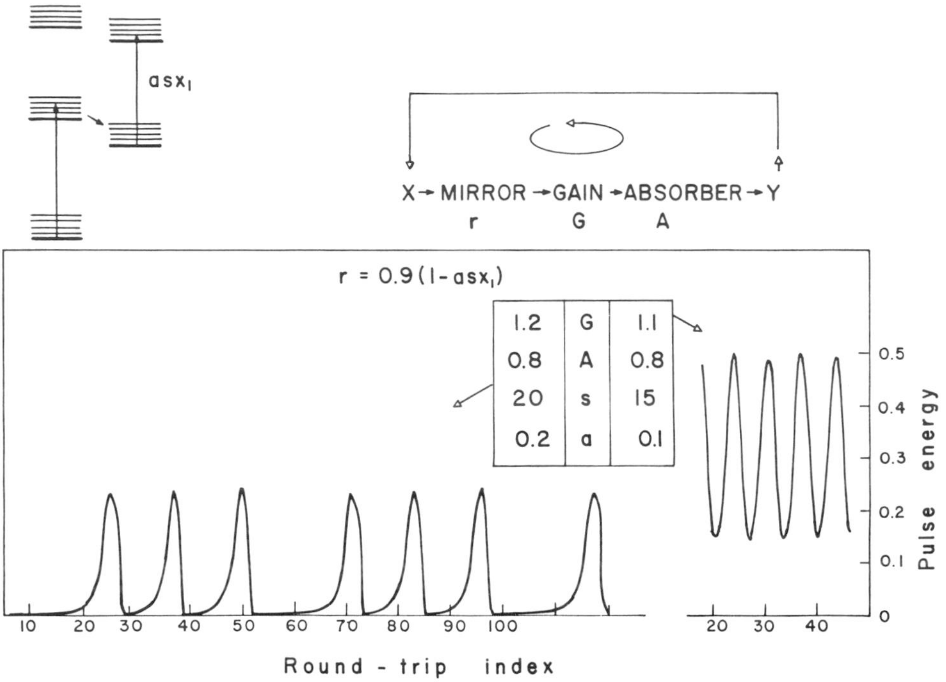

Fig. 3.9 Evolution of the normalized pulse energy x with round-trip index. Nonlinear losses proportional to the energy in the previous round-trip are included in the feedback coefficient $r = r_0(1 - asx_1)$, where sx_1 is the normalized energy in the previous round-trip, and a is the proportionality factor. The pulse energy versus round-trip index is represented for two sets of cavity parameters. The origin of the round-trip index is arbitrary but such that the portion of the evolution represented corresponds to a stationary condition.

saturation alone, is discussed. As in all calculations involving the particular shape of a pulse, a retarded time frame of reference is used.

Sharpening of the leading edge is due to a progressive decrease in attenuation as the saturating pulse enters the medium. Integrating Maxwell's equations (3.14) over time, between $-\infty$ and time t, using Bloch's equation (3.13) and the rate equation (3.18), leads to the saturation equation:

$$\frac{dW(t)}{dz} = -S_a|\alpha_{a0}|\mathcal{T}_{sa}\,[1 - e^{-W(t)/S_a\mathcal{T}_{sa}}]. \tag{3.40}$$

Let us define as $t = t_s$ the instant where the accumulated energy equals the saturation energy $s_a\mathcal{T}_{sa}$. The initial rise of the pulse, for which $t < t_s$, is attenuated by a factor exp $(\alpha_{a0}d) = \exp(A)$. If, for the later part of the pulse, $(t > t_s)$ $W(t) \gg \mathcal{T}S_a$, then the decrease in photon flux is only linear

$[dW/dz \simeq -s_a\alpha_{a0}\mathcal{T}_{sa}]$. For an exponentially rising pulse $I(t) = I_0e^{at}$:

$$j(t) = \int I\,dt = \frac{I}{a}e^{at} \tag{3.41}$$

$$\frac{dW(t)}{dt} = -|\alpha_{a0}|S_a\mathcal{T}_{sa}\,[1 - e^{-e^{at}I_0/aS\mathcal{T}_{sa}}] \tag{3.42}$$

which leads to a rise steeper than exponential.

Saturation of the absorber alone cannot account for an evolution toward short pulses. As pointed out by Haus (1975), after the absorber has bleached, there is a net gain for any disturbance that follows the pulse, before the absorber recovers. As it is being amplified, the pulse will deplete the gain medium itself, leaving a net negative gain in the cavity, in which small perturbations will be damped.

The same equations that were used to describe the saturation of the absorber apply as well to the gain medium. By proper adjustment of the focusing condition, saturation in the gain medium will be reached at a time $t_s' > t_s$. There is a time window, for $t_s < t < t_s'$, during which pulse amplification can occur, since the absorption losses are saturated, and the gain has not yet saturated. This time window will narrow as t_s' recedes, and the pulse leading edge steepens. A detailed analytical and numerical treatment of the pulse evolution due to saturable gain and absorption has been made by New (1972, 1974). Haus (1975) has added a bandwidth-limiting element to this model and found closed-form solutions for the steady-state pulses, using a "small signal" approximation for the absorber and amplifying media (which enables him to expand all exponentials to first order). A simple well-shaped frequency filter is the Gaussian:

$$H(\omega - \omega_d) = e^{-c[1+\omega-\omega_d/\omega_c]^2} \tag{3.43}$$

Provided the filter bandwidth remains broad compared to the pulse spectrum, one can use for $H(\omega - \omega_d)$ its first-order expansion, or, in the time domain, the operator:

$$1 - c\left[1 - \frac{1}{\omega_c^2}\frac{d^2}{dt^2}\right]. \tag{3.44}$$

Assuming a linear gain factor in the amplifier ($\Delta\mathcal{E} = g(t)\mathcal{E}(t)$) and similarly a linear loss factor in the absorber ($\Delta\mathcal{E} = \alpha(t)\mathcal{E}(t)$), the round-trip equation can be written as

$$\mathcal{E}(t+\delta) = \left\{1 - c\left[1 - \frac{1}{\omega_c^2}\frac{d^2}{dt^2}\right] + g(t) - \alpha(t)\right\}\mathcal{E}(t), \tag{3.45}$$

where the time mismatch δ allows for a small temporal drift of the envelope in its frame of reference, after each round-trip. Consistent with the other linearizations, the left-hand side of Eq. (3.45) can be replaced by $(1 + \delta d/dt)\,\mathscr{E}(t)$. The temporal dependence of the gain and absorption has been derived in Section (3.2.2) (Eqs. 3.24 and 3.25):

$$g(t) = g_0 e^{-j_g(t)} \simeq g_0(1 - j_g(t)) \tag{3.46}$$

$$\alpha(t) = \alpha_0 e^{-j_a(t)} \simeq \alpha_0(1 - j_a(t)), \tag{3.47}$$

where $j_g(t)$ and $j_a(t)$ are the accumulated energy densities normalized to the respective saturation energy densities in the gain and absorbing media. Equations (3.45)–(3.47) lead to a differential equation for the steady-state pulse, for the basic passively mode-locked laser cavity, that includes only saturation as pulse-shortening mechanism, and a finite bandwidth element to set a lower limit to the pulse duration. Such an equation is solved by assuming a functional dependence for the pulse shape, and substituting it in the equation to determine the various parameters (pulse energy, duration, temporal shift δ). We refer the interested reader to Haus (1975) for a detailed analysis of the cavity equation.

3.4. The "Colliding-Pulse" Effect

When the saturable absorber is put in close contact with a cavity end mirror (Ruddock and Bradley, 1976) or inside a ring cavity, the resulting standing wave affects in different ways the operation of the mode-locked laser. The space averaging of the standing-wave saturation leads to an apparent reduction of the saturation energy, by up to a factor of 3. On the other hand, the population grating induced in the medium introduces a coupling between the counterpropagating pulses.

3.4.1. Standing Wave versus Traveling Wave

Let us consider an optically thin absorber cell sketched in Fig. 3.10, illuminated from both sides by identical coherent beams.

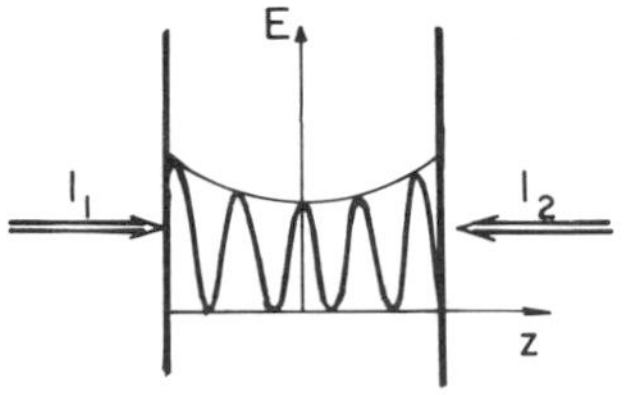

Fig. 3.10 Standing-wave saturation.

In the absence of the beam I_2, the attenuation of the beam I_1 is given by:

$$\frac{dI_1}{dz} = -\frac{\alpha I_1}{1 + I_1/I_s} \text{ or } \frac{d|E_1^2|}{dz} = -\frac{\alpha |E_1|^2}{1 + |E_1|^2/E_s^2}. \tag{3.48}$$

If a second beam of intensity $I_2 = I_1$ is sent counterpropagating to I_1, the medium is not irradiated by the sum of both intensities, but by the superposition of the two fields. The total absorption from both beams will thus be given by

$$2\frac{d|E_1|^2}{dz} = -\left\langle \frac{\alpha_0 \cdot 2E_1^2(1 + \cos 2kz)}{1 + 2E_1^2(1 + \cos 2kz)/E_s^2} \right\rangle_z \tag{3.49}$$

where spatial averaging is symbolized by the brackets $\langle \quad \rangle_z$. By performing the spatial averaging on the series expansion of Eq. (3.49), to first order (assuming for this simple discussion that the intensity of the radiation is small compared with the saturation intensity), we find

$$\begin{aligned} \frac{2d|E_1|^2}{dz} &\sim -2\alpha E_1^2\left[\langle 1 + \cos 2kz\rangle_z - \frac{2E_1^2}{E_s^2}\langle (1 + \cos 2kz)^2\rangle_z\right] \\ &= 2\alpha E_1^2\left(1 - 3\frac{E_1^2}{E_s^2}\right). \end{aligned} \tag{3.50}$$

Since symmetric illumination has been assumed, one can conclude from Eq. (3.50) that the saturation law for each beam of intensity I_1, in presence of a counterpropagating beam of the same intensity, is

$$\frac{dI}{dz} = \frac{\alpha I}{1 + 3I/I_s}. \tag{3.51}$$

The effect of standing-wave saturation is thus to reduce by a factor of 3 the saturation intensity, or the saturation energy density $W_s = I_s T_1 = \mathcal{T}_s/S$.

A more rigorous but tedious derivation can be made, starting from the density matrix equations (3.25) and (3.36), in which two counterpropagating fields are substituted for the total field E. A more general expression than Eq. (3.51), for unequal counterpropagating beams, is

$$\frac{dI_1}{dz} = \frac{-\alpha I_1}{1 + \beta I_1 + \theta I_2}, \tag{3.52}$$

where the mutual saturation coefficient θ equals twice the self-saturation coefficient β in a homogeneously broadened absorber.

The inclusion of a factor $k = 3$ in the "s parameter" (defined in Section 3.3.1.) facilitates the buildup phase of the mode-locking, since lower-energy pulses are required to bring the laser above threshold. It makes it possible for the ring laser to operate with a ratio of spot sizes in the two jets (gain or absorber) close to (or even less than) unity, thereby minimizing the contribution of the Kerr effect (Diels *et al.*, 1985b). Indeed, in a standing-wave configuration, there is only a twofold enhancement of the Kerr effect, of which the importance relative to the dispersive effects of the saturating dye (discussed in Section 3.6) is thus reduced.

In modeling the pulse evolution in the antiresonant ring cavity, the value of the enhancement factor k (of the s parameter defined in Section 2.3.1) should be varied with the pulse duration. Indeed, while it can be assumed that the long pulses at the initial phase of mode-locking overlap in the dye, this may no longer be the case of the ultrashort steady-state pulses. It is neither practical—nor necessary, as shown in Section 3.4.4,—to position the absorbing jet at equal distance, within 10 microns, of the beam splitter. The reduction of the s factor as the pulse energy increases may have a favorable influence on the pulse shortening and stability, by preventing the saturation time t_s to occur too early in the rise of the pulse.

3.4.2. *Induced Grating Effects*

The two counterpropagating beams induce a population grating in the absorbing medium, spanning the full beam overlap volume. Each of the beams will be reflected by this induced grating, resulting in a complex mutual coupling.

In order to derive equations for the mutual coupling, we can either start with the density matrix of Eqs. (3.5) and (3.6) of Section (3.2.1) or use directly the rate-equation approximation obtained by substituting the steady-state solution of Eq. (3.5) into Eq. (3.6):

$$\dot{w} = -\left[\frac{|E|^2}{E_s^2} w + (w - w_{eq})\right] \tag{3.53}$$

where E is the total field:

$$E = \tilde{\mathscr{E}}_1 e^{\mathrm{i}(\omega t - kz)} + \tilde{\mathscr{E}}_2 e^{\mathrm{i}(\omega t + kz)}. \tag{3.54}$$

Substitution of the field (Eq. 3.54) into the rate equation (Eq. 3.53) will produce source terms with spatial modulation $\exp\{2ikz\}$ and $\exp\{-2ikz\}$ for the population difference w. These terms in turn will combine with the field to produce still higher spatial frequencies which we shall neglect here. Let us consider the first-order solution for the population difference, which can be decomposed in a space-independent term w_1, and a sum of two couplex conjugated modulations of amplitude w_2:

$$w = w_1 + w_2 e^{2ikz} + w_2^* e^{-2ikz} \tag{3.55}$$

$$\dot{w}_1 = -\left[\frac{2w_1\tilde{\mathscr{E}}_1\tilde{\mathscr{E}}_2 + w_2^*\tilde{\mathscr{E}}_1^2 + w_2\tilde{\mathscr{E}}_2^2}{E_s^2} + (w_1 - w_0)\right] \tag{3.56}$$

$$\dot{w}_2 = -\left[\frac{2w_2\tilde{\mathscr{E}}_1\tilde{\mathscr{E}}_2 + w_1\tilde{\mathscr{E}}_1^2}{E_s^2} + w_2\right]. \tag{3.57}$$

The total polarization which is the source term in Maxwell's propagation equation is proportional to the population difference w. The equations can thus be grouped according to the propagation direction. Instead of using a retarded time frame of reference, it is more convenient to describe the pulses at each instant in moving spatial coordinates, using the transformation $z' = z \pm ct$, and $t' = t$, as sketched in Fig. 3.11. The propagation equations are of the form:

$$\frac{\partial\tilde{\mathscr{E}}_1}{\partial t'} = A_1\tilde{\mathscr{E}}_1 + A_2\tilde{\mathscr{E}}_2 \tag{3.58}$$

$$\frac{\partial\tilde{\mathscr{E}}_2}{\partial t''} = A_1\tilde{\mathscr{E}}_2 + A_2^*\tilde{\mathscr{E}}_1 \tag{3.59}$$

with

$$A_i = \frac{\mu\omega c^2}{2}\kappa T_2 w_i. \tag{3.60}$$

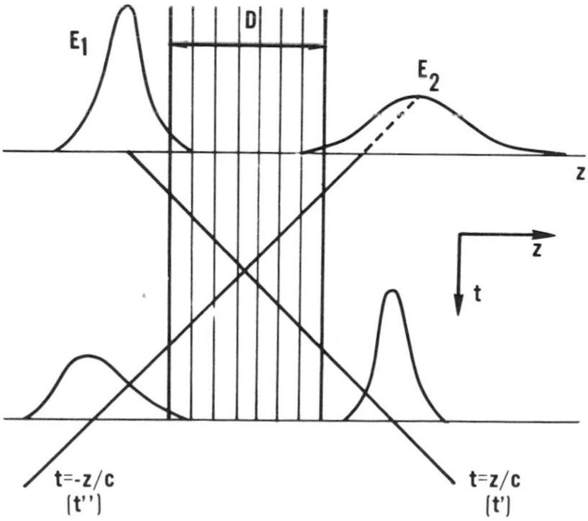

Fig. 3.11 Study of the interaction of two pulses counterpropagating in a thick medium. For each step in time, the interaction is computed as a function of the spatial variable z. Instead of the standard retarded time frame of reference, the appropriate change of variables is $z' = z,\ t' = t - z/c$ for the pulse E_1, and $z'' = z,\ t'' = t + z/c$ for the pulse E_2.

The mutual coupling term A_2 increases during the pulse from zero to a maximum fraction of the initial w_0, (a fraction that is determined by the ratio of the energy density accumulated at time t, $[\mathcal{T}(t)]$) to the saturation energy density $\mathcal{T}_s$. Given a strong pulse $\mathcal{E}_1$, the effect of the coupling term A_1 is generally to induce a gain on a weak counterpropagating pulse. This effect has been shown to be at the origin of the generation of satellite pulses in some mode-locked lasers (Diels *et al.*, 1982b). The calculations leading to the generation of such satellites are straightforward, since they can be made in the thin jet approximation.

Only numerical calculations can evaluate the influence of a finite jet thickness on the dye-laser operation through the mutual coupling mechanism of Eqs. (3.58)–(3.60). The interaction of the two pulses in the absorber jet is computed by successive steps in time, solving at each iteration the system of Eqs. (3.58)–(3.60) by a predictor corrector method. It was shown (McMichael, 1984) that the finite jet thickness has a pulse-broadening influence. In calculations of the pulse evolution in the cavity, the pulse duration was seen to stabilize at a minimum value, slightly shorter than the jet thickness, even in the absence of any bandwidth-limiting element in the cavity. Figure 3.12 shows the steady-state pulse shape that developed in this calculation.

3.4.3. *Phase-Conjugated Coupling*

Saturation does not account fully for the coupling between counterpropagating fields. The magnitude of the coupling term A_2 of Eqs. (3.58) and (3.59), as given by Eq. (3.60), is a lower limit. If we consider the case of DODCI as saturable absorber, the laser wavelength, typically around 620–630 nm, is significantly longer than that of the main resonance. There is however an absorption feature at half the laser wavelength, corresponding to a resonant two-photon transition. The latter has the main single-photon absorption peak as the intermediate level. The large absorption cross section from the ground state to the first excited state, and the even larger excited-state absorption cross section, contribute to a significant two-photon absorption probability. The two-photon resonance induces a coupling between the counterpropagating beams that is comparable to that due to the single-photon resonance. For simplicity, let us assume equal strength for the transitions from the ground and excited states. For the case of counterpropagating pulses of amplitude $\tilde{\mathcal{E}}_1(t)$ and $\tilde{\mathcal{E}}_2(t)$, Eqs. (3.26)–(3.28) are to be replaced by the following system of equations:

$$\dot{\tilde{Q}}_0 = i(\omega_0 - 2\omega + \Delta_2)\tilde{Q}_0 - 2\kappa^2 \frac{\tilde{\mathcal{E}}_1\tilde{\mathcal{E}}_2}{\Delta\omega_1} w - \frac{\tilde{Q}_0}{T_2} \tag{3.61}$$

$$\tilde{Q}_{1,2} = i(\omega_0 - 2\omega + \Delta_2)\tilde{Q}_{1,2} - \frac{\kappa^2\tilde{\mathcal{E}}^2_{1,2}}{\Delta\omega_1} w - \frac{\tilde{Q}_{1,2}}{T_2} \tag{3.62}$$

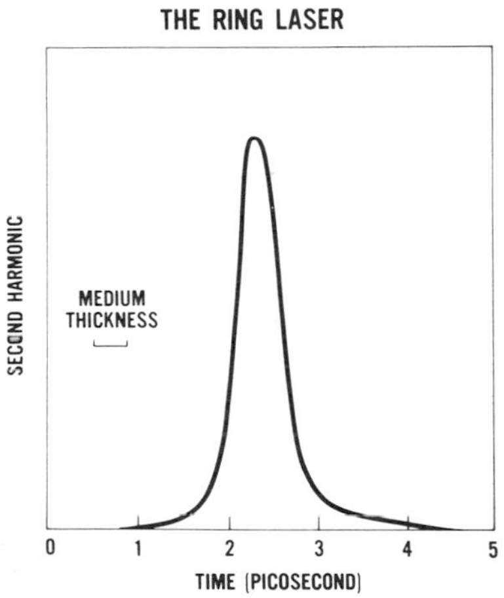

Fig. 3.12 Result of the numerical calculation of the pulse evolution in a ring cavity, without bandwidth-limiting element. Rate equations only are used to model the interaction in the gain and absorber media. Each pulse $E_{1(2)}$ undergoes successive gain, counterpropagating interaction, and saturable absorption in the presence of the counterpropagating pulse $E_{2(1)}$, a 1/4 cavity round-trip time later, dispersion, gain in the amplifier that has recovered for 1/2 cavity round-trip time, again counterpropagating interaction in the absorber that has recovered for a full cavity round-trip. The model was shown to match the experimentally measured pulse shape in the case of a linear cavity (Diels *et al.*, 1982b). Both counterpropagating pulses E_1 and E_2 converge toward the same shape, shown above, in the case of a "thick" jet (the medium thickness/c is indicated on the figure). In the thin-jet approximation, the same model shows continuous pulse shortening, without convergence toward a steady-state shape [from Diels *et al.*, 1982b].

$$\dot{w} = \frac{4\kappa^2}{\Delta\omega_1} \operatorname{Re}\left(\tilde{\mathscr{E}}_1^2 \tilde{Q}_1^* + \tilde{\mathscr{E}}_2^2 \tilde{Q}_2^* + 2\tilde{\mathscr{E}}_1 \tilde{\mathscr{E}}_2 \tilde{Q}_0^*\right). \tag{3.63}$$

To use notations consistent with the single-photon case, the initial condition for w is $w_0 = -p\mathrm{N}_0 = -\hbar\kappa N_0$. Using moving spatial coordinates defined previously, Maxwell's equations for the beam propagating along the positive direction z are the real and imaginary parts of:

$$\frac{\partial \tilde{\mathscr{E}}_1}{\partial t} = -\frac{\mu\omega c^2}{2}\left(Q_1 \frac{\kappa \tilde{\mathscr{E}}_1^*}{\Delta\omega_1} + Q_0 \frac{\kappa \tilde{\mathscr{E}}_2^*}{\Delta\omega_1}\right). \tag{3.64}$$

As for the case of single-photon transition, we can use for Q_0 its steady-state approximation. Assuming exact resonance,

$$Q_0 \simeq 2\frac{\kappa^2 \tilde{\mathscr{E}}_1 \tilde{\mathscr{E}}_2 T_2}{\Delta\omega_1} w \tag{3.65}$$

and the cross-coupling term of Eq. (3.64) is

$$\frac{\partial \tilde{\mathscr{E}}_1}{\partial t} = -\frac{\mu \omega c^2}{2} 2 \frac{\kappa^2 \tilde{\mathscr{E}}_1 \tilde{\mathscr{E}}_2}{\Delta \omega_1^2} \kappa w T_2 \tilde{\mathscr{E}}_2^* . \tag{3.66}$$

If we compare the cross-coupling term given by the two-photon resonance (Eqs. 3.64 and 3.65) to that given by the single-photon saturation, we note that the two-photon term is smaller by a factor 2 $\kappa^2 \tilde{\mathscr{E}}^2/\Delta\omega_1$, but this reduction may be compensated by the larger population difference w_0 and the larger phase-relaxation time (Diels and McMichael, 1986) T_2 of the two-photon transition.

The two-photon cross-conjugated coupling is phase-conjugated. If one can assume that the two counterpropagating pulses remain identical in the cavity, such a phase-conjugated coupling will have a stabilizing influence against phase modulation of the pulse. This chirp compensation effect could explain the very small (or even absence of) phase modulation observed in some ring dye lasers (Diels *et al.*, 1985a).

3.4.4. *Experimental Determination of a "Colliding-Pulse" Effect*

There are conflicting reports as to whether the "colliding-pulse" effect contributes effectively to the mode-locking process. For instance, French and Taylor (1987) report significantly shorter pulses in CPM configuration than in dispersion-compensated linear cavities, in contrast to the results reported by Valdmanis *et al.* (1985). Because of the sensitivity of the pulse-generation mechanism to the various beam parameters, it is difficult to make a quantitative comparison between two completely different cavities. In the case of the antiresonant ring linear laser, however, it is possible to switch from the "colliding" to "noncolliding" configuration without changing the cavity parameters. This is done by changing the relative lengths of the two arms of the antiresonant ring, as sketched in Fig. 3.13, all other cavity parameters remaining unchanged. As the beam splitter is moved away from the position corresponding to the symmetric ring,

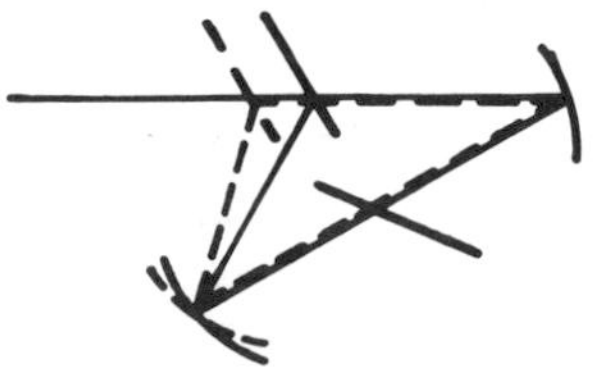

Fig. 3.13 Adjusting the colliding-pulse condition by moving the beam splitter relative to the other components of the antiresonant ring, it is possible to depart from the colliding pulse condition without modifying any of the other cavity parameters [from Jamasbi *et al.*, 1988].

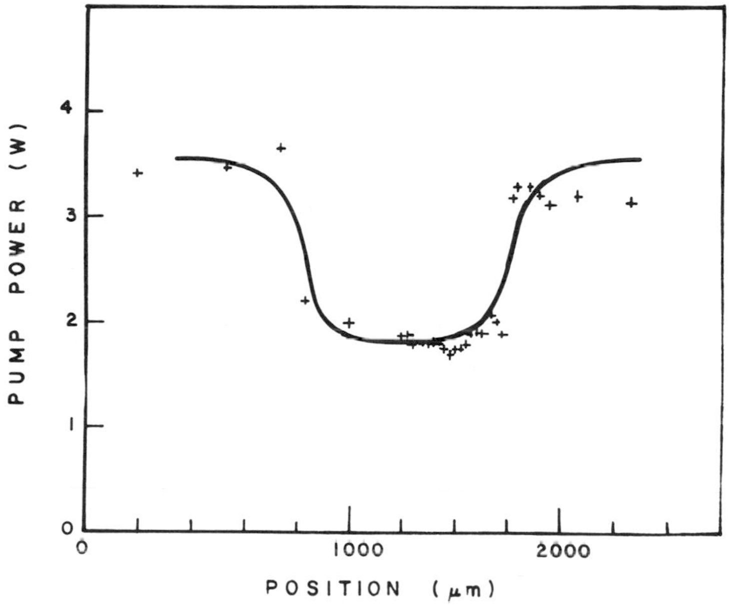

Fig. 3.14 Threshold (argon laser power, in W) for short pulse generation, as a function of the departure from colliding-pulse operation (the "zero" of the position scale is arbitrary). The same intensity autocorrelation is observed for all experimental points shown [from Jamasbi *et al.*, 1988].

the threshold pump power for *short-pulse operation* remains constant in an 800-micron–wide region and then increases from 1.7 to 3.5 watts (Fig. 3.14). However, none of the following parameters of the laser is affected by departing from the symmetric ring: the pulse output power, the intensity autocorrelation, and the losses of the antiresonant ring. The same pulse duration is even obtained when eliminating the beam splitter. However, the stability is reduced, and the threshold is a factor of 2 higher. Figure 3.14 indicates that standing-wave saturation plays an important role, at some point in the evolution of the pulse train, before pulse compression sets in. The width of the "well" indicates that, at this critical point of their evolution, the pulses have a duration of 3 ps.

3.5. Cavity Calculations

3.5.1. Importance of Astigmatism

It was shown in Section 3.3 that the optimum operation for stability and output power of the mode-locked laser corresponds to linear losses (including output coupling) not exceeding 20% for purely passive mode-locking,

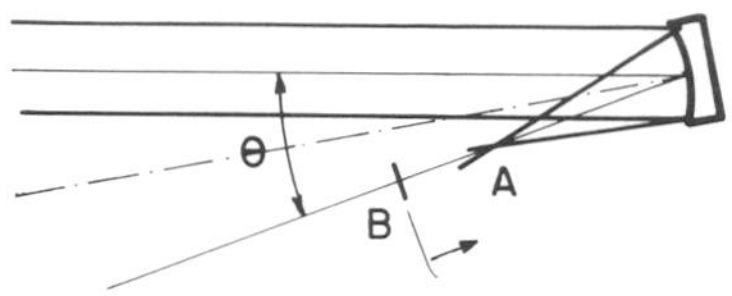

Fig. 3.15 After reflection on a spherical mirror, a collimated beam converges along two orthogonal focal lines A (perpendicular to the plane of the figure) and B. However, if the incident beam is converging in a plane perpendicular to the figure, the focal line B will recede toward A, resulting in a reduction of astigmatism.

or at most 30% in hybrid operation. The performance of the laser will therefore be very sensitive to any misalignment that would increase the losses. In the examples of energy evolution plotted in Fig. 3.7, an "s-parameter" of at least 10 was desirable for optimum stability and pulse energy. It is important to have the flexibility of varying the ratio of beam waists in absorber–amplifier in large proportions, in order to maintain $s > 10$ for the various dye pairs used as absorber–amplifier at different wavelengths (since different pairs of dyes operating at different wavelengths will have different ratios of cross sections and saturation energy). Furthermore, efficient pumping and heat removal of the gain medium calls for the smallest possible spot size. A round spot is desirable in the absorbing medium, in order to have the most uniform possible wavefront across the beam, since that is the region of the cavity where most of the phase modulation takes place.

Tight focusing in the absorber may be difficult to achieve with off-axis reflective optics. Indeed, let us consider a typical focusing configuration in an absorber jet as sketched in Fig. 3.15. The smaller the focal spot in A, the larger the diameter ϕ of the incident beam on the mirrors, hence the larger the clearance angle θ required. However, the astigmatism caused by a large angle of incidence θ will make it impossible to obtain the desired small focal spot. Minimization or suppression of astigmatism is thus an even more pressing problem for mode-locked lasers than for narrow-line dye lasers.

A ring cavity with two focal spots has a very small stability domain. Let us take as example a simple ring cavity, with mirrors of $2f_1 = 3$ cm curvature at the absorber jet and mirrors of $2f_3 = 5$ cm curvature at the amplifier jet. The distances between the two sections are taken to be 3 and 1 m. Assuming a Gaussian beam and parallaxial approximation, the beam characteristics—beam cross section and curvature of the wavefront—can be calculated using the method of ABCD matrices introduced by Kogelnik and Li (1966). Starting from the beam waist in the absorber jet, the various matrices to multiply are, in this particular case:

a propagation by a distance z (**M**1) $\begin{bmatrix} 1 & z \\ 0 & 1 \end{bmatrix}$ (3.67)

a lens of focal distance f_1 (**M**2) $\begin{bmatrix} 1 & 0 \\ -\frac{1}{f_1} & 1 \end{bmatrix}$ (3.68)

propagation for 3 m (**M**3) $\begin{bmatrix} 1 & 3 \\ 0 & 1 \end{bmatrix}$ (3.69)

a lens of focal distance f_3 (**M**4) $\begin{bmatrix} 1 & 0 \\ -\frac{1}{f_3} & 1 \end{bmatrix}$ (3.70)

Two successive propagations by a distance d (**M**5) $\begin{bmatrix} 1 & d \\ 0 & 1 \end{bmatrix}$ (3.71)

followed by a lens (**M**4) again, then

a propagation for a unit distance (**M**6) $\begin{bmatrix} 1 & 1 \\ 0 & 1 \end{bmatrix}$ (3.72)

and finally matrices $\mathbf{M}_2$ and $\mathbf{M}_1$ to close the loop. If we choose $z = f_1$, we find that the round-trip is represented by an **ABCD** matrix of which the elements are:

$$A = \frac{8(d - f_3) - 2(f_3 + f_1)d + f_3^2 + 2f_1f_3}{f_3^2} \tag{3.73}$$

$$B = \frac{2f_1^2(d - f_3)}{f_3^2} \tag{3.74}$$

$$C = ((f_3 - d)[8(1 - f_3 - f_1) + 2(1 - f_1)(f_1 + f_3)] + f_3[4f_3 - d(2f_3 + 2f_1 - 1)])/f_1^2f_3^2 \tag{3.75}$$

$$D = \frac{2(f_3 + f_1 - 1)d - f_3^2 + 2(1 - f_1)(f_3)}{f_3^2}. \tag{3.76}$$

The stability condition for this cavity

$$\left(\frac{A + D}{2}\right)^2 - 1 < 0 \tag{3.77}$$

yields a quadratic equation in d. For the parameters chosen, the two solutions of this equation show that the stability region in this simple cavity is restricted to

$$25.0 \text{ mm} < d < 25.254 \text{ mm}. \tag{3.78}$$

The stability range of this idealized cavity is only 254 microns! Because of the astigmatism, there are two sets of matrices corresponding to the tangential and sagittal focii. The stability range, restricted to the overlap of the stability ranges in the two planes, is therefore much smaller. For instance, if the angle of incidence on the amplifier mirror is 6°, the stability range is the intersection of the two regions:

$$24.863 < d < 25.116 \quad \text{and} \quad 25.137 < d < 25.393; \tag{3.79}$$

i.e., this cavity has no stable region!

Thus, there is clearly a need for minimizing or reducing the astigmatism in mode-locked laser cavities. Standard techniques, as discussed in detail in Chapter 4 of this book, involve the use of plane parallel plates in the regions where the beam is strongly convergent or divergent. Unfortunately, the dye jets used in mode-locked lasers are generally too thin to provide significant astigmatism compensation. Because they are in a portion of the cavity where the beam is practically collimated, the prisms offer no practical possibility of astigmatism compensation. The solution generally used is to minimize the angles on the mirrors, and, if needed, to add a third waist in the cavity where a compensating glass plate can be put at Brewster angle.

Since the astigmatism of successive elements is cumulative, it seems that we are facing a hopeless problem. In such a complex cavity as that of a passively mode-locked laser, ring or linear, common intuition can be misleading. Calculations show that large angles can actually result in astigmatism compensation, and, even with angles of incidence on the focusing mirrors exceeding 10°, large stability ranges are found (Jamasbi *et al.*, 1988). We shall show that, for linear as well as for ring lasers, the cavity geometry (relative location of the components) can be used alone to compensate the astigmatism, without a need for inserting additional "compensating elements." Returning to Fig. 3.15, a collimated beam incident from the left will focus first on a line A perpendicular to the plane of the figure next on a line B in the plane of the figure. However, if the beam incident from the left is collimated in the plane of the figure, but convergent in the orthogonal direction, the focal line B will recede toward the focus A. The incident beam parameters can be adjusted such as to create a tight round focal spot.

Complex cavities can be analyzed analytically, performing multiplications of two sets of matrices (one for the plane of the cavity, the other for the orthogonal plane) as previously shown. Symbolic manipulation languages are needed, but are not a sufficient substitute for human manipulation. The accuracy of numerical calculations of the elements of the final **ABCD** matrix is strongly dependent on the particular form of the analytical expressions and on the order of calculations. In the stability region of the example shown previously, the left-hand side of Eq. (3.77) is at most -0.1. The stable region may not be found if A and D are not calculated with an accuracy better than 0.2%. If no care is taken to reorganize the raw expression from the algebraic manipulation language, the expression for A and D will contain differences of large terms—such as [(cavity length) $\cdot f_3$—(cavity length) $\cdot d$]. Because of the large number of such expressions, even double precision may not be sufficient to calculate the matrix elements A, B, C, and D with an accuracy of only 0.2%. It is possible to eliminate all such differences, factoring out small terms such as $(d - f_3)$.

3.5.2. *Linear Cavities*

Let us first consider the simple linear cavity of Fig. 3.3. Some rough numerical estimates show clearly that the contribution of the folding angles at M_2 and M_3 (and M_4) dominates that of the thin dye jets and that of the four-prism sequence. Because of the symmetric configuration of the antiresonant ring, it can be replaced by combination of one of the curved mirrors and a flat mirror positioned at the absorber jet, for the calculation of the cavity modes. Choosing the absorber jet as origin, the complete round-trip **ABCD** matrix was calculated using the algebraic manipulation language MACSYMA (Jamasbi *et al.*, 1988). The influence of intracavity prisms used in compensating configuration is generally minimal and has therefore been neglected. Inclusion of prisms in the **ABCD** matrix formulation has been made by Duarte (1989). The expressions for the matrix elements were used in a FORTRAN optimization program, which scanned all possible values of the folding mirror angle θ and intermirror distance $M_1 - M_2 = d$, to determine the stability range $z_{max} - z_{min}$ (range of distance between the jet and the antiresonant ring mirror M_3 for which the cavity is stable) and to find the optimum conditions (z, d, θ) which give a *round focal spot* in the absorber, of *minimum size*. Since the absorber jet is the starting point of the calculation, the beam waist is simply given by

$$\left|\frac{\pi w_i^2}{\lambda}\right|^2 = \frac{B_i^2}{1 - A_i^2} \tag{3.80}$$

where the index $i = H$ for the plane of the figure (Fig. 3.3), which is the sagittal plane, and $i = V$ for the orthogonal plane, which is the tangential plane. The beam parameters are subsequently propagated from the beam waist to other parts of the cavity, to determine the important parameters of the beam in the cavity (size of the spots on the mirrors, and in the amplifier jet). Detailed calculations of the stability range and the beam parameters have been published for two sets of focusing optics ($R_1 = R_2 = 7.5$ and 5 cm; and $R_3 = R_4 = 5$ and 3 cm) (Jamasbi *et al.*, 1988). In general, the stability diagram (plots of the stable regions in the θ, d, z space) shows forbidden ranges between stable domains at large angles. The practical consequence is that the laser cannot be tuned continuously from one stable region to the other (for instance by increasing the distance d and optimizing simultaneously the other parameters). Therefore there is a need for an alignment procedure that systematically sets the cavity in a preselected stability region. To illustrate such an alignment tecnique, let us consider a section through the stability diagram of the cavity of Fig. 3.3, with $R_1 = R_2 = 7.5$ cm and $R_3 = R_4 = 5$ cm, corresponding to an angle $\theta = 16°$. The upper and lower limit of the stability range (solid lines), as well as the spot sizes in the absorber (dotted lines) and amplifier medium (dashed lines) are represented in Fig. 3.16. The minimum beam waist in the absorber corresponds to a round spot with $w_{aH} = w_{aV} = 1.2\ \mu$m, at intermirror spacing of $d = 75.8$ mm and $2z = 30.06$ mm. The value of the parameters d and z corresponding to the plotted spot sizes is given by the thin solid line in the stability diagram (shaded region). This example illustrates well the large s parameters that can be achieved through optimization of the cavity geometry, since the ratio of the focal spots of the amplifier to the absorber exceeds 50,000 at that particular point. The beam waist in the absorber being a monotonous function of the mirror spacing d, it is possible to adjust the s parameter of the laser as required for a particular combination of amplifier and absorber dyes.

Complete calculation of the cavity parameters would be only an academic exercise, if it were not possible to accurately align the cavity in predetermined geometry. For a given mirror spacing d, the distance between the point source in the jet and mirror M_2 is taken to be $(d - R_1)$ (this choice limits to two the number of images of the fluorescence, since the source is then imaged upon itself through mirror M_1). Geometric optics is used to calculate the sagittal and tangential images of the fluorescence spot through M_2. For the minimum spot size condition of Fig. 3.16, the correct intermirror spacing d is obtained if the vertical image of the fluorescence is at a point A (Fig. 3.3) 344 mm away from M_2. The pump geometry is thereafter optimized by minimizing the threshold of a simple three-mirror cavity obtained by positioning a flat (parallel faces) mirror at position A. The laser beam from this three-mirror cavity is used to align the antiresonant ring. The spacing between the mirrors of the antiresonant ring is adjusted for maximum feedback into the three-mirror laser.

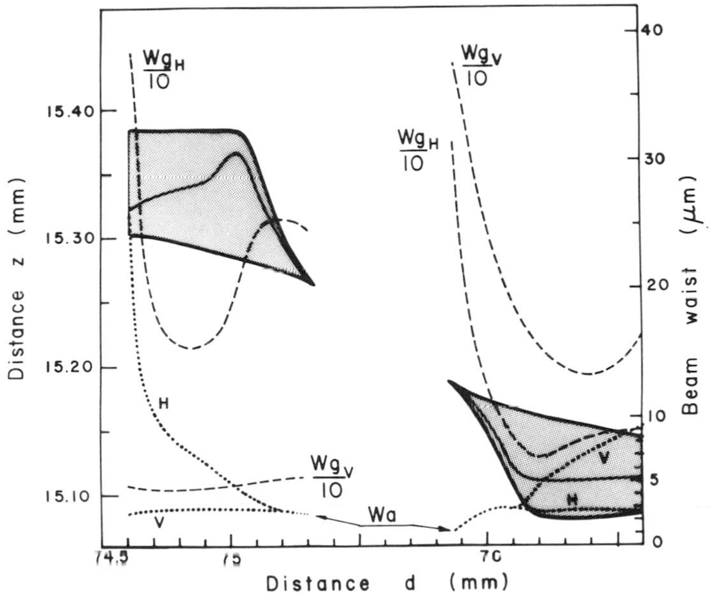

Fig. 3.16 Cross section through the stability diagram of the antiresonant ring dye laser of Fig. 3.3. The angles at the curved mirrors of the antiresonant ring triangle are 16°. The folding angle (at the gain section) is 32°. The stability range is indicated by the shaded region. The radii of the beam waist in the absorber, for which the absolute value of the difference $w_{aH} - w_{aV}$ is minimized, are plotted as a function of the distance d. The corresponding radii of the beam waists in the amplifier section are also shown (dashed lines). The values of the parameters d and z corresponding to the plotted spot sizes are given by the thin solid line in the stability diagram (shaded region) [from Jamasbi *et al.*, 1988]].

It should be noted that the calculations of the cavity of Fig. 3.3 can be applied to other linear cavities. For instance, the antiresonant ring section can be replaced by two curved mirrors or by a curved mirror and a flat mirror.

The astigmatism of successive folding angles is cumulative. Therefore, one would expect that the addition of another fold (Fig. 3.17) would further degrade the quality of the beam emitted by the laser. As in the previous example, simple intuitive rules fail in a cavity of that complexity. A complete mapping of the cavity parameters for different cavity geometries has shown (Jamasbi *et al.*, 1988) that it is possible to have simultaneously a round output beam and a very small (round) waist in the absorber jet. Table 3.1 gives some geometric parameters of a 1.978 m cavity, with folding-mirror radii of 75 mm at the amplifier section and 50 mm at the antiresonant ring section, as a function of the (equal) folding angles (2θ) at the amplifier section, the distance between folding mirrors d, the distance h

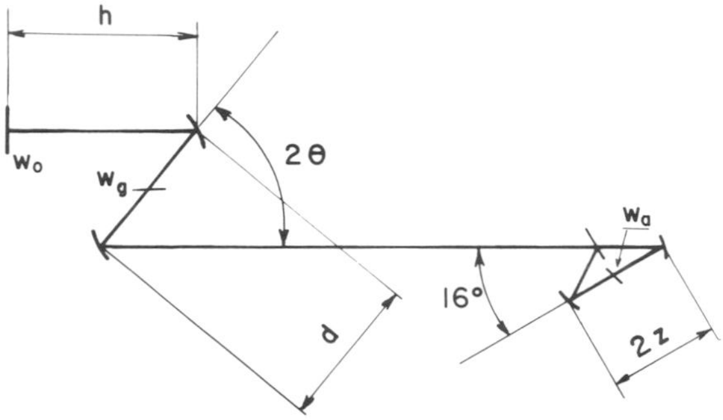

Fig. 3.17 Parameters of the extended antiresonant ring laser cavity.

from the output mirror to the first folding mirror, and the spacing $2z$ between mirrors at the antiresonant ring. s_a is the ratio of the beam cross section in the amplifier jet to that in the absorber. As in the previous example, it can be seen that, despite the large angle of the antiresonant ring triangle (16°), very tight focusing is possible in the absorber jet, without the insertion of any astigmatism-compensating element in the cavity.

3.5.3. Ring Cavities

As shown at the beginning of this section, the focusing optics for the two dye jets will have an overwhelming contribution to the astigmatism of the cavity, as compared to that of the thin dye jets and the pairs of prisms. However, as in the case of the linear laser, a complete cavity calculation may show stable regions where contributions to the astigmatism originating from different parts of the cavity cancel each other, resulting in very sharp focusing in one of the dye jets.

As in the case of the linear cavity, the origin is chosen at midpoint between the absorber mirrors (distance between mirrors, $2z$; radius of curvature, R_1). The angles of incidence on these mirrors are both equal to θ_1. The angles of incidence on the two mirrors of curvature R_3 and spacing $2d$ of the amplifier section are both equal to θ_3. The two sections are connected in one direction by a straight propagation matrix of distance e, in the other direction by a propagation matrix of distance v, a reflection on a mirror of curvature R_2, at an incidence θ_2, and a propagation for a distance w. Since the two arms of the cavity should be in a length ratio of 1 to 3, the distance w can be expressed as a function of the other parameters ($w = 3e + R_1 + R_3 - v$). The resulting **ABCD** matrix can be used to represent various ring configurations as in Fig. 3.1 (where the curved mirror R_2 could be in position M_3 or M_6, and the only acceptable solutions are

Table 3.1

Typical Geometric Parameters of the Antiresonant Ring Laser

θ	s_a	Waist area in ABS $(\mu m)^2$	Spot size in the absorber w_H	w_V	Output eccentricity	d (cm)	z (cm)	h (cm)	Stability range (mm)	Remarks
5.5	540	8.25	2.30	2.28	−0.886	7.801	2.542	42.5	0.002	large eccentricity small stability
8	38.6	30.65	4.04	4.83	0.724	7.643	2.559	40	0.564	
13	95	15.38	3.135	3.124	0.741	7.520	2.558	45	0.052	
14	400	6.26	2.00	1.99	−0.828	7.488	2.557	40	0.006	round spot in both jets
14	150	4.35	1.66	1.66	−0.626	7.706	2.533	40	0.001	tight focusing in absorber
15	8.3	51.50	4.65	7.00	−0.02	7.817	2.551	40	0.27	round output beam
15.42	6.75	40.69	5.09	5.09	0.14	7.752	2.545	45.62	0.22	round colimated beam between amplifier and absorber
15.42	21.09	12.12	2.72	2.83	0.46	7.760	2.530	43.75	0.27	
16.5	2,000	3.03	1.39	1.38	−0.78	7.385	2.562	42.5	0.003	largest s in that range of θ
16.5	15.6	107.	25.41	2.68	0.003	7.618	2.556	42.5	0.09	round output beam

Table 3.2

Typical Geometric Parameters of the Ring Laser

						Spot size in the absorber				Third beam waist					
										H		*V*			
θ_1	θ_3	R_2	*v* (m)	s_a	Waist area in ABS $(\mu m)^2$	w_H	w_V	*d* (cm)	*z* (cm)	*at* (cm)	*w* (mm)	*at* (cm)	*w* (mm)	Stability range (mm)	Remarks
9	10	1	0.5	1.3	520	10.0	33	2.506	1.429	105	0.054	N		0.2	
13	10	1	0.5	1	940	24	25	2.548	1.438	263	0.425	N		0.055	
9	13	1	2	104	7.2	3.8	1.2	2.588	1.528	46	0.085	N		0.01	
9	15	1	2	54	7.8	4.5	1.1	2.495	1.537	38	0.089	132	0.055	0.017	
13	20	1	1.75	26	14.3	1.1	8.3	2.741	1.487	86	0.045	N		0.023	
13	20	1	2.08	620	8	2.3	2.2	2.789	1.486	84	0.112	N		0.005	
17	10	1	2	12	44.5	3.5	8.1	2.633	1.467	61	0.14	N		0.036	
5	10	∞	—	0.15	264	58	2.9	2.524	1.518	N		N		0.011	Smallest gain spot $(40\ \mu m^2)$
5	10	∞	—	7	34	12	1.8	2.534	1.517	N		N		0.011	Smallest absorber spot
9	10	∞	—	2	63.3	12.6	3.2	2.519	1.531	11	0.3	N		0.078	

$\theta_2 = 8.3°$

"N" indicates no third beam waist

those for which there are three beam waists in the plane of the figure) or in Fig. 3.2 (where the radius of curvature R_2 is chosen to be infinite). The same procedure as in the former subsection can be used, namely to optimize the interabsorber mirror spacing $2z$ for best "round spot" (smallest absolute value of the difference between sagittal and tangential beams waists), for every value of the other parameters (distance between amplifier mirrors d, distance v, angles θ_1 and θ_3). Some typical results are summarized in Table 3.2. The cavity of Fig. 3.1 gives a minimum spot of equal dimension in the absorber and amplifier sections. A tighter focusing can be obtained in the case of the cavity of Fig. 3.2 in the absorber jet (60 μm^2). Even smaller spot sizes are possible for smaller angles θ_1, as indicated in Table 3.2. However, the minimum angle of 5° may not be sufficient for beam clearance (at the jet and at the mirrors). Much tighter focusing and largest values of the s parameter are obtained if the curved mirror of focal distance R_2 is located in position M_6 (Fig. 3.1). As in the case of the linear cavity, the smallest spot sizes are not necessarily found for small angles of incidence on the curved mirrors. The very small round focal spot in the absorber obtained for $\theta_1 = 13°$, $\theta_3 = 20°$ (Table 3.2) is an example of geometric astigmatism compensation.

3.6. Phase Modulation and Dispersion

3.6.1. Role of Phase Modulation and Dispersion

Progress in a quantitative assessment of the dispersive properties of the laser cavities and their components have led to dramatic improvements in source performances. It has been shown in Section 3.3 that saturation mechanisms will compress the intracavity pulses. As the pulse duration approaches the 100fs range, phase modulation and dispersive effects become dominant shaping mechanisms.

A somewhat confusing expression has appeared in the literature: that of "chirp compensation" to describe a succession of self-phase modulation and dispersion. These are two different processes which are time-frequency analogs of each other. In the sketch of Fig. 3.18a, a short pulse of amplitude E(t) and of phase $\varphi(t)$ is incident on a medium with an intensity-dependent index of refraction $n_0 + n(t)$. The pulse emerging from a thickness d of that medium will have the same amplitude function, but a phase modulation or "chirp" incremented by

$$\Delta\varphi(t) = -\Delta k \cdot d = \frac{2\pi}{\lambda}\Delta n(t) \cdot d. \tag{3.81}$$

Pulse dispersion can be described by an analogous figure in the frequency domain (Fig. 3.18b). The ultrashort pulse is now described by its spectral

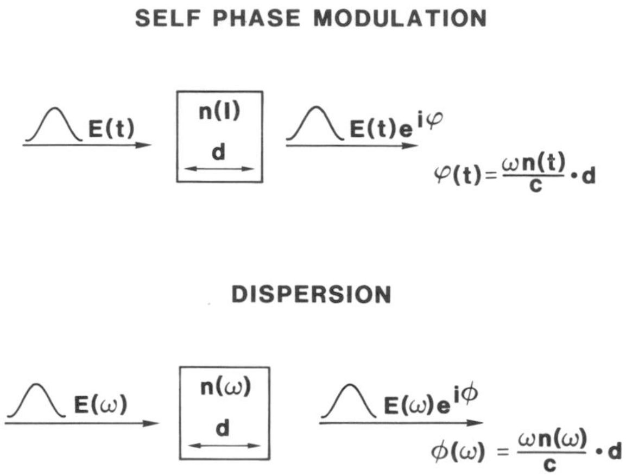

Fig. 3.18 Self-phase modulation (Fig. 3.18a, top) and dispersion (Fig. 3.18b, bottom). In the time domain, a transparent medium with a time- (or intensity-) dependent index of refraction will affect the phase and not the shape (amplitude of the electric field in Fig. 3.18a). In the frequency domain, a transparent dispersive medium (frequency-dependent index of refraction) will affect the phase only and not the shape of the spectrum (Fig. 3.18b).

amplitude function $E(\omega)$, which has a phase function $\phi(\omega)$. If this pulse is incident on a transparent medium with a frequency-dependent index of refraction $n(\omega)$, the spectral intensity distribution of the transmitted pulse will not be affected, but the phase of the spectral component at ω will be modified according to

$$\phi(\omega) = -k \cdot d = -\frac{2\pi}{\lambda} \cdot n(\omega) \cdot d = -\frac{\omega}{c_0} n(\omega) d. \tag{3.82}$$

Both processes of self–phase modulation and dispersion are needed to transform a bandwidth-limited pulse into a shorter bandwidth-limited pulse: the first one to broaden the spectrum, without affecting the temporal profile of the pulse; the second one to change the pulse amplitude in time, without affecting the spectrum. It should be noted that, in a Taylor expansion of the phase function $\phi(\omega - \omega_\ell)$ of Eq. (3.82) around the central pulse frequency ω_ℓ, terms of physical significance start only at the order 2. The zero-and first-order terms reflect respectively the different phase and group velocities of the light in the media traversed, and they will be ignored in the following calculations.

The laser itself is an example of a device where a combination of self-phase modulation and dispersion takes place. Self-phase modulation and dispersion takes place at the intracavity beam waists, where the intensity is sufficient to induce sufficiently large nonlinear effects. Dispersion takes place in all reflective and transmissive elements of the cavity, including the resonator itself. The various cavity designs differ generally by the particular component that has been selected for its dominant dispersion.

Both outputs of a ring laser are not identical. The difference can be significant in the case of large phase modulation–dispersion in each round-trip. Let us compare, for instance, the situation of a large bandwidth-limited pulse, being sent through a medium with a quadratic nonlinearity and through a dispersive medium. Let us assume that the amount of dispersion has been matched to optimally compress a pulse that has traversed the nonlinear medium first. If, instead, the dispersive medium is traversed first, the pulse incident on the nonlinear medium will have been broadened by dispersion, and it will emerge with less phase modulation than would have been produced by the original bandwidth-limited pulse. We refer to a problem at the end of this chapter for an evaluation of the importance of the order of operations, dispersion, and phase modulation, for Gaussian-shaped pulses.

3.6.2. *Dispersion*

Let us consider first a typical ring cavity, where the dominant dispersive components are prisms. The relative importance of the various elements contributing to dispersion will be discussed at the end of this section. For a complete cavity round-trip, we can separate a purely "material" contribution to dispersion, over the total thickness d_G of highly dispersive transparent medium being traversed, and a contribution due to the cavity perimeter P being a function of the frequency. P can also designate a frequency-dependent optical length of a group of components (such as a sequence of prisms), external or internal to a cavity. Ignoring terms of order less than 2, the leading terms of an expansion of the frequency dispersion (3.82) applied to a laser cavity are

$$\phi(\omega-\omega_\ell)=\frac{1}{2c_0}(\omega-\omega_\ell)^2\left\{d_G\left[\frac{d^2(\omega n(\omega))}{d\omega^2}\right]+\left[\frac{d^2(\omega P(\omega))}{d\omega^2}\right]\right\}+\frac{1}{3!c_0}\cdots \tag{3.83}$$

(ω_ℓ being the laser frequency). It is more convenient to express the second derivatives of Eq. (3.83) in terms of derivatives with respect to the wavelength. For any spectral quantity $x(\omega)$, we have the relation:

$$\frac{d^2[\omega x(\omega)]}{d\omega^2}=-\frac{\lambda^3}{2\pi c_0}\frac{d^2x}{d\lambda^2}. \tag{3.84}$$

Substituting in Eq. (3.83) yields an expression for the modification in the phase factor of the Fourier component of the light field at $\Omega = \omega - \omega_\ell$:

$$\phi(\Omega) = -\frac{\lambda^3}{4\pi c_0^2}\left(d_G \frac{d^2 n}{d\lambda^2} + \frac{d^2 P}{d\lambda^2}\right)\Omega^2 + 0(\Omega^3). \tag{3.85}$$

The second derivative of the index of refraction versus wavelength in (3.85) is generally positive. As a result, normal dispersion will only be appropriate to compress pulses with a negative phase modulation or downchirp. Let us consider, for instance, such a Gaussian pulse, with a pulse duration (FWHM of the intensity) of $\tau = \sqrt{2 \ln 2}\, w$; and a linear downchirp characterized by the parameter a:

$$\mathscr{E}(t) = \mathscr{E}_0 \exp\left\{-\left[1 + ia\right]\left(\frac{t}{w}\right)^2\right\}. \tag{3.86}$$

The Fourier transform of the pulse emerging after propagation through a thickness of glass d_G is given by

$$\begin{aligned} e^{-ik \cdot d_G} & \int_{-\infty}^{\infty} \mathscr{E}(t) e^{-i\Omega t}\, dt \\ &= \frac{w\sqrt{\pi}}{\sqrt{1 + ia}} \exp\left\{-\left[\frac{\Omega^2 w^2}{4(1 + a^2)}\right](1 - ia)\right\} \exp[i\phi(\Omega)]. \end{aligned} \tag{3.87}$$

The pulse duration will be reduced by a factor $\sqrt{1 + a^2}$ provided the glass thickness is adjusted for a vanishing value of the imaginary argument of the exponential:

$$d_G = \frac{\pi c_0^2}{\lambda^3} \frac{a w^2}{1 + a^2}\left(\frac{d^2 n}{d\lambda^2}\right)^{-1}. \tag{3.88}$$

For any other value of d_G, the emerging pulses will be longer (and consequently phase-modulated, since they are spanning the same bandwidth as the minimum-duration pulse). Figure 3.19 shows a typical series of autocorrelations taken for increasing amounts of intracavity glass for the laser cavity of Fig. 3.1.

Some values of material dispersion $d^2 n/d\lambda^2$ are given for a few standard glasses in Table 3.3. A 300-fs laser pulse with a downchirp parameter of $a = 3$ can easily be generated by a mode-locked ring laser (Diels *et al.*, 1985a). Optimum compression by a factor of 3 can be obtained by propagating such a pulse through 160 mm of BK7 glass. An adjustable dispersion inside the cavity prevents such a large chirp from developing. Only a few mm of glass is required to compensate the chirp generated at each roundtrip (with partial compensation, the chirp generated by off-resonance saturation of the dye is typically in the range $0.03 < a < 0.15$).

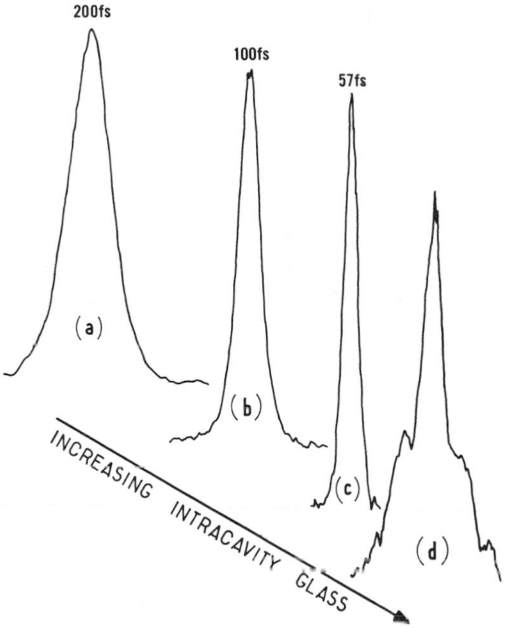

Fig. 3.19 Successive intensity autocorrelation of the output of the ring laser of Fig. 3.1, taken for increasing amounts of intracavity glass. The corresponding pulse durations are indicated (assuming sech^2 pulse shapes) [from Diels *et al.*, 1985a].

Table 3.3

Dispersion of Various Materials Near 0.62 μm

Material	Index	$dn/d\lambda$	$d^2n/d^2\lambda$	Ref.
BK7	1.515548	−0.03635991	0.1388040	Schott
K7	1.509825	−0.03910139	0.2031956	Schott
K10	1.500002	−0.03959791	0.1581152	Schott
BaK I	1.570957	−0.04420354	0.2066466	Schott
F2	1.617473	−0.07357275	0.3433244	Schott
LaSF9	1.847	−0.11808	0.638166	Melles Griot, 1988
SF10	1.724440	−0.10873030	0.5381957	Schott
Quartz	1.457	−0.030509	0.1267	Melles Griot
ZnSe	2.589	−0.355	0.756	Duarte, 1987
CaF_2	1.433	−0.0174	0.1739	Melles Griot

It is only under well-defined saturation conditions that a downchirp can be generated in the laser cavity. The most common source of intra- or extracavity phase modulation is the Kerr effect. A dominant Kerr effect results in a positive frequency modulation or an "upchirp." Gordon and Fork (1984) have shown that the geometric contribution to the dispersion (second term in Eqs. 3.83 and 3.85) can be positive or negative, and can therefore be used to compress upchirped pulses.

Dispersion alone, in the absence of a source of direct phase modulation, will contribute to pulse broadening and a quadratic chirp. Retransforming back the expression (Eq. 3.87) to the time domain, for the particular case of an initially bandwidth limited pulse ($a = 0$), it can be shown (cf. Problem 3 at the end of this chapter) that the pulse is given a linear chirp, or the phase modulation:

$$\phi(t) = a_c t^2 \tag{3.89}$$

while the pulse duration is being multiplied by the factor

$$(1 + a_c^2/2) \tag{3.90}$$

where the chirp coefficient a_c is given by

$$a_c = \frac{2(\ln 2)\lambda^3}{\pi c_0^2} d_G \left(\frac{d^2 n}{d\lambda^2} + \frac{d^2 P}{d\lambda^2} \right) \frac{1}{\tau^2}. \tag{3.91}$$

It should be noted that the pulse broadening due to dispersion is quadratic in distance of dispersive medium traversed, hence a second-order effect as compared to pulse compression or broadening due to propagation of a linearly chirped pulse in a dispersive medium (which is a linear function of the amount of dispersion).

The geometric dispersion is most often associated with a wavelength-dependent deflection angle θ. For the geometric dispersion parameter $d^2P/d\lambda^2$ of Eq. (3.85),

$$\frac{d^2 P}{d\lambda^2} = \frac{d^2\theta}{d\lambda^2}\frac{dP}{d\theta} + \frac{d^2 P}{d\theta^2}\left(\frac{d\theta}{d\lambda}\right)^2. \tag{3.92}$$

The second term of this expression should be small, since it is directly related to the bandwidth $\Delta\lambda$ of the device or cavity:

$$\Delta\lambda = \Delta\theta \left(\frac{\partial\theta}{\partial\lambda} \right)^{-1}. \tag{3.93}$$

Methods to calculate the bandwidth of simple and complex cavities are detailed in Chapter 4. For use with femtosecond dye lasers, it is necessary to select resonators or prism assemblies that are in a compensating configuration (Duarte, 1987) in order to have a bandwidth compatible with the pulses to be generated. On the other hand, the whole expression—hence the first term—of Eq. (3.92) should be a large negative number in order to compress ultrashort pulses with a large positive chirp. Equation (3.92) can be rewritten as (Fork *et al.*, 1984):

$$\frac{d^2P}{d\lambda^2} = \left[\frac{d^2n}{d\lambda^2}\frac{d\theta}{dn} + \left(\frac{dn}{d\lambda}\right)^2 \frac{d^2\theta}{dn^2}\right]\frac{dP}{d\theta} + \left(\frac{dn}{d\lambda}\right)^2 \left(\frac{d\theta}{dn}\right)^2 \frac{d^2P}{d\theta^2}. \tag{3.94}$$

New techniques to calculate this expression for any prism configuration and cavity have been developed by Duarte (1987). The reader is referred to Chapter 4, Section 4.5, for a detailed study of cavity dispersion and bandwidth as it relates to prismatic dispersion and its derivatives.

3.6.3. Mirror Dispersion

Since the laser is operating at a wavelength for which the losses are minimal, one expects the mirrors to have a constant and maximum reflectivity for the wavelength range covered by the pulses. However, the complex reflection coefficient will have in general a frequency-dependent phase, which has to be added to the overall cavity dispersion in Eq. (3.83). Dispersion will ultimately lead to phase modulation (Eq. 3.89) and to pulse broadening (Eq. 3.90), for an initially unmodulated pulse. The modified shape upon reflection of an ultrashort pulse on a dielectric mirror may therefore be used to evaluate the dispersive properties of broadband dielectric mirrors, as shown by Weiner *et al.*, (1984, 1985). The pulse-shaping effect is however a second-order effect of the phase dispersion upon the pulse, and such a measurement requires pulses of less than 30 fs. A method that is both more direct (because it is a direct effect of the phase of the reflection coefficient) and more sensitive is to compare glass and coating dispersion inside a cavity. The importance of mirror dispersion relative to that of the intracavity glass has been measured experimentally (Diels *et al.*, 1984b; Dietel *et al.*, 1984a), by substituting mirrors with different coatings in one cavity position, and noting the change in amount of glass required to compensate for the additional dispersion. The method is very sensitive, because the effect of the sample (mirror) is multiplied by the mean number of cycles of the pulse in the laser cavity. The important result is that, in normal operating conditions, the dispersion of typical laser mirrors (18 quarter wavelength layers) corresponds to less than 0.4 mm of glass, when used within 10% of the central wavelength (full width of the reflection band: 30% of the central wavelength). The sign of the dispersion can be reversed by substituting a $\lambda/2$ layer to the last $\lambda/4$ layer (Fig. 3.20). Not all mirrors can be defined as "good mirrors" consisting of equally spaced quarter-wave dielectric layers. Some mirrors are made on purpose with a broad distribution of layer thicknesses, in order to enhance their

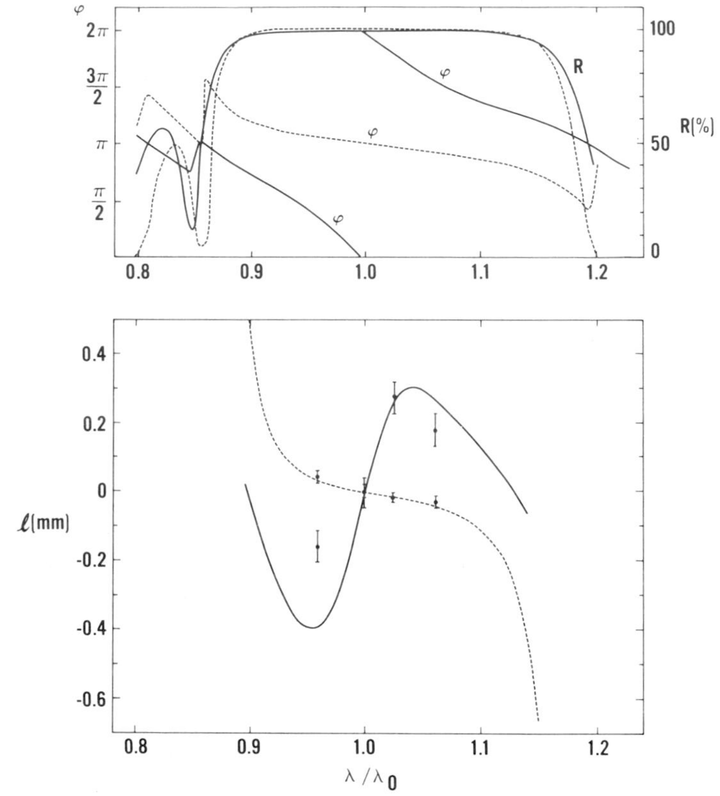

Fig. 3.20 Wavelength-dependence of the reflectivity (top) and of the dispersion (bottom) for two types of mirror coatings. The coatings investigated consist in 18 $\lambda/4$ dielectric layers (dashed lines) or 17 $\lambda/4$ + one $\lambda/2$ layers (solid lines). The calculated reflection coefficient (R) and the phase shift upon reflection are plotted in the top figure. The mirror dispersive properties (second-order) have been converted to equivalent intracavity glass thickness (plots of the lower figure). The data points in the lower figure were obtained by interchanging a cavity mirror [from Diels *et al.*, 1984b].

bandwidth. Calculations by De Silvestri *et al.* (1984b) have shown however that such mirrors can exhibit a large dispersion, even when used far from their band edge. In general, the phase curve of a broadband, multiple-layer dielectric mirror can undergo very rapid changes in the vicinity of certain frequencies (Dobrowski, 1978), referred to as "transition frequencies" by Christodoulides *et al.* (1985). Because of the fast variation of the phase near such transition frequencies, calculations based on a truncated Taylor series expansion as in De Silvestri *et al.* (1984a,b) are no longer adequate. Instead, a simple analytical function with poles at the transition frequencies can be substituted to the real frequency-dependence of the phase of the reflectivity (Christodoulides *et al.*, 1985).

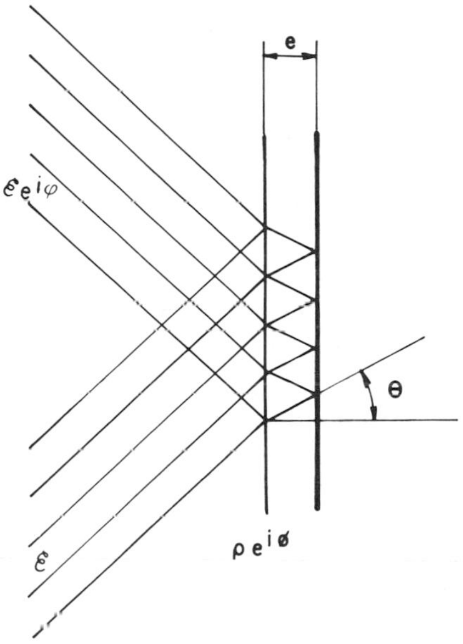

Fig. 3.21 Gires-Tournois interferometer. The light amplitude $\mathscr{E}$ is incident from the lower left. The interferometer consists of two parallel reflecting layers separated by a distance e, with amplitude reflection coefficients of $\rho e^{i\varphi}$ and 1, respectively. θ is the internal angle of incidence.

In view of the importance of dispersive effects with dielectric coatings, the logical next step was to look for a reflective device of which the dispersion could be continuously tuned. Such a multilayer structure was first implemented for femtosecond dye lasers by Heppner and Kühl (1985). The device is a miniaturization of the Gires-Tournois interferometer used by Dugay and Hansen (1969) for compressing picosecond pulses. Gires and Tournois (1964) proposed to adapt the pulse-compression technique used in radar (sending a frequency-modulated pulse through a dispersive line) to optical frequencies. To this effect, they proposed (Gires and Tournois, 1964) the interferometer sketched in Fig. 3.21, consisting of a pair of dielectric-coating flat mirrors with a uniform spacing e of a material of index n. It is essentially a Fabry-Perot resonator, with mirrors of unequal reflectivity: one reflector being ideally 100%, the other with a reflection coefficient $\rho \exp(i\phi)$. This structure is purely dispersive, within the range of angles of incidence for which the total reflector has negligible transmission. Given an incident wave of unit amplitude, the complex field of the reflected wave is given by

$$\mathscr{E}e^{i\varphi} = \frac{\rho e^{i\phi} + e^{-i\omega t_o}}{1 + \rho e^{-(\omega t_o - \phi)}} \tag{3.95}$$

where the constant t_o equals $2en/[C \cos \theta]$, with θ being the internal angle of incidence (Fig. 3.20). The phase factor of the reflected wave is given by

$$\tan \varphi = -\frac{(1-\rho^2)\sin \omega t_o}{2\rho \cos \phi + (1+\rho)^2 \cos \omega t_o} \tag{3.96}$$

In the original proposal of Gires and Tournois (1964), the interferometer consisted of a plane-parallel glass plate on a totally reflecting mirror. Therefore, the phase factor upon reflection ϕ was chosen to be equal to π in their derivation of the reflected phase and amplitude of the Gires-Tournois interferometer. This is no longer the case for the structure used with femtosecond lasers, where both interfaces consist of multilayer coatings, and the phase shift upon reflection ϕ is generally different from π.

The second-and higher-order Taylor expansion of the phase shift upon reflection ρ, around the laser frequency, can be used to calculate the reflective properties of the device, as a function of the parameter t_o and of the reflection amplitude ρ. We have seen that it is desirable to have a large second-order term that can be tuned continuously from positive to negative values, in order to balance the "chirp" coefficient a in Eq. (3.87). Calculation of the second derivative of Eq. (3.96) with respect to frequency yields an expression proportional to $t_0^2(1-\rho^2)\ \rho \sin(\omega t_o)$. The dependence in $\rho(1-\rho^2)$ implies a maximum dispersion for a reflection coefficient of $\sqrt{1/3}$ (or an intensity-reflection coefficient at the partially transmitting interface of 33%). The dependence on the optical pathlength in the spacer t_0 implies that the device can be tuned with the angle of incidence, yielding positive and negative values for the dispersion. There is however an important approximation in deriving Eq. (3.95) that may invalidate the use of the device for the shortest pulses. In order for the reflection after N internal reflections to interfere with the incident beam, the pulse duration should be such that $2Nt_o \ll \tau$ where τ is the pulse duration. For the optimum value of $\rho = \sqrt{1/3}$, N should be at least 5, hence $t_o < \tau/10$. Typically, the Gires-Tournois interferometers are made of a high reflective coating (18–20 quarter-wave layers), followed by the "spacer" consisting of 5 half-wave layers, topped by a quarter-wave layer. The wavelength refers to the laser wavelength corrected for the particular internal angle of incidence (typically 8% larger than the laser wavelength). The Gires-Tournois interferometers are generally used in pairs (Kühl and Heppner, 1986) or groups of four (French *et al.*, 1986a).

Kühlke *et al.* (1986) found a strong correlation between the third-order dispersion and the energy fluctuations of the laser pulses. In the case of the Gires-Tournois, the third-order term of the expansion of the phase (given by Eq. 3.96) in frequency is positive, which implies that the group velocity dispersion decreases with wavelength. For a four-prism sequence, the third-order term has the opposite sign. Kühlke *et al.* (1986a,b) showed that it is possible to find a combination of Gires-Tournois and four-prism sequences for which the two effects cancel each other (i.e., the group velocity dispersion is constant over the bandwidth of the pulse). That particular combination resulted in substantial reduction in energy fluctuations of their laser.

3.6.4. *Phase Modulation*

There are at least two beam waists in a mode-locked dye-laser cavity where nonlinear effects leading to self–phase modulation are likely to occur: in the absorber and in the amplifier dye jet. In the case of the DODCI–Rh-6G absorber–amplifier combination, as well as for most other currently used dye pairs, the laser wavelength is longer than that of the peak absorption of the absorber dye. Therefore, as the dye saturates, its index of refraction will decrease with time. Assuming pure homogeneous broadening for the dye absorption, Maxwell's equation (3.15) for the evolution of the phase of the pulse reduces, in the rate-equation approximation, to

$$\frac{\partial \varphi}{\partial z} = \frac{\mu \omega c}{2} \frac{\kappa T_2 \Delta \omega T_2}{1 + \Delta \omega^2 T_2^2} p \, \Delta N(t) = \frac{\sigma \Delta \omega T_2}{1 + \Delta \omega^2 T_2^2} \Delta N(t) \tag{3.97}$$

where σ is the absorption cross section at the center of the line of the absorbing dye. Linearizing the propagation equation and using the time-dependence (3.24) and (3.22) for the population difference and its time derivative, we find that passage through a thickness d of dye leads to the frequency modulation or chirp:

$$\dot{\varphi}(t, d) = N_o \sigma d \frac{\Delta \omega T_2}{1 + \Delta \omega^2 T_2^2} \frac{\mathrm{I}(t)}{\mathcal{T}_s} e^{-\int_{-\infty}^{t} I(t') dt' / \mathcal{T}_s}$$

$$= A \, \Delta \omega T_2 \frac{\mathrm{I}(t)}{\mathcal{T}_s} e^{\int_{-\infty}^{t} \mathrm{I}(t') dt' / \mathcal{T}_s} \tag{3.98}$$

where A is the (linear) attenuation factor of the absorber jet defined in Section 3.3. Simultaneous numerical integration of Maxwell's equation 3.97 and 3.30 show that, for large energies ($\mathcal{T} > \mathcal{T}_s$), the validity of the approximation (3.98) extends beyond $d = \alpha^{-1}$. Numerical calculations (Fig. 3.22) show that the frequency modulation peaks roughly at a time such that the accumulated energy density equals the saturation energy density $\mathcal{T}_s$. If the pulse focusing in the absorber is such that the saturation energy density is reached during the risetime of the pulse, a uniform downchirp is achieved during most of the pulse. Optimal focusing is such that the total pulse energy density is six times the saturation energy density. The downchirp is nearly linear during most of the pulse. Insertion

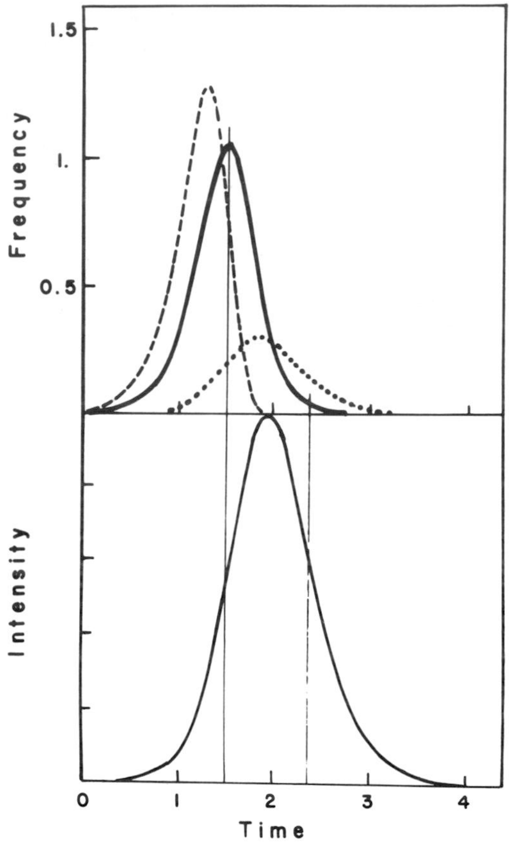

Fig. 3.22 Intensity versus time, and saturation-induced chirp. The exciting pulse (plotted on the lower part of the figure) has the shape

$$I(t) = I_0\left(\frac{2}{e^{(-t/t_r)} + e^{(t/t_f)}}\right)^2$$

with $t_r = 0.75\ \tau$ and $t_f = 1.25\ \tau$. The time is in units of the pulse FWHM ($= 1.72\ \tau$). The instantaneous frequency plotted in the upper part of the figure corresponds to a propagation through one linear absorption length of the absorber ($A = 1$), one linewidth below resonance ($\Delta\omega T_2 = 1$), and it is expressed in units of inverse pulsewidth (FWHM). The pulse energies, normalized to the saturation energy in the absorber, are 0.5 (dotted line), 1 (solid line), and 10 (dashed line).

of the proper amount of intracavity glass leads to a linearly downchirped to bandwidth-limited pulses (Diels *et al.*, 1985a). For a tighter focusing, the main contribution of the dye to the downchirp occurs early in the pulse, and other nonlinearities dominate the pulse chirp. If the pulse-energy density does not reach 1/6 of the saturation-energy density, upchirp may result—provided that the saturation is still sufficient for mode-locked operation.

In a first approximation, the phase modulation due to the electronic Kerr effect can be considered to be instantaneous. Consequently, the Kerr-effect–induced frequency modulation is proportional to (minus) the time derivative of the pulse intensity. The resulting upchirp is maximum and approximately linear near the pulse peak. In a situation where Kerr-effect–induced upchirp dominates, it is preferable to have it sufficiently large to dominate all other sources of phase modulation. Additives can be used to increase the nonlinear susceptibility of the solvent. Several investigators (Yamashita *et al.*, 1988; Jamasbi *et al.*, 1988) have used 2-methyl-4-nitroaniline (MNA) mixed in the DODCI solution in ethylene glycol. MNA has a higher nonlinear index than the solvent, and its absorption peak at 410 nm is sufficiently removed from the laser wavelength so as not to affect significantly the cavity losses. Higher concentrations can be reached by using a mixture of ethylene glycol and benzyl alcohol as solvent (Yamashita *et al.*, 1988). The significant improvement in stability and a slight decrease in minimum pulse duration should induce a search for more efficient and less toxic additives to enhance the Kerr effect.

3.6.5. Merging These Elements in the Theory

It has been shown in Section 3.6.1 that a succession of frequency modulation and dispersion can lead to pulse compression. The frequency modulation and dispersion should be treated, respectively, in the time and frequency domains. This rigorous approach however is not adequate for analytical calculations of the steady-state pulse. In the time domain, the dispersion function can be described by a series of derivative operators resulting from the Fourier transform of a Taylor-series expansion of the dispersion function. This series is then incorporated in closed-form equations of the laser cavity, such as Eq. (3.45) (Kühlke *et al.*, 1983; Diels *et al.*, 1983; Martinez *et al.*, 1984; Haus, 1986). The steady-state pulse shape is thereafter determined by a parametric approach (an ansatz envelop shape is chosen, depending on several parameters that are determined by solving the cavity equation).

3.7. Coherent Effects

In the preceding considerations, the dye was approximated by a pure two-level system, with a Lorentzian line shape, consistent with the rate-equation approximation made. Several investigations have been made to determine whether this simple model describes adequately the experimental situation, and to which extent a departure from these assumptions would modify the theoretical results.

As shorter and shorter pulses are being produced, one might question the validity of the rate equation used in all aforementioned models. In order to assess the influence of a finite phase relaxation rate of the two-level system, a perturbative approach has been used (Rudolph and Wilhelmi, 1984; Diels *et al.*, 1985b). Instead of inserting the steady-state solution of the off-diagonal matrix element equation (3.7) into the population equation (3.8) to derive the rate equation, the formal solution of Eq. (3.7) is written as a power series of source terms:

$$Q = \left(- \frac{T_2}{1 - i\,\Delta\omega T_2} \right)^{n+1} \frac{d^n \kappa \mathcal{E} w}{dt^n} \tag{3.99}$$

with $\Delta\omega = \omega_0 - \omega$. This solution is inserted in the population equation (3.8) to form a coupled set of differential equations, which can be solved iteratively (starting with the solution for $n = 0$, which is the rate-equation approximation). Applying this perturbative approach to correct the rate-equation approximation, it has been shown that, as the pulse duration becomes comparable to the phase-relaxation time, the magnitude of the induced downchirp decreases (Rudolph and Wilhelmi, 1984). An important feature resulting from the inclusion of coherent effects (to first order) is that the correct lasing frequency is found (Rudolph and Wilhelmi, 1984).

Another appproximation that has been questioned is that of modeling the absorbing dye molecule with only two levels. As noted by Petrov *et al.* (1987a), the assumption of two-level systems implying equal profiles for the absorption and stimulated emission bands is justified as long as the relaxation from the highly excited vibronic levels is considerably slower than the pulse duration. Petrov *et al.* (1987a) have studied the interaction of light with quasi-resonant four-level systems (two vibronic states in each of the ground and excited electronic levels considered). That model is valid for vibronic and phase-relaxation times comparable to the pulse duration. As compared to the two-level system, the result of the theory of Petrov *et al.* shows a significant increase of the saturation-induced downchirp, for particular laser frequencies between the emission and absorption resonance.

Consistent with the perturbative treatment mentioned previously, mode-locked lasers should not be able to operate with pulses shorter than the phase-relaxation time of the dye. This conclusion however may simply be the reflection of the use of an inappropriate model. The perturbation approach is no longer valid if pulses of "area" (the angle of rotation of the pseudopolarization vector defined in Section 3.2.1) $\theta \gg 1$ are produced in the absorber. Such a situation is conceivable with the extreme focusing that can be achieved in the astigmatism-compensated cavity discussed in Section 3.5. Is the absorber bandwidth (or its phase-relaxation rate) an intrinsic limit to the pulse duration of a laser? To answer this question, let us consider a very simplified laser model, with an extreme s parameter (cf. Section 3.3, Eq. 3.32) such that the gain can be assumed to be linear. There are essentially two components acting on the field $\mathscr{E}e^{i\varphi}$ at each round-trip; a linear one with a net gain $g = re^{iA}$, and a nonlinear one consisting in a two-level system in a standing-wave excitation (the absorber). Steady-state condition for a pulse propagating at a velocity V implies the system of equations:

$$g\mathscr{E} = \frac{\mu\omega c}{2} v \tag{3.100}$$

$$\left[1 - \frac{n^2V^2}{c^2}\right]\mathscr{E} = \mu V^2 u. \tag{3.101}$$

The second equation should be used to describe a steady state instead of the first-order approximation (3.15) of the real part of Maxwell's propagation equation. Equation (3.101) essentially states that each component of a chirped steady-state pulse has to propagate at the same velocity or reproduce itself after each round-trip with the same temporal shift. Therefore, a constant phase velocity V,

$$V = \frac{(\omega + \dot{\varphi})}{\left(k - \frac{\partial\varphi}{\partial z}\right)}, \tag{3.102}$$

must be associated with the chirped steady-state pulse. In the system of equations (3.100) and (3.101), the functions u and v are given by Bloch's equation for counter propagating identical pulses:

$$\dot{u} = (\Delta\omega - \dot{\varphi})v - u/T_2 \tag{3.103}$$

$$\dot{v} = (\Delta\omega - \dot{\varphi})u - \kappa\mathscr{E}w - v/T_2 - \kappa R\mathscr{E} \tag{3.104}$$

$$\dot{w} = 2\kappa\mathscr{E}v \tag{3.105}$$

$$\dot{R} = \kappa\mathscr{E}v. \tag{3.106}$$

Equations (3.101) and (3.105) imply that E, u, and v should have the same functional dependence. For instance

$$\mathscr{E}(t) = E_0 \operatorname{sech}\left[\frac{t}{\tau}\right] \tag{3.107}$$

$$u(t) = A \operatorname{sech}\left[\frac{t}{\tau}\right] \tag{3.108}$$

$$v(t) = B \operatorname{sech}\left[\frac{t}{\tau}\right] \tag{3.109}$$

A solution is found by substituting the ansatz (Eq. 3.107) into the system of Eqs. (3.100)–(3.106) (cf. Problem 4 at the end of this chapter). There is no constraint on the pulse duration τ of the steady-state solution, except for the general condition that $\tau \ll T_1$. The steady-state field amplitude E_0 should be larger than $1/\sqrt{3}\kappa\tau$. There is an unchirped solution below resonance and a downchirped solution above resonance. The detuning of the upchirped solution is

$$\Delta\omega = \frac{1}{T_2}\sqrt{3\kappa^2 E_0^2 \tau^2 - 1} \tag{3.110}$$

and its frequency modulation is

$$\dot{\varphi} = \Delta\omega T_2 \tanh\left[\frac{t}{\tau}\right] \tag{3.111}$$

(the downchirped solution corresponds to the opposite sign in Eq. 3.110.

This approach gives a new type of solution that cannot be found with the standard perturbative approaches (Haus, 1975–1986; Diels *et al.*, 1985b; Martinez *et al.*, 1984).

The analytical solution was found off-resonance, including the exact quantum mechanical interaction of counterpropagating standing waves in a thin saturable absorber, and assuming linear unsaturated gain (Eq. 3.100). Similar solutions were found numerically by Petrov *et al.* (1987b), assuming however exact resonance, but using Bloch's equations for both the absorber and amplifier (neglecting the counterpropagating interaction in the absorber). The pulses are seen to evolve in the numerical simulation toward exact "π pulses" ($\theta = \pi$) for the amplifying medium and toward "2π pulses" ($\theta = 2\pi$) for the absorbing medium, for the particular value of the s parameter $s = 4$. The pulse area in the amplifier and the pulse duration decrease with increasing value of s. This dependence of the pulse area (in the amplifier) on the parameter s maintains a value close to 2π for the pulse area in the absorber. These "coherent" stationary solutions are characterized by their extreme stability and reproducibility.

The types of solutions discussed previously (analytical and numerical) have not included the dispersive effect of the cavity, or the effects of other levels (for instance, two-photon absorption) that are likely to play a role at these large intensities (Diels and McMichael, 1986). Nevertheless, the fact that such solutions exist may be an indication that we have not yet reached the full potential of the fs dye laser, and there are still some unexplored modes of operation to be investigated. The required energy densities correspond to that of standard fs dye lasers, with the tightest focusing obtainable by geometric astigmatism compensation (cf. Section 3.5 and Tables 3.1 and 3.2;. The use of more than one gain jet or a high gain configuration with more than one gain spot (Blit and Tittel, 1978) may lead to the required intracavity power.

3.8. Solitons

The perturbative approach taken in Sections 3.3 and 3.6 leads to the most comprehensive description of the operation of the mode-locked laser, and is the model chosen by most investigators. It mixes a little bit of everything (saturation, modulation, coherence, dispersion, and coupling of counterpropagating waves), and there is always room for the addition of other effects in this already thick brew. Some strong nonlinearities cannot be included, as illustrated in the preceding section for the case of the "strong coherent absorber" in a mode-locked laser. More than the inclusion of the exact quantum mechanical response of a two-level system, it was the *exclusion* of most other parameters that made the elegance of that theory. The individual influences of the numerous parameters associated with the operation of the laser become buried in their sheer number, in the case of the perturbative treatment. In a similar way, as we have shown in the preceding section the existence of steady-state pulse due solely to the coherent interaction with the absorbing medium, we shall show next that the dispersive effects alone—in the absence of saturation or absorptive mechanism—can also lead to steady-state pulses. The treatment of phase modulation and dispersion effects is identical to the one of Section 3.6. The only difference with the previous analysis is that all other effects will be neglected. The resulting equation for the field envelope is the Schrödinger equation with cubic nonlinearity. This second-order equation is one of the few that can be solved analytically (Zakharov and Shabat, 1972, 1973).

As in most analytical treatment, a continuity assumption is made—namely that the pulse passes through a continuous succession of infinitesimal Kerr media, followed by an infinitesimal disperser, each of thickness Δz. The first step of the sequence sketched in Fig. 3.18 leads to the electric field amplitude:

$$\tilde{\mathscr{E}}(z_0 + \Delta z, t) = \tilde{\mathscr{E}}(z_0, t)e^{-iK_r\Delta z} \simeq \tilde{\mathscr{E}}(z_0, t)[1 - iK_r\Delta z] \tag{3.112}$$

where we have expanded the exponential to first order, assuming a small intensity-induced phase modulation $K_r\Delta z$, with K_r given by

$$K_r = \frac{2\pi n_2 \mathscr{E}^2}{\lambda} \tag{3.113}$$

The corresponding approximation for the dispersive medium (Fig. 3.18b) is

$$\tilde{\mathscr{E}}(z_0 + \Delta z, \Omega) = \tilde{\mathscr{E}}(\Omega)e^{-ik\Delta z} \simeq \tilde{\mathscr{E}}(\Omega)[1 - ik\,\Delta z] \tag{3.114}$$

where the argument of the Fourier amplitude $\mathscr{E}(\Omega)$ is

$$\Omega = \omega - \omega_\ell \tag{3.115}$$

(ω_ℓ being the laser frequency). Consistent with the approximation of an infinitesimal disperser, we can use a Taylor-series expansion of the wave vector around the laser frequency, up to second order in the frequency difference Ω. The expansion of (3.115) can be converted back to the time domain, using the operation $\partial/\partial t$ to replace the product by $i\Omega$. The equation for a succession of infinitesimal modulation-dispersion reduces to

$$\tilde{\mathscr{E}} + \Delta\tilde{\mathscr{E}} = \left(1 - k'\,\Delta z\partial_t + i\frac{k''}{2}\Delta z\partial_{tt}^2\right)\left(1 - iK_\mathrm{r}\Delta z\right)\tilde{\mathscr{E}} \tag{3.116}$$

or

$$i\left(\frac{\partial\tilde{\mathscr{E}}}{\partial z} + k'\,\frac{\partial\tilde{\mathscr{E}}}{\partial t}\right) = -\frac{k''}{2}\frac{\partial^2\tilde{\mathscr{E}}}{\partial t^2} + K_\mathrm{r}\tilde{\mathscr{E}}. \tag{3.117}$$

In most media ("normal dispersion"), the second derivative of the index of refraction versus frequency is a positive quantity. Hence, the second derivative (Eq. 3.85) of the wave vector versus frequency is negative. With the exception of some organic materials (as investigated for instance by Prasad (1986), the Kerr effect K_r is a positive quantity. The Kerr-effect–induced upchirp mentioned in Section 3.6.4 can be derived directly from the real part of Eq. (3.117). Neglecting the dispersion, using the explicit field-dependence for the Kerr effect and $\mathscr{E}\exp(i\varphi)$ for the complex electric field envelope, yields the following expression for the induced phase modulation:

$$\frac{\partial\varphi}{\partial z} = -\frac{2\pi}{\lambda}n_2\mathscr{E}^2. \tag{3.118}$$

The time derivative of this expression indicates that, indeed, the instantaneous frequency is proportional to (minus) the signal intensity, showing an upchirp near the peak of the pulse. Using a retarded time frame of reference, and with appropriate normalization (cf. Problem 5 at the end of the chapter), Eq. (3.117) takes the form:

$$i\frac{\partial\tilde{E}}{\partial z} + \frac{\partial^2\tilde{E}}{\partial t^2} + |\tilde{E}|^2\tilde{E} = 0 \tag{3.119}$$

where z, t and $\tilde{E}$ are no longer the physical quantities, but the appropriately reduced variables or function. Equation (3.119) is encountered in many areas of physics, in particular in nonlinear optics. Zakharov and Shabat (1972) have shown that it can be solved exactly by the inverse problem method (Gardner *et al.*, 1967). The problem is reduced to a search for the eigenvalues ζ_j of coupled differential equations. These eigenvalues are generally complex and are called the poles of the solutions for $\tilde{E}(z, t)$ which are themselves designated as "solitons." The "order" of the soliton is the number of poles associated with that particular solution. For the particular problem that Eq. (3.119) describes, the real part of a pole ξ_j is a measure of the corresponding propagation velocity of the corresponding soliton envelope, and the imaginary part of the pole η_j is associated with the soliton energy. The solutions of interest for the study of the femtosecond laser are those for which the poles are all lined up on the imaginary axis (or on a parallel to this axis). These solutions are periodic steady-state pulses. The simplest one is the soliton of order 1, a sech-shaped pulse:

$$\tilde{E}(z,t) = \sqrt{2}\eta e^{4i\eta^2 z} \text{ sech } (2\eta t). \tag{3.120}$$

This is a stable pulse shape, propagating without distortion, with a continuous slippage of phase with distance. More generally, solitons are periodic solutions. For instance, a symmetrical soliton (i.e., with a symmetrical envelope $E(z, t)$) will have an envelope changing with distance with a period given by

$$T = \frac{\pi}{\eta_2^2 - \eta_1^2}. \tag{3.121}$$

Soliton pulse propagation propagation due to Kerr-effect phase modulation and dispersion has been observed in optical fibers (Mollenauer *et al.*, 1980). Can the steady-state pulse of a femtosecond dye laser also be labeled a soliton? Several investigators (Dietel *et al.*, 1984b; Diels *et al.*, 1985b) felt that it was justified because the nonlinear Schrödinger equation (3.119) is a particular simplification (neglecting for instance saturation effects) of the more complete perturbative treatment. Salin *et al.* (1986) associated the periodic deformation of the intracavity pulse with a higher-order soliton. Periodic evolution of the pulse train has been reported earlier (Diels and Sallaba, 1980) and exploited to extract (by boxcar averaging) the shortest pulses from the train (Diels *et al.*, 1980). Salin *et al.* (1986) interpret the triple-peaked autocorrelation trace associated with a 35 and 70 kHz periodicity in the pulse train as a soliton of order 3. It is clear however that the nonlinear Schrödinger equation is only a crude approximation of the complex pulse evolution in a dye laser. As shown in Section 3.3.1, saturation effects alone can also account for complex periodicities in the evolution of the pulse train. Avramopoulos *et al.* (1988) have made a numerical model of the cavity, including gain and absorption saturation, linear loss, noise injection, spectral filtering, group velocity dispersion and self-phase, modulation, dispersion and bandwidth filtering. As mentioned

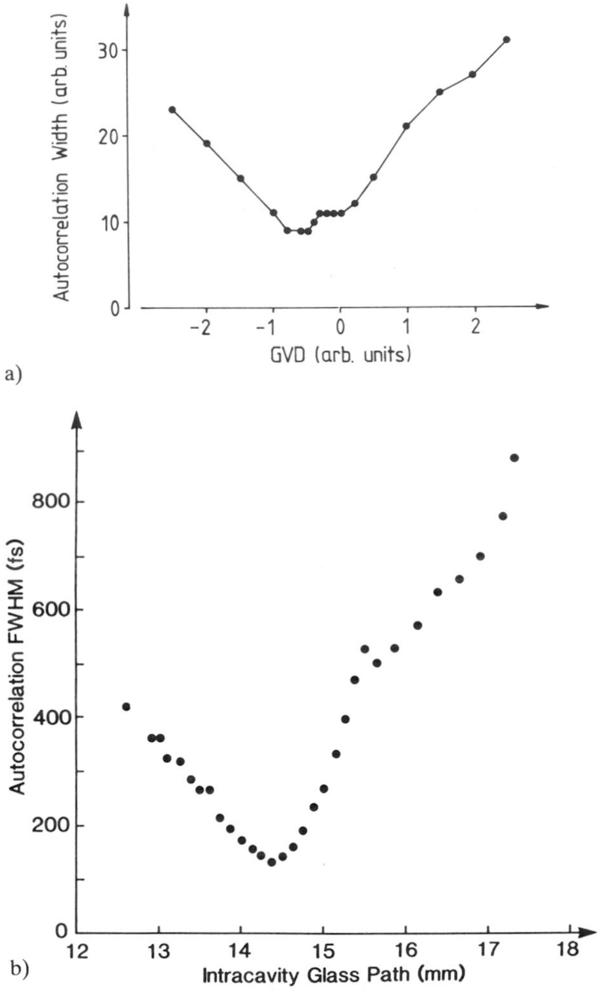

Fig. 3.23 FWHM of the autocorrelation versus group velocity dispersion, obtained for a numerical model (top: Fig. 3.23a) with both Kerr-effect and absorption-saturation contributions to the intracavity frequency chirp as compared to the measurement (bottom: Fig. 3.23b) [from Avramopoulos *et al.*, 1988]. © 1988 IEEE.

in the introduction to this section, such a numerical approach, where the theoretical pulse interacts successively with each discrete element of the cavity, has the advantage of following closely the behavior of the real system. It still does not lead to a *quantitative* comparison with the real system because, in order to keep computing time to reasonable limits, the strengths of the nonlinearities and the amounts of intracavity group velocity dispersion used exceed those values that would be found in the real system. Nevertheless, the comparison that is made with autocorrelation measurements on a ring-dye laser shows convincing qualitative agreement (Avramopoulos *et al.*, 1988). The laser used for the experimental comparison is a Rhodamine 110 dye laser, passively mode-locked using 1,1′,3,3,3′,3′-hexamethylindocarbocyanine iodide (HICI) and 2-(p-dimethylamynostyryl)-benzothiacarbocyanine iodide (DASBTI) as saturable absorbers. Figure 3.23 compares theoretically predicted autocorrelation widths and measurements as a function of group velocity dispersion. The only fitting parameter is the relative strength of the two sources of intracavity self–phase modulation. The shapes of the computed autocorrelation and the measured ones are also found to be in qualitative agreement (Avramopoulos *et al.*, 1988). A more stringent test yet to be performed is an experimental verification of the theoretically predicted pulse shapes, in amplitude and phase, using the techniques described in the next section.

In the theory of Avramopoulos *et al.* (1988), the bandwidth-limiting filter plays a dominant role. Generally, though not in all systems, such a filter is avoided in order to obtain the shortest possible pulse. It has been observed (Diels *et al.*, 1983; Fontaine *et al.*, 1983) that, in the absence of a bandwidth-limiting element, the laser frequency can be tuned between the amplifier and absorber resonances by changing the concentration of absorber. The bandwidth-limiting effect arising from the saturable absorber and the amplifier through their phase memory has been calculated by the techniques mentioned in Sections 3.3.2 and 3.6.4 (Rudolph and Wilhelmi, 1984; Diels *et al.*, 1985b). A numerical model (Petrov *et al.*, 1989) of the Rh 6G-DODCI absorber (excluding an intracavity filter, taking into account coherent interactions, the photoisomer of DODCI, Kerr effect, and dispersion) shows similar periodic oscillations of the envelope in the kHz range, as observed experimentally.

More stringent tests have to be performed to make an accurate assessment of the various theories. An example of such a test (yet to be performed) is an experimental verification of the theoretically predicted pulse *shapes*, in amplitude and phase, using the techniques described in the next section.

4. DIAGNOSTIC TECHNIQUES

Since electronic-detection techniques are still limited at around a psec (streak camera detection), there is a need for methods with a resolution in the femtosecond scale. The need arises not only for the characterization of the ultrashort pulses themselves, but also for research methods that can fully exploit the capability of the source. The various tools available to the femtosecond spectroscopist are presented in this section.

4.1. Second-Order Autocorrelations

In principle, successive optical correlations of increasing order (2, 3, . . . n) would provide a complete characterization of ultrashort pulses. However, high-order correlations require simultaneously a high intensity and a large bandwidth (the bandwidth of the nonlinear process has to exceed that of the short pulse). Therefore, for the detection of the relatively weak output of continuously mode-locked dye lasers, one chooses generally to remain at order 2 with second harmonic detection. The crystal length determines the bandwidth of second harmonic generation. The crystal thickness—or equivalently the confocal parameter of the beam focused in the crystal—should be shorter than the coherence length at all the frequency components of the light pulse. The differences in group velocity between the fundamental and second harmonic should also be negligible over the crystal thickness (Weber, 1966).

The basic elements of a second-order autocorrelation setup are sketched in Fig. 3.24. The pulse train is split into two beams which are recombined after having passed through a fixed and an adjustable optical delay. The average power of the second harmonic of the recombined beam (generated, for instance, in a KDP crystal) is recorded as a function of the variable delay. What distinguishes particular designs are various schemes to split and recombine the beams, the method to scan the optical delay, and the choice of detector.

A standard Michelson interferometer is an example of autocorrelator design that is generally avoided, because part of the energy is reflected

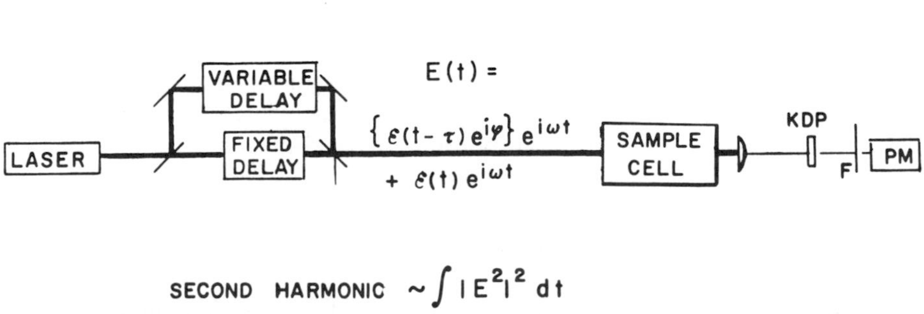

Fig. 3.24 Sketch of a basic autocorrelator setup. Optical delay lines are used to split the original laser pulse into a sequence of two pulses of relative phase ϕ and delay τ. The second harmonic generated in a nonlinear crystal (for instance, KDP) is filtered from the remaining fundamental by the filter F, and recorded by the detector PM [from Diels *et al.*, 1985a).

back into the laser, resulting in unstable operation. However, if corner cube reflectors are used in each arm, feedback into the laser can be avoided by lateral translation of the return beam. A polarizing beam splitter and quarter-wave retardation plates in each arm not only prevent reflection into the laser, but also makes it possible to utilize all the incoming light for the nonlinear detection. The use of urea crystals phase-matched for second harmonic generation type II leads to background-free measurement (Diels *et al.*, 1987a). Another arrangement that does not reflect light back into the source is the Mach-Zehnder interferometer. However, as is the case for the Michelson without polarizing beam splitter, half of the available light power is wasted.

For weak signals for which long integration times are needed, the scanning of the variable arm of the correlator will be performed with translation stages and synchronous or stepping motors. It is however more convenient for laser-optimization purposes to have a "real-time" (i.e., in less than 20 ms) display of the correlation. Several commercial instruments use the periodic modulation in optical path, obtained by inserting a rotating slab of glass in one arm of the correlator, to scan the delay. The disadvantage of this method is that it is limited to picosecond pulses for which the dispersion of glass is negligible. For fs pulses, significant pulse broadening or compression may occur, depending on the eventual chirp of these pulses (Diels *et al.*, 1985a). A simple scanning method that does not have this shortcoming uses corner cube reflectors mounted on audio speakers. Such speakers can be scanned over a range of 60 ps. The position of the corner cube (i.e., the parameter of the autocorrelator) can be monitored with an accelerometer followed by a pair of integrators (Jain *et al.*, 1982).

Frequency doubling can be used for the nonlinear detection required for second-order autocorrelation for all sources of wavelength longer than 280 nm (see Fig. 3.24). For shorter wavelength, surface second harmonic generation (Kintzer and Rempel, 1987) or multiphoton ionization (Bourne and Alcock, 1986) can be used. A more sensitive and simpler technique proposed and demonstrated by Noack *et al.* (1988) involves the use of two-photon–induced luminescence in scintillator crystals used for VUV detection. Na doped CsI crystals were selected because two-photon absorption is efficient in these crystals and is not followed by the formation of color centers. The luminescence induced by two-photon absorption of 308 nm is centered at 440 nm, and it decays in several μs.

In the arrangement of Fig. 3.24, the function that is recorded is proportional to $\int |(E_t)^2|^2 dt$, where $E_t = E(t) + E(t - \tau)$ is the electrical field of the light entering the frequency doubler KDP crystal. If the measurement sketched in Fig. 3.23 is performed with interferometric accuracy, the second harmonic recording is proportional to

$$I_t = \int |\{\mathscr{E}(t) \exp i(\omega t + \varphi) + \mathscr{E}(t-\tau) \exp i[\omega(t-\tau) + \varphi(t-\tau)]\}^2|^2 \, dt$$

$$= \int \{2\mathscr{E}^4 + 4\mathscr{E}^2(t)\mathscr{E}^2(t-\tau)$$

$$+ 4\mathscr{E}(t)\mathscr{E}(t-\tau)[\mathscr{E}^2(t) + \mathscr{E}^2(t-\tau)] \cos[\omega\tau + \varphi(t) - \varphi(t-\tau)]$$

$$+ 2\mathscr{E}^2(t)\mathscr{E}^2(t-\tau) \cos 2[\omega\tau + \varphi(t) - \varphi(t-\tau)]\} \, dt. \qquad (3.122)$$

If the interference fringes in the recombined beam are averaged out, which is generally the case, all terms in cos ( ) vanish. The resulting limit of (3.122) is the intensity autocorrelation:

$$I_{c_1} = \frac{1 + 2\int I(t)I(t-\tau)\, dt}{|\int I^2 \, dt|} \qquad (3.123)$$

where I is the intensity of the light pulse. The function I_{c_1} has a peak-to-background ratio of 3 to 1. In some experimental arrangements only the function

$$I_c = \frac{\int I(t)I(t-\tau)\, dt}{|\int I^2 \, dt|} \qquad (3.124)$$

is measured (Ippen and Shank, 1975; Diels *et al.*, 1987a). The function I_c (or I_{c_1}) does not carry any phase information and therefore cannot be used to distinguish between coherent and incoherent pulses. Yet the use of the expression *coherent spike*, associated with the intensity autocorrelation, has sometimes created confusion by making it appear as if a pulse free of coherent spike (in its autocorrelation) is a coherent pulse. We summarize therefore some of the properties and information contained in the intensity autocorrelation to clarify the distinction with the interferometric autocorrelation.

Any noise on top of a cw signal will be identifiable in the function I_c by a small "bump" riding on an infinite background. The width of the bump is a measure of the temporal width of the fluctuations, and the contrast ratio (peak-to-background ratio of I_c) is a measure of the modulation depth. A 100% modulation depth results in a peak-to-background ratio of 2 to 1 for I_c (Weber and Danielmeyer, 1970). Any signal of a finite duration results in a function I_c of finite width. If that signal has some fine structure (amplified modulation, noise), a narrow spike will appear in the middle of the correlation function. This is the coherence spike, typical of a signal consisting of a burst of noise (Bradley and New, 1974).

An autocorrelation free of coherence spike is a necessary condition for a coherent pulse. However, it is not a sufficient condition. A white-light-incoherent pulse (free of amplitude modulation) will have an autocorrelation I_c without coherence spike. That is also the case of a pulse with random or deterministic phase modulation.

In the case of the interferometric autocorrelation given by Eq. (3.122), the signal at zero delay being a coherent superposition of the field from each arm is

$$I_t(0) = 2^4 \int E^4(t)\,dt. \tag{3.125}$$

The peak-to-background ratio of the interferometric autocorrelation is thus 16:2 or 8:1. The envelope of the constructive and destructive interferences will merge into the intensity autocorrelation for delays exceeding the pulse phase coherence time. Since it involves the fourth power of the fields combining in phase, the upper envelope will be more sensitive to the pulse shape than the intensity autocorrelation. The interferometric autocorrelation can provide very useful information about pulse chirp, because various types of chirps have characteristic signatures. Let us consider, for instance, the interferometric autocorrelation of a pulse that has been chirped by self–phase modulation (for example, by propagation through a Kerr-like nonlinearity). Because of the frequency sweep that is largest in the center of the pulse, there will be a narrowing of the upper and lower envelopes. However, since the pulse tail and pulse front remain coherent with each other, the interferences in the wings of the interferometric autocorrelation will extend to delays as large as those for an unchirped pulse of the same duration. In the case of self–phase modulation, the narrowing of the upper and lower envelopes is a much more sensitive indication of the phase modulation than the spectral broadening. Indeed, in the case of a Gaussian pulse with self-modulation ($\varphi \propto I$), a chirp that reduced the interferometric autocorrelation width to 2/3 of its value (for unchirped pulses) broadens the spectrum by only 10%. Another typical phase modulation is that caused by a linear chirp. The case of Gaussian pulses linearly chirped can be treated analytically (Diels *et al.*, 1985a). The meeting point of the lower and upper envelopes of the interferometric autocorrelation can be used to determine the chirp magnitude.

Amplitude and phase information on fs pulses can thus be extracted from intensity and interferometric autocorrelation measurements. Since these measurements are basically symmetric, they can only lead to a pulse-shape reconstruction under the limiting assumption that the pulses are exactly symmetric in amplitude and phase. Since there is no physical justification for such an assumption, additional information is needed to resolve the symmetry of the signal. This information can be provided by

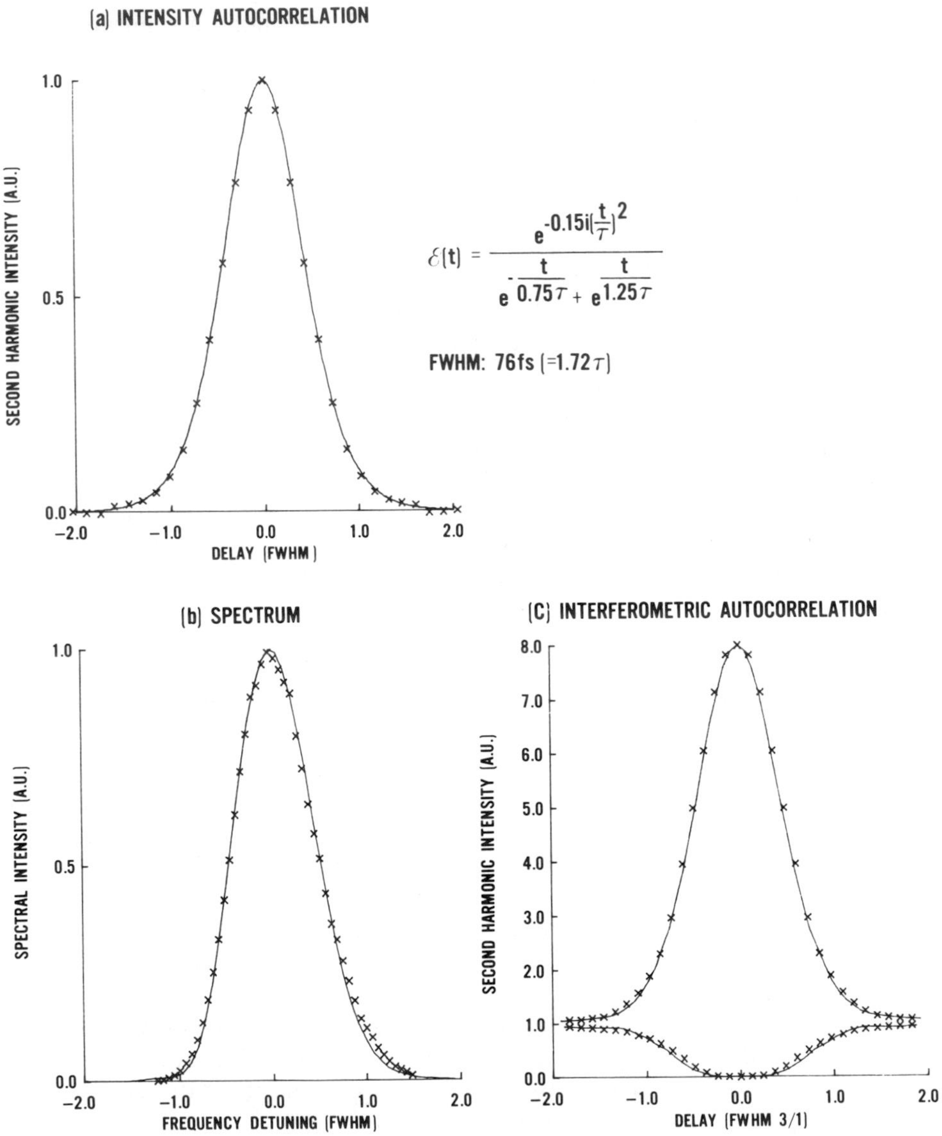

Fig. 3.25 Example of pulse-shape determination through (a) iterative fitting of the intensity autocorrelation, (b) the pulse spectrum, and (c) the interferometric autocorrelation. The crosses are the experimental data points. The solid lines are the corresponding functions calculated for the final iteration of the pulse shape, which is given in the figure [from Diels *et al.*, 1985a].

the pulse spectrum. Simple pulse shapes can be fitted to the autocorrelations and pulse spectra, as shown in Fig. 3.25. The method is accurate: a change in any of the parameters of the pulse (risetime, falltime, or chirp coefficient) by 5% results in a significant discrepancy between data and fit in one of the three curves of Fig. 3.25. As a guide to such fitting procedures, analytical expressions have been derived for the three functions for standard symmetric and nonsymmetric pulse shapes (Table 3.4).

The method of iterative fitting is quite accurate but tedious, particularly for complex pulse shapes. In addition, the method fails in the limit of vanishing phase modulation, since in that case all data are symmetric. Rothenberg and Grischkowsky (1985) have proposed the use of cross-correlation, combined with interferometry, to determine the phase of a relatively long pulse, when a shorter one is available. This method however does not solve the problem of finding the shape of the shorter pulse.

4.2. Convolutions

Montgomery *et al.* (1986) have proposed the use of self-convolution (time-reversed reflection) as an alternative to autocorrelations. The Fourier transform of the self-convolution of a function is the square of the Fourier transform of that function (as opposed to the absolute value squared as in the case of the autocorrelation). The self-convolution thus carries all the information on the pulse shape. It has been proposed to use degenerate four-wave mixing (DFWM) for time-reversed reflection. However, it has been shown (Diels *et al.*, 1981) that most of the shape information of an ultrashort probe pulse can no longer be found in the signal if the medium thickness is not negligibly small compared to the pulse length (a medium of thickness L stretches out a DFWM reflection by a length $2L$). It is only under the simultaneous condition of negligible medium thickness and weak reflection that DFWM provides a true time-reversed reflection (Diels *et al.*, 1981).

4.3. The Asymmetric Correlator

The basic technique is to transform the pulse to be measured by an operation involving a well-known transfer function, to a shape that is easily measured. The simplest operation is propagation through glass, which causes the pulse to be broadened by normal dispersion. The broadened pulse is then cross-correlated with the original (shorter) pulse. The intensity cross-correlation yields information on the amplitude shape of the signal, while its phase can be extracted from the interferometric cross-correlation. The basic experimental setup is thus a standard interferometric autocorrelator, in the arm of which a block of glass (typically 5–10 cm of

Table 3.4

Correlations and Spectra for typical pulse shapes

$I(t)$	Δt	$I(\omega)$	$\Delta\omega$	$\Delta\omega\,\Delta t$	$g_1^{(\tau)}$	$\Delta\tau$
e^{-t^2}	1.665	$e^{-\omega^2}$	1.665	2.772	$1+e^{-\tau^2/4}$	3.330
$\operatorname{sech}^2 t$	1.763	$\operatorname{sech}^2 \frac{\pi\omega}{2}$	1.122	1.978	$1 \pm \frac{\tau}{\sinh\tau}$	4.355
$\frac{1}{(e^{t/(1+A)}+e^{-t/(1-A)})^2}$ $A=\frac{1}{4}$	1.715	$\frac{1+1/\sqrt{2}}{\cosh\frac{15\pi}{16}\omega+1/\sqrt{2}}$	1.123	1.925	$1 \pm 4\frac{\sinh\frac{1}{3}\tau}{\sinh\frac{4}{3}\tau}$	3.405
$A=\frac{1}{2}$	1.565	$\operatorname{sech}\frac{3\pi}{4}\omega$	1.118	1.749	$1 \pm 2\frac{\sinh\tau}{\sinh 2\tau}$	2.634
$A=\frac{3}{4}$	1.278	$\frac{1-1/\sqrt{2}}{\cosh\frac{7\pi}{16}\omega-1/\sqrt{2}}$	1.088	1.391	$1 \pm \frac{4}{3}\frac{\sinh 3\tau}{\sinh 4\tau}$	1.957
$\frac{1}{(e^t+e^{-rt})^2}$	—	$\frac{2(1-y)x}{x^2-2yx+1}$ where $x=\exp\left(\frac{2\pi\omega}{1+r}\right)$ $y=\cos\left(\frac{2\pi r}{1+r}\right)$	—	—	$1 \pm \frac{r+1}{r-1}\frac{\sinh\left(\frac{r-1}{2}\tau\right)}{\sinh\left(\frac{r+1}{2}\tau\right)}$	

Source: Diels *et al.*, 1985a.

BK7 or SF5 glass) has been inserted. The second harmonic is digitized and stored on a minicomputer, at a rate of up to 100 kHz. A high data rate, such as to record a complete set of data in less than one second, is desirable in order to maximize the accuracy of the measurement.

Let $E_1(t)$ $[I_1(t)]$ and $E_2(t)$ $[I_2(t)]$ be the amplitude (intensity) of the original laser pulse and that of the pulse broadened by glass. The intensity cross-correlation $A(\tau)$ is by definition

$$A(\tau) = \int I_1(t-\tau)I_2(t)\,dt. \tag{3.126}$$

$\Delta\tau/\Delta t$	$G_2(\tau)$	$\Delta\tau$	$\Delta\tau/\Delta t$	$g_2(\tau)$
2	$e^{-\tau^2/2}$	2.355	1.414	$1 + 3G_2(\tau) \pm 4e^{-(3/8)\tau^2}$
2.470	$\dfrac{3(\tau\cosh\tau - \sinh\tau)}{\sinh^3\tau}$	2.720	1.543	$1 + 3G_2(\tau) \pm \dfrac{3(\sinh 2\tau - 2\tau)}{\sinh^3\tau}$
1.985	$\dfrac{1}{\cosh^3\dfrac{8}{15}\tau}$	2.648	1.544	$1 + 3G_2(\tau) \pm 4\dfrac{\cosh^3\dfrac{4}{15}\tau}{\cos^3\dfrac{8}{15}\tau}$
1.683	$\dfrac{3\sinh\dfrac{8}{3}\tau - 8\tau}{4\sinh^3\dfrac{4}{3}\tau}$	2.424	1.549	$1 + 3G_2(\tau) \pm 4\dfrac{\tau\cosh 2\tau - \dfrac{3}{2}\cosh^2\dfrac{2}{3}\tau\sinh\dfrac{2}{3}\tau\left(2 - \cos\dfrac{4}{3}\tau\right)}{\sinh^3\dfrac{4}{3}\tau}$
1.531	$\dfrac{2\cosh\dfrac{16}{7}\tau + 3}{5\cos^3\dfrac{8}{7}\tau}$	2.007	1.570	$1 + 3G_2(\tau) \pm 4\dfrac{\cosh^3\dfrac{4}{7}\tau\left(6\cosh\dfrac{8}{7}\tau - 1\right)}{5\cosh^3\dfrac{8}{7}\tau}$
—	—	—	—	—

A rough estimate of I_1 can be used to compute the Fourier transform $I_2(\Omega)$ of the broadened pulse

$$I_2(\Omega) = \frac{A(\Omega)}{I_1^*(\Omega)}, \tag{3.127}$$

where $A(\Omega)$ is the Fourier transform of the measured intensity cross-correlation. The amplitudes $E_1(t)$ and $E_2(t)$ ($\propto \sqrt{I_{1,2}(t)}$) are used to compute the upper and lower envelopes of an interferometric cross-correlation assuming no chirp:

$$F^{\pm}(\tau) = \int [E_1(t - \tau) \pm E_2(t)]^4\, dt. \tag{3.128}$$

The phase $\varphi_2(t)$ of the pulse broadened by glass can be determined from the difference $\Delta(\tau) = [F^+(\tau) - G^+(\tau)] + [F^-(\tau) - G^-(\tau)]$, where $G^+(\tau)$ and $G^-(\tau)$ are the upper and lower envelopes of the measured interferometric cross-correlation. Indeed, if the pulse E_1 is much shorter than the pulse E_2,

$$\Delta(\tau) = 4\int E_1^2(t)E_2^2(t-\tau)[1 - \cos 2\varphi_2(t-\tau)]\,dt. \tag{3.129}$$

A similar decorrelation procedure as in Eq. (3.127) is used to extract cos $(2\varphi_2)$, hence the phase of the broadened pulse E_2 from Eq. (3.129). This procedure leads, thus, to a complete determination of the complex electric-field amplitude of the pulse broadened by glass. Taking the Fourier transform of this pulse and then multiplying by the transfer function of the glass traversed leads to the Fourier transform of the original pulse. A last inverse Fourier transform leads to the complex amplitude, hence the amplitude and phase of the original laser pulse. It should be noted that the procedure just outlined computes the spectrum of the laser pulse, which can be used as an independent test of the method by comparing it to a measured pulse spectrum.

An example of application of this method is shown in Figs. 3.26a and 3.26b. The measurements pertain to two different modes of operation of the antiresonant ring (ARR) dye laser described in the previous section (Jamasbi *et al.*, 1988). This laser exhibits practically indistinguishable intensity autocorrelation when pumped continuously or synchronously (hybrid mode locking). The propagation of the laser pulses through dispersive media is drastically different in the two modes of operation, and the pulse spectra have opposite asymmetry. Applying our diagnostic technique to the output of the ARR laser, we find that, indeed, the frequency modulation of the passively mode-locked laser (Fig. 3.26a) shows much more structure than that of the synchronously mode-locked laser (Fig. 3.26b). The very complex frequency modulation of the passively mode-locked laser results from the competition between the resonant (saturation of the dye near resonance) and nonresonant (Kerr effect in the solvent) contributions. At the higher intracavity powers in the case of hybrid mode locking, the contribution from the Kerr effect alone dominates. In the case of the ring dye laser for which a pulse analysis is illustrated in Fig. 3.25, the focal spot (see Table 3.2 in Section 3.5) in the absorber jet is much larger, resulting in a negligible Kerr-effect contribution to the frequency modulation.

4.4. Single-Shot Detection

All the correlation techniques discussed previously are designed for a train of identical pulses. There are modes of operation of the passively

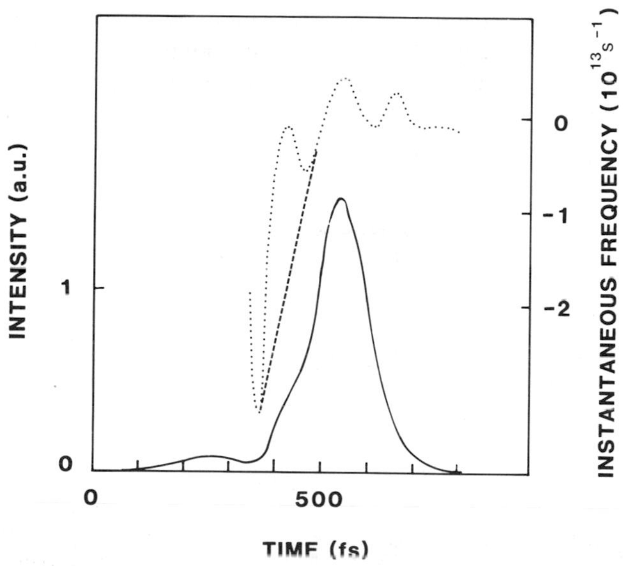

a)

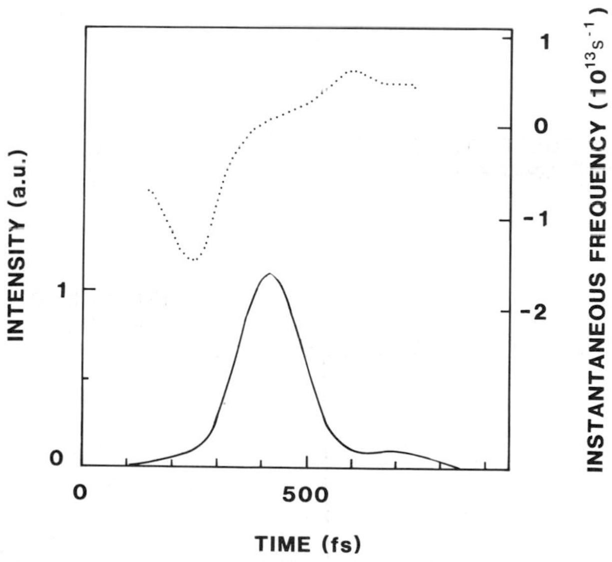

b)

Fig. 3.26 Measurement of the intensity versus time (solid line, arbitrary units) and the instantaneous frequency (dotted line) versus time for a pulse from a passively mode-locked laser (26a, upper figure) and the same laser in hybrid operation (26b, lower figure). The dashed line indicates the contribution from the Kerr effect to the chirp [from Jamasbi *et al.*, 1988].

mode-locked dye laser where the pulse shape evolves periodically along the pulse train (cf. Section 3.7). It has been pointed out by Van Stryland (1979) that the exponential wings of the intensity autocorrelation of synchronously pumped dye laser are the result of statistical fluctuations in pulse duration. The intensity autocorrelation for a series of Gaussian

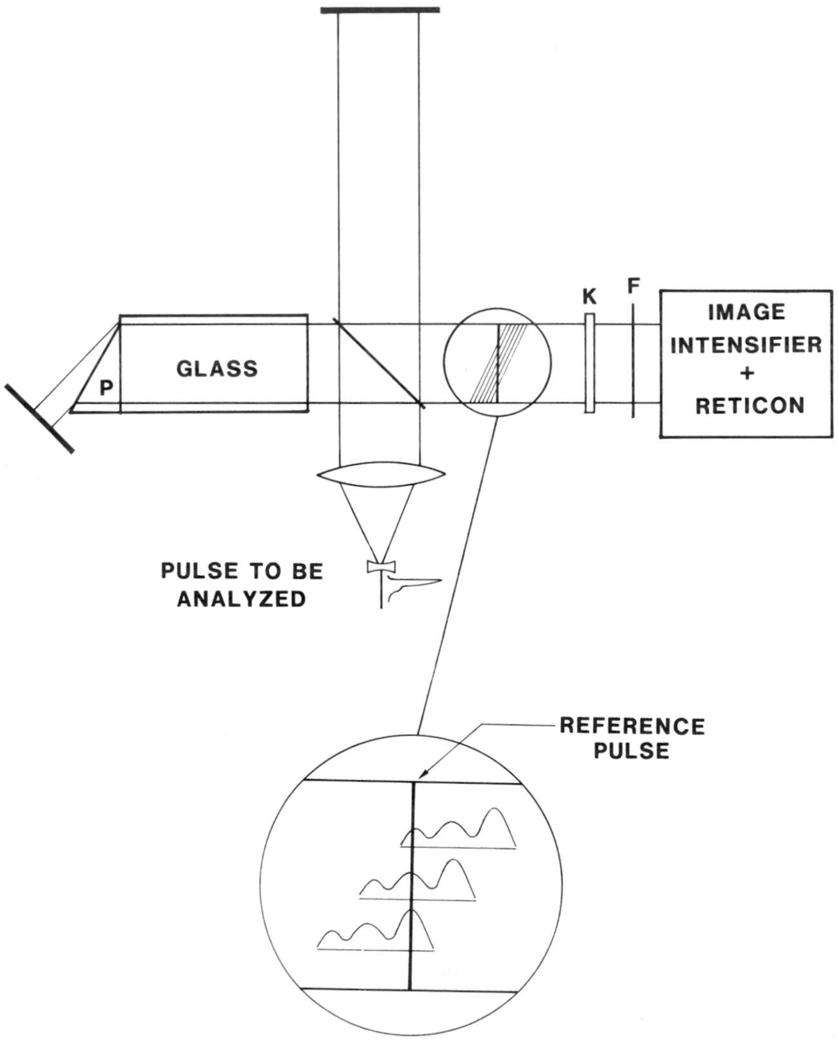

Fig. 3.27 Single-shot pulse diagnostic instrumentation. K is a thin KDP frequency doubling crystal; F is a UV transmitting filter, P is a Brewster-angle prism. The function of the glass prism is to introduce group velocity dispersion across the beam. The block of glass introduces pulse broadening.

pulses statistically distributed in pulse width has exponential wings. Catherall and New (1984) demonstrated that one indeed expects a jitter in pulse duration for this type of laser. A similar type of ambiguity exists for the phase function. The interferometric autocorrelations for a train of unchirped Gaussian pulses, with a Gaussian statistical distribution of frequencies, is identical to that of a single linearly chirped Gaussian pulse (Diels *et al.*, 1985a). Therefore, for all the correlation techniques described in the preceding paragraphs, care has to be taken that all pulses from the train are identical. Since it involves linear (spectrum) and quadratic measurements, the iterative fitting technique (Section 4.1) will simply not converge if all pulses are not identical. In the case of the asymmetric correlator (Section 4.3), comparison of the computed and measured spectra provide a method of verifying and uniformity of the pulse train.

Nevertheless, since not all fs dye lasers produce a train of identical pulses, and pulse-to-pulse fluctuation may be enhanced in amplifier chains, there is clearly a need for single-shot diagnostic techniques. Gyuzalian *et al.* (1979) have used the spatial distribution of the second harmonic in a noncolinear arrangement to determine the pulse duration. Salin *et al.* (1987) applied this method to the measurement of femtosecond pulses. Unfortunately, this method provides only an intensity autocorrelation, rather than a complete pulse characterization.

Szabo *et al.* (1988) used group velocity dispersion in a glass prism to transfer the delay to the transverse dimension in a Michelson-type interferometer. In addition, an interferometric autocorrelation can be obtained by slightly tilting one of the mirrors of the interferometer. That single-shot correlator can be modified into providing a full complete single-shot diagnostic of the pulse by the addition of a known disperser in one arm of the interferometer, as shown in Fig. 3.27 (Diels *et al.*, 1988). In one arm of the interferometer, a prism is used to transfer the delay information to a transverse coordinate. In the arrangement sketched in Fig. 3.27, group velocity dispersion in the prism P causes the delay of the reference signal to be dependent on the transverse coordinate. Therefore, the spatial recording of the second harmonic is a cross-correlation of the pulse broadened by glass with a reference pulse. An interferometric cross-correlation is obtained by introducing a fringe pattern along the plane of the figure through a slight misalignment of one of the interferometer mirrors.

5. HYBRID MODE LOCKING

Active mode locking, which involves an externally imposed loss or gain modulation in the cavity, is commercially the most readily available source of short pulses. The technique is not specific to dye lasers and therefore will not be covered here. Most often, an actively mode-locked (by loss modulation with an acousto-optic modulator) pump laser is used (after eventual pulse compression) to synchronously pump a dye laser, thereby modulating the gain at or near the cavity round-trip frequency. Synchronously mode-locked lasers operate with very large output coupling losses (the reflectivity of the output coupler is typically of the order of 60%). Unless strong intracavity bandwidth filtering is used, the intensity autocorrelation of the output pulses (which are longer than for the passively mode-locked laser) have a typical cusp shape. Since the shape of that autocorrelation can be fitted with a double exponential, many authors have considered the synchronously mode-locked laser to be a source of single-side exponential pulses, of duration half that of the full width at half-maximum (FWHM) of the autocorrelation recording. However, it has been pointed out by Van Stryland (1979) that a double-sided exponential autocorrelation is a typical signature for a random distribution of pulse duration in the train. The interferometric autocorrelation of synchronously mode-locked pulse trains exhibits also a shape characteristic of a random (Gaussian) distribution of pulse frequencies (Diels *et al.*, 1985a). A reduction, but not a complete elimination, of the background of the intensity autocorrelations of the pulse train can be obtained by active stabilization of the cavity length and mode-locker frequency. In this stabilization scheme, the error signal was the change in the ratio $P_F{}^2/P_{SH}$, where P_F and P_{SH} are the fundamental and second harmonic powers, respectively (Rotman *et al.*, 1980).

The fluctuations in pulse duration and frequency of the synchronously mode-locked laser have been vertified by theoretical (numerical) calculations of New and Catherall (1986). The stochastic nature of spontaneous emission, which is a source term in the equation of the pulse evolution, is shown in the computer simulation of Fig. 3.28; fluctuations in the spontaneous emission induce severe fluctuations in the mode-locked pulses. The synchronously mode-locked lasers are very sensitive to cavity mismatch. New and Catherall (1986) have shown that the conditions for eliminating the large-scale fluctuations and for obtaining short pulses are mutually exclusive.

A logical step to improve the characteristics of the synchronous mode-locked dye laser, one that has been implemented commercially since 1985, is the addition of a saturable absorber in the cavity (Ryan *et al.*, 1978; Couillaud *et al.*, 1985). The hybrid mode-locked laser is thus either a synchronously mode-locked laser perturbed by the addition of saturable absorption or else a passively mode-locked laser pumped synchronously. The distinction is important, since it is associated with drastically different modes of operation. The most common hybrid mode-locked lasers use a large output coupling and a low concentration of saturable absorber. The

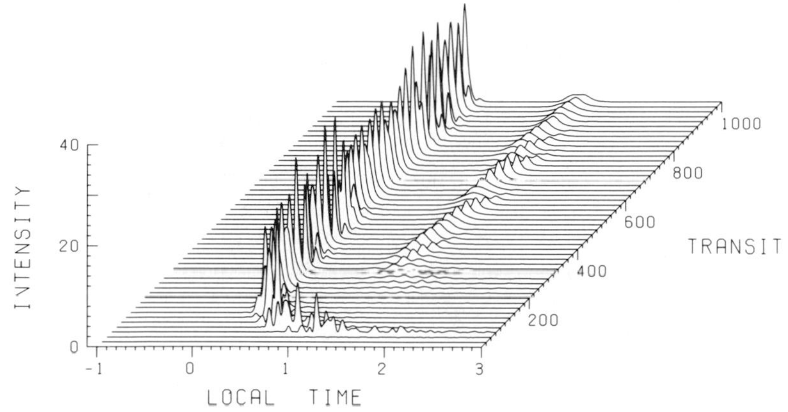

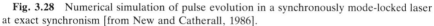

Fig. 3.28 Numerical simulation of pulse evolution in a synchronously mode-locked laser at exact synchronism [from New and Catherall, 1986].

output characteristics are very similar to that of the synchronously mode-locked laser (large tunability and intensity autocorrelations with exponential wings), but the pulse duration is less than one ps (Ishida *et al.*, 1980). Since the saturable absorber is not here the primary mode-locker, its position in the cavity is not as critical as for passive mode-locked lasers (or hybrid mode-locked lasers with small linear losses). The most common location is at the end of the cavity, although single jet configurations (with a mixture of saturable absorber and gain dyes) are also used (Mourou and Sizer, 1982; Ishida *et al.*, 1985). As a consequence of the higher intracavity power resulting from the pulsed gain, the phase modulation is dominated by the Kerr effect. This upchirp and the associated asymmetric spectra of the pulses have been analyzed by Ishida *et al.* (1985). The shortest pulses are thus obtained by incorporating a negative dispersion line inside the cavity. The arrangement most commonly used in hybrid modelocking (cf. Section 6.2) consists of a two-prism negative dispersion line incorporated in a linear cavity (cf. Section 6.6).

As the concentration of the saturable absorber is increased, the sensitivity of the laser to cavity detuning decreases. The absorber introduces an additional timing mismatch which partially compensates the pulse-advancing influence of the gain and spontaneous emission (Petrov *et al.*, 1989). The reduced sensitivity to cavity mismatch is illustrated by calculations of the steady-state pulses, for three values of the cavity mismatch, for synchronously (Fig. 3.29 a) and hybrid (Fig. 3.29 b) mode-locked laser (linear losses equal to nonlinear losses) (Petrov *et al.*, 1989). However, moderate concentrations of saturable absorber can lead to complex self-pulsing solutions, as shown by Catherall and New (1984).

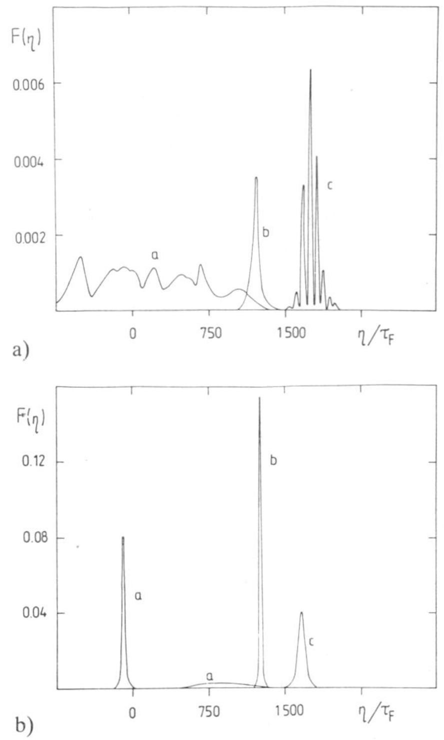

Fig. 3.29 Normalized pulse intensity in the case of synchronous (Fig. 3.29a) or hybrid (Fig. 3.29b) mode locking for cavity-length mismatch of $-1.25\ \tau_F$(a), 0(b), and $1.25\ \tau_F$(c), where τ_F is the filter response time [from Petrov *et al.*, 1989].

Most synchronously pumped systems are linear lasers, with the exception of a hybrid system demonstrated by Nuss *et al.* (1985). The antiresonant ring laser offers the possibility of combining the standing-wave saturation of the ring laser with the ease of synchronization of a liner laser (Vanherzeele *et al.*, 1984a; Norris *et al.*, 1985). For high concentrations of saturable absorber, such that the nonlinear losses dominate the linear losses, the mode-locking behavior at exact synchronism is similar to that of a passively mode-locked laser (Jamasbi *et al.*, 1988). This is also the type of hybrid mode-locked operation that was discussed in Section 3.3. The antiresonant ring-dye laser can be operated with either continuous or synchronous pumping. The intensity autocorrelations are identical in the two modes of operation. Because of the higher intracavity power, the phase modulation, in the case of hybrid mode-locking, is dominated by the

Kerr effect (Jamasbi *et al.*, 1988; and this chapter, Section 4). This type of operation can lead to as short pulse duration as the passively mode-locked dye laser, but with higher output powers. Kubota *et al.* (1988) reported 29-fs pulses from a hybrid mode-locked Kiton red S dye laser, using DODCI and diethyl-quinoloxacarbocyanine iodide (DQOCI) as saturable absorbers, providing an average output power of 60 mW at 76 MHz (peak power of 27 kW). The parameters for the shortest pulses are critically dependent on the cavity dispersion (the prism position should be controlled to within 10 μm). A steady-state dispersion and phase modulation can only be defined in the amplifying medium if the gain is accurately timed at each round-trip. Therefore, it is not surprising to find the shortest pulse operation to be as sensitive to an eventual cavity-length mismatch as the synchronously mode-locked laser.

6. EXTENDING THE PALETTE OF FS DYE LASERS

The search for a better understanding and control of the mode-locking process in the 70s' and the early 80s' has taken precedence over attempts to generate short pulses with other dye combinations that Rh 6G and DODCI (or DQOCI). There are a variety of techniques to generate short pulses at various wavelengths with a source around 620 nm. These are generation of a continuum with amplified pulses, harmonic generation, and parametric conversion.

6.1. Continuum Generation

Continuum generation involves a relatively complex setup, since the femtosecond pulses have to be amplified to the μJ level in order to generate a broad continuum in a liquid. The technique is used for pump-probe experiments that require an intense pump pulse.

The high peak-power densities associated with fs pulses allow the use of slightly different techniques of continuum generation than with more energetic pulses. For instance, a 500-μm–thick jet of ethylene glycol was shown to be sufficient for continuum generation with 100-μJ pump pulses of 80-fs duration (Fork *et al.*, 1983). The dominant mechanism for continuum production appears to be self-phase modulation. Four-photon parametric mixing is expected to contribute to a lesser extent, since the emission spectrum does not appear to depend on the emission angle. Using a copper-vapor laser pumped amplifier instead of the standard 10-Hz Nd:YAG laser pumped amplifier, continuum generation was achieved at 5 kHz with pulses of only one-μJ energy (Knox *et al.*, 1984). Self-focusing, multiphoton ionization, and plasma production appear to be the mechanisms responsible for continuum generation in gaseous media (Corkum and Rolland, 1987; Corkum *et al.*, 1986). The continuum spectra produced with 620-nm pulses extend approximately from 400 to 800 nm (Corkum *et al.*, 1986).

The continuum generated is most intense in the wavelength region close to that of the exciting pulse, and its intensity decreases by several orders of magnitude away from the pump wavelength. Therefore, even for those experiments that generate a continuum probe, there is a need for source lasers in various spectral regions.

6.2. Harmonic Generation

Because most biological materials absorb in the UV, there is a need for short-wavelength fs sources. The technique generally used is frequency doubling and tripling. However, here again, amplified pulses are generally needed to achieve the required efficiency and the desired power level in the UV. There is a compromise between the conversion efficiency and the power level that one desires to achieve. Intracavity frequency doubling has been demonstrated (Focht and Downer, 1988), using a one-mm–thick KDP crystal at a beam waist of a ring cavity, with an efficiency approaching 0.2% and only minor perturbation of the operation of the laser. The most noticeable effect on the laser characteristics is an increase of the hysteresis in the laser threshold from 0.3 watts of pump power to 1 to 1.5 watts. More efficient nonlinear crystals have become available such as urea (Halbout *et al.*, 1979) and β-BaB_2O_4 (Chen *et al.*, 1985). For the latter, the parameters applicable to femtosecond applications have been determined by Ishida and Yajima (1987).

6.3. Cascade Synchronous, Double Mode Locking

The main pump sources for the dye lasers are the argon ion laser, the frequency-doubled Nd:YAG laser, and the krypton ion laser. Multiple-wavelength femtosecond sources have the pump source in common. The various gain dyes should have their peak absorption near the pump-laser wavelength. The range of wavelengths available appears to be limited by the Stokes shift of the various dye molecules that absorb at a given wavelength (for instance 514.5 nm). One possibility to generate synchronized trains at various wavelengths is to cascade several lasers. The technique has been applied by Heritage and Jain to synchronously pumped dye lasers (Heritage and Jain, 1978). With this technique, the prices to pay for the simultaneous availability of laser pulses at two wavelengths are low conversion efficiency and a complex setup. The corresponding technique applied to passively mode-locked lasers is called "double mode locking" (Arjavalingam *et al.*, 1982). In the latter method, the saturable absorber of the primary laser is the gain medium for the secondary laser. One would expect this method, pioneered by Yasa *et al.* (1977), to generate shorter pulses than simple passive mode locking, because the lifetime of the primary absorber is shortened by the secondary laser action. Unfortunately, the increased complexity has not paid off by a significant reduction of pulse duration. An economical aspect of the technique has been to eliminate the need for active mode locking of the argon laser in synchronous or hybrid pumping. Double mode-locking of an argon laser and a dye laser in a ring cavity was reported by Wang *et al.* (1985). The particular performance edge of this configuration over passive mode locking is the overall efficiency: the average power output was 50 mW, for an argon pump power of one W.

6.4. Energy Transfer

An alternative technique for extending the lasing range of an oscillator is the use of excitation transfer between a donor and acceptor. The energy transfer can be intramolecular, as in the case of bifluorophoric dyes (Lüttke and Schäfer, 1983). In the latter, the excitation energy is transported from a short-wavelength–absorbing chromophore of the bifluorophoric molecule to a long-wavelength–emitting one by intramolecular energy transfer. In this case, lasing can only be achieved in the spectral region of the long-wavelength–emitting chromophore. Another possibility is the use of excitation transfer dye mixtures, where the excitation is transferred radiatively from the donor to the acceptor. Such mixtures have made it possible to extend the emission range of dye laser to the red and near-infrared, even with green pumping sources (Jain, 1982). As opposed to bifluorophoric dyes, excitation transfer dye mixtures can be operated simultaneously at two wavelengths (Selfridge and Dienes, 1985; Kaschke *et al.*, 1986). Synchronous pumping was made either with a Q-switched, mode-locked Nd:YAG laser or with a mode-locked argon laser. Using a mixture of Rh6G (2×10^{-3} M) as donor and Cresyl violet (from 10^{-4} to 10^{-3} M depending on the pump source and power) as acceptor, the double-wavelength operation (580 nm for the donor, 660 nm for the acceptor) resulted in a pulse shortening and improved stability at the donor wavelength. Mode-locking performances (stability and pulse duration) were generally poorer in the acceptor region (Kaschke *et al.*, 1986). Excitation transfer dye mixtures, combined with extensive testing of dyes, has contributed to finding dye combinations for passive and hybrid mode-locking over the visible and near-infrared range of wavelengths.

Table 3.5

Passive Mode Locking

Gain dye	Absorber[1]	Wavelength range		Minimum Width (fs)	at (nm)	Remarks	References
Coumarin 102	DOCI	493	502	93	497	Ring laser,	French and Taylor, 1988b
Coumarin 102	D9MOCI	488	512			UV argon laser	
Coumarin 102	DPQI	494	512			pumped	
Coumarin 102	DQTI	492	512				
Coumarin 6H	DOCI	492	507	110	497	Ring laser	French and Taylor, 1988a
Rh 110	DASBTI	553	570	210	561	Linear laser	French and Taylor, 1986a, b
	HICI	553	570	70	573	ring	French *et al.*, 1987a
Rh 6G	DASTBI	570	600	500		Linear laser	
	DODCI	550	570	250			French *et al.*, 1986a
	DQTCI						
Rh B	DQTCI	616	658	220	635	Linear cavity	French and Taylor, 1986c
Rh6G/SRh101	DQTCI	652	682	120	666	Linear cavity,	French and Taylor, 1986d
	DCCI	652	694	240	650	energy transfer	
DCM	DQTCI	655	673	680	670	Linear cavity	French and Taylor, 1986e
Rh 700	DOTCI + DCI	727	740	350	740	Linear cavity,	Smith *et al.*, 1985
	HITCI	762	778			krypton laser pump	
Rh 700/DCM	DDI	742	761	110	754	Ring laser,	
	cryptocyanine	732	743	260h12		energy transfer pumped	French *et al.*, 1987a, b

[1] See appendix for abbreviations.

Table 3.6

Hybrid Mode Locking

Gain dye	Absorber[1]	Wavelength range		Minimum Width (fs)	at (nm)	Remarks	References
Disodium fluo-	Rh B	535	575	450	545		Ishida *et al.*, 1982
Rh. 110 rescein	Rh B	545	585	250	560		
Rh 6G	DODCI	574	611	300	603		Ishida *et al.*, 1980
Rh 6G	DODCI			110	620	Ring laser	Nuss *et al.*, 1985
Rh 6G	DODCI			60	620	Antiresonant ring	Diels *et al.*, 1986b
Kiton red S	DODCI, DQOCI			29	615		Kubota *et al.*, 1988
Rh B	Oxazine 720 (perchlorate)	616	658	190	650		Dawson *et al.*, 1987a
SRh101	DQTCI	652	682	55	675	Direct pumping with doubled Nd: YAG	Dawson *et al.*, 1987a
	DCCI	652	694	240	650		
Pyridine 1 (solvent: propylene carbonate and ethylene glycol)	DDI			103	695		Dawson *et al.*, 1987b
Rhodamine 700	DOTCI	710	718	470	713		Langford *et al.*, 1986
Pyridine 2 (solvent: propylene carbonate and ethylene glycol)	DDI, DOTCI			263	733		Dawson *et al.*, 1987b
Rhodamine 700	HITCI	770	781	550	776		Langford *et al.*, 1986
LDS–751	HITCI	790	810	100			Dawson *et al.*, 1987b
Styryl 8	HITCI			70	800		
Styryl 9 (solvent: propylene carbonate and ethylene glycol)	IR 140 in benzylalcohol	840	880	65	865	Ring laser	Dobler *et al.*, 1986 Knox, 1987
Styryl 14	DaQTeC	—	—	228	974		Dawson *et al.*, 1987c

[1] See appendix for abbreviations.

6.5. Passive Mode Locking

The results of an extensive search of amplifying-dye–saturable-absorber combinations to cover the visible and near-IR range with femtosecond pulses are summarized in Table 3.5. The shortest pulse duration quoted should be viewed as an upper limit, since part of the testing was made with a linear cavity that was not dispersion-compensated. It should be noted that all the visible spectrum between 490 and 780 nm is covered, except for a gap between 512 and 553 nm.

6.6. Hybrid Mode Locking

The dye combinations shown in Table 3.5 can also be applied to hybrid mode locking. The range of mode locking covered by hybrid mode-locked lasers is however larger, as can be seen from Table 3.6.

One can reasonably expect the range to expand further in the UV and into the infrared. Synchronous mode locking has been demonstrated in the 412–438-nm wavelength range with stilbene 420 and in the 432–460-nm wavelength range with coumarin 450 (Ure *et al.*, 1986). These lasers were pumped by the third harmonic of the Nd:YAG laser. Substituted diphenyl-benzo-bisoxazoles appear promising for the wavelength range of 350 to 400 nm (Gusten *et al.*, 1986).

At the long-wavelength end of the spectrum, several dyes are now available for synchronous pumping by the Nd:YAG laser. Infrared laser dyes differ from the visible laser dyes by a much smaller quantum efficiency (in the range of 10^{-3} to 10^{-4}) and shorter fluorescence lifetimes (in the picosecond range). Generation of bandwidth-limited pulses with dyes having a fluorescence lifetime shorter than the pump pulse is difficult (Roskos *et al.*, 1986). The use of compressed Nd:YAG laser pulses for synchronously pumping infrared dye lasers leads to more stable operation, with a pulse duration of 400 fs (Beaud *et al.*, 1986). One can reasonably expect that this laser is the seed for extending further the range of fs sources into the infrared.

PROBLEMS

1. Evolution of the pulse energy

1a. Compare the sequence of elements of Fig. 3.5 with that leading to Eq. (3.33). Show that the two sequences can be made to correspond to two pulses propagating in opposite directions in a ring cavity. Are the two equations (3.32) and (3.33) identical? Find the common limits and differences.

1b. Study the case $G = 1.1$, $A = 0.8$, $r = 0.7$, $x_o = 0.1$, $s = 5$. Compare with the results of Fig. 3.8.

2. ***Effect of excited-state absorption***

Calculate the intracavity pulse evolution for nonlinear losses proportional to the pulse energy. Choose in Eq. (3.33) $r - r_o(1 - 10 \cdot sx)$. Find periodic solutions.

3. ***Phase modulation and dispersion***

3a. Consider a Gaussian pulse of FWHM $\sqrt{2 \ln 2} = 1.177$ and intensity 1. After passage through a nonlinear device, its phase is incremented by $\phi = -aI_o t^2$. Thereafter it is sent through a dispersive medium with $k'' = \alpha\Omega^2$. Find the value of α providing optimum compression in the sequence modulator-disperser, as well as the parameters of the compressed pulse.

3b. Using the same parameters as in 3a, find the pulse resulting from the sequence disperser-modulator.

4. ***Coherent fs laser***

4a. Find the steady-state solutions on-resonance and off-resonance for the linear amplifier and the colliding-pulse resonant absorber (in particular, find expressions for the parameters A, B, and E_o).

4b. Use the preceding solution for the colliding-pulse coherent absorber off-resonance. Find the conditions for the amplifying medium, on resonance, that would satisfy the steady-state condition for the combination 2-level resonant amplifier, colliding-pulse off-resonant absorber, and linear losses $(1 - r)$.

5. ***Solitons***

Derive Eq. 3.119

REFERENCES

Arecchi, F.T., and Bonifacio, R. (1965). Theory of optical laser amplifiers. *IEEE J. Quantum Electron.* **QE-1**, 169–178.

Arjavalingam, G., Dienes, A., and Whinnery, J.R. (1982). Highly synchronous mode-locked picosecond pulses at two wavelengths. *Opt. Lett.* **7**, 193–195.

Avramopoulos, H., French, P.M.W., Williams, J.A.R., New, G.H.C., and Taylor, J.R. (1988). Experimental and theoretical studies of complex pulse evolutions in a passively mode-locked ring dye laser. *IEEE J. Quantum Electron.* **24**, 1884–1892.

Beaud, P., Zyssed, B., and Weber, H.P. (1986). 1.3 m subpicosecond pulses from a dye laser pumped by compressed Nd:YAG laser pulses. *Opt. Lett.* **11**, 24–27.

Besnainou, S., Diels, J.-C., and Stone, J. (1984). Molecular multiphoton excitation of phase coherent pulse pairs. *J. Chem. Phys.* **81,** 143–149.

Blit, S., and Tittel, F.K. (1978). A new laser configuration for high power cw dye laser operation. CLEO '78, San Diego, Calif.

Bourne, O.L., and Alcock, A.J. (1986). An ultraviolet and visible single shot autocorrelator based on multiphoton ionisation. *Rev. Sci. Instruments* **57,** 2979–2982.

Bradley, D.J., and New, G.H.C. (1974). Ultrashort pulse measurements. *Proc. IEEE* **62,** 313.

Catherall, J.M., and New, G.H.C. (1984). Theoretical studies of active, synchronous, and hybrid mode-locking. In *Ultrafast Phenomena IV* (Auston, D.H., and Eisenthal, K.B., eds.). Springer-Verlag, Berlin, pp. 75–77.

Catherall, J.M., and New, G.H.C. (1986). Role of spontaneous emission in the dynamics of mode-locking by synchronous pumping. *IEEE J. Quantum Electron.* **QE-22,** 1593–1599.

Chekalin, S.V., Kryukov, P.G., Matveetz, Yu. A., and Shatberashvili, O.B. (1974). The processes of formation of ultrashort laser pulses. *Opto-electronics* **6,** 249–261.

Chen, C., Wu, B., Jiang, A., and You, G. (1985). A new type ultraviolet SHG crystal. *Scientia Sinica (Ser. B)* **28,** 235–243.

Christodoulides, D.N., Bourkoff, E., Joseph, R.I., and Simos, T. (1985). Reflection of femtosecond optical pulses from multiple-layer dielectric mirrors: Analysis. *IEEE J. Quantum Electron.* **QE-22,** 186–191.

Corkum, P.B., Rolland, C., and Srinivasan-Rao, T. (1986). Supercontinuum generation in gases: a high order nonlinear optics phenomenon. In *Ultrafast Phenomena V* (Fleming, G.R., and Siegman, A.E., eds.). Springer-Verlag, Berlin, pp. 149–152.

Corkum, P.B., and Rolland, C. (1987). Ultrashort pulse multiphoton ionization of Xenon at moderate pressures. In *Atomic and Molecular Processes with Short and Intense Laser Pulses* (Bandrauk, A.D., ed.). NATO ASI series. Plenum Press, New York, pp. 157–163.

Couillaud, B., Fossati-Bellani, V., and Mitchell, G. (1985). Ultrashort pulse spectroscopy and applications. *SPIE Proceedings*, **vol. 533,** pp. 46–51.

Dawson, M.D., Boggess, T.F., Garvey, D.W., and Smirl, A. (1987a). Femtosecond pulse generation in the red/deep red spectral region. *IEEE J. Quantum Electron.* **QE-23,** 290-292.

Dawson, M.D., Boggess, T.F., and Smirl, A. (1987b). Femtosecond synchronously pumped pyridine dye lasers. *Opt. Lett.* **12,** 254–256.

Dawson, M.D., Boggess, T.F., and Smirl, A. (1987c). Picosecond and femtosecond pulse generation near 1000 nm from a frequency-doubled Nd : YAG-pumped cw dye laser. *Optics Lett.* **12,** 590–592.

De Silvestri, S., Laporta, P., and Svelto, O. (1984a). Analysis of quarter-wave dielectric-mirror dispersion in femtosecond dye-laser cavities. *Opt. Lett.* **9,** 335–337.

De Silvestri, S., Laporta, P., and Svelto, O. (1984b). The role of cavity dispersion in cw mode-locked lasers. *IEEE J. Quantum Electron.* **QE-20,** 533–539.

De Silvestri, S., Laporta, P., and Svelto, O. (1984c). Effects of cavity dispersion on femtosecond mode-locked dye lasers. In *Ultrafast Phenomena IV* (Auston, D.H., and Eisenthal, K.B., eds.). Springer-Verlag, New York, pp. 23–26.

Diels, J.-C., and Hahn, E.L. (1976). Pulse propagation stability in absorbing and amplifying media. *IEEE J. Quantum Electron.* **QE-12,** 411–416.

Diels, J.-C., and Sallaba, H. (1980). Black magic with red dyes. *JOSA* **70,** 629–630.

Diels, J.-C., and McMichael, I.C. (1986). Degenerate four wave mixing of femtosecond pulses in an absorbing dye jet. *JOSA* **B3,** 535–543.

Diels, J.-C., Van Stryland, E.W., and Benedict, G. (1978a). Generation and measurements of pulses of 0.2 psec duration. *Opt. Comm.* **25,** 93–96.

Diels, J.-C., Van Stryland, E.W., and Gold, D. (1978b). Parameters affecting picosecond and subpicosecond pulse duration in passively mode-locked dye lasers. In *Picosecond Phenomena I* (Shank, C.V., Ippen, E.P., and Shapiro, S.V., eds.). Springer Series in Chemical Physics, **vol. 4.** Springer-Verlag, Berlin, pp. 117-120.

Diels, J.-C., Menders, J., and Sallaba, H. (1980). Generation of coherent pulses of 60 optical cycles through synchronization of the relaxation oscillations of a mode-locked dye laser. In *Picosecond Phenomena II* (Hochstrasser, R.M., Kaiser, W., and Shank, C.V., eds.). Springer Series in Chemical Physics, **vol. 14.** Springer-Verlag, Berlin, pp. 1–45.

Diels, J.-C., Wang, W.C., and Winful, H. (1981). Dynamics of the nonlinear four-wave mixing and interaction. *Appl. Phys.* **B26,** 105–110.

Diels, J.-C., Fontaine, J.J., McMichael, I.C., and Wang, C.Y. (1982a). Experimental and theoretical study of the mutual interaction of subpicosecond pulses in absorbing and gain media. Paper presented at XIIth Int. Quantum Electron. Conf., Munich.

Diels, J.-C., McMichael, I.C., Fontaine, J.J., and Wang, C.Y. (1982b). Subpicosecond pulse shape measurement and modeling of a passively mode-locked dye laser including mutual interaction in a dye jet. In *Picosecond Phenomena III* (Eisenthal, K.B., Hochstrasser, R.M., Kaiser, W., and Lauberean, A., eds.). Springer Series in Chemical Physics, **vol. 23.** Springer-Verlag, Berlin, pp. 116–119.

Diels, J.-C., Fontaine, J.J., McMichael, I.C., and Wilhelmi, B. (1983). Experimental and theoretical study of a femtosecond laser. *Kvantovaya Elektron.* **10,** 2398–2410 (translation in *Sov. J. Quantum Electron.* **13,** 1562–1569).

Diels, J.-C., McMichael, I.C., and Vanherzeele, H. (1984a). Degenerate four wave mixing of picosecond pulses in the saturable amplification of a dye laser. *IEEE J. Quantum Electron.* **QE-20**, 630–636.

Diels, J.-C., Dietel, W., Döpel, E., Fontaine, J., McMichael, I.C., Rudolph, W., Simoni, F., Torti, R., Vanherzeele, H., and Wilhelmi, B. (1984b). Colliding pulse femtosecond lasers and applications to the measurement of optical parameters. In *Ultrafast Phenomena IV* (Auston, D., and Eisenthal, K.B., eds.). Springer Series in Chemical Physics, **vol. 38.** Springer-Verlag, Berlin and Heidelberg, pp. 30–34.

Diels, J.-C., Vanherzeele, H., and Torti, R. (1984c). Femtosecond pulse generation in a linear cavity terminated by an antiresonant ring. Paper WC5, presented at Conf. on Lasers and Electro-optics, Anaheim, California.

Diels, J.-C., Fontaine, J.J., McMichael, I.C., and Simoni, F. (1985a). Control and measurement of ultrashort pulse shapes (in amplitude and phase) with femtosecond accuracy. *Appl. Opt.* **24,** 1270–1282.

Diels, J.-C., Dietel, W., Fontaine, J.J., Rudolph, W., and Wilhelmi, B. (1985b). Analysis of a mode-locked ring laser: chirped-solitary-pulse solutions. *JOSA B* **2,** 680–686.

Diels, J.-C., Jamasbi, N., and Sarger, L. (1986a). Engineering of a five mirror femtosecond laser, using geometrical astigmatism compensation. Presented at Conf. on Lasers and Electro-optics CLEO'86, San Francisco, Calif.

Diels, J.-C., Jamasbi, N., and Sarger, L. (1986b). Passive and hybrid fsec operation of a linear astigmatism compensated dye laser. In *Ultrafast Phenomena V* (Fleming, G.R., and Siegman, A.E., eds.). Springer-Verlag, Berlin, pp. 2–4.

Diels, J.-C., Fontaine, J.J., and Rudolph, W. (1987a). Ultrafast diagnostics. *Rev. Phys. Appl.* **22,** 1605–1611.

Diels, J.-C., Fontaine, J.J., Jamasbi, N., and Lai, M. (1987b). The femtonitpicker. Presented at Conf. on Lasers and Electro-optics CLEO'87, Baltimore, Md.

Diels, J.-C., Jamasbi, N., Yan, C., and Lai, M. (1988). Ultrafast detection of weak signals. SPIE Conference on Advances in Semiconductors and Superconductors, Newport Beach, Calif. *SPIE* **942,** 190–194.

Dietel, W., Döpel, E., Kühlke, D., and Wilhelmi, B. (1982). Pulses in the femtosecond range from a cw dye ring laser in the colliding pulse mode-locking regime with down chirp. *Opt. Comm.* **43,** 433–436.

Dietel, W., Fontaine, J.J., and Diels, J.-C. (1983). Intracavity pulse compression with glass: a new method of generating pulses shorter than 60 femtoseconds. *Opt. Lett.* **8,** 4.

Dietel, W., Döpel, E., Hehl, K., Rudolph, W., and Schmidt, E. (1984a). Multilayer dielectric mirror generated chirp in femtosecond dye ring lasers. *Opt. Comm.* **50,** 179–182.

Dietel, W., Rudolph, W., Wilhelmi, B., Diels, J.-C., and Fontaine, J.J. (1984b). Formation of solitary femtosecond light pulses with chirp in passively mode-locked dye lasers. *Izv. Akad. Nauk. SSSR Se Fiz* **48,** 480–491.

Dobler, J., Schulz, H.H., and Zinth, W. (1986). Generation of femtosecond light pulses in the near infrared around 850 nm. *Opt. Comm.* **57,** 407–409.

Dobrowski, J.A. (1978). Coatings and filters. In *Handbook of Optics* (Driscoll, W.G., and Vaughan, W., eds.). McGraw-Hill, New York, pp. 8.67–8.69.

Duarte, F.J. (1987). Generalized multiple prism dispersion theory for pulse compressions in ultrafast dye lasers. *Opt. Quantum Electron.* **19**, 223–229.

Duarte, F.J. (1989). Ray transfer matrix analysis of multiple-prism dye laser oscillators. *Opt. Quantum Electron.* **21,** 47–54.

Dugay, M.A., and Hansen, J.W. (1969). Compression of pulses from a mode-locked He–Ne laser. *Appl. Phys. Lett.* **14,** 14–15.

Feynman, R.P., Vernon, F.L., and Hellwarth, R.W. (1957). Geometrical representation of the Schrödinger equation for solving maser problems. *J. Appl. Phys.* **28,** 49–52.

Focht, G., and Downer, M.C. (1988). Generation of synchronized ultraviolet and red femtosecond pulses by intracavity frequency doubling. *IEEE J. Quantum Electron*, special issue on ultrafast optics and electronics (Feb. 1988).

Fontaine, J.J., Dietel, and Diels, J.-C. (1983). Chirp in a mode-locked ring dye laser. *IEEE J. Quantum Electron.* **QE-19,** 1467.

Fork, R.L., and Shank, C.V. (1981). Generation of optical pulses shorter than 0.1 ps by colliding pulse mode-locking. *Appl. Phys. Lett.* **38,** 671–673.

Fork, R.L., Shank, C.V., Hirlimann, C., Yen, R., and Tomlinson, W.J. (1983). Femtosecond white-light continuum pulses. *Opt. Lett.* **8,** 1–3.

Fork, R.L., Martinez, O.E., and Jordan, J.P. (1984). Negative dispersion using pairs of prisms. *Opt. Lett.* **9,** 150–152.

Fork, R.L., Cruz, C.H., Becker, P.C., and Shank, C.V. (1987). Compression of optical pulses to six femtoseconds by using cubic phase compensation. *Opt. Lett.* **12,** 483–485.

French, P.M.W., and Taylor, J.R. (1986a). Femtosecond pulse generation from passively mode-locked continuous wave dye lasers 550–700 nm. In *Ultrafast Phenomena V* (Fleming, G.R., and Siegman, A.E., eds.). Springer Series in Chemical Physics, **vol. 46.** Springer-Verlag, Berlin, New York, London, pp. 11–13.

French, P.M.W., and Taylor, J.R. (1986b). Passively mode-locked continuous-wave Rhodamine 110 dye laser. *Opt. Let.* **11,** 297–299.

French, P.M.W., and Taylor, J.R. (1986c). The passive mode-locking of the continuous wave Rhodamine B dye laser. *Opt. Comm.* **58,** 53–55.

French, P.M.W., and Taylor, J.R. (1986d). Passive mode-locking of an energy transfer continuous wave dye laser. *IEEE J. Quantum Electron.* **QE22,** 1162–1164.

French, P.M.W., and Taylor, J.R. (1986e). Passive mode-locking of the continuous wave DCM dye laser. *J. Appl. Phys.* **B41, 53–55.**

French, P.M.W., and Taylor, J.R. (1987). The passively mode-locked and dispersion compensated Rh110 dye laser. *Opt. Comm.* **63,** 124–129.

French, P.M.W., and Taylor, J.R. (1988a). The passively mode-locked coumarin 6H ring dye laser. *Opt. Comm.* **67,** 51–52.

French, P.M.W., and Taylor, J.R. (1988b). Generation of sub-100 fs pulses tunable around 497 nm from a colliding pulse mode-locked ring dye laser. *Opt. Lett.* **13,** 470–472.

French, P.M.W., Chen, G.F., and Sibbett, W. (1986a). Tunable group velocity dispersion interferometer for intracavity and extracavity applications. *Opt. Comm.* **57,** 263–268.

French, P.M.W., Dawson, M.D., and Taylor, J.R. (1986b). Passive mode-locking of a cw dye laser in the yellow spectral region. *Opt. Comm.* **56,** 430–432.

French, P.M.W., Williams, J.A.R., and Taylor, J.R. (1987a). Passively mode-locked dye lasers operating from 490 nm to 800 nm. *Rev. Phys. App.* **22,** 1651–1655.

French, P.M.W., Williams, J.A.R., and Taylor, J.R. (1987b). Passive mode-locking of a c.w. energy transfer dye laser operating in the near infrared around 750 nm. *Opt. Lett.* **12,** 684–686.

Gardner, C.S., Green, G., Kruskal, M., and Miura, R. (1967). *Phys. Rev. Lett.* **19,** 1095.

Gires, F., and Tournois, P. (1964). Interferomètre utilisable pour la compression d'impulsions lumineuses modulées en fréquence. *C.R. Acad. Sc. Paris* 6112–6115.

Glenn, W.H. (1975). The fluctuation model of a passively mode-locked laser. *IEEE J. Quantum Electron.* **QE-11,** 8–17.

Gordon, J.P., and Fork, R.L. (1984). Optical resonator with negative dispersion. *Opt. Lett.* **9,** 153–156.

Gusten, H., Rinke, M., Kao, C., Zhou, Y., Wang, M., and Pan, J. (1986). New efficient laser dyes for operation in the near UV range. *Opt. Comm.* **59,** 379–384.

Gyuzalian, R.N., Sogomonian, S.B., and Horvath, Z. Gy. (1979). Background-free measurement of time behaviors of an individual picosecond laser pulse. *Opt. Comm.* **29,** 239–242.

Halbout, J.M. (1979). Efficient phase matched second harmonic generation and sum frequency in urea. *IEEE J. Quantum Electron.* **15,** 1176–1180.

Haus, H.A. (1975). Theory of mode-locking with a slow saturable absorber. *IEEE. J. Quantum Electron.* **QE-11,** 736–746.

Haus, H.A., and Silberberg, Y. (1986). Laser mode-locking with addition of nonlinear index. *IEEE J. Quantum Electron.* **QE–22,** 325–331.

Heppner, J., and Kühl, J. (1985). Intracavity chirp compensation in a colliding pulse mode-locked laser using thin-film interferometers. *Appl. Phys. Lett.* **47,** 453–456.

Heritage, J.P., and Jain, R.K. (1978). Subpicosecond pulses from a cw. mode locked dye laser. *Appl. Phys. Lett.* **32,** 101–103.

Ippen, E.P., and Shank, C.V. (1975). Dynamic spectroscopy and subpicosecond pulse compression. *Appl. Phys. Lett.* **27,** 488–491.

Ippen, E.P., Shank, C.V., and Dienes, A. (1972). Passive mode-locking of the cw dye laser. *Appl. Phys. Lett.* **21,** 348–350.

Ishida, Y., and Yajima, T. (1987). Characteristics of a new type SHG crystal β-BaB_2O_4 in the femtosecond region. *Opt. Comm.* **62,** 197–200.

Ishida, Y., Yajima, T., and Naganuma, K. (1980). Generation of broadly tunable subpicosecond light pulses from a synchronously and passively mode-locked cw dye laser. *Jap. J. App. Phys.* **19,** L717–L720.

Ishida, Y., Naganuma, K., and Yajima, T. (1982). Tunable subpicosecond pulse generation from 535 to 590 nm by a hybridly mode-locked dye laser. *Jap. J. App. Phys.* **21,** L312–L314.

Ishida, Y., Naganuma, K., and Yajima, T. (1985). Self-phase modulation in hybridly mode-locked cw dye lasers. *IEEE J. Quantum Electron.* **QE-21,** 69–77.

Jain R.K. (1982). Near infrared picosecond pulse generation in a cw. mode-locked dye laser pumped directly by an argon ion laser. *Appl. Phys. Lett.* **40,** 295–297.

Jain, R.K., Brown, J.E., and Robinson, W.P. (1982). Simple distortion-free real-time optical pulse correlator. *Appl. Opt.* **21,** 4073–4076.

Jamasbi, N., Diels, J.-C., and Sarger, L. (1988). Study of a linear femtosecond laser in passive and hybrid operation. *J. Modern Opt.* **35,** 1891–1906.

Kaschke, M., Stamm, U., and Volger, K. (1986). Subpicosecond pulse generation in synchronously pumped energy transfer dye lasers. *Appl. Phys.* **B39,** 183–186.

Kintzer, E.S., and Rempel, C. (1987). Near-surface second harmonic generation for autocorrelation measurements in the UV. *Appl. Phys.* **B42,** 91–95.

Knox, W.H. (1986). Femtosecond vibrational relaxation of the F_2^4 center in LiF. In *Ultrafast Phenomena V* (Fleming, G.R., and Siegman, A.E., eds.). Springer Series in Chemical Physics, **vol. 46.** Springer-Verlag, Berlin, pp. 277–299.

Knox, W.H. (1987). Generation and kilohertz-rate amplification of femtosecond optical pulses around 800 nm. Conference on Lasers and Electro-optics, Baltimore, Md., April 1987.

Knox, W.H., Downer, M.C., Fork, R.L., and Shank, C.V. (1984). Amplified femtosecond optical pulses and continuum generation at 5 kHz repetition rate. *Opt. Lett.* **9,** 552–554.

Kogelnik, H., and Li, T. (1966). Laser beams and resonators. *Appl. Opt.* **9,** 1550–1566.

Kubota, H., Kurokawa, K., and Nakazawa, M. (1988). 29 fs pulse generation from a linear cavity synchronously pumped dye laser. *Opt. Lett.* **13,** 749–751.

Kühl, J., and Heppner, J. (1986). Compression of femtosecond optical pulses with dielectric multilayer interferometers. *IEEE J. Quantum Electron.* **QE-22,** 2070–2074.

Kühlke, D., Rudolph, W., and Wilhelmi, B. (1983). Influence of transient absorber grating on the pulse parameters of passively mode-locked ring lasers. *Appl. Phys. Lett.* **42,** 325–327.

Kühlke, D., Bonkhofer, T., and von der Linde, D (1986a). Pulse fluctuations and chirp compensation in colliding-pulse mode-locked dye laser. *Opt. Comm.* **59,** 208–212.

Kühlke, D., Bonkhofer, T., Herpers, U., and von der Linde, D. (1986b). Fluctuations and chirp in colliding-pulse mode-locked dye laser. In *Ultrafast Phenomena V* (Fleming, G.R., and Siegman, A.E., eds.). Springer-Verlag, Berlin, pp. 17–19.

Kühlke, D., Herpers, V., and von der Linde, D. (1985). Characteristics of a hybridly mode locked cw dye laser. *Appl. Phys.* **B38,** 233–240.

Kühlke, D., Rudolph, W., and Wilhelmi, B. (1982). Pulses in the femtosecond range from a cw ring laser in the colliding pulse mode locking (CPM) regime with downchirp. *Opt. Comm.* **43,** 433.

Langford, N., Smith, K., Sibbett, W., and Taylor, J.R. (1986). The hybrid mode-locking of a cw Rhodamine 700 dye laser. *Opt. Comm.* **58,** 56–58.

Lüttke, W., and Schäfer, F.P. (1983). Neue eutwicklungen bei laserfarbstoffen. *Laser und Optoelektr.* **15,** 127–136.

Martinez, O.E., Fork, R.L., and Gordon, J.P. (1984). Theory of passively mode-locked lasers including self-phase modulation and group velocity dispersion. *Opt. Lett*, **9,** 156–158.

McMichael, I.C. (1984). Degenerate four wave mixing of short and ultrashort light pulses. Ph.D. thesis. North Texas State University, Denton, Texas.

Melles Griot, *Optics Catalogue* Irvine, Calif. *1988.*

Mollenauer, L.F., Stolen, R.H., and Gordon, J.P. (1980). Experimental observation of picosecond pulse narrowing and solitons in optical fibers. *Phys. Rev. Lett.* **45,** 1095–1098.

Montgomery, S., Pederson, D., and Salamo, G. (1986). Intensity profiles of short optical pulses via temporarily reversed pulses. *Appl. Phys. Lett.* **49,** 620–621.

Mourou, G.A., and Sizer II, T. (1982). Generation of pulses shorter than 70 fs with a synchronously pumped cw dye laser. *Opt. Comm.* **41,** 47–48.

New, G.H.C. (1972). Mode-locking of quasi-continuous lasers. *Opt. Comm.* **6,** 188–192.

New, G.H.C. (1974). Pulse evolution in mode-locked quasi-continuous lasers. *IEEE J. Quantum Electron.* **QE-10**, 115–124.

New, G.H.C., and Catherall, J.M. (1986). Advances in the theory of mode-locking by synchronous pumping. In *Ultrafast Phenomena V* (Fleming, G.R., and Siegman, A.E., eds.). Springer-Verlag, Berlin, pp. 24–26.

Noack, P., Rudolph, W., Deijch, R., and Postovalos, W.B. (1988). Measurement of UV femtosecond light pulses using two-photon luminescence in CsI crystals doped with Na. Annual meeting of the Am. Electrochem. Soc., Chicago, 1988, proceedings to be published.

Norris, T., Sizer II, T., and Mourou, G. (1985). Generation of 85-fs pulses by synchronous pumping of a colliding-pulse mode-locked dye laser. *JOSA* **B2,** 613–614.

Nuss, M.C., Leonhart, R., and Zinth, W. (1985). Stable operation of a synchronously pumped colliding-pulse mode-locked ring dye laser. *Opt. Lett.* **10,** 16–18.

Petrov, V., Rudolph, W., and Wilhelmi, B. (1987a). Chirping of femtosecond light pulses passing through a four-lever absorber. *Opt. Comm.* **64,** 398–402.

Petrov, V., Rudolph, W., and Wilhelmi, B. (1987b). Evolution of chirped light pulses and the steady state regime passively mode-locked femtosecond dye lasers. *Rev. Phys. Appl.* **22,** 1639–1650.

Petrov, V., Rudolph, W., Stamm, U., and Wilhelmi, B. (1989). Limits of ultrashort pulse generation in cw mode-locked dye lasers. *Phys. Rev. A* **40,** 1474–1483.

Prasad, P.N. (1986). Nonlinear optical interactions in polymer thin films. Society for Photo-Optical Instrumentation Engineers, SPIE **vol. 682.** Molecular and Polymeric Optoelectronic Materials: Fundamentals and Applications, pp 120–124. SPIE 29th Int. Symposium, San Diego, Calif., Aug. 1986.

Risken, H., and Nummedal, K. (1968). Self-pulsing in lasers. *J. Appl. Phys.* **39,** 4662–4672.

Roskos, H., Optiz, S., Seilmeier, A., and Kaiser, W. (1986). Operation of an infrared dye laser synchronously pumped by a mode-locked cw Nd:YAG laser. *IEEE J. Quantum Electron.* **QE-22,** 697–703.

Rothenberg, J.E., and Grischkowsky, D. (1985). Measurement of the phase of a frequency-swept ultrashort optical pulse. *JOSA* **B2,** 626–633.

Rotman, S.R., Roxlo, C.B., Bebelaar, D., and Salour, M.M. (1980). Pulse-width stabilization of a synchronously pumped mode-locked dye laser. *Appl. Phys. Lett.* **36,** 886–888.

Ruddock, I.S., and Bradley, D.J. (1976). Bandwidth-limited subpicosecond pulse generation in mode-locked cw dye lasers. *Appl. Phys. Lett.* **29,** 296–297.

Rudolph, W., and Wilhelmi, B. (1984). Calculation of light pulses with chirp in passively mode-locked lasers taking into account the phase memory of absorber and amplifier. *Appl. Phys.* **B35,** 37–44.

Ryan, J.P., Goldberg, L.S., and Bradley, D.J. (1978). Comparison of synchronous pumping and passive mode-locking of cw dye lasers for the generation of picosecond and subpicosecond light pulses. *Opt. Comm.* **27,** 127–132.

Salin, F., Grangier, P., Roger, G., and Brun, A. (1986). Observation of high-order solitons directly produced by a femtosecond ring laser. *Phys. Rev. Lett.* **56,** 1132–1135.

Salin, F., Georges, P., Roger, G., and Brun, A. (1987). Single shot measurement of a 52 fs pulse. *Appl. Opt.* **26,** 4528–4531.

Schott, "Optical Glass Pocket Catalogue," from the Catalogue *Optical Glass No. 3111*, Duryea, Pa.

Selfridge, R., and Dienes, A. (1985). Synchronously pumped double mode-locking of a rhodamine 6G-cresyl violet mixture. *Appl. Phys. B.* **37,** 7.

Shank, C.V. (1982). Program in ultrashort optical pulse generation. Paper WE.3 presented at the XIIth Int. Quantum Electronics Conf., Munich, Germany, June 22.

Shank, C.V., Becker, P.C., Fragnito, H.L., and Fork, R.L. (1988). Femtosecond photon echoes. In *Ultrafast Phenomena VI* Yajima, T., Yoshihara, K., Harris, C.B., and

Shionoya, S., eds.). Springer Series in Chemical Physics, **vol. 48.** Springer-Verlag, Berlin, pp. 344–348.

Siegman, A.E. (1973). An antiresonant ring interferometer for coupled laser cavities, laser output coupling, mode-locking, and cavity dumping. *IEEE J. Quantum Electron.* **QE-9,** 247–250.

Siegman, A.E. (1981). Passive mode-locking using an antiresonant ring laser cavity. *Opt. Lett.* **6,** 334–335.

Smith, K., Langford, N., Sibbet, W., and Taylor, J.R. (1985). Passive mode-locking of a continuous wave dye laser in the red-near infrared spectral region. *Opt. Lett.* **10,** 559–561.

Szabo, G., Bor, Z., and Müller, A. (1988). Phase sensitive single-pulse autocorrelator for ultrashort laser pulses. *Opt. Lett.* **13,** 746–748.

Ure, K.A., Hanna D.C., and Pointer, D.J. (1986). A high power synchronously pumped dye laser operating in the blue and green spectral region. *Opt. Comm.* **60,** 229–231.

Valdmanis, J.A., Fork, R.L., and Gordon, J.P. (1985). Generation of optical pulses as short as 27 fs directly from a laser balancing self-phase modulation, group velocity dispersion, saturable absorption, and saturable gain. *Opt. Let.* **10,** 131–133.

Van Stryland, E.W. (1979). The effect of pulse to pulse variation on ultrashort pulsewidth measurements. *Opt. Comm.* **31,** 93–94.

Vanherzeele, H., Diels, J.-C., and Torti, R. (1984a). Tunable passive colliding pulse mode-locking in a linear dye laser. *Opt. Lett.* **9,** 549–551.

Vanherzeele, H., Torti, R., and Diels, J.-C. (1984b). Synchronously pumped dye laser passively mode-locked with an antiresonant ring. *Appl. Opt.* **23,** 4182–4183.

Wang, C.Y., Xing, Q.-R., and Zhao, Xin-miao (1985). The double and colliding pulse mode-locking of an Art and dye laser in a ring cavity. *Opt. Comm.* **55,** 135–137.

Wang, C.Y., Zhang, R., Xiang, W., and Liu, H. (1986). Generation of optical pulses as short as 30 fs from a simple CPM dye laser. Presented at the 1986 Int. School of Laser Applications, Beijing.

Weber, H.P. (1966). Method for pulsewidth measurement of ultrashort light pulses generated by phase-locked lasers using nonlinear optics. *J. Appl. Phys.* **38,** 2231–2234.

Weber, H.P., and Danielmeyer, H.G. (1970). Multimode effects in intensity correlations measurements. *Phys. Rev.* **A2,** 2074–2078.

Weiner, A.M., Fujimoto, J.G., and Ippen, E.P. (1984). Compression and shaping of femtosecond pulses. In *Ultrafast Phenomena IV* (Auston, D.J., and Eisenthal, K.B., eds.). Springer-Verlag, New York, pp. 11–15.

Weiner, A.M., Fujimoto, J.G., and Ippen, E.P. (1985). Femtosecond time-resolved reflectometry measurements of multiple-layer dielectric mirrors. *Opt. Lett.* **10,** 71–73.

Wilhelmi, B., Rudolph, W., Döpel, E., and Dietel, W. (1986). Chirp production and its compensation. Paper WK42, CLEO, June 1986, San Francisco, Calif.

Yamashita, M., Torizuka, K., and Sato, T. (1988). Intracavity femtosecond pulse compression with the addition of highly nonlinear organic material. *Opt. Lett.* **13,** 24–26.

Yasa, Z.A., Dienes, A., and Whinnery, J.R. (1977). Subpicosecond pulses from a cw double mode-locked dye laser. *Appl. Phys. Lett.* **30,** 24–26.

Yoshizawa, M., and Kobayashi, T. (1984). Experimental and theoretical studies on colliding pulse mode-locking. *IEEE J. Quantum Electron.* **QE-20,** 797-803.

Zakharov, V.E., and Shabat, A.B. (1972). Exact theory of two-dimensional self-focusing and one-dimensional self-modulation of waves in nonlinear media. *Sov. Phys. JETP* **34,** 62–69.

Zakharov, V.E., and Shabat, A.B. (1971). Exact theory of two-dimensional self-focusing and *Sov. Phys. JETP* **37,** 823–828.

Chapter 4

NARROW-LINEWIDTH PULSED DYE LASER OSCILLATORS

F.J. Duarte

Photographic Research Laboratories
Photographic Products Group
Eastman Kodak Company
Rochester, New York

ISBN 0-12-222700-X

1. INTRODUCTION

Narrow-linewidth pulsed-dye lasers have become a very successful, widely used class of laser sources spanning the spectrum from the near-UV to the near-IR. Applications of these tunable devices range from numerous types of spectroscopy to large-scale laser isotope separation. In this chapter we focus on the development and physics of narrow-linewidth pulsed-dye–laser oscillators. In this regard, the dispersion characteristics of multiple-prism grating techniques are discussed in detail. At this introductory stage it should be made explicitly clear that the dispersive configurations considered here can be applied to a wide variety of lasers.

Early efforts to produce narrowband tunable laser emission from pulsed dye lasers led to the demonstration of a grating-mirror cavity by Soffer and McFarland (1967). Shortly afterward, Bradley *et al.* (1968) reported on a refinement of the grating-mirror cavity that incorporated an intracavity etalon to achieve linewidths of about 0.05 nm. A mirror resonator in conjunction with an intracavity four–isosceles-prism tuning arrangement was used by Strome and Webb (1971) to obtain a linewidth of 0.17 nm. Thus, in a relatively short period all the basic components of wavelength selectivity had been applied to dye-laser cavities.

However, until 1971, dye lasers were mostly known as light sources offering tunability over the broadband gain spectrum of the dye utilized. In the early pulsed devices, narrow-linewidth emission, an essential component of the laser concept, was not fully realized. In that year, Hänsch (1972) demonstrated that the use of an intracavity telescope enhanced considerably the dispersion of the tuning element, a grating used in Littrow configuration. Thus, the use of intracavity beam expansion yielded linewidths of about 0.003 nm at $\lambda \sim 600$ nm. Further linewidth refinement in the telescopic device was achieved by employing an intracavity etalon (Hänsch, 1972).

During the same period, the principle of augmenting the grating dispersion via intracavity beam expansion was independently utilized by other authors employing single-prism expanders and gratings in Littrow configuration to achieve linewidths less than 0.1 nm (Myers, 1971; Stokes *et al.*, 1972). Considerable refinement in the performance of the single-prism grating design was reported by Hanna *et al.* (1975). These authors achieved linewidths in the 0.003- 0.007-nm range.

In the late '70s, the development of the grazing-incidence dye laser (Shoshan *et al.*, 1977; Littman and Metcalf, 1978) provided a further alternative for the design of compact narrow-linewidth dye lasers. In this case full illumination of the grating was accomplished using the grating in non-Littrow configuration, at a high angle of incidence, in conjunction with a tuning mirror. The reported linewidth was 0.003 nm at 600 nm (Littman and Metcalf, 1978). Single-mode oscillation in this type of laser was demonstrated soon after (Littman, 1978; Saikan, 1978).

The interest in prismatic cavities was mainly sustained by the prospects of developing compact, easy-to-align resonators (Hanna *et al.*, 1975). Although those ingredients were present in the single-prism devices, the use of the prism at a high angle of incidence implied either a high level of amplified spontaneous emission (ASE) or low efficiencies, depending on the output beam coupling configuration.

Multiple-prism expanders in dye-laser cavities were introduced independently in several publications (Novikov and Tertyshnik, 1975; Klauminzer, 1978; Wyatt, 1978a; Duarte and Piper, 1980). The dual function of the multiple-prism expander permitting the attainment of very low ASE levels, in a closed-cavity configuration, at significantly improved conversion efficiencies was reported by Duarte and Piper (1980). Certainly, using a multiple-prism Littrow (MPL) configuration (with the grating used in fifth order), linewidths of about 0.0014 nm (at $\lambda \sim 510$ nm) and conversion efficiencies of 14% were achieved at ASE levels reduced more than two orders of magnitude as compared with the single-prism open-cavity design (Duarte and Piper, 1980).

Although grazing-incidence dye lasers provide narrow-linewidth emission, their operation is associated with intrinsically low efficiencies due to the high angle of incidence required at the grating. By characterizing the efficiency of the holographic grating as a function of angle of incidence, Duarte and Piper (1981) calculated the optimum prism intracavity beam expansion necessary to minimize the losses of the grating and the dispersive system as a whole. Thus, using prismatic preexpansion, efficiencies of about 7% were reported at 0.001-nm linewidths for $\lambda \sim 510$ nm.

The dispersive theory of multiple-prism systems was a parallel development which includes generalized expressions for the dispersion of multiple-prism assemblies and the development of close-form formulae applicable to the geometry utilized in intracavity laser-beam expanders (Duarte and Piper, 1982a). Consideration of multipass dispersive effects in multiple-prism systems were also incorporated in the description (Duarte and Piper, 1984a). Interest in the exact design of multiple-prism expanders yielding zero dispersion at a wavelength of interest led to the formulation of further criteria on the subject (Barr, 1984; Duarte, 1985a; Trebino, 1985).

The success of grazing-incidence and multiple-prism grating configurations in laser-pumped dye lasers led to further application of these ideas to produce narrow-linewidth emission in other laser systems such as excimer lasers (McKee *et al.*, 1979; Lyutskanov *et al.*, 1980; Sze, 1984), CO_2 lasers (Duarte, 1985b), and long-pulse flashlamp-pumped dye lasers (Duarte and Conrad, 1987).

Alternative techniques of linewidth narrowing include the use of a

Cassegrain beam expander in conjunction with a Littrow grating (Beiting and Smith, 1979). A detailed assessment of this method in comparison to other telescopic designs, grazing-incidence resonators, and multiple-prism oscillators was provided by Trebino *et al.* (1982). Hnilo *et al.* (1980) considered a reflection beam expander, Koning *et al.* (1980) described a folded mirror beam expansion system, and Meyer and Nenchev (1981) incorporated a Fizeau interferometer. Also, techniques related to the grazing-incidence configuration include beam preexpansion by a telescope (Gallagher *et al.*, 1982) and intracavity lenses (Lisboa *et al.*, 1983; Yodh *et al.*, 1984; Smith and DiMauro, 1987; Kong *et al.*, 1987).

For historical reasons, in this introduction quoted linewidths are given in wavelength units. In the rest of the chapter frequency units (GHz or MHz) are used when appropriate. Also, quoted linewidths are at full-width, half-maximum (FWHM) unless stated otherwise.

2. BASIC CONCEPTS

This section is provided mainly to review some of the basic concepts associated with wavelength tuning and linewidth narrowing in pulsed dye lasers.

The intrinsic emission bandwidth of a mirror-mirror pulsed dye laser is rather broad and can be of the order of a few nm. Linewidth characteristics are considerably improved in the type of grating-mirror resonator illustrated in Fig. 4.1a. In pulsed dye laser cavities incorporating dispersive elements, the single-pass linewidth is given to a first approximation by

$$\Delta\lambda \approx \Delta\theta(\partial\Theta/\partial\lambda)_C^{-1} \tag{4.1}$$

where $\Delta\theta$ is the beam divergence and $(\partial\Theta/\partial\lambda)_C$ is the total dispersion provided by the optical components in the cavity. From this equation it is clear that $\Delta\lambda$ is minimized, reducing $\Delta\theta$ and increasing the dispersion. For the case of the grating-mirror cavity the wavelength is determined by the grating equation

$$m\lambda = a(\sin\Theta \pm \sin\Theta') \tag{4.2}$$

where *m is the order, a* is the groove spacing, Θ the angle of incidence, and Θ' the angle of diffraction. For a grating utilized in Littrow configuration, $\Theta = \Theta'$ and the dispersion becomes

$$(\partial\Theta/\partial\lambda)_G = (2\tan\Theta)/\lambda \tag{4.3}$$

Using this type of grating configuration, Soffer and McFarland (1967) demonstrated a linewidth of 0.06 nm by employing a frequency-doubled ruby laser as excitation source and rhodamine 6G as gain medium. Notice

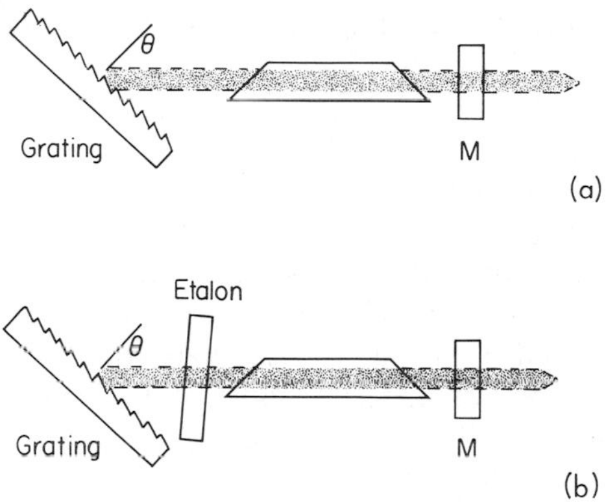

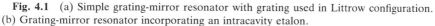

Fig. 4.1 (a) Simple grating-mirror resonator with grating used in Littrow configuration. (b) Grating-mirror resonator incorporating an intracavity etalon.

that for the broadband mirror-mirror cavity they reported a linewidth of 6 nm. Thus, the simple use of a grating in a Littrow mounting instead of one of the broadband reflectors led to a continuously tunable dye laser that provided bandwidth narrowing. Additional linewidth refinement in a grating-mirror resonator can be obtained by utilizing an intracavity etalon as shown in Fig. 4.1b (see Section 2.2 for further details).

2.1. Early Prism Cavities

The early use of prism arrangements in dye lasers involved isosceles prisms in conjunction with a tuning mirror in a flashlamp-excited device as illustrated in Fig. 4.2 (Strome and Webb, 1971). It can be shown that for several identical isosceles prisms, arranged so that the exit angle from one

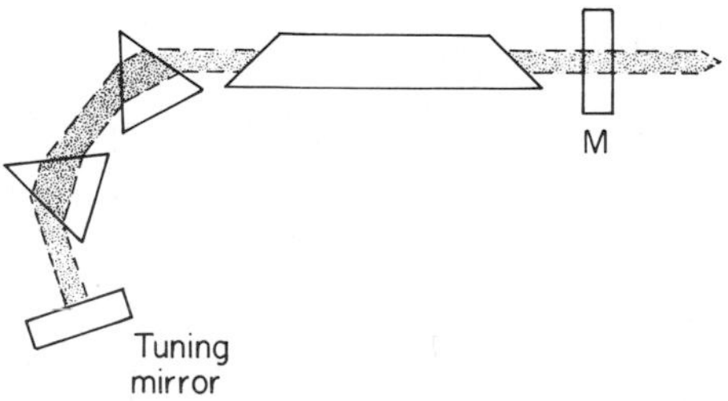

Fig. 4.2 Prismatic cavity utilizing an isosceles or equilateral prism tuner.

prism is equal to the entrance angle at the next prism, the overall dispersion is given by

$$(\partial\Theta/\partial\lambda)_P = r(\partial\phi/\partial\lambda) \tag{4.4}$$

where r is the total number of prisms and $(\partial\phi/\partial\lambda)$ is the individual dispersion contribution from each prism (here assumed to be identical). In other words, prism dispersions added, as is known to those familiar with prism spectrometers (Meaburn, 1976). In the system of Strome and Webb (1971) four prisms were used in an additive configuration. Thus, for their long-pulse flashlamp-pumped device a linewidth of 0.17 nm was measured using rhodamine 6G dye. At about the same time Schäfer and Müller (1971) introduced a six-prism ring device.

Notice that both Soffer and MacFarland (1967) and Strome and Webb (1971) reported measured linewidths narrower than calculated values. That is due to multipass effects and it is discussed later in this chapter.

2.2. Telescopic Dye Laser

The approach of Hänsch (1972) introduced the concept of beam expansion in a transversely excited device (see Fig. 4.3). That is, by increasing the number of grooves being illuminated at the grating, the dispersion term in Eq. (4.1) is multiplied by the beam-magnification factor M provided by the intracavity telescope. Thus, for the expander-grating system the linewidth equation is modified to

$$\Delta\lambda = \Delta\theta[M(\partial\Theta/\partial\lambda)_G]^{-1} \tag{4.5}$$

where the dispersive term of the Littrow grating is given by Eq. (4.3). An alternative way to express Eq. (4.5) is $\Delta\lambda = (\Delta\theta/M)\,(\partial\Theta/\partial\lambda)_G^{-1}$, and con-

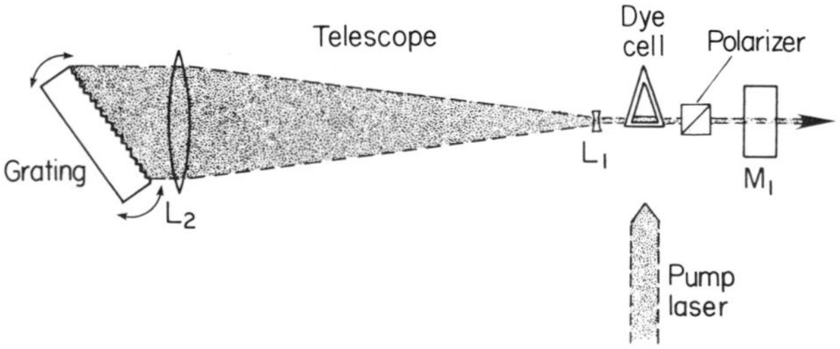

Fig. 4.3 Telescopic dye-laser resonator. In this particular configuration, a Galilean telescope is utilized to expand the beam. Lens L_1 has focal length $|f_1|$ and lens L_2 has a focal length f_2. For improved linewidth performance, a Fabry-Perot etalon is inserted between the telescope and the grating (see text).

ceptual differences between these two alternatives are discussed in Section 5. The Hänsch telescopic device produced linewidths of about 0.003 nm at a near diffraction limited beam divergence of 2.5 mrad and a power conversion efficiency of 20%. Insertion of an etalon between the telescope and the grating reduced the linewidth further to 0.0004 nm but power conversion efficiency declined to the 2–4% range (Hänsch, 1972).

Telescopic expanders utilized in pulsed dye lasers provide two-dimensional beam magnification, typically in the X20–X50 range. The length of these telescopes depends on the focal lengths of the lenses employed. Practical expanders, for the magnification range given previously, are often available in lengths in the 10–20-cm range and provide output apertures of up to 50 mm. Usually, the lenses employed in these telescopes are multielement, are designed for minimum spherical aberration, and have antireflection coatings. In order to avoid parasitic oscillations it is convenient to utilize these telescopes in a slightly off-axis configuration.

Design parameters and beam-propagation characteristics in telescopic beam expanders depend on the A and B elements of the ray transfer matrix. For a well-adjusted Newtonian telescope composed of two lenses, with focal lengths f_1 and f_2 (where $f_2 \gg f_1$), the matrix element A provides the beam-magnification factor $(-)\ M$ and is given by the ratio $(-)(f_2/f_1)$. The matrix element B depends on the separation between the two lenses given by $(f_1 + f_2)$. Similarly, for a well-adjusted Galilean telescope (where $f_2 \gg |f_1|$) the magnification factor M is given by the matrix element A (that is, $(f_2/|f_1|)$) and the B element is given by $(f_2 - |f_1|)$ (see, for example, Siegman (1986)). Further information on telescopic resonators can be found in Kogelnik (1965), Hanna *et al.* (1981), and Routledge *et al.* (1986).

As mentioned previously, intracavity etalons provide a further avenue to linewidth narrowing. For instance, Bradley *et al.* (1986) utilized a grating in Littrow configuration in conjunction with an intracavity etalon to obtain linewidths of about 0.05 nm that were continuously tunable over 10 nm. Magyar and Schneider-Muntau (1972) and then Maeda *et al.* (1975) employed multi-etalon systems to produce linewidths of ~0.01 and 0.005 nm, respectively.

A basic parameter of interest in resonators incorporating intracavity etalons is the free spectral range (FSR) of the etalon:

$$\begin{aligned} FSR &= \lambda^2/(2nd_e) && \text{in wavelength units (m)} \\ &= c/(2nd_e) && \text{in frequency units (Hz)} \\ &= 1/(2nd_e) && \text{in wave numbers (m}^{-1}\text{)} \end{aligned}$$

where n is the refractive index and d_e is the separation distance of the reflecting surfaces. An additional basic quantity is the finesse (F) which depends on parameters such as surface flatness and reflectivity (see, for example, Born and Wolf (1975) and Steel (1967)). For instance, the reflectivity finesse is given by $F_R = (\pi\sqrt{R})/(1 - R)$, where R is the surface reflectivity. Also, an estimate of the minimum bandwidth provided by the etalon is the ratio $[(FSR)/F]$.

In dye lasers, intracavity etalons are mainly used to provide additional linewidth refinement beyond that offered by other dispersive elements in the cavity (Bradley *et al.*, 1968; Hänsch, 1972). Thus, the FSR for a particular intracavity etalon is determined in relation to the dispersive bandwidth of the cavity. An approach used by many authors is to choose an FSR greater than (often about twice) the dispersive linewidth established by Eq. (4.5) in the case of the telescope-grating resonator. Then, the finesse of the etalon and, thus, its reflectivity are determined by the ultimate desired linewidth (itself restricted by the Fourier transform limit). For those interested in single-mode lasing the finesse should be sufficiently high so that the bandwidth provided by the etalon is less than the separation of two longitudinal modes of the cavity (see Section 7 for further details). Tuning and dispersive characteristics of etalons are considered in Chapter 6.

2.3. Single-Prism Devices

The principle of operation of prismatic devices is the same as that of the telescopic cavity. That is, the single prism acts as a compact one-dimensional intracavity beam expander. As mentioned in the introduction, single-prism dye lasers suffered from two limitations when operating in a narrow linewidth configuration. The use of an open cavity, as shown in Fig. 4.4a, provided a laser output with a large component of ASE. On the

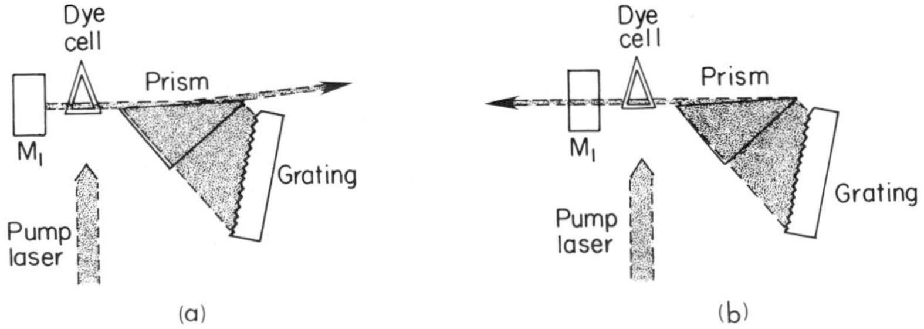

Fig. 4.4 (a) Open-cavity dye laser incorporating single-prism beam expander and grating in Littrow configuration. (b) Closed-cavity version of single-prism design.

other hand, utilizing a closed cavity (see Fig. 4.4b) minimized the ASE level but at the same time reduced the efficiency of the cavity substantially. This effect was due to the use of a high angle of incidence at the prism which was needed to provide the necessary beam expansion required for narrow linewidth operation. For a single-prism the reflection losses for the component of polarization parallel to the plane of incidence is given by

$$l_p = \frac{\tan^2(\phi_1 - \psi_1)}{\tan^2(\phi_1 + \psi_1)} \tag{4.6}$$

where ϕ_1 is the angle of incidence and ψ_1 is the angle of refraction (Born and Wolf, 1975). The beam expansion coefficient, for a right-angle prism designed for orthogonal beam exit, can be written as

$$M = \frac{\cos \psi_1}{\cos \phi_1} \tag{4.7}$$

which can be expressed more explicitly as

$$M = \left[\frac{(n^2 - \sin^2 \phi_1)}{(n^2 - n^2 \sin^2 \phi_1)} \right]^{1/2} \tag{4.8}$$

(Hanna *et al.*, 1975). Using Eqs. (4.6) and (4.7), it is easy to illustrate the conflict of interest between large beam expansion and the requirement for low reflection losses.

Linewidth calculations applicable to single-prism Littrow grating dye lasers are discussed by Hanna *et al.* (1975), Wyatt (1978b), and Zhang *et al.* (1981). Overall dispersion values of single-prism grating combinations are given by Krasinski and Sieradzan (1979) for different prism materials. In this chapter the single-prism case is considered as a special result of the multiple-prism grating dispersion theory discussed in Section 4.

2.4. Grazing-Incidence Designs

In grazing-incidence dye lasers the resolution of the cavity is enhanced by illuminating the whole grating with the unexpanded beam (which is widened by natural beam divergence only) using the grating at a large angle of incidence.

The efficiency and ASE limitations experienced with grazing-incidence cavities are similar to those observed in single-prism Littrow dye lasers. In the case of an open cavity, as described by Shoshan *et al.* (1977) (see Fig. 4.5a), a large component of ASE is present since the output is coupled from the losses of the grating utilized at a high angle of incidence. The closed-cavity alternative, shown in Fig. 4.5b (Littman and Metcalf, 1978),

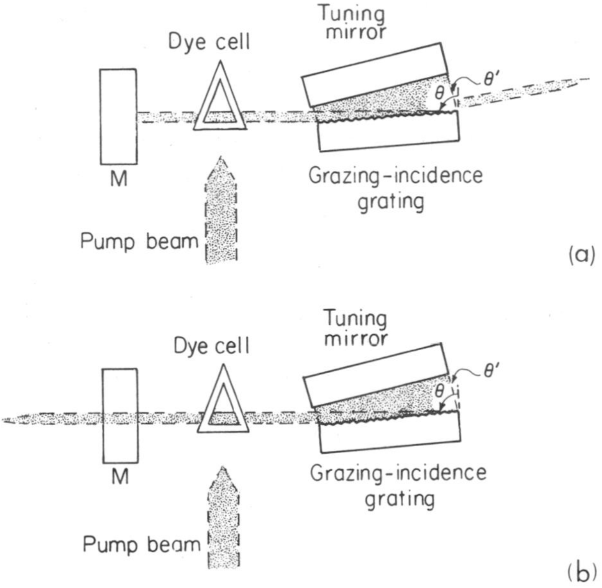

Fig. 4.5 (a) Open-cavity grazing-incidence dye laser. (b) Closed-cavity configuration of grazing-incidence design.

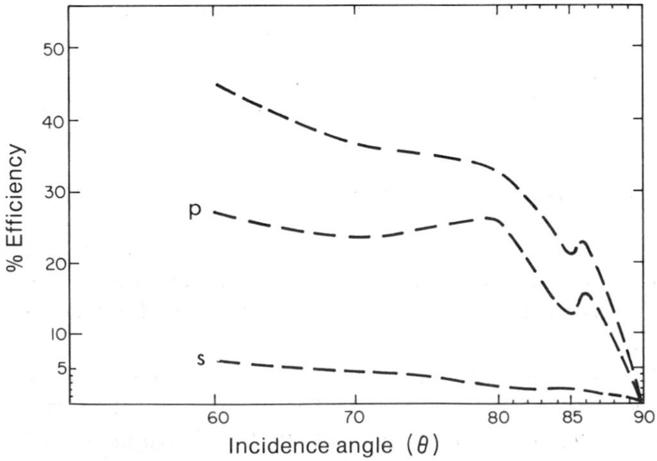

Fig. 4.6 Grating-efficiency curve as function of angle of incidence for a typical holographic grating at $\lambda = 632.8$ nm. Both polarization components are included [from Duarte and Piper, 1981].

reduces ASE but yields much lower efficiencies. This is due to the fact that grating efficiency tends to deteriorate rapidly at high angles of incidence required for narrow-linewidth operation (see Fig. 4.6).

In their discussion of the grazing-incidence configuration, Littman and Metcalf (1978) derived an expression for the grating dispersion in that geometry:

$$(\partial\Theta/\partial\lambda)_G = 2(\sin\Theta + \sin\Theta')/(\lambda\cos\Theta) \tag{4.9}$$

where Θ is the angle of incidence and Θ' is the angle of diffraction. Notice that by using the grating equation it can also be shown that an equivalent expression is

$$(\partial\Theta/\partial\lambda)_G = 2m/(a\cos\Theta) \tag{4.10}$$

(see for example Shoshan *et al.* 1977; Iles, 1981) where m is the grating order, and a is the groove spacing. For a given set of parameters it is easy to see that the dispersion of a grating in grazing-incidence configuration is intrinsically superior to the dispersion of a grating in Littrow mounting. However, it is well known that the efficiency of a Littrow grating is much better than the efficiency obtained from a grating in grazing-incidence configuration. An additional alternative to increase the dispersion in a grazing-incidence cavity even further is to replace the tuning mirror by a second grating utilized in Littrow configuration (Shoshan and Oppenheim, 1978; Littman, 1978; Iles, 1981; Lebedev and Plyasulya, 1981). Mory *et al.* (1981a) describe a double-grating design where the second grating is used in non-Littrow configuration in conjunction with a tuning mirror. The use of a Fabry-Perot etalon inserted between the grating and the tuning mirror has been reported by Mory *et al.* (1981b). Utilizing a method of excitation that allows the use of multiple active regions in a single dye cell, Hung and Brechignac (1985a) report a 6% conversion efficiency for a 1.3-GHz linewidth in a grazing-incidence cavity.

The use of grazing-incidence configurations in flashlamp-pumped dye lasers is discussed by Godfrey *et al.* (1980), Mazzinghi *et al.* (1981), and Zhang *et al.* (1982).

2.5. Double-Wavelength Dye Lasers

Double-wavelength lasing in grazing-incidence and prism expanded Littrow cavities has been considered by several authors. Dual-wavelength grazing-incidence cavities are utilized in several geometric arrangements. For example, Prior (1979) employed two mirrors in conjunction with two successive diffraction orders from the grazing-incidence grating. Additional methods to achieve double-wavelength lasing include interception of one diffracted order with two mirrors (Chandra and Compaan, 1979;

Burlamacchi and Ranea Sandoval, 1979; Dinev *et al.*, 1980a), the use of a second grating-mirror assembly in series (Nair and Dasgupta, 1980), and the excitation of two vertically displaced regions at the dye cell in conjunction with a single grating and two mirrors (Williams and Heddle, 1983).

Double-wavelength prismatic cavities incorporating gratings in Littrow configuration include interception of the prism-expanded beam by two gratings (Chandra and Compaan, 1979) and the use of two single-prism grating assemblies in series (Nair, 1979). Two independent single-prism devices excited by a single source via a beam splitter were used by Kong and Lee (1981) to rotate the polarization of the beam by rotating the cavities about the axis of propagation. A further method to produce an additional wavelength has been demonstrated in telescopic dye lasers and employs a wedge to intercept the expanded beam illuminating the Littrow grating (Lotem and Lynch, 1975). A similar technique in a prismatic oscillator is described by Alden *et al.* (1984). Melikechi and Allen (1987) describe double-wavelength emission in a telescopic device that utilizes two dye cells, separated (by ~1.5 mm) from one another, to produce two slightly displaced dye-emission beams in the plane of incidence. A three-wavelength device using interference filters is described by Saito *et al.* (1985). A review of multiple-wavelength methods, including early references, is provided by Nair (1982).

3. MULTIPLE-PRISM GRATING OSCILLATORS

The efficiency and ASE problems associated with single-prism Littrow devices are reduced substantially with the introduction of designs incorporating further stages of prism expansion. For the closed cavity multiple-prism Littrow oscillator introduced by Duarte and Piper (1980) and shown in Fig. 4.7, conversion efficiency was about 14% while the level of ASE ($\sim 7.5 \times 10^{-6}$) was reduced by more than two orders of magnitude as compared with a single-prism open-cavity design (see Fig. 4.8). For a given beam expansion, yielding $\Delta\lambda \sim 0.0014$ nm, the single-pass prismatic losses in the single-prism device were ~90% as compared to ~70% for the MPL device.

The employment of prismatic preexpansion in grazing-incidence cavity configurations (Duarte and Piper, 1981, 1984b) led to the development of hybrid multiple-prism grazing-incidence (HMPGI) oscillators similar to that shown in Fig. 4.9. This provided improvements in efficiency and ASE characteristics. The approach here consists of identifying a suitable choice of angles of incidence for the grating and prisms that would minimize the losses while producing the same linewidth as the pure grazing-incidence design. For example, by using the grating at a reduced angle of incidence

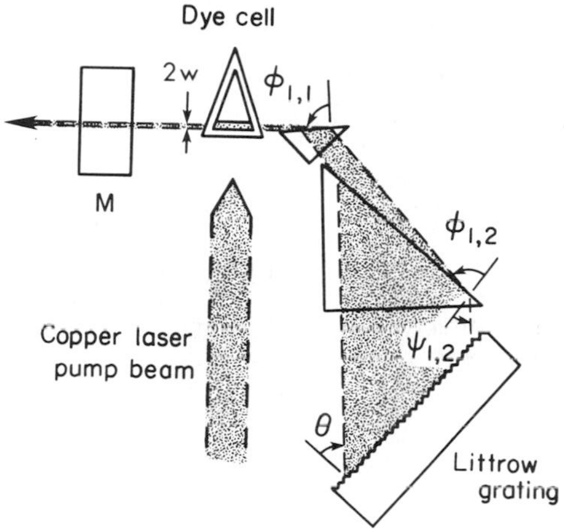

Fig. 4.7 Schematic of multiple-prism Littrow (MPL) oscillator.

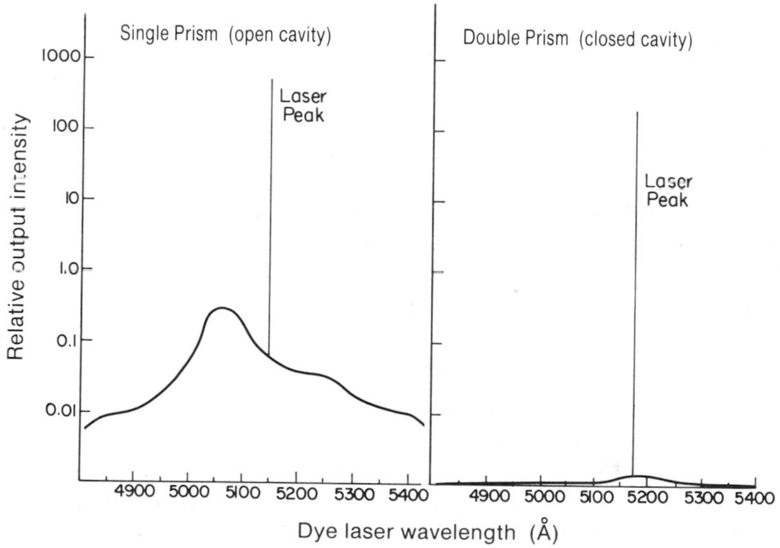

Fig. 4.8 ASE characteristics of single-prism open-cavity design and closed-cavity MPL oscillator [from Duarte and Piper, 1980].

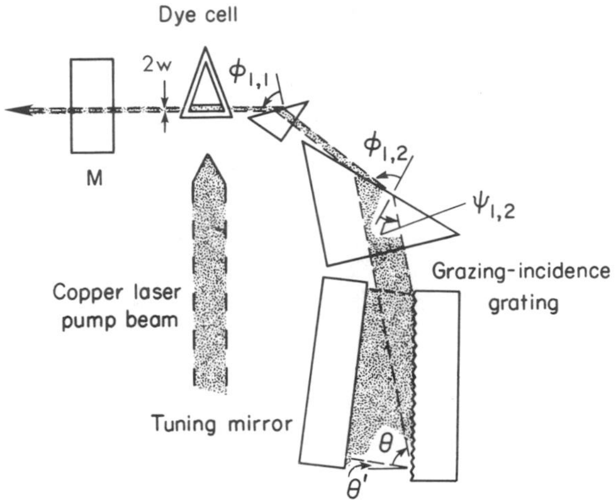

Fig. 4.9 Schematic of hybrid multiple-prism grazing-incidence (HMPGI) oscillator.

(~85 °), in conjunction with prismatic preexpansion, linewidths of about 0.001 nm were obtained at efficiencies in the 7–10% range (Duarte and Piper, 1981). By comparison, the simple grazing-incidence cavity offered less than 2% efficiency at the same linewidth. Again, the level of ASE of the closed-cavity–prism preexpanded oscillator was substantially less than that provided by the equivalent open-cavity configuration. Further evidence on the efficiency advantages of this type of configuration was given by Racz *et al.* (1981) and Elizarov (1985). A discussion on the use of this type of design in an oscillator–amplifier system is provided by Dupre (1987).

Inherent advantages of intracavity multiple-prism beam magnification over telescopic expansion include compactness, ease of alignment, and the availability of a polarized output beam. In this regard, MPL and HMPGI dye-laser oscillators have been shown to provide laser emission almost 100% *p*-polarized (Duarte and Piper, 1982b, 1984b). In addition, matching the polarization of the pump laser to the preferred polarization of the prisms, gain medium, and grating can enhance the photon efficiency of the system (see, for example, Duarte and Piper (1984b)). This topic is considered in more detail in Chapter 6.

The main disadvantage of prismatic expansion is the detraction to some degree from the achromaticity offered by designs where dispersion is purely dependent on the grating characteristics. Further discussion of this topic is provided later in this chapter.

Transversely excited multiple-prism grating dye-laser oscillators are compatible with a variety of pump sources. For instance, the oscillators

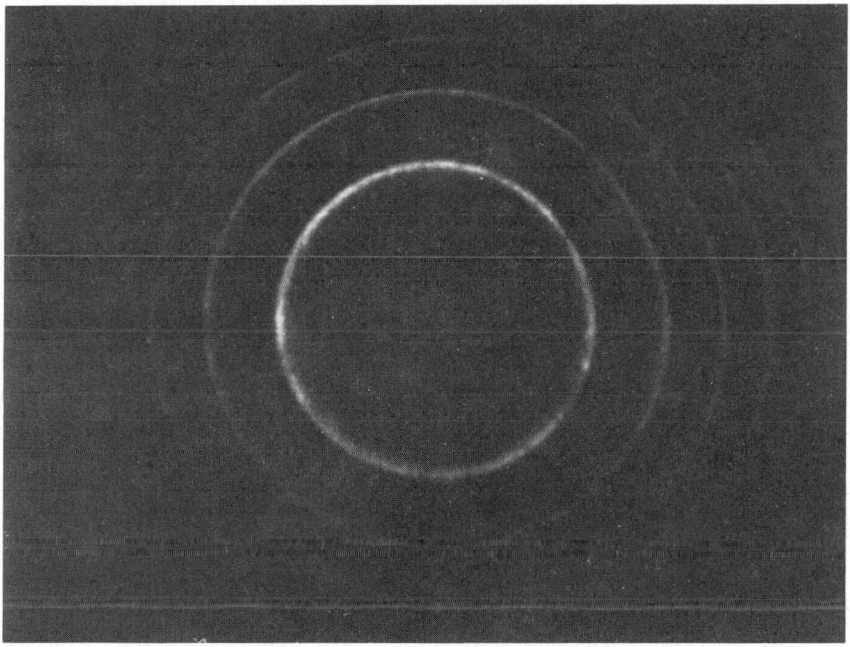

Fig. 4.10 Interferogram from high-prf copper-vapor laser-pumped HMPGI dye-laser oscillator. The measured linewidth ($\Delta\nu \sim 650$ MHz) corresponds to an integrated value from more than 2 million laser pulses.

described in Figs. 4.7 and 4.9 have been utilized in conjunction with low-pulse repetition frequency (prf) ultraviolet lasers (such as N_2 and excimer lasers) and high-prf copper vapor lasers. Indeed, the HMPGI oscillators have yielded very stable single-longitudinal-mode oscillation at $\Delta\nu \sim 650$ MHz at a prf of 10 kHz for a conversion efficiency of about 5% (Duarte and Piper, 1984b). An interferogram from these experiments is shown in Fig. 4.10. The main feature of these Fabry-Perot rings is that they provide a time-integrated linewidth corresponding to more than two million laser pulses. Thus, the linewidth quoted involves frequency jitter and drift. In Chapter 6 some of the technical requirements for high-prf operation in MPL and HMPGI oscillators are discussed.

Using a copper-laser pumped MPL design in conjunction with an intracavity etalon, Bernhardt and Rasmussen (1981) reported linewidths as low as 60 MHz. High-prf copper-vapor lasers have also been utilized in the excitation of open-cavity grazing-incidence dye lasers yielding linewidths of about 3 GHz (Broyer *et al.*, 1984). A systematic comparison of grazing-incidence, MPL, and HMPGI oscillators excited by a TE CuBr laser was performed by Duarte and Piper (1982b).

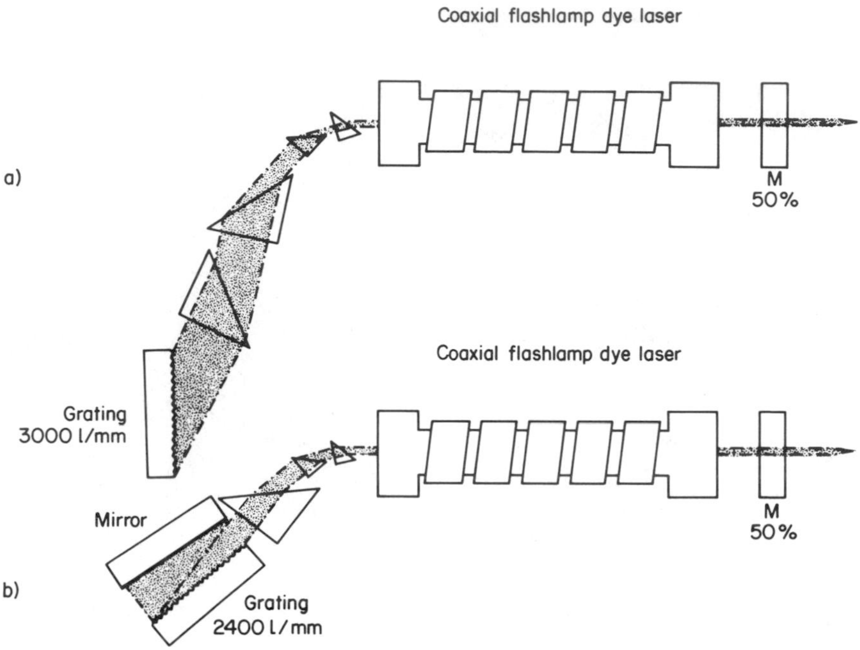

Fig. 4.11 Schematics of (a) MPL and (b) HMPGI flashlamp-pumped dye-laser oscillators. In the MPL design, the first three prisms are utilized in additive configuration while the fourth prism is arranged in compensating configuration.

Table 4.1

Performance of Narrow-Linewidth Dye Laser Oscillators

Cavity type	Linewidth	% Eff	Pump source	Reference
Telescopic	2.5 GHz	20	N_2 laser	Hänsch (1972)
	300 MHz[1]	2–4	N_2 laser	Hänsch (1972)
Grazing-incidence	2.5 GHz	~4	Nd: YAG laser[5]	Littman and Metcalf (1978)
	420 MHz[2]	~6	N_2 laser	Saikan (1978)
	300 MHz[3]	2	Nd: YAG laser[5]	Littman (1978)
	150 MHz	3	Nd: YAG laser[5]	Littman (1984)
MPL	1.61 GHz	14	N_2 laser	Duarte and Piper (1980)
HMPGI	1.15 GHz	7–10	N_2 laser	Duarte and Piper (1981)
MPL	60 MHz[1]	5	Cu laser	Bernhart and Rasmussen (1981)
HMPGI	400–650 MHz	4–5	Cu laser	Duarte and Piper (1984b)
HMPGI and MPL	250–375 MHz[4]		Flashlamp	Duarte and Conrad (1987)

[1] Incorporates intracavity etalon.
[2] Open-cavity configuration.
[3] Single-pulse linewidth. Tuning mirror is replaced by a second grating.
[4] Single-pulse linewidth
[5] Frequency-doubled.

Usually multiple-prism oscillators incorporate more than two prisms in the cavity. An example of such an arrangement is given in Fig. 4.11 for the case of flashlamp operation (Duarte and Conrad, 1987). These oscillators yielded single-mode oscillation with linewidths in the 250–375 MHz range, utilizing a 140-mm–wide, 3000-ℓ/mm holographic grating. As discussed later in the chapter, these configurations are particularly useful when a dispersionless expander, at a given wavelength, is required. For comparison purposes, Table 4.1 provides some reported performances of the main types of cavities considered here.

4. MULTIPLE-PRISM AND INTRACAVITY DISPERSION

A parameter of fundamental importance in determining frequency characteristics in laser cavities, incorporating prisms and gratings, is dispersion. For instance, from the expression $\Delta\lambda = \Delta\theta\,(\partial\Theta/\partial\lambda)_C^{-1}$, it is clear that laser linewidth is inversely proportional to the overall cavity dispersion. In this section we examine in detail the dispersive properties of resonators incorporating multiple-prism grating assemblies. It should be emphasized that the multiple-prism theory given here has a number of applications beyond frequency selectivity in laser resonators. Further applications of the formulae are related to the use of prismatic expanders in novel electro-optical devices and the utilization of prisms in pulse compression in femtosecond lasers.

4.1. The Single-Prism Case

This subsection describes the usual case of dispersion from a single prism, as treated in most textbooks. The aim of this exercise is to introduce the notation and basic ideas.

It is well known (Born and Wolf, 1975) that dispersion from a single prism (Fig. 4.12) can be derived from the following set of relations:

$$\phi_1 + \phi_2 = \epsilon + \alpha \tag{4.11}$$

$$\psi_1 + \psi_2 = \alpha \tag{4.12}$$

$$n \sin \psi_1 = \sin \phi_1 \tag{4.13}$$

$$n \sin \psi_2 = \sin \phi_2 . \tag{4.14}$$

As indicated in Fig. 4.12, ϕ_1 and ϕ_2 are the angles of incidence and emergence, ψ_1 and ψ_2 are the corresponding angles of refraction, α is the apex angle, ϵ is the angle of deviation, and n is the refractive index of the prism.

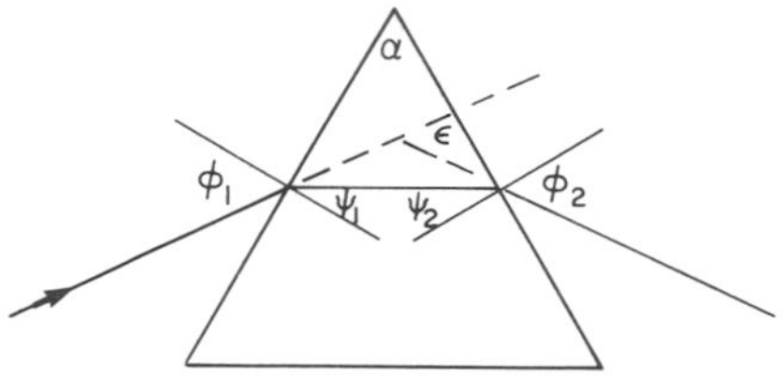

Fig. 4.12 Ray path through single prism.

From Eq. (4.12) we obtain $d\psi_1/dn = -d\psi_2/dn$; differentiating Eqs. (4.13) and (4.14) with respect to n we find that

$$\frac{d\phi_2}{dn} = \frac{\sin\psi_2}{\cos\phi_2} + \frac{\cos\psi_2}{\cos\phi_2}\tan\psi_1 \tag{4.15}$$

which is the result given by Born and Wolf (1975) (in this equation the (+) sign is replaced by a (−) sign if the apex angle is less than ψ_1). For the case of a right-angle prism designed for orthogonal beam exit (see Fig. 4.4), so that $\phi_2 = \psi_2 \approx 0$, Eq. (4.15) is simplified to

$$(d\phi_2/dn) \approx \tan\psi_1 \tag{4.16}$$

which is the single-prism result employed by Wyatt (1978b).

4.2. Single-Pass Multiple-Prism Dispersion

The development of multiple-prism grating dye lasers highlighted the necessity to extend the dispersion equation to the general case involving any arbitrary number of prisms. To that effect generalized multiple-prism arrays, as shown in Fig. 4.13, were investigated (Duarte and Piper, 1982a, 1983). Considering the mth prism of the arrangement, the angular relations are written as

$$\phi_{1,m} + \phi_{2,m} = \epsilon_m + \alpha_m \tag{4.17}$$

$$\psi_{1,m} + \psi_{2,m} = \alpha_m \tag{4.18}$$

$$n_m \sin\psi_{1,m} = \sin\phi_{1,m} \tag{4.19}$$

$$n_m \sin\psi_{2,m} = \sin\phi_{2,m}. \tag{4.20}$$

Using these equations it was shown (Duarte and Piper, 1982a, 1983) that the single-pass dispersion following the mth prism is given by

$$\frac{d\phi_{2,m}}{d\lambda} = \frac{\sin\psi_{2,m}}{\cos\phi_{2,m}}\frac{dn_m}{d\lambda} + \frac{\cos\psi_{2,m}}{\cos\phi_{2,m}}\frac{\cos\phi_{1,m}}{\cos\psi_{1,m}}\left(\frac{\sin\psi_{1,m}}{\cos\phi_{1,m}}\frac{dn_m}{d\lambda} \pm \frac{d\phi_{2,(m-1)}}{d\lambda}\right). \tag{4.21}$$

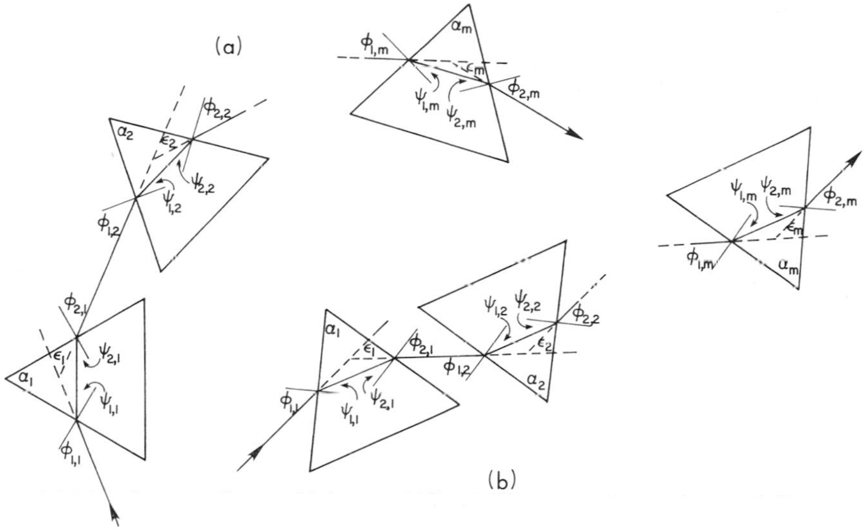

Fig. 4.13 Generalized multiple-prism assemblies in (a) additive and (b) compensating configurations.

Equation (4.21) is a recursion relation which permits $(d\phi_{2,m}/d\lambda)$ to be determined for ascending values of m, the initial value being $(d\phi_{2,1}/d\lambda)$. The $(\pm)$ sign provides the alternative between an additive configuration $(+)$ (Fig. 4.13a) and a compensating arrangement $(-)$ (Fig. 4.13b).

This equation is general and as such is applicable to any prismatic array regardless of the prism's geometry or material. From Eq. (4.21) one can consider the case of an array of identical isosceles (or equilateral) prisms arranged symmetrically so that the angles of incidence and emergence at each prism are the same. Under those circumstances it is not difficult to prove that the total dispersion is the result of the addition of the individual dispersions. Thus the cumulative dispersive value following the light passage of r prisms can be written as

$$(d\phi_{2,r}/d\lambda) = r\,(d\phi_{2,1}/d\lambda) \tag{4.22}$$

which is the result discussed earlier in Section 2. It should be noticed, however, that this notion of straight dispersion addition does not extend to the case of multiple-prism expanders composed of right-angle prisms designed for orthogonal beam exit. A recent frequency-selectivity application of equilateral prisms in additive configuration has been reported by van Hoek and van Wijk (1987).

An alternative way to express the generalized dispersion equation is given by Duarte (1989a)

$$\frac{d\phi_{2,r}}{d\lambda}=\left(\prod_{j=1}^{r}k_{1,j}\prod_{j=1}^{r}k_{2,j}\right)^{-1}\sum_{m=1}^{r}(\pm 1)\frac{\tan\phi_{2,m}}{n_m}\frac{dn_m}{d\lambda}\left(\prod_{j=1}^{m}k_{1,j}\prod_{j=1}^{m}k_{2,j}\right)$$
$$+\sum_{m=1}^{r}(\pm 1)\frac{\tan\phi_{1,m}}{n_m}\frac{dn_m}{d\lambda}\left(\prod_{j=m}^{r}k_{1,j}\prod_{j=m}^{r}k_{2,j}\right)^{-1} \tag{4.23}$$

where $k_{1,j}=(\cos\psi_{1,j}/\cos\phi_{1,j})$ and $k_{2,j}=(\cos\phi_{2,j}/\cos\psi_{2,j})$.

For multiple-prism assemblies incorporating prisms designed for orthogonal beam exit ($\phi_{2,1}=\phi_{2,2}=\ldots=\phi_{2,m}=\cdots=0$) Eq. (4.21) (or Eq. (4.23)) reduces to

$$(d\phi_{2,r}/d\lambda)=\beta\sum_{m=1}^{r}(\pm 1)\left(\prod_{j=1}^{m}k_{1,j}\right)\tan\psi_{1,m}(dn_m/d\lambda) \tag{4.24}$$

where

$$(1/\beta)=\prod_{m=1}^{r}k_{1,m}. \tag{4.25}$$

In this equation $k_{1,m}=(\cos\psi_{1,m}/\cos\phi_{1,m})$ is the individual beam expansion factor (notice that the subscripts m and j are interchangeable (see, for example, Duarte (1987)).

In practice, Eq. (4.24) can be simplified further since all prisms are made of the same material ($n_m=n$). Moreover, if the expander is composed of prisms with identical apex angle oriented at the same angle of incidence (so that $\phi_{1,1}=\phi_{1,2}=\cdots=\phi_{1,m}=\cdots$), then Eq. (4.24) can be written as

$$(d\phi_{2,r}/d\lambda)=\tan\psi_{1,1}\sum_{m=1}^{r}(\pm 1)\,(1/k)^{m-1}\,(dn/d\lambda) \tag{4.26}$$

with the k factor identical for all prisms. Explicitly, Eq. (4.26) is written as

$$(d\phi_{2,1}/d\lambda)=\tan\psi_{1,1}(1)(dn/d\lambda) \tag{4.27}$$

$$(d\phi_{2,2}/d\lambda)=\tan\psi_{1,1}(1\pm k^{-1})(dn/d\lambda) \tag{4.28}$$

$$(d\phi_{2,3}/d\lambda)=\tan\psi_{1,1}(1\pm k^{-1}+k^{-2})(dn/d\lambda) \tag{4.29}$$

$$(d\phi_{2,4}/d\lambda)=\tan\psi_{1,1}(1\pm k^{-1}+k^{-2}\pm k^{-3})(dn/d\lambda) \tag{4.30}$$

for $r=1$, 2, 3, and 4 (for further details see Duarte and Piper, (1982a, 1983)). The results for $r=1$ and $r=2$ can be written as those given by Wyatt (1978b) and Kasuya *et al.* (1978), respectively.

For the important case of incidence at Brewster's angle that minimizes reflection losses, Eq. (4.26) takes the very simple form of

$$(d\phi_{2,r}/d\lambda)=\sum_{m=1}^{r}(\pm 1)(1/n)^m(dn/d\lambda). \tag{4.31}$$

4.3. Return-Pass Multiple-Prism Grating Dispersion

In a laser cavity, the magnified beam from the multiple-prism expander is returned, as shown in Fig. 4.14, on the same optical path by reflection from a mirror or diffraction from a grating. Thus, we are now interested in determining the overall multiple-prism grating dispersion of the return-pass at the first prism. As indicated by Duarte and Piper (1982a, 1984a), that calculation can be done thinking of the return-pass as a mirror image of the initial light passage. If we indicate the return-pass angles as prime quantities, then the overall double-pass dispersion can be written as

$$\frac{d\phi'_{1,m}}{d\lambda} = \frac{\sin\psi'_{1,m}}{\cos\phi'_{1,m}}\frac{dn_m}{d\lambda} + \frac{\cos\psi'_{1,m}}{\cos\phi'_{1,m}}\frac{\cos\phi'_{2,m}}{\cos\psi'_{2,m}} \times \left(\frac{\sin\psi'_{2,m}}{\cos\phi'_{2,m}}\frac{dn_m}{d\lambda} \pm \frac{d\phi'_{1,(m+1)}}{d\lambda}\right) \quad (4.32)$$

where the $(d\phi'_{1,(m+1)})/d\lambda)$ term provides the cumulative single-pass prism dispersion plus the grating contribution; that is,

$$(d\phi'_{1,(m+1)}/d\lambda) = [(\partial\Theta/\partial\lambda)_G \pm (d\phi_{2,r}/d\lambda)] \quad (4.33)$$

where $(\partial\Theta/\partial\lambda)_G$ is the grating dispersion, either in grazing-incidence or Littrow configuration, and $(d\phi_{2,r}/d\lambda)$ is the cumulative single-pass dispersion given by Eq. (4.21). In practice, Eq. (4.32) is utilized to provide the cumulative return-pass dispersion of the whole round-trip relative to the

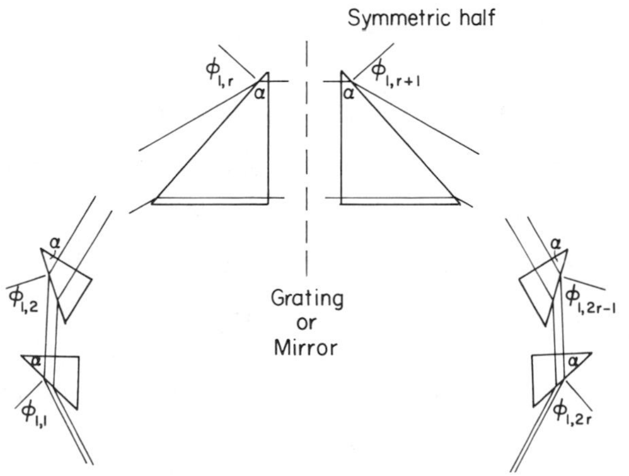

Fig. 4.14 Schematics of multiple-prism expander, in additive configuration, and its mirror image. Analysis of the beam path through the mirror image is equivalent to a return-pass analysis [from Duarte and Piper, 1982a].

first prism. Since the quantity of interest involves also the grating contribution, for notational convenience we label it $(d\Theta/d\lambda)_C$.

The explicit version of the generalized return-pass dispersion for a multiple-prism mirror system (that is, $(d\Theta/d\lambda)_G = 0)$) is given by

$$(d\Theta/d\lambda)_p = 2\left(\prod_{j=1}^{r} k_{1,j}\prod_{j=1}^{r} k_{2,j}\right)\sum_{m=1}^{r}(\pm 1)\frac{\tan\phi_{1,m}}{n_m}\frac{dn_m}{d\lambda}\left(\prod_{j=m}^{r} k_{1,j}\prod_{j=m}^{r} k_{2,j}\right)^{-1} + 2\sum_{m=1}^{r}(\pm 1)\frac{\tan\phi_{2,m}}{n_m}\frac{dn_m}{d\lambda}\left(\prod_{j=1}^{m} k_{1,j}\prod_{j=1}^{m} k_{2,j}\right). \quad (4.34)$$

For the case of right-angle prisms designed for orthogonal beam exit $(\phi_{2,1} = \phi_{2,2} = \cdots = \phi_{2,m} = 0 = \cdots)$, Eq. (4.32) reduces to

$$(d\Theta/d\lambda)_C = \prod_{m=1}^{r} k_{1,m}(\partial\Theta/\partial\lambda)_G + 2\sum_{m=1}^{r}(\pm 1)\left(\prod_{j=1}^{m} k_{1,j}\right)\tan\psi_{1,m}(dn_m/d\lambda) \quad (4.35)$$

(Duarte and Piper, 1982a; Duarte, 1985a). Here, the overall beam expansion is given by

$$M = \prod_{m=1}^{r} k_{1,m}. \quad (4.36)$$

Equation (4.35) explains very succinctly why multiple-prism grating oscillators yield very narrow linewidths. For instance, in MPL oscillators the total intracavity beam expansion can be in the 100–200 range, thus augmenting the grating dispersion significantly. For the special case of a prismatic expander composed of identical prisms $(n_1 = n_2 = n_3 = \cdots = n)$, Eq. (4.35) becomes

$$(d\Theta/d\lambda)_C = k(\partial\Theta/\partial\lambda)_G + 2\tan\psi_{1,1}(k)(dn/d\lambda) \quad (4.37)$$

$$(d\Theta/d\lambda)_C = k^2(\partial\Theta/\partial\lambda)_G + 2\tan\psi_{1,1}(k \pm k^2)(dn/d\lambda) \quad (4.38)$$

$$(d\Theta/d\lambda)_C = k^3(\partial\Theta/\partial\lambda)_G + 2\tan\psi_{1,1}(k \pm k^2 + k^3)(dn/d\lambda) \quad (4.39)$$

$$(d\Theta/d\lambda)_C = k^4(\partial\Theta/\partial\lambda)_G + 2\tan\psi_{1,1}(k \pm k^2 + k^3 \pm k^4)(dn/d\lambda) \quad (4.40)$$

for $r = 1$, 2, 3, and 4 (Duarte and Piper, 1982a).

If the grating is replaced by a mirror, so that $(\partial\Theta/\partial\lambda)_G = 0$, then the overall dispersion is reduced to the prismatic return-pass prismatic component:

$$(d\Theta/d\lambda)_P = 2\sum_{m=1}^{r}(\pm 1)\left(\prod_{j=1}^{m} k_{1,j}\right)\tan\psi_{1,m}(dn_m/d\lambda). \quad (4.41)$$

For an expander incorporating right-angle prisms of identical material, designed for orthogonal beam exit at Brewster's angle of incidence, Eq. (4.41) reduces to

$$(d\Theta/d\lambda)_P = 2 \sum_{m=1}^{r} (\pm 1)(n)^{m-1}(dn/d\lambda). \tag{4.42}$$

Here the individual beam expansion at each prism becomes $k_{1,1} = k_{1,2} = \cdots = k_{1,m} = \cdots = n$, and the overall beam expansion is given by $M = n^r$.

In Eqs. (4.33) and (4.35) the term $(\partial\Theta/\partial\lambda)_G$ was referred to as the grating dispersion either in Littrow or grazing-incidence configuration. The dispersion for the grating in the Littrow mounting is given by Eq. (4.3) and the dispersion for the grating in grazing-incidence configuration is given by Eq. (4.9).

4.4. Design of Discrete-Wavelength Zero-Dispersion Multiple Prism Expanders

The equation for multiple-prism return-pass dispersion can be used to illustrate the design of prismatic expanders yielding zero dispersion at a given wavelength (also referred in the literature as achromatic N-prism expanders). The subject of achromatic expanders has been well treated in papers by Barr (1984), Duarte (1985a), and Trebino (1985). Barr (1984) considers numerical solutions to the design of zero-dispersion double-prism expanders, and Trebino (1985) discusses optimal configurations for multiple-prism expanders as applied to MPL and HMPGI oscillators.

It is important to state explicitly that an achromatic multiple-prism expander of any geometry or material(s) can be designed utilizing any of the following Eqs.: (4.21), (4.23), (4.32), or (4.34). This type of design requires that either $(d\phi_{2,r}/d\lambda) = 0$ or $(d\Theta/d\lambda)_P = 0$, depending on the choice of equations. Numerical solutions for nonorthogonal double-prism expanders with $M = 6$ have been provided by Niefer and Atkinson (1988).

In an example given by Duarte (1985a), a four-prism expander (prisms of identical material) required to provide an overall magnification of about 100 is discussed. Here, we consider the same example with a slightly different prism material. The first three prisms are chosen to be identical and in additive configuration at the same angle of incidence. Therefore the last prism (of different geometry) must be oriented in the compensating mode at a different angle of incidence. Thus, for $(d\Theta/d\lambda)_P = 0$, Eq. (4.41) can be written as

$$(k_{1,1} + k_{1,1}k_{1,2} + k_{1,1}k_{1,2}k_{1,3}) \tan\psi_1 = (k_{1,1}k_{1,2}k_{1,3})k_{1,4} \tan\psi_{1,4} \tag{4.43}$$

where $\psi_{1,1} = \psi_{1,2} = \psi_{1,3} = \psi_1$. For a total beam expansion of ~100 we adopt $k_{1,1} = k_{1,2} = k_{1,3} = 4$. Considering borosilicate crown glass (BK 7) at 510 nm ($n = 1.5207$), we find that for prisms 1 to 3, $\phi_1 = 78.99°$ and $\psi_1 = 40.20°$ (which becomes the common apex angle). Then we solve

$$(4 + 16 + 64)\tan(40.20) = 64(\cos\psi_{1,4}/\cos\phi_{1,4})\tan\psi_{1,4}$$

which means that $\tan\phi_{1,4} = 1.68$. Therefore, $\phi_{1,4} = 59.33°$, $\psi_{1,4} = 34.44°$ (which is the apex angle for the fourth prism), and the total beam magnification becomes $M = k_{1,1}k_{1,2}k_{1,3}k_{1,4} = 103.49$. The total reflection loss factor for p-polarization can be calculated using the expression for cumulative losses at the mth prism:

$$\mathrm{L}_m = \mathrm{L}_{m-1} + (1 - \mathrm{L}_{m-1})\mathrm{l}_m \tag{4.44}$$

Duarte and Piper, 1980; Duarte and Conrad, 1986) where the individual loss at the mth prism is

$$\mathrm{L}_m = \tan^2(\phi_{1,m} - \psi_{1,m})/\tan^2(\phi_{1,m} + \psi_{1,m}) \tag{4.45}$$

and is based on Eq. (4.6). Using the cumulative loss expression the total reflection loss for the expander considered here is ~0.49.

Notice that the same zero-dispersion configuration can be obtained using the single-pass approach (Eq. (4.24)). Although the example considered is quite specific, the method can be applied to design an infinite number of quasi-achromatic expanders composed of two or more prisms. Transmission-optimized configurations considered by Trebino (1985) include several designs containing three to six prisms offering magnification factors from 4.42 to 57.73. Further numerical examples are provided by Trebino *et al.* (1985).

The compensating multiple-prism expander considered here provides $(d\Theta/d\lambda)_{\mathrm{P}} = 0.00\ m^{-1}$ at 510 nm for $M = 103.49$. On the other hand, if the last prism is oriented in an additive mode, then for the same beam expansion we obtain $(d\Theta/d\lambda)_{\mathrm{P}} = 1.78 \times 10^7\ m^{-1}$. Moreover, if for the compensating design the wavelength is shifted to 490 nm, the overall dispersion becomes $-1.52 \times 10^4\ m^{-1}$ which should be compared to $-1.98 \times 10^7\ m^{-1}$ for the additive configuration. As indicated by Duarte (1985a) operation away from the design wavelength implies that the beam does not exit the prisms at $\phi_{2,m} = 0$ thus causing deviations from prism to prism. Under these conditions the generalized dispersion expressions (Eqs. (4.21) or (4.32)) are utilized.

An important question relates to the effect of these dispersive variations on the return-pass linewidth. Considering the present four-prism expander, in conjunction with a 600-ℓ/mm echelle grating used in sixth order $(M(\partial\Theta/\partial\lambda)_{\mathrm{G}} \sim 9.39 \times 10^8\ m^{-1})$, the full-width return-pass laser linewidth becomes 3.75 GHz for the compensating design and 3.68 GHz for the additive configuration at 510 nm (in this example we assumed a cavity length equal to the Rayleigh length, $w = 0.15$ mm, and a grating length ~ 80 mm).

The aim of this example was to illustrate that at a given wavelength, the differences in $\Delta\nu$ between the two configurations are relatively small. However, for wavelength scanning around the wavelength of design (510 nm in this case) the compensating design can be more advantageous than the additive alternative; also, the quasi-achromatic configuration is less susceptible to changes in the refractive index due to thermal variations. The question of thermal effects on wavelength drift in prismatic cavities has been considered by several authors (Duarte, 1983; Zhuo, 1984; Trebino, 1985). The single-pass angular displacement as a function of temperature can be written as

$$(d\phi_{2,r}/dT) = (d\phi_{2,r}/d\lambda)(dn/d\lambda)^{-1}(dn/dT) \tag{4.46}$$

where $(d\phi_{2,r}/d\lambda)$ is given by Eq. (4.23), $(dn/d\lambda) \sim -7.06 \times 10^4\ m^{-1}$ for BK 7 glass (at 490 nm), and $(dn/dT) \sim n \times 10^{-6}\ ^\circ\mathrm{C}^{-1}$ (Hanna *et al.*, 1975). For a quasiachromatic multiple-prism expander $(d\phi_{2,r}/d\lambda) = 0$ at the wavelength of design; thus, thermal variations have a minimum effect under these circumstances. It is clear from Eq. (4.46) that $(d\phi_{2,r}/dT)$ has a nonzero value either if $(d\phi_{2,r}/d\lambda)$ is of an additive nature or if a quasiachromatic multiple-prism expander is utilized away from its design wavelength. If the achromatic four-prism expander considered here is utilized away from its design wavelength (at 490 nm) then we estimate that $(d\phi_{2,4}/dT) \sim 1.6 \times 10^{-9}$ rad $^\circ\mathrm{C}^{-1}$. This value becomes $\sim 0.32 \times 10^{-6}$ rad $^\circ\mathrm{C}^{-1}$ for the return-pass case. In Section 9 it is indicated that angular deviations of about 1 μrad can cause a wavelength detuning of the order of one linewidth in certain long-pulse lasers.

4.5. Applications of the Multiple-Prism Dispersion Theory to Pulse Compression in Ultrafast Lasers

An extension of the multiple-prism dispersion theory discussed here is its application to pulse-compression schemes in femtosecond lasers (which are discussed in detail in Chapter 3). Here, a symmetric array of four isosceles prisms configured as shown in Fig. 4.15 was introduced by Fork *et al.* (1984) to provide negative dispersion. In this subsection we illustrate how to perform the basic Fork calculation utilizing the dispersion theory discussed previously.

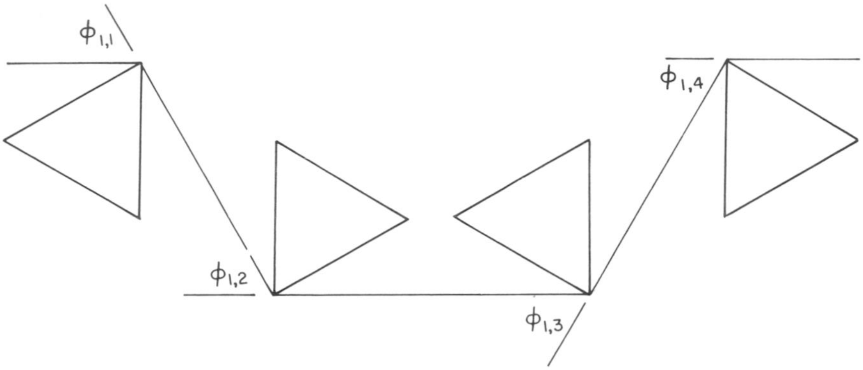

Fig. 4.15 Compensating prismatic array as utilized in pulse-compression arrangements in femtosecond dye lasers.

Briefly, in the work of Fork *et al.* (1984) it was shown that the dispersion constant for pulse compression depends on the second derivative of the relevant optical path ($d^2P/d\lambda^2$) which itself is a function of the prismatic dispersion ($d\phi_{2,m}/dn$) and its derivative ($d^2\phi_{2,m}/dn^2$). The prismatic dispersion needed can be obtained by multiplying Eq. (4.21) by ($d\lambda/dn_m$) so that

$$\frac{d\phi_{2,m}}{dn} = \frac{\sin\psi_{2,m}}{\cos\phi_{2,m}} + \frac{\cos\psi_{2,m}}{\cos\phi_{2,m}}\frac{\cos\phi_{1,m}}{\cos\psi_{1,m}}\left(\frac{\sin\psi_{1,m}}{\cos\phi_{1,m}} \pm \frac{d\phi_{2,(m-1)}}{dn}\right) \tag{4.47}$$

and the derivative of the single-pass dispersion becomes

$$\begin{aligned}\frac{d^2\phi_{2,m}}{dn^2} &= \tan\phi_{2,m}\left(\frac{d\phi_{2,m}}{dn}\right)^2 + (k_{2,m})^{-1}\left(\frac{d\psi_{1,m}}{dn} + \frac{d\psi_{2,m}}{dn}\right) \\ &\quad + (k_{2,m})^{-1}\tan\psi_{1,m}\tan\phi_{1,m}\frac{d\phi_{1,m}}{dn} \pm (k_{1,m}k_{2,m})^{-1}\frac{d^2\phi_{2,(m-1)}}{dn^2} \\ &\quad + \left(\frac{\sin\psi_{1,m}}{\cos\phi_{1,m}} \pm \frac{d\phi_{2,(m-1)}}{dn}\right)\left[(k_{1,m}k_{2,m})^{-1}\tan\psi_{1,m}\frac{d\psi_{1,m}}{dn}\right. \\ &\quad \left. - (k_{2,m})^{-1}\frac{\sin\phi_{1,m}}{\cos\psi_{1,m}}\frac{d\phi_{1,m}}{dn} - (k_{1,m})^{-1}\frac{\sin\psi_{2,m}}{\cos\phi_{2,m}}\frac{d\psi_{2,m}}{dn}\right]\end{aligned} \tag{4.48}$$

where $k_{1,m} = (\cos\psi_{1,m}/\cos\phi_{1,m})$ and $k_{2,m} = (\cos\phi_{2,m}/\cos\psi_{2,m})$.

For a single isosceles prism Eq. (4.47) becomes

$$\frac{d\phi_{2,1}}{dn} = \frac{\sin\psi_{2,1}}{\cos\phi_{2,1}} + \frac{\cos\psi_{2,1}}{\cos\phi_{2,1}}\tan\psi_{1,1} \tag{4.49}$$

and Eq. (4.48) reduces to

$$\frac{d^2\phi_{2,1}}{dn^2} = \tan\phi_{2,1}\left(\frac{d\phi_{2,1}}{dn}\right)^2 - \frac{\tan^2\psi_{1,1}}{n}\frac{d\phi_{2,1}}{dn}. \tag{4.50}$$

For minimum deviation and Brewster's angle of incidence, we get $(d\phi_{2,1}/dn) = 2$ and $(d^2\phi_{2,1}/dn^2) = [4n - (2/n^3)]$, which is the result utilized by Fork *et al.* (1984).

An isosceles prism can be considered to be composed of two identical right-angle prisms back-to-back. Thus the passage of light through an isosceles prism can be treated using the return-pass analysis as applied to right-angle prisms. Thus from Eq. (4.41) we get

$$(d\phi'_{1,1}/dn) = 2\sum_{m=1}^{r}(\pm 1)\left(\prod_{j=1}^{m} k'_{1,j}\right)\tan\psi'_{1,m} \tag{4.51}$$

and its derivative is

$$\begin{aligned}\frac{d^2\psi'_{1,1}}{dn^2} = {} & \pm(1/\beta)\sum_{m=1}^{r}\left(\prod_{j=m}^{r} k'_{1,j}\right)^{-1}\tan\phi'_{1,m}\left(\frac{d\psi'_{1,m}}{dn}\right)^2 \\ & \pm\sum_{m=1}^{r}\left(\prod_{j=1}^{m} k'_{1,j}\right)\tan\psi'_{1,m}\frac{d\psi'_{1,m}}{dn}\frac{d\phi'_{1,(m+1)}}{dn} \\ & \pm\left(\prod_{m=1}^{r} k'_{1,m}\right)\frac{d^2\phi'_{1,(m+1)}}{dn^2}\end{aligned} \tag{4.52}$$

(Duarte, 1987). At the prism boundary (where light is sent back on its original path), $(d\phi'_{1,(m+1)}/dn) = d\phi_{2,m}/dn$. Thus using Eqs. (4.51) and (4.52), in conjunction with the expressions

$$(d\psi'_{1,m}/dn) = n^{-1}[(k'_{1,m})^{-1}(d\phi'_{1,m}/dn) - \tan\psi'_{1,m}] \tag{4.53}$$

and

$$(d\psi_{1,m}/dn) = -n^{-1}[\pm k_{1,m}^{-1}(d\phi_{2,(m-1)}/dn) + \tan\psi_{1,m}] \tag{4.54}$$

it can be shown that $d\phi'_{1,1}/dn = 2$, and $d^2\phi'_{1,1}/dn^2 = [4n - (2/n^3)]$, which again is Fork's result. Although the single-pass analysis appears to be simpler, the use of Eqs. (4.51) and (4.52) is more suitable to calculations involving large number of prisms. This is particularly true if the case to be considered requires a return-pass analysis.

It follows from the perspective presented here and other discussions in the literature (see for example Fork *et al.* (1984), Duarte (1987), Kafka and Baer (1987) that geometric as well as prism material properties (n, $dn/d\lambda$, and $d^2n/d\lambda^2$) can affect the amount of negative dispersion necessary to achieve pulse compression. For instance, Kafka and Baer (1987) discuss the use of flint glass prisms at the grazing exit angle. Indeed, it can be shown via Eqs. (4.47) and (4.48) (or the simpler versions of Eqs. (4.49) and (4.50)) that the dispersion and its derivative is augmented as the angle of incidence in an isosceles prism is decreased, so that the exit angle is increased. This is for a fixed apex angle. Alternatively, one can augment dispersion by increasing the angle of incidence, while at the same time *adjusting* (by design) the apex angle, to maintain minimum deviation. Also, dense materials such as LaSF 9 and ZnSe reduce requirements on prism separation length (Bor and Racz (1985) provide a material survey). These statements can be best illustrated by an example using the parameters utilized by Fork *et al.* (1984). In that work, negative dispersion is achieved for a prism separation $\geqslant$138.4 mm using quartz prisms ($n = 1.457$), at $\phi_{1,1} = 55.53°$. If, instead, the material becomes ZnSe ($n = 2.586$), negative dispersion could be achieved at a separation $\geqslant$11.2 mm at $\phi_{1,1} = 68.85°$ (the corresponding Brewster angle).

An additional comment in this area relates to the configuration of the prismatic pulse compressor. In this regard, it should be stated that the use of the simplified single-prism dispersive expressions (Eqs. (4.49) and (4.50)) is justified only in the case of perfectly balanced prism pairs. Deviations from this type of compensating configuration require the use of the generalized expressions (Eqs. (4.47) and (4.48)) to determine the overall dispersion of the prism chain.

5. MULTIPASS DISPERSION ANALYSIS AND THE LINEWIDTH EQUATION

It is well established that measured linewidths in pulsed dye lasers are narrower than predicted values (Soffer and McFarland, 1967; Strome and Webb, 1971; Hänsch, 1972; Littman and Metcalf, 1978; Iles, 1981; Duarte and Piper, 1981, 1982b, 1984b). This phenomenon is particularly pronounced in long-pulse flashlamp-pumped dye lasers (Strome and Webb, 1971; Duarte and Conrad, 1987). Normally, this discrepancy is attributed to multipass effects. In this subsection we consider multipass dispersive effects that can lead to partial reduction of calculated linewidths. In addition we discuss an appropriate conceptual approach to the use of the linewidth equation in resonators involving intracavity multiple-prism expanders in conjunction with gratings.

Utilizing the generalized dispersion Eqs. (4.21) and (4.32) as the starting point, Duarte and Piper (1984a) analyzed mathematically the multi–return-pass dispersion of an intracavity multiple-prism grating arrangement. Thus it was shown that the emission linewidth for multiple-prism grating dye-laser oscillators, employing right-angle prisms designed for orthogonal beam exit, is given by

$$\Delta\lambda = \Delta\theta\left[R \prod_{m=1}^{r} k_{1,m}(\partial\Theta/\partial\lambda)_G + 2R \sum_{m=1}^{r} (\pm 1)\left(\prod_{j=1}^{m} k_{1,j}\right) \tan\psi_{1,m}(dn_m/d\lambda)\right]^{-1} \tag{4.55}$$

which can be stated more succinctly as

$$\Delta\lambda = \Delta\theta[RM(\partial\Theta/\partial\lambda)_G + R(d\Theta/d\lambda)_P]^{-1} \tag{4.56}$$

where R is the number of intracavity return passes and all other parameters have their usual meaning (Duarte and Piper, 1984a; Duarte, 1985b). The value of R is determined experimentally and for laser-excited oscillators is often less than 5. Independently, at about the same time, Sze (1984) provided an intuitive argument for linewidth reduction due to multipass dispersive effects. In this description the multipass coefficient R is considered outside the dispersive term and as such it divides the divergence factor. In his argument Sze (1984) added that $(\Delta\theta/R)$ can only decrease until the corresponding diffraction limit is reached.

Allowing multipass effects to reduce $\Delta\theta$ toward its diffraction limit or using the measured value of $\Delta\theta$ in laser-pumped MPL and HMPGI oscillators is not always sufficient to ensure agreement between measured and calculated $\Delta\nu$, via Eq. (4.55). This effect is even more pronounced in long-pulse flashlamp-pumped narrow-linewidth oscillators where the measured $\Delta\nu$ is considerably lower than the estimated linewidth (Duarte and Conrad, 1987). This is mainly due to the fact that multipass effects are only partly geometric and the dynamics of the excitation mechanism should be incorporated in the discussion. In this regard we point out that the overall solution should incorporate time, space, and wavelength-dependent population and intensity equations where the multipass-intensity equations incorporate frequency-selective mirrors whose reflectivity is a function of Eq. (4.55).

A topic that has received little attention in the literature, is the mode of applicability of Eq. (4.55) (for $R = 1$). For cavities incorporating a dispersionless beam expander, such as a telescope, the second term in parentheses of the linewidth equation vanishes and

$$\Delta\lambda \approx \Delta\theta[M(\partial\Theta/\partial\lambda)_G]^{-1} \tag{4.57}$$

Under these circumstances one may be allowed to utilize the value of the divergence corresponding to the expanded beam multiplied by the reciprocal of the grating dispersion (Hänsch, 1972). Thus, the preceding expressions may be written as

$$\Delta\lambda \approx (\Delta\theta/M)[(\partial\Theta/\partial\lambda)_G]^{-1} \tag{4.58}$$

where $(\Delta\theta/M)$ can be written as $[\lambda/(\pi w M)]$ for the far field case (see, for example, Hänsch (1972)). This approach is also applied to pure grazing-incidence resonators (see, for example, Littman and Metcalf (1978) and Iles (1981)) where the beam width interacting with the grating is expressed in terms of $(l \cos \Theta)$ where l is the grating length being illuminated and Θ is the angle of incidence.

The situation is altered considerably in cavities incorporating multiple-prism grating assemblies where the multiple-prism expander contributes to the overall dispersion of the cavity. In this case, Eq. (4.55) can be used with the overall measured value of $\Delta\theta$. Alternatively, it can be utilized with the calculated single-pass convolution value of the beam divergence prior to expansion by the multiple-prism arrangement (see, for example, Duarte and Piper, 1982a). Further, the overall cavity convolution value of $\Delta\theta$ (as given by Eq. (4.69)) can be employed. In all cases the lower limit of this value of $\Delta\theta$ is the diffraction limit $[\lambda/(\pi w)]$. It is implicit in this approach that Eq. (4.55) should not be used in conjunction with a beam-divergence value corresponding to the expanded beam (that is, for example, $\lambda/(\pi w M)$). Also, the use of the reduced beam-divergence value, resulting from prismatic expansion, in conjunction with the reciprocal of the grating dispersion alone, implies neglect of the prismatic dispersion and may lead to conceptual difficulties.

6. BEAM DIVERGENCE

In addition to output power, emission wavelength, and linewidth, beam divergence is a parameter of basic importance to lasers. In this regard, the output quality of a laser is considerably enhanced by the ability of the device to provide relatively low beam divergence. Further, laser linewidth, via Eq. (4.1), is directly related to intracavity dispersion and $\Delta\theta$.

Following a number of previous authors (Hänsch, 1972; Littman and Metcalf, 1978; Kasuya *et al.*, 1978; Corney *et al.*, 1979; Iles, 1981; Buffa *et al.*, 1986; Dupre, 1987) the discussion assumes a Gaussian beam profile. Although this may be somewhat idealized, it is known that, at least for single-mode operation, beam profiles approach Gaussian behavior. However, we should emphasize that $\Delta\nu$ is best determined using the measured value of $\Delta\theta$.

A preliminary discussion on the subject involves the concept that the minimum beam divergence attainable is determined by the well-known diffraction limit of a Gaussian beam, $\Delta\theta = \lambda/(\pi w)$. Here, w is the waist, or radius, of the beam in the active dye region. This occurs when far-field conditions are satisfied so that the cavity length is much greater than the Rayleigh length, $L_R = (\pi w^2)/\lambda$. Thus the use of a very small beam waist ($w \leq 0.2$ mm) is advantageous since beam divergences approaching the diffraction limit are possible in compact cavities. Further, a short cavity length in cavities incorporating dispersive elements (such as grazing-incidence, MPL, and HMPGI resonators) enhances conditions favorable for single-mode operation.

6.1. Geometric Approach

In this subsection we review the approach used by Littman and Metcalf (1978) which has also been discussed by Iles (1981). In this case the overall angular divergence is considered to be the result of a convolution of entrance ($\Delta\theta_1$) and exit ($\Delta\theta_2$) slit functions, so that

$$\Delta\theta = [(\Delta\theta_1)^2 + (\Delta\theta_2)^2]^{1/2}. \tag{4.59}$$

In their treatment, Littman and Metcalf (1978) assigned the following values to the relevant functions: $\Delta\theta_1 = (l\cos\Theta)/(2d)$ and $\Delta\theta_2 = w/d$, where d was the dye-cell grating length, l the grating length, and Θ the incidence angle at the grating.

For the far-field case ($d \gg L_R$) it was argued that $\Delta\theta_1 \approx \lambda/(\pi w)$, so that Eq. (4.59) becomes

$$\Delta\theta = (w/d)[1 + (d/L_R)^2]^{1/2}. \tag{4.60}$$

which for the far-field conditions can be simplified further to $\Delta\theta \approx \lambda/(\pi w)$. For the case of $d \ll L_R$, it is shown by Littman and Metcalf (1978) that

$$\Delta\theta = [(w/d)^2 + (w/d)^2]^{1/2} = [(\lambda\sqrt{2})/(\pi w)](L_R/d) \tag{4.61}$$

which, for $L_R = d$, reduces to $\Delta\theta = (\lambda\sqrt{2})/(\pi w)$.

6.2. Ray Matrix Description of Multiple-Prism Configurations

In this section a ray matrix description for tilted interfaces is applied to multiple-prism assemblies, and expressions for $\Delta\theta$ are subsequently derived. Initially, we consider the propagation matrix for two tilted interfaces (Turunen, 1986; Tache, 1987) as applied to a single prism (see Fig. 4.12):

$$\begin{bmatrix} A & B \\ C & D \end{bmatrix} = \begin{bmatrix} (kk') & \ell k'/(nk) \\ 0 & (k'k)^{-1} \end{bmatrix} \tag{4.62}$$

where $k = (\cos\psi_1/\cos\phi_1)$, $k' = (\cos\phi_2/\cos\psi_2)$, l is the prism's path length, and n is the refractive index. For a right-angle prism designed for orthogonal beam exit $\psi_2 = \phi_2 = 0$; thus, Eq. (4.62) simplifies to

$$\begin{bmatrix} A & B \\ C & D \end{bmatrix} = \begin{bmatrix} k & \ell/(nk) \\ 0 & 1/k \end{bmatrix}. \tag{4.63}$$

Practical intracavity beam expanders are composed of two or more right-angle prisms designed for orthogonal beam exit as illustrated in Figs. 4.7 and 4.11. For the case of a multiple-prism expander composed of r identical prisms separated at a uniform distance D and oriented at the same angle of incidence, the corresponding matrix at the plane of incidence is given by (Duarte, 1989b)

$$\begin{bmatrix} A & B \\ C & D \end{bmatrix} =$$

$$\begin{bmatrix} \prod_{m=1}^{r} k_{1,m} & (\ell/n)\left[\left(\prod_{m=1}^{r} k_{1,m}\right) \sum_{m=1}^{r} \left(\prod_{j=1}^{m} k_{1,j}\right)^{-2}\right] + D\left[\left(\prod_{m=1}^{r} k_{1,m}\right) \sum_{m=1}^{r-1} \left(\prod_{j=1}^{m} k_{1,j}\right)^{-2}\right] \\ 0 & \left(\prod_{m=1}^{r} k_{1,m}\right)^{-1} \end{bmatrix}. \tag{4.64}$$

Here, $k_{1,m} = (\cos\psi_{1,m}/\cos\phi_{1,m})$ is the individual beam expansion factor, where $\phi_{1,m}$ is the angle of incidence and $\psi_{1,m}$ the corresponding angle of refraction. Again, in this notation the overall beam magnification factor is given by

$$M = \prod_{m=1}^{r} k_{1,m}.$$

The transfer matrix for the more general case of r prisms with non-orthogonal beam exit can be derived, extending the result given in Eq. (4.62). For instance, the A and D terms of Eq. (4.64) are multipled by

$$\left(\prod_{m=1}^{r} k_{2,m}\right) \tag{4.65a}$$

and

$$\left(\prod_{m=1}^{r} k_{2,m}\right)^{-1}, \tag{4.65b}$$

respectively, where $k_{2,m} = (\cos\phi_{2,m}/\cos\psi_{2,m})$. The B term for this case is rather lengthy and thus is not included in the text.

For a typical multiple-prism resonator there is a finite distance L between the dye cell and the entrance of the expander (see Figs. 4.7 or 4.9). Also, there is a distance x between the exit of the expander and the reflective end of the cavity (either a mirror or a grating). Thus, the propagation matrix can be written as

$$\begin{bmatrix} A' & B' \\ C' & D' \end{bmatrix} = \begin{bmatrix} M & ML + B + (x/M) \\ 0 & 1/M \end{bmatrix} \tag{4.66}$$

where the B term is given by that in Eq. (4.64). Notice that B' for a double prism expander ($r = 2$) is given by

$$B' = ML + D + (\ell/n) + (\ell/nM) + (x/M). \tag{4.67}$$

The Gaussian beam at distance x from the expander is given by (Turunen, 1986).

$$\mathrm{w}(x) = w[M^2 + (B'/L_R)^2]^{1/2}. \tag{4.68}$$

Thus, it can be shown that the overall single-pass divergence can be written as

$$\Delta\theta = (w/L_R)[1 + (L_R/B')^2 + (L_R M/B')^2]^{1/2}. \tag{4.69}$$

In the absence of prisms $L = D = \ell = 0$ and $n = M = 1$. Hence, if we write $x = d$, we find that $\Delta\theta = \lambda/(\pi w)$ for $d \gg L_R$, $\Delta\theta = [(\lambda\sqrt{2})/(\pi w)](L_R/d)$ for $d \ll L_R$, and

$$\Delta\theta = (\lambda\sqrt{3})/(\pi w) \tag{4.70}$$

for $d = L_R$. Thus, this ray matrix approach yields the well-known geometric results for the limiting case of a free-space cavity with a slight difference for the case of $d = L_R$. It should be noticed that Eq. (4.69) can be written as Eq. (9) of Dupre (1987) which was derived for the free-space case exclusively.

A case of interest relates to a cavity with a finite distance between the dye cell and the prism expander which is configured to provide a large beam-magnification factor. Under those circumstances $B' \approx ML$, and Eq. (4.69) reduces to

$$\Delta\theta = (w/L_R)[1 + (L_R/L)^2 + (L_R/ML)^2]^{1/2} \tag{4.71}$$

which is further simplified for the case of $ML \gg L_R$. Hence, for this set of parameters $\Delta\theta \approx \lambda/(\pi w)$ for $L \gg L_R$, $\Delta\theta = [\lambda/(\pi w)](L_R/L)$ for $L \ll L_R$, and $\Delta\theta = (\lambda\sqrt{2})/(\pi w)$ for $L = L_R$.

It should be noticed that Eq. (4.64) in conjunction with the ray matrix for the reverse passage through the multiple prism expander, namely,

$$\begin{bmatrix} \left(\prod_{m=1}^{r} k_{1,m}\right)^{-1} & B \\ 0 & \prod_{m=1}^{r} k_{1,m} \end{bmatrix} \tag{4.72}$$

where the B term is given in Eq. (4.64), can be used to perform a round-trip analysis of the cavity to determine stability conditions.

As discussed in Section 5, it is well known that measured laser linewidths are narrower than those predicted. This is partly due to the fact that the measured half-angle beam divergence (typically in the 1–2-mrad range for optimized cavities) can also be smaller than those calculated. In his telescopic Littrow cavity, Hänsch (1972) assumed a far-field situation for which the calculated diffraction-limited divergence ($\lambda/(\pi w)$) was very close to the measured value. In prismatic cavities, where appropriate calculations were performed using the ideas outlined in this section, the measured divergence is lower than the predicted value (see, for example, Dupre (1987) and Duarte and Conrad (1987)). It is believed that the main source of refinement in $\Delta\theta$ is due to multipass effects as outlined in Section 5. An additional factor that may affect the value of $\Delta\theta$ is the possible presence of thermal lensing in the active region (thermal effects in this type of lasers are discussed in Peters and Mathews (1980) and Duarte (1983)). The effect of thermal lensing can be incorporated in Eq. (4.69) by describing the active dye region as an aperture in conjunction with a thin lens (see, for example, Duarte (1989b) and Siegman (1986)).

Finally, it should be stated that, so far, the best estimate of $\Delta\nu$ (via Eq. (4.55)) can be obtained by using the measured value of $\Delta\theta$. Alternatively, $\Delta\theta$ can be estimated using the methods described in this section with the understanding that the measured $\Delta\theta$ will most likely be lower. Although it is clear that the lowest $\Delta\theta$ will be obtained in a long cavity (far-field case, $\Delta\theta \approx \lambda/(\pi w)$), it is also evident that for single-mode operation, the shortest possible cavity is desired. Thus the main usefulness of the Gaussian treatment provided is a description of the interaction of parameters such as beam waist, cavity length, and beam expansion, which determine the beam-divergence characteristics of different cavity configurations.

7. SINGLE–LONGITUDINAL-MODE PULSED DYE LASER OSCILLATORS

In this section we outline some of the basic criteria utilized in the design of single longitudinal-mode pulsed dye lasers.

An important parameter to consider when contemplating the design of a single-mode dye laser is the cavity-mode spacing which depends on $(c/2L)$ where L is the cavity length. Thus, for cavity lengths of 5, 10, or 15 cm the corresponding free spectral range of the cavity becomes 2.99, 1.49, and 0.99 GHz, respectively. Therefore, one should ensure the shortest possible cavity length and a dispersive assembly capable of providing a return-pass dispersive linewidth comparable or smaller than $(c/2L)$. In this regard, it should be mentioned that in the case of relatively short-pulse (ns-regime) laser-pumped dye lasers, where it is necessary to consider only a few cavity round trips, it is found that the return-pass dispersive linewidth provides a useful upper limit estimate of the observed laser linewidth (Littman and Metcalf, 1978; Iles, 1981; Duarte and Piper, 1981, 1984b).

In practice the principles just outlined can be applied utilizing the various approaches described in this chapter. As our first example we consider the short grazing-incidence cavity described by Littman (1984). The reported cavity length for this laser is 4–5 cm which results in a longitudinal-mode separation of 2.99 to 3.74 GHz. Besides this relatively wide mode spacing, the short cavity allows a greater number (~10) of intracavity round-trips (Littman, 1984), thus permitting multipass effects to offer a reduced linewidth. The single–longitudinal-mode linewidth reported for this short-cavity oscillator is 150 MHz (Littman, 1984). Wavelength tuning of this type of laser is considered in Chapter 6. Other single-mode grazing-incidence pulsed dye lasers are described by Saikan (1978), Dinev *et al.* (1980b), Chang and Li (1980), Beltyugov *et al.* (1981), Koprinkov *et al.* (1982), and Hung and Brechignac (1985b). Designs providing single-mode oscillation using double-grating arrangements are discussed by Littman (1978) and Lebedev and Plyasulya (1981). Dinev *et al.* (1980a) discuss a single-mode grazing-incidence cavity ($\Delta\nu \sim 310$ MHz) designed to lase at two independently tunable wavelengths.

In addition to the grazing-incidence design single–longitudinal-mode lasing has been achieved with a high-prf copper-laser-pumped HMPGI oscillator (Duarte and Piper, 1984b). In this case the return-pass dispersive linewidth (via Eq. (4.55)) can be estimated to be in the 1.3–1.9-GHz range and the cavity mode separation is ~1 GHz. Thus, only a few round-trips (~3) are necessary to account for the observed time-integrated ~650-MHz linewidth (see Fig. 4.10).

As mentioned previously, Bernhardt and Rasmussen (1981) utilized a high-prf copper-laser-pumped MPL oscillator incorporating an intracavity etalon to achieve single–longitudinal-mode lasing at a 60-MHz linewidth. In this case the free spectral range of the cavity was ~1 GHz and the linewidth obtained with the multiple-prism grating system was 9 GHz. Thus an etalon with a FSR of 20 GHz and a finesse of 13 was inserted in the path from the expander to the Littrow grating. Bernhardt and Rasmussen (1981) indicate that only two round-trips are necessary to restrict oscillation to a single axial mode. Bos (1981) reports on the use of a telescopic dye laser, incorporating an intracavity etalon, to achieve single-mode oscillation at ~320 MHz and 6% conversion efficiency. This efficiency was obtained using a pump energy of 12 mJ, at 532 nm, and a dye concentration of 4.5×10^{-4} M of rhodamine 6G (Bos, 1981) (see Section 9 of Chapter 6 for further details).

Implicit in this discussion is the fact that to achieve narrow-linewidth characteristics we require good beam quality (near TM_{00} mode) in a near-diffraction-limited beam. In addition, as indicated by Bernhardt and Rasmussen (1981), mode discrimination is a function of dye concentration with better discrimination achieved at lower concentrations. Also, for high-prf operation, it is absolutely necessary to ensure a high-quality laminar dye flow and the elimination of turbulence (Duarte and Piper, 1984b).

As demonstrated by several authors working on flashlamp-pumped dye-laser oscillators (Magyar and Schneider-Muntau, 1972; Maeda *et al.*, 1975; Flamant *et al.*, 1984), an effective option to reduce linewidth further is the use of multiple etalons. However, the introduction of additional etalons is accompanied by a decline in output energy.

Single– and double–longitudinal-mode lasing in long-pulse flashlamp-pumped MPL and HMPGI dye-laser oscillators has been reported by Duarte and Conrad (1987) (see Fig. 4.11). In this case a 140-mm–wide grating (3000 ℓ/mm) is fully illuminated by multiple-prism expansion and the return-pass dispersive linewidth is estimated to be 2.26 GHz. Here, multipass effects play an important part in reducing the linewidth to $\leqslant$375 MHz.

8. GRAZING-INCIDENCE AND MULTIPLE-PRISM GRATING TECHNIQUES IN GAS LASERS

In this section a brief review of the use of grazing-incidence grating and multiple-prism grating techniques in long-cavity lasers and pulsed gas lasers is provided. In long-cavity gas lasers, where oscillation occurs in high-gain and relatively broad molecular transitions, several techniques are usually employed to provide linewidth narrowing. Among the most common of these techniques are mode selection by injection from narrow-linewidth oscillators such as compatible cw lasers (see, for example, Lachambre *et al.* (1976), Harrison *et al.* (1983), Bigio and Slatkine (1982), and Bourne and Alcock (1983)) and the use of single and multiple intracavity etalons (see, for example, Weiss and Goldberg (1972), Lee and Aggarwal (1977), Pacala *et al.* (1984a, b)). Frequency discrimination available from systems involving injection-seeding from cw lasers can readily provide single-mode operation (a review on this method is provided by Tratt *et al.* (1985)). However, this technique requires an additional compatible seeding laser which may not be readily available for lasers operating at unusual wavelengths.

In the case of the intracavity etalon approach, a serious limitation originates in the susceptibility to optical damage in coatings due to high intracavity energy fluxes. A further disadvantage is losses due to walk-off effects which tend to limit the tunability.

An alternative to the injection-seeding and etalon approaches is the use of the grazing-incidence configuration. In the case of the excimer laser this technique has been used in the XeCl laser (McKee *et al.*, 1979; Lyutskanov, 1980; Pacala *et al.*, 1982; Buffa *et al.*, 1983; Armandillo and Proch, 1983; Sugii *et al.*, 1987; Chaltakov *et al.*, 1988), KrF laser (Caro *et al.*, 1982), and the XeF laser (Armandillo and Giuliani, 1983). This technique has also been applied to TEA CO_2 lasers (Duarte, 1985c; Bobrovskii *et al.*, 1987). Advantages offered by grazing-incidence configurations include a reduction of the intracavity energy flux incident on the grating and a wide tunability range (Chalkatov *et al.*, 1988; Bobrovskii *et al.*, 1987). However, limitations analogous to those present in dye lasers related to efficiency and ASE are present. For instance, McKee *et al.* (1979) report that in excimer lasers increased output energy is observed in an open-cavity configuration; however, the background ASE was about half the intensity of the narrow linewidth emission. The use of grazing-incidence configurations has also been reported in solid-state lasers by German (1981), Gladkov and Kuznetsov (1985), and Kangas *et al.* (1989).

An additional method of frequency narrowing utilizes the concept of multipass grating interferometers (MGI) (Armandillo *et al.*, 1984; Partanen, 1986; Giuliani *et al.*, 1987). Table 4.2 lists reported performance for some of these lasers.

As in the case of dye lasers, isosceles (or equilateral) prism tuning in gas lasers has been known for some time (see, for example, Loree *et al.* (1978)). Similarly, the use of multiple-prism beam expansion in conjunction with a Littrow or near–grazing-incidence grating can provide improved efficiency as compared to the pure–grazing-incidence alternative. Utilizing a combination of specifically designed right-angle ZnSe prisms (hypotenuses of 25 and 60 mm) and a 10-cm (150 ℓ/mm) grating, Duarte (1985b) reported linewidths less than 140 MHz in MPL and HMPGI TEA CO_2 laser oscillators. In these oscillators the uncoated right-angle prisms were utilized in a compensating configuration at near Brewster's angle of incidence (see Fig. 4.16). Thus the transmission of the multiple-prism expanders was $\geqslant 90\%$ for p-polarization. The emission from these oscillators varied stochastically, from pulse to pulse, between near- and single-longitudinal-mode ($\Delta\nu \leqslant 140$ MHz). The beam divergence was diffraction-limited (see Table 4.2 for further details). The performance of multiple-prism and grating assemblies in excimer lasers has been studied by Sze *et al.* (1986). Also some commercial excimer lasers incorporate this type of dispersive cavity.

Table 4.2

Narrow-Linewidth Gas Laser Oscillators

Laser	Cavity	λ (nm)	$\Delta\nu$	E_0	Reference
ArF	MPL	193	10 GHz	150 μJ	Ludewigt *et al.* (1987)
KrF	GI	248	≤9 GHz	15 μJ	Caro *et al.* (1982)
XeCl	GI[1]	308	~31 GHz	50 mJ	Buffa *et al.* (1983)
XeCl	GI[2]	308	~1.5 GHz	~1 mJ	Sze *et al.* (1986)
XeCl	GI	308	~1 GHz	4 mJ	Sugii *et al.* (1987)
XeCl	3 etalons	308	≤ 150 MHz	2–5 μJ	Pacala *et al.* (1984b)
XeF	MGI	351	~40 MHz	~0.1 μJ	Armandillo *et al.* (1984)
CO_2	GI[1]	10,591	117 MHz	140 mJ	Duarte (1985c)
CO_2	GI[1]	10,591	400–700 MHz	230 mJ	Bobrovskii *et al.* (1987)
CO_2	MPL	10,591	≤140 MHz	200 mJ	Duarte (1985b)
CO_2	HMPGI	10,591	107 MHz	85 mJ	Duarte (1985b)

[1] Open-cavity configuration.
[2] Incorporates Michelson interferometer.

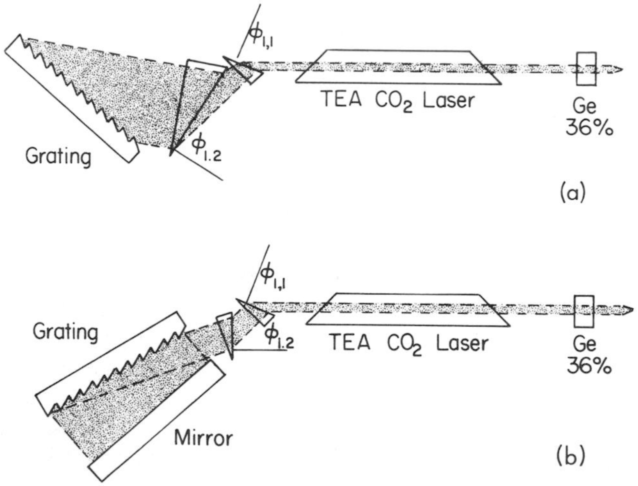

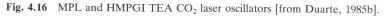

Fig. 4.16 MPL and HMPGI TEA CO_2 laser oscillators [from Duarte, 1985b].

An important feature of MPL and HMPGI oscillators is their inherent ability to operate at high intracavity energies since the energy density incident on the grating is substantially reduced by expansion. In this regard, provided that a single transverse mode is selected (this can be accomplished by the aperturing offered by the prisms themselves), the dispersive return-pass linewidth given by Eq. (4.55) provides an upper limit (depending on pulse length) for expected laser linewidth. A particular design applicable to a large-scale long-pulse e-beam pumped CO_2 laser involves the use of three ZnSe prisms ($\phi_{1,1} = \phi_{1,2} = 67.40°$, $\psi_{1,1} = \psi_{1,2} = 22.59°$, $\phi_{1,3} = 69.51°$, and $\psi_{1,3} = 22.94°$) and a 25-cm grating (150 ℓ/mm) to obtain a return-pass linewidth of ~456 MHz for an MPL configuration involving an total expansion of 15.18, assuming an unexpanded beam of 5 mm in radius and a diffraction-limited divergence (here, $\lambda = 10.59\ \mu m$, P20 (00°1–10°0) transition). Transmission for *p*-polarization radiation in this case is 99.79%. For the same beam characteristics and grating a HMPGI oscillator incorporating two prisms yielding an expansion factor of 5 can provide a return-pass linewidth of ~228 MHz (Duarte and Conrad, 1986). Notice that this type of design involving ZnSe prisms could be applied to other infrared lasers, such as free electron lasers, operating in the 9–20-μm region, provided the intracavity power density is suitable for ZnSe material.

At this stage we should reemphasize that single-mode and near–single-mode oscillation in long-cavity–length lasers incorporating multiple-prism grating assemblies is facilitated by the long-pulse characteristics of some of these lasers (see Section 5). For instance, in the case of the flashlamp-pumped MPL and HMPGI oscillators (described in Fig. 4.11) for which double–and single–longitudinal-mode oscillation was observed in the $250 < \Delta\nu < 375$-MHz range, the corresponding pulse length was in the 200–300-ns range. In the case of the TEA CO_2 MPL and HMPGI oscillators ($107 < \Delta\nu < 140$ MHz), pulse length was in the 60–80-ns range at FWHM.

The use of prismatic techniques in solid-state lasers is discussed by Kravchenko and Anohov (1974) and Brosnan and Byer (1979).

9. LINEWIDTH INSTABILITIES IN MPL AND HMPGI OSCILLATORS

In this section we discuss linewidth instabilities in flashlamp-excited MPL and HMPGI oscillators. For a detailed discussion of dynamic instabilities in lasers, readers should refer to Abraham *et al.* (1988).

Several reports in the literature provide useful background information on this subject. Pease and Pearson (1977) discuss the axial mode structure in a copper-laser-pumped telescopic oscillator incorporating an intracavity etalon, and they provide evidence of pulse-to-pulse fluctuations in the mode structure. Duarte and Piper (1984b) reported on the influence of dye-flow turbulence on linewidth variations and frequency jitter in high-prf copper-laser-pumped MPL and HMPGI dye-laser oscillators. Westling and

Raymer (1986) discuss statistical processes in multimode grazing-incidence dye lasers, and Yang (1988) considers the effects of turbulence on intensity and gain distribution in liquid systems. Berik and Davidenko (1988) made a series of statistical measurements on the linewidth performance of multimode prism-grating excimer-laser-pumped dye lasers. Also, pulse-to-pulse mode-intensity–ratio variations for double-longitudinal-mode lasing in a homogeneous gaseous medium are reported by Duarte (1985b).

The time-dependent linewidth instabilities observed in long-pulse flashlamp-pumped MPL and HMPGI oscillators are distinguished by a lower frequency modulation overimposed on the frequency beating characteristic of double–longitudinal-mode oscillation (250 MHz $\leq \Delta\nu \leq$ 375 MHz) (Duarte *et al.*, 1988).

A type of lower frequency modulation affecting the mode-beating oscillation is illustrated in Fig. 4.17 for $\Delta\nu \sim 250$ MHz. This nonuniform pattern of oscillation is due to a time-dependent variation in the relative intensity of the two longitudinal laser modes. For the case considered in Fig. 4.17 there is an initial mode-intensity ratio of about 200:1; then, as time progresses an intensity-ratio evolution takes place at a modulation frequency of about 20 MHz. Notice that changes in experimental condi-

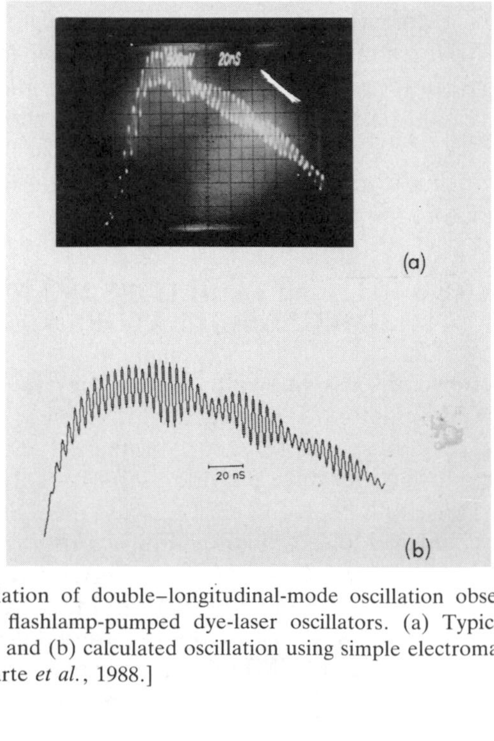

Fig. 4.17 Modulation of double–longitudinal-mode oscillation observed in MPL and HMPGI long-pulse flashlamp-pumped dye-laser oscillators. (a) Typical single-pulse experimental response and (b) calculated oscillation using simple electromagnetic wave representation [from Duarte *et al.*, 1988.]

tions, such as alignment, can produce severe changes in the pattern of the instability.

This time-dependent linewidth variation may occur due to several possible mechanisms. In order to discuss the mechanics of this effect it is useful to emphasize that the linewidth equation discussed in Section 5, namely

$$\Delta\lambda = (\Delta\theta/R)\left[\prod_{m=1}^{r} k_{1,m}(\partial\Theta/\partial\lambda)_G + 2\sum_{m=1}^{r}(\pm 1)\left(\prod_{j=1}^{m} k_{1,j}\right)\tan\psi_{1,m}(dn_m/d\lambda)\right]^{-1},$$

is dispersive in origin. Thus, this is a frequency transmission function that allows one or more laser modes to oscillate. As such, the linewidth of a single–longitudinal-mode can be much narrower than the width of the dispersive linewidth function. Since the dispersive contribution of the multiple-prism expander can be made equal to zero by appropriate design, the resonant frequency can be determined by the grating alone. In the case of a Littrow configuration ($m\lambda = 2a \sin\Theta$), it can be shown that very slight angular displacements ($\sim 10^{-6}$ rads) of the expanded beam toward the grating can cause a frequency shift ~250 MHz (Duarte *et al.*, 1988). In addition, it was found that changes in the beam-expansion coefficient of the dispersive linewidth function (Eq. (4.55)) influence the amplitude and frequency of the instability modulation. This is partly due to the fact that broadening of the linewidth function (by reducing M) allows an increase of the intensity of the secondary mode.

Interferometric observations indicate that the temporal modulation of the double-mode frequency beating was accompanied by frequency jitter. This observation tends to suggest that a dual mechanism involving frequency displacement of the dispersive function in conjunction with an out-of-phase frequency jitter of the laser modes may be a likely vehicle for the linewidth instability. However, it is only necessary to use the simpler concept of a frequency displacement of the two laser modes relative to the central frequency of the dispersive function to obtain agreement with the temporal measurements by using a double-mode electromagnetic wave representation of the effect (see Fig. 4.17b). This simple wave representation considers the resulting field from two modes of frequencies, ω_1 and ω_2, and amplitudes, E_1 and E_2, so that

$$E = E_1\cos(\omega_1 t - k_1 z) + E_2\cos(\omega_2 t - k_2 z). \tag{4.73}$$

For incidence at $z = 0$ on a square law detector the output signal can be expressed as

$$\begin{aligned} E^2 = \tfrac{1}{2}(E_1^2 + E_2^2) + \tfrac{1}{2}E_1^2\cos 2\omega_1 t + \tfrac{1}{2}E_2^2\cos 2\omega_2 t \\ + E_1E_2\cos(\omega_1+\omega_2)t + E_1E_2\cos(\omega_1-\omega_2)t. \end{aligned} \tag{4.74}$$

As indicated by Pacala *et al.* (1984b), detectors with response in the nanosecond regime respond only to the first and last terms of Eq. (4.74). Thus, the beat frequency in Fig. 4.17 is derived from the frequency difference $(\omega_1 - \omega_2)$ term. The calculated version of the modulated double–longitudinal-mode frequency beating was obtained using a non-Gaussian temporal representation derived from experimental data in the form of a rational function

$$f(t) = (a_2 t^2 + a_1 t + a_0)(b_1 t + b_0)^{-1}. \tag{4.75}$$

The ratio of frequency jitter $\delta\omega$ to cavity mode spacing $(\Delta\omega = \omega_1 - \omega_2)$ was represented by a sinusoidal function at 20 MHz (initial mode intensity ratio is 200:1).

When these instabilities were first observed (Duarte and Conrad, 1987) it was postulated that they were the result of mode competition created or exacerbated by inhomogeneities in the dye medium due to thermal gradients and instabilities in the pumping photon flux.

In practice, thermal gradients occur due to temperature differences between the dye solution and the cooling water at the wall of the dye cell ($\Delta T \leqslant 1$ °C). Under ideal conditions of equilibrium, thermal characteristics can be described using cylindrical coordinates so that the conduction equation becomes

$$\partial v/\partial t = \kappa[\partial^2 v/\partial r^2 + (1/r)\,\partial v/\partial r] \tag{4.76}$$

where r is the radius of the active region (Carslaw and Jaeger, 1973). A solution of this equation is of the form

$$v(r, t) = \sum_{n=1}^{\infty} A_n J_0(\alpha_n r)\, e^{-\alpha_n^2 \kappa t} \tag{4.77}$$

where J_0 is a Bessel function of the first kind. The exact form for the solution of Eq. (4.76) depends on boundary conditions. In practice, it is difficult to achieve a deterministic radial temperature profile as described by Eq. (4.77). The situation is severely altered by the presence of turbulence which causes a nonuniform distribution of dye solution from different temperature regions, thus seriously affecting the optical homogeneity of the active medium. Thus, dye-flow turbulence appears to contribute to the cause of linewidth instabilities by providing a nonuniform optical path which results in slight propagation deviations of the intracavity beam. Experimentally, it is found that neutralization of dye-flow turbulence leads to near-stable linewidth oscillation in the long-pulse regime (these results are illustrated in Fig. 4.18). A detailed study of the effects of thermal

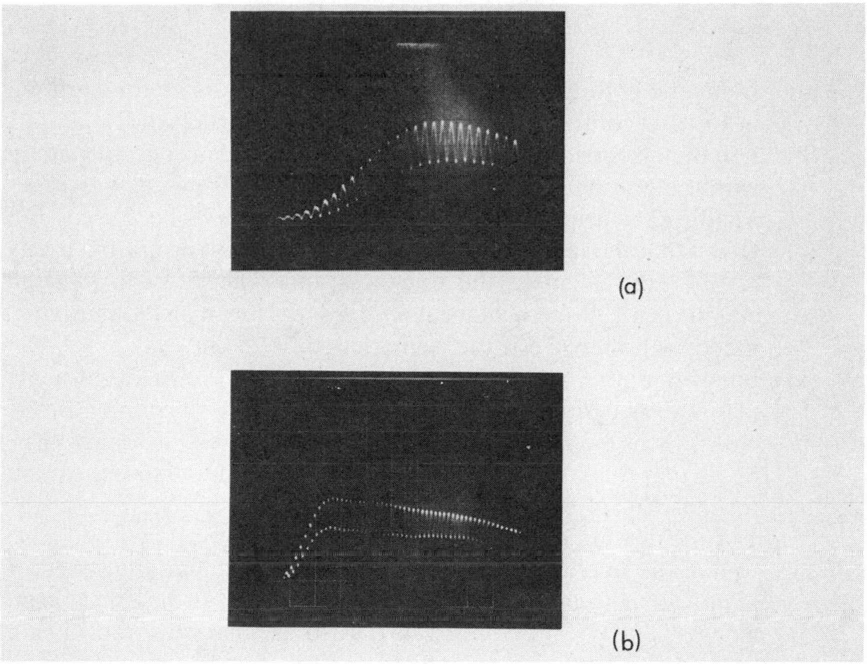

Fig. 4.18 Typical single-pulse unmodulated double–longitudinal-mode oscillation (at different excitation energies) at reduced, or absent, dye flow turbulence: (a) 10 ns/div. and (b) 20 ns/div.

gradients on output energy in flashlamp-pumped dye lasers is provided by Fletcher *et al.* (1986).

In addition to thermal and dye-flow control, it was found that a decrease in the excitation energy of the laser produces shorter pulses (~150-ns full width) characterized by near-uniform double–longitudinal-mode oscillation, under optimum beam alignment.

Contribution from stochastic processes, such as those present in shot-to-shot variations in the mode intensity ratio (see, for example, Duarte, 1985b) are believed to contribute a fraction of the total variability. We also note that acoustical waves are known to cause optical distortions in lasers operating in the μs regime (Gavronskaya *et al.*, 1977; Aristov *et al.*, 1978).

Hence, it is suggested that stable narrow-linewidth oscillation in flashlamp-pumped dye lasers requires thermal uniformity and optical homogeneity of the active medium. Besides, it is clear that other parameters such as variations in the pump photon flux also affect the outcome. In this regard, further measurements are needed to quantify the influence of these variables.

PROBLEMS

1. (a) Using the grating equation, show that the dispersion for a grating in Littrow configuration is given by $(\partial\Theta/\partial\lambda) = (2\tan\Theta)/\lambda$.
   (b) Use the dispersion equation for a grazing-incidence grating along with the grating equation to show that an alternative expression for dispersion in this configuration is $(\partial\Theta/\partial\lambda) = 2m/(a \cos\Theta)$.
   (c) Given that the laser linewidth provided by a mirror-grating cavity is 10 GHz, calculate the thickness and reflectivity for a solid etalon ($n = 1.46$) to achieve $\Delta\nu \sim 1$ GHz. How many longitudinal modes will there be if the cavity length is 25 cm?
2. (a) Show that the beam-expansion factor for a single right-angle prism designed for orthogonal beam exit is given by $k_{1,1} = \cos\psi_{1,1}/\cos\phi_{1,1}$.
   (b) Show that this expansion factor can be written in a more explicit form as Eq. (4.8).
   (c) Use the cumulative reflection loss expression given by Eq. (4.44) to find the overall loss of a double-prism expander composed of identical right-angle prisms designed for near-orthogonal beam exit ($\phi_{1,1} = \phi_{1,2} = 75°$, $n = 1.46$). Also calculate the total beam magnification of such an expander.
3. Show that for identical right-angle prisms designed for near-orthogonal beam exit, and beam incidence at Brewster's angle, Eq. (4.42) follows from Eq. (4.41).
4. Design a three-prism beam expander providing an overall beam magnification of about 80. This beam expander should be made of quartz and must provide zero dispersion at $\lambda = 540$ nm. If the unexpanded beam diameter is 0.15 mm, estimate the return-pass linewidth, at the given wavelength, if a 50-mm holographic grating (3600 ℓ/mm) is to be used in the first order (assume a dye-cell-grating length equal to the Rayleigh length). What would be the linewidth if the prisms are utilized in an additive configuration? What would be the linewidth if we assume a multipass reduced beam divergence approaching the diffraction limit? (Use the quasi-achromatic design for this estimate.)
5. In reference to the design of question 4, assume that a diffraction-limited beam divergence is obtained. Given that the dye cell is about 10 mm wide and that the distance of the output coupler to the cell is also 10 mm, is it possible to achieve single-mode operation with the multiple-prism grating assembly in place? Here, assume that the smallest prism has a hypotenuse of 15 mm. Discuss your options.
6. (a) Using the same beam parameters as in Question 4, calculate the optimum return-pass linewidth using the 3600 ℓ/mm grating in

grazing-incidence configuration. Since the cavity can now be made very short, comment on the conditions for single-mode operation.

(b) If more efficiency is required and two small prisms (hypotenuses of 15 mm) are introduced to provide an expansion factor of 25, does the increase in cavity length exclude single-mode operation?

7. For the case of isosceles or equilateral prisms in femtosecond dye lasers, show that dispersion can be increased by either reducing the angle of incidence on the prism or, alternatively, by increasing the angle of incidence while the apex angle is adjusted (by design) so that $\alpha_1 = 2\psi_{1,1}$ (and $\psi_{1,1} = \psi_{2,1}$). Discuss the implications of each of these approaches for the design of prism chains for pulse compression.

8. (a) Use the generalized ray matrix transfer to write the corresponding transfer matrices for single- and a double-prism expanders.

(b) For a cavity composed of an output coupler mirror, dye cell, double-prism expander, and a total reflector, perform a round-trip ray matrix analysis to determine the stability of the cavity (consider the dye cell as a thin convex lens).

REFERENCES

Abraham, N.B., Mandel, P., and Narducci, L.M. (1988). Dynamical instabilities and pulsations in lasers. In *Progress in Optics*, **vol. XXV** (Wolf, E., ed.). North-Holland, New York, pp. 3–190.

Alden, M., Fredricksson, K., and Wallin, S. (1984). Application of a two-color dye laser in CARS experiments for fast determination of temperatures. *Appl. Opt.* **23,** 2053–2055.

Aristov, A.V., Kozlovskii, D.A., Staselko, D.I., and Strigun, V.L. (1978). Acoustooptical distortions induced by pump lamp radiation in ethanol and water dye solutions. *Opt. Spectrosc.* **45,** 683–687.

Armandillo, E., and Giuliani, G. (1983). Estimation of the minimum laser linewidth achievable with a grazing-grating configuration. *Opt. Lett.* **8,** 274–276.

Armandillo, E., and Proch, D. (1983). Highly efficient, high-quality phase-conjugate reflection at 308 nm using stimulated Brillouin scattering. *Opt. Lett.* **8,** 523–525.

Armandillo, E., Lopatriello, P.V.M., and Giuliani, G. (1984). Single-mode, tunable operation of a XeF excimer laser employing an original interferometer. *Opt. Lett.* **8,** 327–329.

Barr, J.R.M. (1984). Achromatic prism beam expanders. *Opt. Commun.* **51,** 41–46.

Beiting, E.J., and Smith, K.A. (1979). An on-axis reflective beam expander for pulsed dye laser cavities. *Opt. Commun.* **28,** 355–358.

Beltyugov, V.N., Nalivaiko, V.I., Plekhanov, A.I., and Safonov, V.P. (1981). Single-mode pulsed dye laser. *Sov. J. Quantum Electron.* **11,** 837–839.

Berik, E., and Davidenko, V. (1988). Statistical properties of pulsed dye laser radiation. *Opt. Commun.* **67,** 129–132.

Bernhardt, A.F., and Rasmussen, P. (1981). Design criteria and operating characteristics of a single-mode pulsed dye laser. *Appl. Phys. B* **26,** 141–146.

Bigio, I.J., and Slatkine, M. (1982). Transform-limited-bandwidth injection locking of an XeF laser with an Ar-ion laser at 3511 Å. *Opt. Lett.* **7,** 19–21.

Bobrovskii, A.N., Branitskii, A.V., Zurin, M.V., Kozhevnikov, A.V., Mishchenko, V.A., and Mylnikov, G.D. (1987). Continuously tunable TEA CO_2 laser. *Sov. J. Quantum Electron.* **17,** 1157–1159.

Bor, Z., and Racz, B. (1985). Dispersion of optical materials used for picosecond spectroscopy. *Appl. Opt.* **24,** 3440–3441.

Born, M., and Wolf, E. (1975). *Principles of Optics*, 5th ed. Pergamon, New York.

Bos, F. (1981). Versatile high-power single-longitudinal-mode pulsed dye laser. *Appl. Opt.* **20,** 1886–1890.

Bourne, O.L., and Alcock, A.J. (1983). A high-power, narrow linewidth XeCl oscillator. *Appl. Phys. Lett.* **42,** 777–779.

Bradley, D.J., Gale, G.M., Moore, M., and Smith, P.D. (1968). Longitudinally pumped, narrow-band continuously tunable dye laser. *Phys. Lett.* **26A,** 378–379.

Brosnan, S.J., and Byer, R.L. (1979). Optical parametric oscillator threshold and linewidth studies. *IEEE J. Quantum Electron.* **QE-15,** 415–431.

Broyer, M., Chevaleyre, J., Delacretaz, G., and Woste, L. (1984). CVL-pumped dye laser for spectroscopic application. *Appl. Phys. B* **35,** 31–36.

Buffa, R., Burlamacchi, P., Salimbeni, R., and Matera, M. (1983). Efficient spectral narrowing of a XeCl TEA laser. *J. Phys. D: Appl. Phys.* **16,** L125–L128.

Buffa, R., Cavalieri, S., Matera, M., and Mazzoni, M. (1986). Analysis and performance of a nitrogen pumped dye laser optimized for narrowband operation. *Opt. Commun.* **58,** 255–258.

Burlamacchi, P., and Ranea Sandoval, H.F. (1979). Characteristics of a multicolor dye laser. *Opt. Commun.* **31,** 185–188.

Caro, R.G., Gower, M.C., and Webb, C.E. (1982). A simple tunable KrF laser system with narrow bandwidth and diffraction-limited divergence. *J. Phys. D: Appl. Phys.* **15,** 767–773.

Carslaw, H.S., and Jaeger J.C. (1973). *Conduction of Heat in Solids*, 2nd ed. Oxford University Press, New York.

Chaltakov, I.V., Minkovski, N.I., and Tomov, I.V. (1988). A widely tunable XeCl excimer laser. *Opt. Commun.* **65,** 437–439.

Chandra, S., and Compaan, A. (1979). Double-frequency dye lasers with a continuously variable power ratio. *Opt. Commun.* **31,** 73–75.

Chang, T., and Li, F.Y. (1980). Pulsed dye laser with grating and etalon in a symmetric arrangement. *Appl. Opt.* **19,** 3651–3654.

Corney, A., Manners, J., and Webb, C.E. (1979). A narrow bandwith, pulsed, ultra-violet dye laser. *Opt. Commun.* **31,** 354–358.

Dinev, S.G., Koprinkov, I.G., Stamenov, K.V., Stankov, K.A., and Radzewicz, C. (1980a). Two-wavelength single mode grazing incidence dye laser. *Opt. Commun.* **32,** 313–316.

Dinev, S.G., Koprinkov, I.G., Stamenov, K.V., and Stankov, K.A. (1980b). A novel double grazing-incidence single-mode dye laser. *Appl. Phys.* **22,** 287–291.

Duarte, F.J. (1983). Thermal effects in double prism dye laser cavities. *IEEE J. Quantum Electron.* **QE-19,** 1345–1347.

Duarte, F.J. (1985a). Note on achromatic multiple-prism beam expanders. *Opt. Commun.* **53,** 259–262.

Duarte, F.J. (1985b). Multiple-prism Littrow and grazing-incidence pulsed CO_2 lasers. *Appl. Opt.* **24,** 1244–1245.

Duarte, F.J. (1985c). Application of dye laser techniques to frequency selectivity in pulsed

CO_2 lasers. In *Proceedings of the International Conference on Lasers '84* (Corcoran, K.M., Sullivan, D.M., and Stwalley, W.C., eds.). STS Press, McLean, Va., pp. 397–403.

Duarte, F.J. (1987). Generalized multiple-prism dispersion theory for pulse compression in ultrafast dye lasers. *Opt. Quantum Electron.* **19,** 223–229.

Duarte, F.J. (1989a). Transmission efficiency in achromatic nonorthogonal multiple-prism laser beam expanders. *Opt. Commun.* **71,** 1–5.

Duarte, F.J. (1989b). Ray transfer matrix analysis of multiple-prism dye laser oscillators. *Opt. Quantum Electron.* **21,** 47–54.

Duarte, F.J., and Conrad, R.W. (1986). Evaluation of multiple-prism techniques for linewidth narrowing in large scale CO_2 lasers. In *Proceedings of the International Conference on Lasers '85* (Wang, C.P., ed.). STS Press, McLean, Va. pp. 145–152.

Duarte, F.J., and Conrad, R.W. (1987). Diffraction-limited single–longitudinal-mode multiple-prism flashlamp-pumped dye laser oscillator: linewidth analysis and injection of amplifier system. *Appl. Opt.* **26,** 2567–2571.

Duarte, F.J., Ehrlich, J.J., Patterson, S.P., Russell, S.D., and Adams, J.E. (1988). Linewidth instabilities in narrow-linewidth flashlamp-pumped dye laser oscillators. *Appl. Opt.* **27,** 843–846.

Duarte, F.J., and Piper, J.A. (1980). A double-prism beam expander for pulsed dye lasers. *Opt. Commun.* **35,** 100–104.

Duarte, F.J., and Piper, J.A. (1981). Prism preexpanded grazing-incidence grating cavity for pulsed dye lasers. *Appl. Opt.* **20,** 2113–2116.

Duarte, F.J., and Piper, J.A. (1982a). Dispersion theory of multiple-prism beam expanders for pulsed dye lasers. *Opt. Commun.* **43,** 303–307.

Duarte, F.J., and Piper, J.A. (1982b). Comparison of prism-expander and grazing-incidence grating cavities for copper laser pumped dye lasers. *Appl. Opt.* **21,** 2782–2786.

Duarte, F.J., and Piper, J.A. (1983). Generalized prism dispersion theory. *Am. J. Phys.* **51,** 1132–1134.

Duarte, F.J., and Piper, J.A. (1984a). Multi-pass dispersion theory of prismatic pulsed dye lasers. *Opt. Acta* **31,** 331–335.

Duarte, F.J., and Piper, J.A. (1984b). Narrow linewidth, high prf copper laser-pumped dye-laser oscillators. *Appl. Opt.* **23,** 1391–1394.

Dupre, P. (1987). Quasiunimodal tunable pulsed dye laser at 440 nm: theoretical development for using a quad prism beam expander and one or two gratings in a pulsed dye laser oscillator cavity. *Appl. Opt.* **26,** 860–871.

Elizarov, A.Y. (1985). A telescope for dye lasers, consisting of a prism and a diffraction grating. *Sov. Phys. Tech. Phys.* **30,** 1222–1223.

Flamant, P.H., Josse, D., and Maillard, M. (1984). Transient injection frequency-locking of a microsecond-pulsed dye laser for atmospheric measurements. *Opt. Quantum Electron.* **16,** 179–182.

Fletcher, A.N., Bliss, D.E., Pietrak, M.E., and McManis, G.E. (1986). Improving the output and lifetime of flashlamp-pumped dye lasers. In *Proceedings of the International Conference on Lasers '85* (Wang, C.P., ed.). STS Press, McLean, Va., pp. 797–804.

Fork, R.L., Martinez, O.E., and Gordon, J.P. (1984). Negative dispersion using pairs of prisms. *Opt. Lett.* **9,** 150–152.

Gallagher, T.F., Kachru, R., and Gounand, F. (1982). Simple linewidth reducing modification for a Hansch dye laser. *Appl. Opt.* **21,** 363–364.

Gavronskaya, E.A., Groznyi, A.V., Staselko, D.I., and Strigun, V.L. (1977). Dynamics of thermooptical inhomogeneities in the active medium of an organic dye laser with flash-lamp pumping. *Opt. Spectrosc.* **42,** 213–215.

German, K.R. (1981). Grazing angle tuner for cw lasers. *Appl. Opt.* **20,** 3168–3171.

Giuliani, G., Palange, E., and Salvetti, G. (1987). Spectral characteristics of laser cavities employing multipass grating interferometers as output couplers. *J. Opt. Soc. Am. B.* **4,** 1781–1789.

Gladkov, S.M., and Kuznetsov, V.I. (1985). Tunable YAG:Nd^{3+} laser with a grazing-incidence diffraction grating. *Sov. J. Quantum Electron.* **15,** 141.

Godfrey, L.A., Egbert, W.C., and Meltzer, R.S. (1980). Grazing incidence grating techniques for flashlamp-pumped tunable dye lasers. *Opt. Commun.* **34,** 108–110.

Hanna, D.C., Karkkainen, P.A., and Wyatt, R. (1975). A simple beam expander for frequency narrowing of dye lasers. *Opt. Quantum Electron.* **7,** 115–119.

Hanna, D.C., Sawyers, C.G., and Yuratich, M.A. (1981). Telescopic resonators for large-volume TEM_{OO}-mode operation. *Opt. Quantum Electron.* **13,** 493–507.

Hänsch, T.W. (1972). Repetitively pulsed tunable dye laser for high resolution spectroscopy. *Appl. Opt.* **11,** 895–898.

Harrison, R.G., Kar, A.K., Tratt, D.M., Wright, E.M., Firth, W.J., and Smith, S.D. (1983). Longitudinal mode selection in TEA CO_2 lasers by injection locking. In *Proceedings of the International Conference on Lasers '82* (Powell, R.C., ed.). STS Press, McLean, Va., pp. 627–631.

Hnilo, A.A., Manzano, F.A., and Burgos, A.H. (1980). A reflection beam expander for short cavity dye laser. *Opt. Commun.* **33,** 311–314.

Hung, N.D., and Brechignac, P. (1985a) A narrowband grazing incidence pulsed dye laser improved by using an active multiregion. *Opt. Commun.* **53,** 405–408.

Hung, N.D., and Brechignac, P. (1985b). A single-mode single-grating grazing incidence pulsed dye laser. *Opt. Commun.* **54,** 151–154.

Iles, M.K. (1981). Unified single-pass model of linewidths in the Hansch single- and double-grating grazing-incidence dye lasers. *Appl. Opt.* **20,** 985–988.

Kafka, J.D., and Baer, T. (1987). Prism-pair dispersive delay lines in optical pulse compression. *Opt. Lett.* **12,** 401–403.

Kangas, K.W., Lowenthal, D.D., and Muller, C.H. (1989). Single–longitudinal-mode, tunable, pulsed Ti:sapphire laser oscillator. *Opt. Lett.* **14,** 21–23.

Kasuya, T., Suzuki, T., and Shimoda, K. (1978). A prism anamorphic system for Gaussian beam expander. *Appl. Phys.* **17,** 131–136.

Klauminzer, G.K. (1978). Optical beam expander for dye laser. U.S. Patent. no. 4, 127, 828.

Kogelnik, H. (1965). Imaging of optical modes—resonators with internal lenses. *Bell. Syst. Tech. J.* **44,** 455–494.

Kong, H.J., and Lee, S.S. (1981). Dual wavelength and continuously variable polarization dye laser. *IEEE J. Quantum Electron.* **QE-17,** 439–441.

Kong, X., Pan, Z., and Shi, B. (1987). Effect of the intracavity lens on the linewidth of a dye laser. *Appl. Opt.* **26,** 1366–1367.

Koning, R., Minkwitz, G., and Christov, B. (1980), Nanosecond dye laser of short constructional length with a folded mirror beam expansion system. *Opt. Commun.* **32,** 301–305.

Koprinkov, I.G., Stamenov, K.V., and Stankov, K.A. (1982). High-efficiency single-mode dye laser. *Opt. Commun.* **42,** 264–266.

Krasinski, J., and Sieradzan, A. (1979). A note on the dispersion of a prism used as a beam expander in a nitrogen laser pumped dye laser. *Opt. Commun.* **28,** 14.

Kravchenko, V.I., and Anohov, S.P. (1974). A Nd:glass sweep laser for spectroscopic research. *Opt. Commun.* **12,** 248–251.

Lachambre, J.L., Lavigne, P., Otis, G., and Noel, M. (1976). Injection locking and mode selection in TEA-CO_2 laser oscillators. *IEEE J. Quantum Electron.* **QE-12,** 756–764.

Lebedev, V.V., and Plyasulya, V.M. (1981). Single-frequency tunable pulsed dye laser with diffraction grating. *Opt. Spectros.* **50,** 408–410.

Lee, N., and Aggarwal, R.L. (1977). Single longitudinal mode TEA CO_2 laser with tilted intracavity etalon. *Appl. Opt.* **16,** 2620–2621.

Lisboa, J.A. Teixeira, S.R., Gunha, S.L.S., Francke, R.E., and Grieneisen, H.P., (1983). A grazing-incidence dye laser with an intracavity lens. *Opt. Commun.* **44,** 393–396.

Littman, M.G. (1978). Single-mode operation of grazing-incidence pulsed dye laser. *Opt. Lett.* **3,** 138–140.

Littman, M.G. (1984). Single-mode pulsed tunable dye laser. *Appl. Opt.* **23,** 4465–4468.

Littman, M.G., and Metcalf, H.J. (1978). Spectrally narrow pulsed dye laser without beam expander. *Appl. Opt.* **17,** 2224–2227.

Loree, T.R., Butterfield, K.B., and Barker, D.L. (1978). Spectral tuning of ArF and KrF discharge lasers. *Appl. Phys. Lett.* **32,** 171–173.

Lotem, H., and Lynch, R.T. (1975). Double-wavelength laser. *Appl. Phys. Lett.* **27,** 344–346.

Ludewigt, K., Pfingsten, W., Mohlmann, C., and Wellegehausen, B. (1987). High-power vacuum-ultraviolet anti-Stokes Raman laser with atomic selenium. *Opt. Lett.* **12,** 39–41.

Lyutskanov, V.L., Khristov, K.G., and Tomov, I.V. (1980). Tuning of the emission frequency of a gas-discharge XeCl laser. *Sov. J. Quantum Electron.* **10,** 1456–1457.

Maeda, M., Uchino, O., Okada, T., and Miyazoe, Y. (1975). Powerful narrow-band dye laser forced oscillator. *Jap. J. Appl. Phys.* **14,** 1975–1980.

Magyar, G., and Schneider-Muntau, H.J. (1972). Dye laser forced oscillator. *Appl. Phys. Lett.* **20,** 406–408.

Mazzinghi, P., Burlamacchi, P., Matera, M., Ranea-Sandoval, H.F., Salimbeni, R., and Vanni, U. (1981). A 200 W average power, narrow bandwidth, tunable waveguide dye laser. *IEEE J. Quantum Electron.* **QE-17,** 2245–2249.

McKee, T.J., Banic, J., Jares, A., and Stoicheff, B.P. (1979). Operating and beam characteristics, including spectral narrowing, of a TEA rare-gas halide excimer laser. *IEEE J. Quantum Electron.* **QE-15,** 332–334.

Meaburn, J. (1976). *Detection and Spectrometry of Faint Light.* Reidel, Boston.

Melikechi, N., and Allen, L. (1987). A two wavelength dye laser with broadband tunability. *J. Phys. E: Sci. Instrum.* **20,** 558–559.

Meyer, Y.H., and Nenchev, M.N. (1981). Single-mode dye laser with a double-action Fizeau interferometer. *Opt. Lett.* **6,** 119–121.

Mory, S., Rosenfeld, A., and Konig, R. (1981a). Generation of subnanosecond dye laser pulses of narrow linewidth by grazing-incidence on two gratings. *Opt. Commun.* **38,** 416–418.

Mory, S., Rosenfeld, A., Polze, S., and Korn, G. (1981b). Nanosecond dye laser with a high-efficiency holographic grating for grazing incidence. *Opt. Commun.* **36,** 342–346.

Myers, S.A. (1971). An improved line narrowing technique for a dye laser excited by a nitrogen laser. *Opt. Commun.* **4,** 187–189.

Nair, L.G. (1979). A double-wavelength nitrogen-laser-pumped dye laser. *Appl. Phys.* **20,** 97–99.

Nair, L.G. (1982). Dye lasers. *Prog. Quantum Electron.* **7,** 153–268.

Nair, L.G., and Dasgupta, K. (1980). Double wavelength operation of a grazing incidence tunable dye laser. *IEEE J. Quantum Electron.* **QE-16,** 111–112.

Niefer, R.J., and Atkinson, J.B. (1988). The design of achromatic prism beam expanders for pulsed dye lasers. *Opt. Commun.* **67,** 139–143.

Novikov, M.A., and Tertyshnik, A.D. (1975). Tunable dye laser with a narrow emission spectrum. *Sov. J. Quantum Electron.* **5,** 848–849.

Pacala, T.J., McDermid, I.S., and Laudenslager, J.B. (1982). A wavelength scannable XeCl oscillator-ring amplifier laser system. *Appl. Phys. Lett.* **40,** 1–3.

Pacala, T.J., McDermid, I.S., and Laudenslager, J.B. (1984a). Ultranarrow linewidth, magnetically switched, long pulse, xenon chloride laser. *Appl. Phys. Lett.* **44,** 658–660.
Pacala, T.J., McDermid, I.S., and Laudenslager, J.B. (1984b). Single longitudinal mode operation of an XeCl laser. *Appl. Phys. Lett.* **45,** 507–509.
Partanen, J.P. (1986). Multipass grating interferometer applied to line narrowing in excimer lasers. *Appl. Opt.* **25,** 3810–3815.
Pease, A.A., and Pearson, W.M. (1977). Axial mode structure of a copper vapor pumped dye laser. *Appl. Opt.* **16,** 57–60.
Peters D.W., and Mathews, C.W. (1980). Temperature dependence of the peak power of a Hansch-type dye laser. *Appl. Opt.* **19,** 4131–4132.
Prior, Y. (1979). Double frequency narrow-band grazing incidence pulsed dye laser. *Rev. Sci. Instrum.* **50,** 259–260.
Racz, B., Bor, Z., Szatmari, S., and Szabo, G. (1981). Comparative study of beam expanders used in nitrogen laser pumped dye lasers. *Opt. Commun.* **36,** 399–402.
Routledge, P.A., Berry, A.J., and King, T.A. (1986). A flashlamp-pumped dye laser in a telescopic resonator configuration. *Opt. Acta* **33,** 445–451.
Saikan, S. (1978). Nitrogen-laser-pumped single-mode dye laser. *Appl. Phys.* **17,** 41–44.
Saito, Y., Teramura, T., Nomura, A., and Kano, T. (1985). Simultaneously tunable three-wavelength dye laser. *Appl. Opt.* **24,** 2477–2478.
Schäfer, F.P., and Müller, H. (1971). Tunable dye ring-laser. *Opt. Commun.* **2,** 407–409.
Shoshan, I., Danon, N.N., and Oppenheim, U.P. (1977). Narrowband operation of a pulsed dye laser without intracavity beam expansion. *J. Appl. Phys.* **48,** 4495–4497.
Shoshan, I., and Oppenheim, U.P. (1978). The use of a diffraction grating as a beam expander in a dye laser cavity. *Opt. Commun.* **25,** 375–378.
Siegman, A.E. (1986). *Lasers.* University Science Books, Mill Valley, Ca.
Smith, R.S., and DiMauro, L.F. (1987). Efficiency and linewidth improvements in a grazing incidence dye laser using an intracavity lens and spherical mirror. *Appl. Opt.* **26,** 855–859.
Soffer, B.H., and McFarland, B.B. (1967). Continuously tunable, narrow-band organic dye lasers. *Appl. Phys. Lett.* **10,** 266–267.
Steel, W.H. (1967). *Interferometry.* Cambridge University Press, London.
Stokes, E.D., Dunning, F.B., Stebbings, R.F., Walters, G.K., and Rundel, R.D. (1972). A high efficiency dye laser tunable from the UV to the IR. *Opt. Commun.* **5,** 267–270.
Strome, F.C., and Webb, J.P. (1971). Flashtube-pumped dye laser with multiple-prism tuning. *Appl. Opt.* **10,** 1348–1353.
Sugii, M., Ando, M, and Sasaki, K. (1987). Simple long-pulse XeCl laser with narrow-line output. *IEEE J. Quantum Electron.* **QE-23,** 1458–1460.
Sze, R.C. (1984). Prism tuning characteristics of a long-pulsed XeCl laser. In *Technical Digest, Conference on Lasers and Electro-Optics.* Paper THP6. Optical Society of America, Washington, D.C.
Sze, R.C., Kurnit, N.A., Watkins, D.E., and Bigio, I.J. (1986). Narrow band tuning with small long-pulse excimer lasers. In *Proceedings of the International Conference on Lasers '85* (Wang, C.P., ed.). STS Press, McLean, Va., pp. 133–144.
Tache, J.P. (1987). Ray matrices for tilted interfaces in laser resonators. *Appl. Opt.* **26,** 427–429.
Tratt, D.M., Kar, A.K., and Harrison, R.G. (1985). Spectral control of gain-switched lasers by injection-seeding: applications to TEA CO_2 systems. *Prog. Quantum Electron.*, **10,** 229–266.
Trebino, R. (1985). Achromatic N-prism beam expanders: optimal configurations. *Appl. Opt.* **24,** 1130–1138.
Trebino, R., Barker, C.E., and Siegman, A.E. (1985). Achromatic N-prism beam expanders:

optimal configurations II. In *SPIE vol. 540, South West Conference on Optics*. Society of Photo-Optical Instrumentation Engineers, Bellingham, Wash., pp. 104–109.

Trebino, R., Roller, J.P., and Siegman, A.E. (1982). A comparison of the Cassegrain and other beam expanders in high-power pulsed dye lasers. *IEEE J. Quantum Electron.* **QE-18,** 1208–1213.

Turunen, J. (1986). Astigmatism in laser beam optical systems. *Appl. Opt.* **25,** 2908–2911.

van Hoek, A., and van Wijk, F..G.H. (1987). Tunable broadband pulsed dye laser. *Appl. Opt.* **26,** 1164–1166.

Weiss, J.A., and Goldberg, L.S. (1972). Single longitudinal mode operation of a transversely excited CO_2 laser. *IEEE J. Quantum Electron.* **QE-8,** 757–758.

Westling, L.A., and Raymer, M.G. (1986). Intensity autocorrelation measurements and spontaneous FM phase locking in a multimode dye laser. *J. Opt. Soc. Am. B,* **3,** 911–917.

Williams, S.W., and Heddle, D.W.O. (1983). Simultaneous two-wavelength tunable dye laser with no mode competition and with a wavelength separation of more than 200 nm. *Opt. Commun.* **45,** 112–114.

Wyatt, R. (1978a). Narrow linewidth, short pulse operation of a nitrogen-laser-pumped dye laser. *Opt. Commun.* **26,** 429–431.

Wyatt, R. (1978b). Comment on "On the dispersion of a prism used as a beam expander in a nitrogen-laser pumped dye laser." *Opt. Commun.* **26,** 9–11.

Yang, C.C. (1988). Average intensity and contrast of light amplified by a partially homogeneously broadened, slightly saturated laser amplifier with stochastic gain distributions. *Opt. Lett.* **13,** 366–368.

Yodh, A.G., Bai, Y., Golub, J.E., and Mossberg, T.W. (1984). Grazing-incidence dye lasers with and without intracavity lenses: a comparative study. *Appl. Opt.* **23,** 2040–2042.

Zhang, G.W., Grethen, H., Kronfeldt, H.D., and Winkler, R. (1981). The influence of the beam expanding prism in a dye laser resonator on the linewidth and its dependence on the expansion ratio. *Opt. Commun.* **40,** 49–53.

Zhang, J.H., Hilty, B.R., and Schuessler, H.A. (1982). Narrow linewidth operation of a flashlamp-pumped dye laser. *Appl. Opt.* **21,** 3065–3067.

Zhou, C.S. (1984). Design of a pulsed single-mode dye laser. *Appl. Opt.* **23,** 2879–2885.

Chapter 5

CW DYE LASERS

Leo Hollberg

National Institute of Standards and Technology
(formerly National Bureau of Standards)
Boulder, Colorado

1. INTRODUCTION

The cw dye laser is a well-established tool of optical science. It plays the premier role as a source of tunable radiation in the visible and near-visible regions of the spectrum. This laser has a unique set of capabilities that include broad tunability, high power, and the potential for extremely high

* I gratefully acknowledge the helpful discussions and contributions made to this chapter by Dr. J.C. Bergquist, Dr. T.F. Johnston, Jr., Dr. J.L. Hall, Dr. S.L. Gilbert, and Dr. M. Young.

 ISBN 0-12-222700-X

resolution. Because of these capabilities we overlook its high cost, sometimes complicated design, and the requirement of a high-power optical pump. No other optical source can provide a comparable combination of tunability, resolution, and power.

1.1. Overview of Characteristics

The most important attribute of the dye laser is its tunability, which gives the user access to essentially any wavelength in the visible and near-visible spectrum. The spectral range of ion-laser-pumped cw dye lasers is essentially complete coverage from 365 to 1000 nm. It is even possible to extend their cw tuning range by using nonlinear optical methods to generate wavelengths further into the ultraviolet and infrared. With the addition of these methods of nonlinear optical synthesis, the spectral range for complete coverage with cw radiation has been extended to 250 nm. Radiation at even shorter wavelengths has been generated for special applications. In addition nonlinear methods have been used to generate radiation in the 2–5 μm infrared region. This range of spectral coverage will certainly be extended in the future.

The power levels available from cw dye lasers are generally more than adequate for spectroscopic applications. The output power of cw dye lasers varies with the type of dye, but typical cw systems produce between .1 and 1 watt of output power. As with most things, there are some applications with almost unlimited demand for power. Special high-power cw-dye-laser systems have been developed that can produce tens of watts of tunable visible radiation.

With broad tuning ranges and narrow linewidths, single-mode cw dye lasers can provide an impressively large number of resolution elements. For example, a standard single-frequency dye laser (linewidth $\approx$ 1 MHz) operating with rhodamine 6G dye (tuning range $\approx$ 100 nm or $\approx 10^{14}$ Hz) is capable of resolving $\approx 10^8$ spectral elements across its tuning curve. If we considered one of the special high-resolution lasers that have been demonstrated (linewidths less than 1 kHz) the number of resolution elements is $\approx 10^{11}$, probably more than any application can actually use and certainly one of the highest of any electromagnetic source.

1.2. Brief History of the CW Dye Laser

Historically, the development of the cw dye laser was an outgrowth of research into the photophysics of dye-laser action in pulsed dye lasers (Schäfer *et al.*, 1966; Snavely and Schäfer, 1969). The first flashlamp-pumped dye lasers were developed in 1967 (Schmidt and Schäfer 1967; Sorokin and Lankard, 1967), just one year after the first successful pulsed dye lasers, which were achieved using ruby lasers as the pump source (Sorokin and Lankard, 1966; Spaeth and Bortfeld, 1966; Stepanov *et al.* 1967). With flashlamp pumping, sufficient population inversion could be attained with surprisingly long output-pulse lengths. The pulse lengths were longer than expected because the trapping of molecular excitation in the unwanted triplet states was less probable than expected. The other important contribution that allowed cw operation was the development of a system to flow the dye rapidly through the pumping region in order to reduce the optical inhomogeneities induced in the dye by heating from the cw pump laser. In 1970 Peterson *et al.* (1970) demonstrated the first cw lasing of an organic dye. That laser used the dye, rhodamine 6G, with an argon-ion laser as the pump. Two decades later, the combination of the argon-ion laser and the rhodamine 6G dye is the most common, and in some ways, the best cw dye-laser system that exists.

That first cw dye laser appears extremely simple, but the design is very clever in solving some of the serious technical problems. Figure 5.1 is a simplified diagram of the type of laser design that was used in that first successful demonstration of a cw dye laser. This laser has the dye flowing between two dichroic mirrors that form the laser resonator as well as part of the containment cell for the dye. The argon-ion laser pumping beam was injected collinearly through one of the dichroic mirrors. The Peterson *et al.* (1970) laser of this type produced an output power of 30 mW in a spectral width of 3 nm when pumped with 1 W of 514-nm (argon ion) laser light. We can get some idea of the technological evolution of dye lasers by

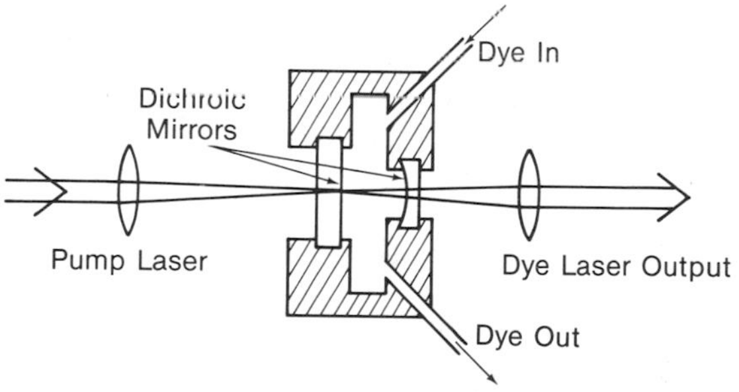

Fig. 5.1 Cross section of the type of dye laser used for the first successful cw operation. The two dichroic mirrors serve the dual role of containing the dye fluid and acting as the dye-laser resonator. The dichroic mirrors transmit the blue-green pump laser beam but are high reflectors for the reddish fluorescence from the dye. The pump-laser beam is derived from an Ar^+ laser. A lens is used to focus the pump beam into the dye cell at the dye-laser resonator waist which is on the flat dichroic mirror. The dye flows rapidly through the cell to remove the excess heat and triplet state population. With this type of laser Peterson *et al.* (1970) achieved $\simeq$30 mW of cw dye-laser output power at 597 nm when the laser was pumped with $\simeq$1 W of 514-nm light.

comparing these results with the performance characteristics of modern lasers, which can produce 1 W of output in a spectral width of about 10^{-6} nm ($\approx$1 MHz) with a pump power of 6 W. More detailed discussion of the early cw dye lasers can be found in Tuccio and Strome (1972).

Development of the cw dye laser since 1970 has been extensive and continues in many labs around the world today. As in most technologies, progress in the development of the cw dye laser has involved a gradual improvement in the design of each of its components, combined with a few important discoveries that have significantly altered the basic system. The important achievements will be discussed in later sections, but it is useful to have some historical perspective of the development. One of the important developments that improved the performance of the dye laser was the introduction in 1972 (Runge and Rosenberg, 1972) of the jet-stream dye-circulation system, which gave more reliable high-power operation. About that same time techniques were developed that allowed frequency stabilization of these lasers and thus greatly enhanced their spectral resolving power (Barger *et al.*, 1973, 1975; Liberman and Pinard, 1973; Schröder *et al.*, 1973, 1975; Wu *et al.*, 1974, Hall and Lee, 1976b). The widespread use of traveling-wave ring lasers started in the late 70s and improved both the spectral characteristics and the power output from the cw dye lasers (Schröder *et al.*, 1977; Jarret and Young, 1979; Johnston *et al.*, 1982). The incorporation of computer control has provided automatic long-range tuning and enhanced data-collection capabilities (Marshall *et al.*, 1980; Williams *et al.*, 1983; Clark *et al.*, 1986). In addition to the continual improvement of laser dyes the '80s have seen the application nonlinear optical techniques that extended the spectral coverage that was possible with cw lasers into the UV and IR regions (Blit *et al.*, 1977, 1978; Couillaud, 1981; Couillaud *et al.*, 1982; Bergquist *et al.*, 1982; Hemmati *et al.*, 1983; T. F. Johnston, Jr., 1987; T. Johnston, 1988; Pine, 1974, 1976; Oka, 1988).

1.3. Other Related Laser Systems

Dye lasers are often the only choice or the best choice of laser for a given application, particularly if it is necessary to tune to a specific wavelength. In the dye laser's region of operation, only a few laser sources are competitive and those are only in limited parts of the dye laser's spectral region. A few new tunable lasers are beginning to threaten some of the traditional domains of the dye laser. These include semiconductor, color-center, Ti-sapphire, alexandrite, and a few optically pumped solid-state lasers. Other nontunable lasers exist that oscillate in the visible and near-visible region of the spectrum; for example, the Kr-ion and Ar-ion lasers have many useful lasing lines, and the optically pumped dimer lasers (sodium, lithium, iodine, etc.) have hundreds of lines, but the lines are relatively narrow and, even when combining all of these lasers, the actual fractional coverage of the available spectrum is small.

Semiconductor lasers are now available in some parts of the near-IR and red regions of the spectrum. As of the late 1980s, in comparison with dye lasers the semiconductor lasers generally have large linewidths and low powers. But the characteristics of semiconductor lasers are improving and their future looks quite bright.

The broad class of optically pumped lasers shares many of the characteristics of the cw dye laser–for example, color-center lasers (Mollenauer, 1985; Mollenauer and White, 1987). Two important, and relatively new optically pumped solid-state lasers are the Ti-sapphire laser (Schulz, 1988) and the LNA (Lanthamide hexa-aluminate) laser (Schearer *et al.*, 1986). The designs of these lasers, as well as that of color center lasers, are very similar to the designs of cw dye lasers. The Ti-sapphire laser is an interesting example of a laser competing for some of the domain dominated by the dye laser (Ti-sapphire tuning range ≃700–1000 nm). These lasers use the same pumping source and essentially the same cavity design, but they have the dye medium replaced by a solid-state material. Most of the design features discussed here for dye lasers apply equally well to other optically pumped lasers. True, some types of lasers produce higher power, and some yield higher spectral purity, but the dye laser combines both of these traits with broad tunability and the ability to change dyes for even broader spectral coverage.

This chapter is not intended to be a comprehensive review of cw dye lasers but rather a description of the basic principles of some common cw dye lasers. Keep in mind that the descriptions are simply examples and the possible variations and extensions are large and often unexplored. This material is but a glimpse at the field from one perspective, where I have additionally chosen to cover in more detail those aspects that have been neglected in other sources, and to give less emphasis to material that is covered well elsewhere. Effort is made to give reference to other sources of material on cw dye lasers but the reference list is not complete. It is rather a place to start for further study. In this light it is important to note some of the review articles and books that are particularly relevant to the subject of cw dye lasers, including Shank (1975), Snavely (1977), Schäfer (1977), Peterson (1979), and Mollenauer and White (1987). Of paramount interest is the excellent article by T. F. Johnston (1987) which is a review of dye lasers. Applications of dye lasers are discussed by Hänsch (1976) and Hall (1978). In addition, a great deal of useful information about the design and characteristics of cw dye lasers can be obtained from the instruction manuals of commercial laser systems.

2. BASIC SYSTEM DESIGN

2.1. Example Systems

The years since 1970 have seen the development of a large number of cw dye-laser designs. Two examples of dye-laser systems that have seen widespread use are the three-mirror folded-linear cavity and the four-mirror unidirectional-ring cavity. These systems are diagrammed in Fig. 5.2. We see some features that are common to both designs; for example, both use

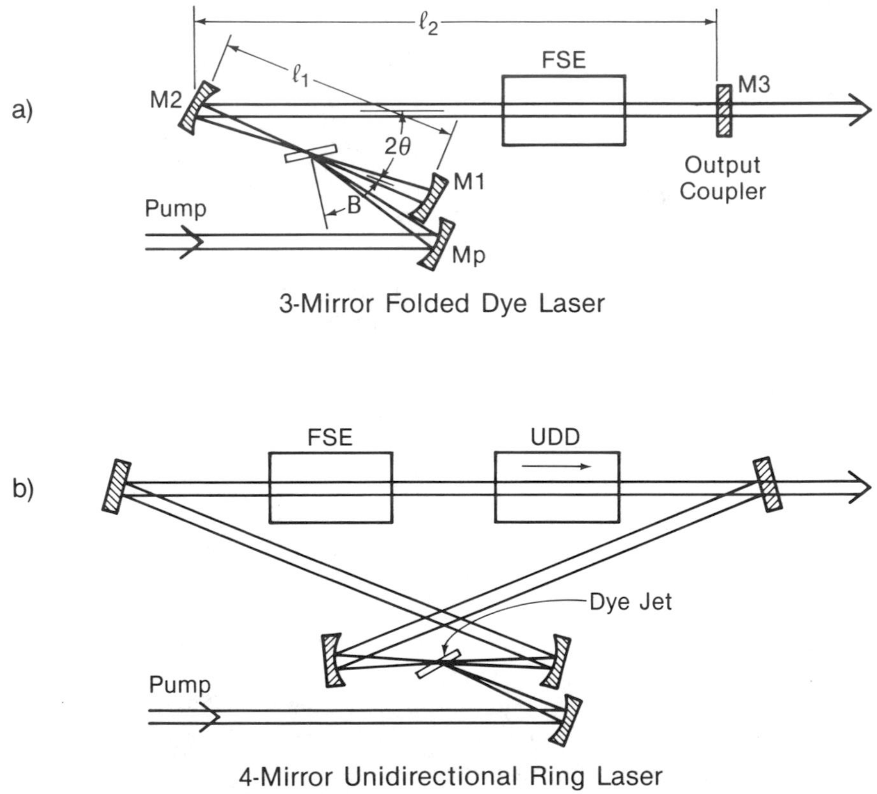

Fig. 5.2 Two popular cw dye-laser resonator configurations. The standing wave three-mirror resonator shown in (a) has a folded geometry with two short-radius curved mirrors (M1 and M2) and a flat output coupler, M3. The curved pump mirror, Mp, focuses the pump beam on the dye jet, which is placed between M1 and M2 at the waist of the dye-laser resonator. FSE represents any frequency-selective elements that might be used inside the laser's resonator. In (b), another flat mirror has been added and the resonator is transformed into a figure-8–shaped ring. This resonator also has a unidirectional device, which ensures that the laser oscillates as a traveling wave, propagating in only one direction around the ring.

a free-flowing jet stream for the dye gain medium and a tightly focused ion-laser beam for the pump source. The two mirrors with short radii of curvature, which surround the dye jet, match the dye-laser cavity mode to the small pump spot. This strongly pumped region of the dye provides a high gain over a very short interaction length. The high gain allows flexibility in designing the other parts of the dye-laser resonator. The remaining components of these two resonators are chosen for specific attributes or applications and many variations of these basic systems are found. The frequency-selective elements, FSE, and the unidirectional device, UDD, in the ring laser, indicated schematically in Fig. 5.2, are components that control the frequency and lasing direction, respectively, and will be discussed in detail in a later section.

Some of the early cw dye-laser designs (which have not survived the test of time) used glass cells to contain the gain medium. In almost all modern cw systems the dye cells have been replaced by free-flowing dye jets. The introduction of the dye jet (Runge and Rosenberg, 1972) and its subsequent development (for example, Harri *et al.*, 1982) has been a major contribution to the success of the cw dye laser. The jets do not have the problem of the pump laser burning the dye on the cell windows. Also, much higher output powers can be obtained with dye jets, rather than cells, because more rapid flow rates are possible. Dye-cell lasers could have some advantage in low-power applications because they might have less frequency noise than the high-power dye-jet lasers. But in general the problems outweigh the advantages of the dye cells.

2.2. CW Output Powers

In the beginning it is useful to have a rough idea of the performance characteristics that can be expected from a modern cw dye laser. Curves for laser output power as a function of laser wavelength are shown in Fig. 5.3. These outstanding results are obtained for single-frequency cw-laser oscillation using a number of dyes, pump-laser lines, and changes of laser optics. One of the important things to notice here is that it is possible to have reasonably high output powers and single-frequency operation anywhere in the visible and near-visible spectrum (in fact from about 365 to 1000 nm). The broad spectral coverage demonstrates the versatility of the dye laser. With different ion-laser pump sources and 11 different dyes it is possible to generate high-intensity, coherent radiation with powers of up to several watts. In exceptional cases output powers as high as 33 W cw have been obtained from dye lasers (Anliker *et al.*, 1977).

The portion of the spectrum covered by the cw dye laser extends for more than an octave. The requirement of a large number of dyes is usually

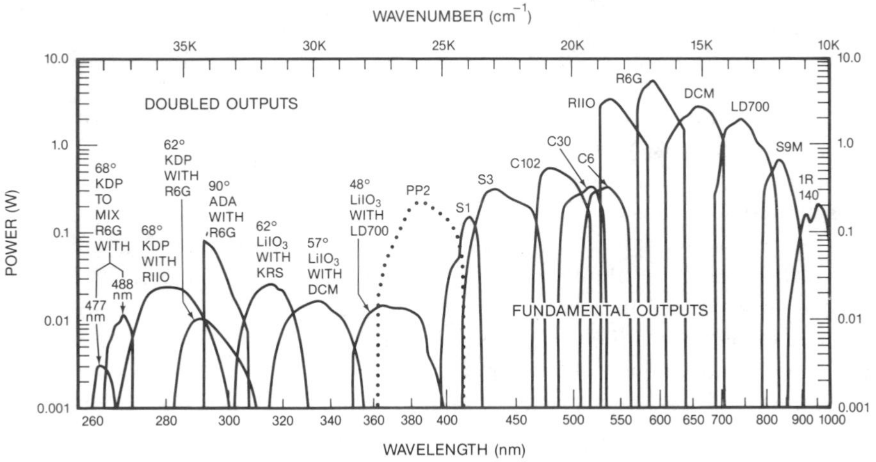

Fig. 5.3 Tuning curves of single-frequency dye-laser radiation, including frequency-doubled and sum frequency mixed outputs. The vertical axis is single-frequency power output displayed as a function of wavelength. The direct dye-laser output spans the spectral region from the blue end of the dye PP2 at ≃365 nm to the red end of the dye IR 140 at ≃1 μm. The outputs from the intracavity frequency doubled radiation extends from ≃395 to 270 nm. The two curves at the extreme blue and end show the single-frequency power obtained by intracavity sum-frequency-mixing the R6G dye laser radiation with argon-ion laser lines at 488 and 477 nm. These curves are taken with permission from Johnston, T. F. (1987) and have been modified slightly to include the results from Johnston, T. (1988) for the dye PP2 (polyphenl 2).

not a problem for most applications because spectroscopic studies usually have spectral windows that are easily covered by one dye. In fact, with the high resolution (approximately one MHz) of a single-frequency dye laser, there is often more spectroscopic resolution than is needed. Most of the UV outputs indicated in this figure (260–400 nm) are generated by using the strong visible dye-laser outputs and nonlinear optical-mixing methods.

2.3. Dye Characteristics

2.3.1. Absorption and Emission Spectra

The basic design of the cw dye laser is naturally constrained in fundamental ways by the photophysics and chemistry of the dye molecules. In Fig. 5.4 we see the representative absorption and emission spectra of a common laser dye. One of the important things to notice here is that the emission band is lower in frequency (red shifted) and is nearly the mirror image of the absorption band. In addition both the absorption and emission spectra are broadband features without sharp lines, indicative of a multitude of broadened mechanisms and overlapping energy levels. This is

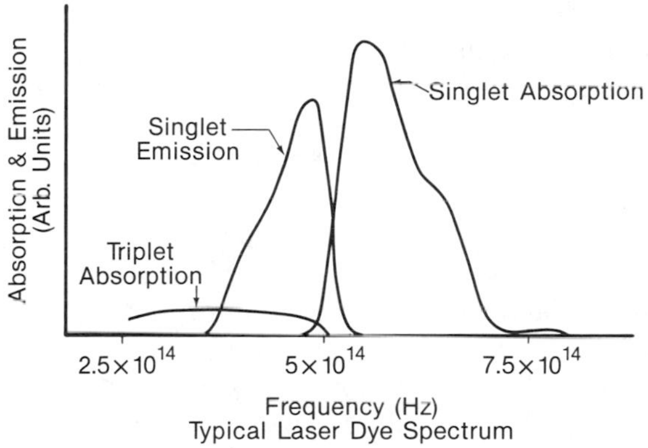

Fig. 5.4 Representation of typical absorption and emission spectra of a laser dye. The singlet emission spectrum is to the red and nearly the mirror image of the singlet absorption spectrum. Triplet absorption is smaller and typically extends even farther to the red than the singlet emission spectrum.

characteristic of large organic molecules in the liquid state and is depicted by a plausible energy-level diagram shown in Fig. 5.5. Here we see two manifolds of states, the singlets and the triplets, as designated by the net

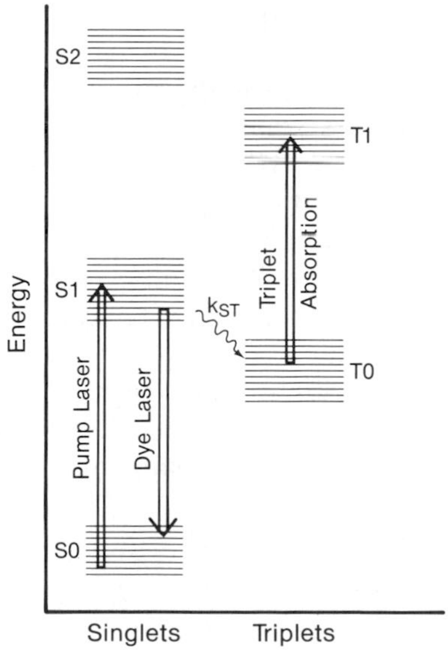

Fig. 5.5 Schematic picture of the energy levels of a generic laser dye, showing the singlet and triplet manifolds of states. The double lined arrows represent the optical redistribution of excitation by the pump laser, by the dye-laser emission (and dye fluorescence), and by triplet-state absorption. The arrow labeled k_{st} represents the transfer of excitation from the singlet to the triplet manifolds.

electronic spin. The transitions from the singlet manifold to the triplet manifold have low probability because they require an electron spin flip. The large organic-dye molecules have many internal degrees of freedom (vibration and rotation) that give rise to broad overlapping energy levels and spectra. The transfer of excitation within these energy bands can be very rapid, with typical time scales on the order of a picosecond. This energy transfer within a band is much faster than the spontaneous decay rates for interband transitions which have nanosecond time scales.

For laser oscillation the dye molecule's absorption band is excited by the intense pump laser from the lower part of the ground state, S0, to higher energy regions of the first excited singlet state S1 (Fig. 5.5). This excitation energy is then rapidly redistributed within the S1 state and reappears predominantly as fluorescence photons as the molecule decays by an electric-dipole transition from the S1 state to the S0 state. This transition in the singlet manifold from the first excited-electronic state S1 to the upper regions of the ground state S0 is the lasing transition. It is obviously advantageous to have efficient conversion of the pump photons into fluorescence photons and to avoid the loss of the pump energy to other channels. The primary loss mechanism (indicated on Fig. 5.5) is the transition from the S1 state in the singlet manifold to the T0 state in the triplet manifold. This excitation, which reaches the triplet states, is detrimental for two reasons. First, it is a loss of available energy from the lasing singlet system. Second, the population in the T1 state can absorb photons by making transitions to higher levels in the triplet manifold, thereby creating additional loss for the laser system. The characteristic time scale for the transfer of population from the S1 state to the T0 state is typically 100 ns. This needs to be compared with the time scales for other processes that affect the excited-state population. These are the fluorescence lifetime for the S1–S0 transition (typically a few nanoseconds) and the time scale for transfer of population within the S1 state (typically one ps). With the good laser dyes available today, the conversion efficiency, in terms of fluorescence photons (S1 decays to S0) for pump photons, is in the range of 0.8 to 1.0. Note that this efficiency is not the energy-conversion efficiency because of the difference in color between the pump and fluorescence photons.

2.3.2. *Dye-Laser Gain and Threshold*

Following the treatments of Shank (1975) and Johnston (1987) and making some simplifying assumptions that ignore the actual spatial distribution of the pump and the dye laser intensities we can write the small-signal gain of the laser as

$$\alpha_0 \ell = \frac{N \ell \beta_p I_p (\sigma_e - k_{st} \tau \sigma_t - \sigma_a)}{h\nu_p + \beta_p I_p (1 + k_{st} \tau)}. \tag{5.1}$$

Here, $\beta_p = \sigma_p \tau \phi$, h is Planck's constant, ν_p is the pump frequency, τ is the spontaneous emission lifetime and ϕ is fluorescence efficiency. This efficiency is the ratio of the fluorescence decay rate to the total decay rate for the S1 state and is usually near 1 for most good laser dyes. The quantities σ_p, σ_a, and σ_e are the singlet pumping, absorption, and emission cross sections, respectively. N is the density of dye molecules, ℓ is the dye-jet interaction length, and I_p is the intensity of the pump laser. The factor k_{st} is the rate of transfer of population from the singlet to the triplet manifold and multiplies the triplet absorption cross section σ_t. Above threshold the dye laser will show laser-intensity–dependent saturation as is usual for a homogeneously broadened system; that is,

$$\alpha\ell = \frac{\alpha_0 \ell}{1 + I_e/I_s}. \tag{5.2}$$

The saturation intensity I_s is a characteristic of the dye and is a function of the pump intensity and limits the gain as the dye laser intensity I_e increases.

For example, we can make the reasonable assumptions that the loss terms σ_a (which represents absorption of the dye-laser radiation by the dye itself) and k_{st} (the singlet-to-triplet transfer rate) are small enough to be ignored. Then Eq. (5.1) becomes

$$\alpha_0 \ell = \frac{N\ell \sigma_p I_p \tau \sigma_e}{h\nu_p + \sigma_p I_p \tau \phi}. \tag{5.3}$$

The pump intensity at threshold can be estimated from Eq. (5.3) by equating the gain to the resonator losses. We shall represent the total resonator loss as T, which is the sum of the output coupling transmission and the actual scattering and absorption losses. The threshold pump intensity is then

$$I_{p,th} = \left(\frac{h\nu_p}{\sigma_p \tau \phi}\right)\left(\frac{T}{N\ell\sigma_e - T}\right). \tag{5.4}$$

The optimum transmittance of the output coupler for most cw dye lasers ranges from 1 to 15% and is almost always the dominant loss term for the laser resonator.

By using typical values for the dye rhodamine 6G and the parameters in Eq. (5.4), we can estimate the minimum required pump intensity for laser oscillation. For rhodamine 6G we expect $T \simeq 15\%$, $\ell = .01$ cm, $N \simeq 10^{18}$ cm^{-3}, $\sigma_e = \sigma_p \simeq 1 \times 10^{-16}$ cm^2, $\nu_p = 5.8 \times 10^{14}$ Hz, $\tau \simeq 3$ ns, and $\phi \simeq 1$. This gives a rough value for the pump intensity at threshold of $I_{p,th} \simeq 2 \times 10^5$ W/cm^2. Such high intensities can be obtained continuously only at the focus of a strong pump laser. Actual operation of a cw dye laser will often be at pump intensities that are several times the threshold value.

In these simple considerations we have neglected many factors that strongly influence the gain and tuning characteristics of dye lasers. Two of the important omissions here are the spatial dependence of the pump and dye laser modes (Tuccio, 1971), and the wavelength dependence of the gain and loss cross sections. A more complete discussion of dye-laser gain can be found in Chapters 1, 3, and 7 as well as in articles by Peterson *et al.* (1971), Peterson (1979), Danielmeyer (1971), Snavely (1977), Shank (1975), and Kuhlke and Dietel (1977).

Two conclusions that we can glean from this calculation are that with practical pump powers the useful gain in laser dyes will be achieved in a very small active region and that the gain will be large. High gain is an asset in that it allows design flexibility because the laser is relatively insensitive to small amounts of extra loss. This, in part, is the reason that there are so many different dye-laser designs. Almost anything works if you have enough gain. Thus dye lasers can be designed with characteristics optimized for specific applications. Even though high pump powers are often needed to reach threshold with dye lasers, the high slope efficiencies that are possible result in laser-conversion efficiencies from pump power to dye-laser power that can be quite acceptable (even as high as 50% in some cases). However, at the highest pump powers the effects of gain saturation and dye heating limit the ultimate powers that are achieved.

2.3.3. *Broadening—Homogeneous versus Inhomogeneous*

The most important aspect of the gain of organic dyes is that it is spectrally broad. The processes that determine the spectral width are categorized according to whether they are homogeneous or inhomogeneous. Operationally there are important distinctions between the two classes of broadening. Homogeneous broadening means that all of the gain medium can contribute power to the oscillating laser mode. In contrast, for inhomogeneously broadened systems, only a fraction of the total gain is available for a specific oscillating mode. The physics of a specific broadening mechanism will determine whether the broadening is homogeneous or inhomogeneous. In the case of homogeneous broadening, all of the excited molecules (or atoms, or excited carriers, etc.) are effectively equivalent and can emit radiation at any frequency within the fluorescence bandwidth. In inhomogeneously broadened systems, on the other hand, the emission spectrum of a specific group of excited molecules is different from the emission spectrum of the system as a whole. When the broadening is inhomogeneous the molecules can be divided into groups that have distinct absorption and emission spectra. The Doppler shift is the best example of inhomogeneous broadening; atoms with different velocities have different emission spectra and hence contribute gain to a laser oscillator at different frequencies. On the other hand, the broadening due to collisions (for example, in a dense liquid) is homogeneous broadening.

The homogeneous broadening of the gain of organic dyes is both a blessing and a curse. It is a blessing in the sense that most of the available gain can be used for a single oscillation frequency and because the broad emission spectrum provides the laser's tunability. It is a curse because the broad spectral width means that the excited-state lifetime is short and hence intense pump powers are required in order to achieve sufficient population inversion for laser oscillation.

2.3.4. Dye Lifetimes

An annoying characteristic of organic dyes is that the dyes have limited productive lifetimes. (Unfortunately this is also the case for laser scientists!) The factors that limit the lifetime of laser dyes are thought to be the chemical and photochemical degradation of the dye in solution. The lifetime of the gain of a dye is often specified in terms of watt-hours, based on empirical data. This power-lifetime product is a measure of the pump-laser energy that has been used to excite the dye. Thus the degradation must be at least in part due to thermally activated chemistry (resulting from the pump-laser heating of the dye) and/or actual laser-induced photochemistry in the dye solution. The chemistry of the dyes can be quite complicated as is evidenced by the fact that the lifetime of the dyes' gain can even be affected by the type of metal plumbing components used in the dye-circulating system. The lifetimes of the typical cw dyes range from ≃75 W · hr (coumarin 480) to several 1000 W · hr (Rh6G). Obviously the predicted lifetimes are estimates of typical performance and cannot be depended upon without further specification of the conditions. For example, the lifetimes usually assume a typical dye-pumping system with approximately one liter of dye solution.

2.4. Pump Sources

We require strong, broad fluorescence from the dye for broadly tunable cw laser operation. This in turn requires an intense optical-pump source tuned to some part of the absorption band of the dye. Incoherent optical pumping of a cw dye laser to threshold has been demonstrated by Thiel *et al.* (1987) using high-pressure arc lamps; but in order to have sufficient intensity for a practical system, the pump sources are almost always cw ion-lasers. In practice, we are presently limited to the argon and krypton ion lasers. The obvious reason is that they are the only cw lasers that can produce high enough power (>4 W, say) in a good single-spatial mode and in the visible or UV region of the spectrum. Recall that the dye's emission band is to

Table 5–1

Powers Available from High-Power Ion Lasers

	Wavelength (nm)	Power (W)
Argon ion	528.7	1.5
	514.5	10.0
	501.7	1.5
	496.5	2.5
	488.0	7.0
	476.5	2.8
	472.7	1.2
	465.8	0.75
	457.9	1.4
	454.5	1.0
	Multiline visible	18–25
	Multiline UV	5.0
	351.1–385.8	3.0
	333.6–363.8	5.0
	275.4–305.5	0.6
Krypton ion	Multiline	—
	752.5–799.3	1.6
	647.1–676.4	4.6
	520.8–568.2	3.6
	468.0–530.9	2.5
	406.7–422.6	1.3
	337.5–365.4	2.0

the red of the absorption band; this means that for visible dye lasers we need strong blue and UV pump lasers. As of 1989, we are stuck with the very inefficient high-power ion-lasers for dye-laser pumping. Some of the strongest lines that are available from high-power ion-lasers are listed in Table 5-1. To be useful as a laser dye, a compounds-absorption band must overlap one of these strong ion laser lines. Other pump sources are certainly possible. For example, dye lasers can be pumped with a HeNe laser at 633 nm (Runge and Rosenberg 1972; Thiel *et al.*, 1986) and also the future holds promise for pumping with high-power green light (532 nm) obtained from frequency doubling the output of cw YAG lasers. There has also been some progress in designing dye lasers that could be pumped with concentrated solar radiation (Lee *et al.*, 1988).

2.5. Resonator Design

A good way to understand the performance characteristics of modern cw dye lasers is to study an example laser that is typical of modern designs.

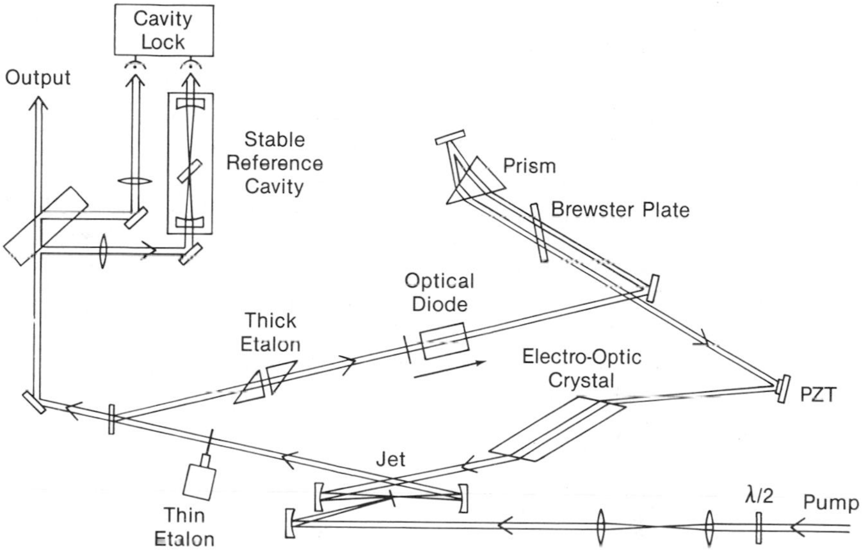

Fig. 5.6 Diagram of a frequency-stabilized ring dye laser. The basic laser resonator is formed with four flat mirrors and two short radius curved mirrors (≃5 cm). The two curved mirrors form a tight waist that overlaps the focused pump spot in the dye jet. The optical diode is made from a Faraday rotator and a thin quartz plate. It forces the ring laser to oscillate in one direction as a traveling wave. The hierarchy of frequency-selective elements are the prism (double passed for added dispersion) the thin etalon (driven with a galvo motor) and the air-spaced thick etalon (constructed from two Brewster prisms and scanned with a PZT, piezo-electric ceramic). The frequency of the laser is lock to the resonance of a stable reference cavity using the cavity side lock method. The resonance frequency of the stable reference cavity is scanned using an intracavity Brewster plate that is rotated by a galvo motor. To stabilize the laser's frequency the cavity lock feeds back to the galvo-driven Brewster plate, the PZT translated mirror, and the intracavity electro-optic crystal. This laser system can scan continuously for about 80 Ghz and has a linewidth of about one kHz. Figure reprinted with permission from Hollberg (1984).

The goal of this reverse engineering is to understand how and why dye lasers have evolved to their present state. We choose here a laser design that has very good performance and is in regular use today.

Figure 5.6 shows a schematic diagram of a ring dye laser based on a six-mirror ring-cavity resonator. The two mirrors close to the dye jet have short radii of curvature (≃5 cm) in order to produce the small laser-cavity mode that is necessary in the dye jet. The remaining mirrors are flat. They close the optical cavity in a manner that also provides space within the resonator for other optical elements. The organic dye is dissolved in a liquid solvent (for example, ethylene glycol) and squirted as a small ribbon of liquid dye into the space between the two short-radius mirrors. The pump radiation from ion laser is focused by the pump mirror to a small spot size ($w_o \simeq 15\ \mu m$) on the flowing jet of dye. Because of the heat produced in the dye by the focused pump laser and because of triplet-state trapping, it is necessary to have the dye molecules traverse the pump spot very rapidly, where "rapidly" in this case is determined by the heating rate and by the rate of transfer of excitation to the triplet state (typically about 1×10^{-7} s). To avoid significant loss into the triplet manifold the dye fluid needs to traverse the pumping region at a velocity of about 10 m/s. Usually the dye jet is a nearly rectangular ribbon with a cross section of 0.1 by 3 mm and is oriented at Brewster's angle (to minimize reflection loss) relative to the laser-cavity mode and near Brewster's angle to the pump beam. The laser mode passes through the thin dimension of the dye jet. The concentration of the organic dye in the transporting fluid is chosen to be high enough that the dye absorbs about 85% of incident pump radiation.

2.5.1. Spatial Hole Burning

The first question we need to address about the design of the laser cavity is why have we abandoned the linear three-mirror cavity of Fig. 5.2 in favor of the ring resonator of Fig. 5.6. The problem with the linear cavity is spatial hole burning. This is easily understood by thinking about the longitudinal mode structure of a linear dye-laser cavity. The dye laser's oscillating mode traverses the dye jet twice, in opposite directions; this sets up a standing wave field in the gain medium. The standing wave is a strong periodic spatial modulation of the laser intensity across the gain medium. Because of the intensity-dependence of stimulated emission, a single-frequency laser mode is only able to use the gain that is available where the field strength is high. But we recall that the pumping field is a traveling wave that provides gain throughout the jet. Thus gain at the nodes of the standing wave cannot be used by the single-mode oscillatory field. The laser can, and will, use this available gain by oscillating simultaneously at a different frequency so that the nodes of the new oscillating mode overlap the antinodes of the original mode. The effect of spatial hole burning is to cause the linear standing-wave cavity to oscillate with more than one frequency (longitudinal mode).

2.5.2. Unidirectional Ring Laser

The problem of spatial hole burning can be alleviated by designing a laser resonator that oscillates in a traveling wave rather than a standing wave. Both the effect of spatial hole burning and its solution have been known for quite some time (Tang *et al.*, 1963; Danielmeyer, 1971; Pike, 1974; Green *et al.*, 1973; Kuhlke and Dietel, 1977), but it was not until about 1977 (Schröder *et al.*, 1977) that the travelling-wave ring dye lasers

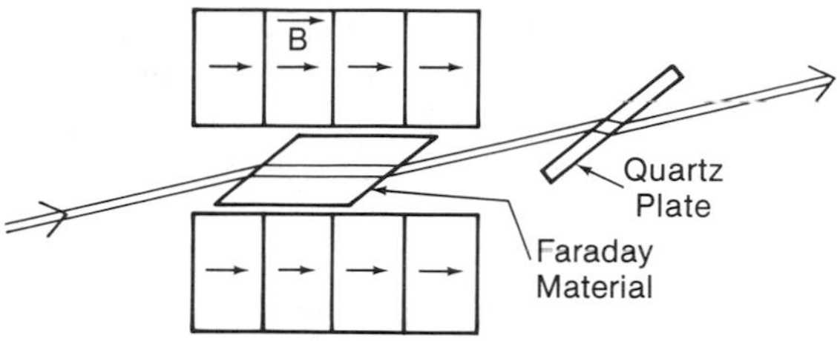

Fig. 5.7 Cross-section diagram of a unidirectional device (optical diode). A stack of permanent magnets creates a strong magnetic field (a few kG) within its central bore which contains Faraday material. The quartz "rotate-back" plate uses optical activity to rotate the polarization of the laser field. Thus, the laser-propagation direction inside the quartz plate must be along the z-axis of the crystal. The configuration shown here is designed for minimum loss with both the Faraday material and the quartz plate oriented at Brewster's angle relative to the laser-propagation direction. For the beam propagating in the direction shown, the net rotation of the polarization is 0°, whereas the other direction has a net rotation of about 4°.

came into prominence (Jarrett and Young, 1979; Johnston *et al.*, 1982). An interesting discussion of using prisms to create ring resonators for dye lasers is given by Marowsky and Zaraga (1974).

To avoid lasing in both directions, the ring laser's cavity must have more loss for one of the two propagation directions. This increased loss for one direction is provided by a nonreciprocal loss element, depicted in Fig. 5.6 as an optical diode. This optical diode (or unidirectional device) provides the directionally specific loss by using the axial vector nature of magnetic fields. A typical unidirectional device is diagrammed in Fig. 5.7. The device works by rotating the polarization of the propagating laser mode for only one of the two propagation directions. The rotated polarization encounters extra loss in the remaining Brewster-angle elements of the ring cavity. The rotation is made nonreciprocal by the Faraday effect of a crystal or glass in a magnetic field (applied parallel to the direction of propagation of the laser mode). The direction of polarization rotation due to the Faraday effect is independent of the direction of propagation of the light through the material. This Faraday rotation can then be combined with a reciprocal polarization rotation to produce an optical diode that rotates the direction of polarization for a beam from one direction but has no net rotation for a beam from the other direction. The reciprocal rotation is easily obtained from the optical activity in a thin quartz plate. A small differential loss of about $\simeq 0.5\%$ is sufficient for unidirectional operation of most ring dye lasers. This loss is achieved from a one-way polarization rotation of about 4° in a laser cavity that has six Brewster surfaces or more. The design criteria for unidirectional devices are discussed by Biraben (1979) and by Johnston and Proffitt (1980).

Unidirectional ring lasers with traveling wave fields have advantages over standing-wave lasers because of more efficient use of the available gain and better spectral characteristics. On the other hand, with ring lasers there is reduced gain per pass because the radiation field only traverses the gain medium once for each incidence on the output coupler. The net result is that with sufficient pump power the traveling-wave ring dye lasers can provide almost twice the single-mode output power of the standing-wave lasers. For this extra power and mode stability ring lasers have the disadvantage of slightly higher cavity loss because of the extra optical components. This means that the thresholds are typically higher for ring cavities. Some of the very low gain dyes actually work better in linear cavities because of their lower threshold.

From the gain arguments just given and knowledge of the pump powers that are available, we know that the gain medium will occupy a small volume (about 10^{-7} cm^3). To optimally use the available gain the dye laser's spatial mode is approximately matched to the pump-laser spatial mode in the dye medium. "Approximately matched" here indicates that the optimal size of the pump and dye laser modes is actually a function of the saturation of the gain medium and is thus a weak function of the pump power (Snavely, 1977; Johnston, 1987). In practical cases, where the laser operates well above threshold, this dependence is weak and is usually not worth the trouble of readjusting the laser mode when the pump power is changed. We usually have a pump-laser spot size of about $w_0 = 15$ μm, which is consistent with the available pump powers and the threshold intensity of the dye. We can then use Gaussian mode analysis to calculate the transverse mode structure for a laser resonator (Yariv, 1975; Siegman, 1986). This analysis can give us the information that is necessary to select the mirror spacings, orientations, and radii of curvature that will approximately match the dye-laser mode to the gain region.

2.5.3. Resonator Aberrations

Up to now we have naively assumed that the natural fundamental transverse mode structure of the dye laser is the round TEM_{00} mode. But in fact the orientation of the curved mirrors in the laser cavity relative to the incidence angle of the laser beam alters the transverse structure of the TEM_{00} mode. Having a laser beam incident on spherical mirrors at an angle other than normal incidence causes the beam to be aberrated by astigmatism and coma. The dye jet is also a source of astigmatism and coma because it intercepts the strongly focused Gaussian laser mode at an angle. These aberrations distort the laser mode from the desired TEM_{00} mode, and this degrades the resonator's stability and reduces the laser's conversion efficiency. In most modern designs the astigmatism is compensated by the appropriate choice of angles of the resonator's mode at the curved mirrors or by the addition of other optical elements that introduce an astigmatism of the opposite sign. It is even possible to simultaneously compensate for both astigmatism and coma. Details of the compensation methods can be found in Kogelnik *et al.* (1972), Johnston and Runge (1972), and Dunn and Ferguson (1977).

Using the geometry shown in Fig. 5.2a, we can follow the analysis of Kogelnik *et al.* (1972), and Mollenauer and White (1987) to calculate the astigmatism introduced by the two curved mirrors and the dye jet. The dye jet with thickness t is located between the two short-radius curved mirrors (M1, M2) and is tilted at Brewster's angle B relative to the beam propagation direction. In this geometry there are two sources of astigmatism; the first is that due to the dye jet crossing the laser mode at an angle and the second due to the curved mirror M2 that is tilted at the angle θ relative to the laser-mode—propagation direction. Fortunately, the sign of the astigmatism is opposite for these two sources, and the angles can be chosen so that the mirror astigmatism cancels that of the jet. The condition for zero astigmatism is

$$f(\sin\theta)(\tan\theta) = t(n^2 - 1)\frac{\sqrt{n^2+1}}{n^4}, \tag{5.5}$$

where f is the focal length of mirror M2 and n is the index of refraction of the gain medium (Kogelnik *et al.*, 1972). Since the dye jets are quite thin ($\approx$0.1 mm) and the mirrors have radii of about 5 cm, the angle θ must be small. These small angles are not always practical, so other methods have been used to compensate the astigmatism. These includes lenses or additional curved mirrors that are titled off normal incidence, a Brewster plate between the dye jet and the short-radius mirrors, and rhombi with Brewster-cut surfaces that are placed in an auxiliary beam waist in the laser cavity.

3. EXTENDED TUNING

3.1. UV Generation

Nonlinear optical methods can be used with dye lasers to extend the already broad tuning range of cw dye lasers. If we look back to Fig. 5.3 we see the UV-tuning curves that were obtained by T. F. Johnston (1987) and T. Johnston (1988) using a high-power single-mode cw dye laser. This cw single-frequency UV output is generated either by frequency doubling the output of the dye laser or by sum-frequency mixing the output of the dye laser with the output of a single-frequency argon-ion laser. Using these methods, they have generated single-frequency cw UV radiation across the frequency range between 260 and 400 nm. The excellent UV powers obtained in these experiments were typically about 10 mW. Both the doubling and the frequency mixing were done inside the resonator of a ring dye laser. For the sum-frequency mixing experiments the two laser resonators were actually arranged to overlap at the nonlinear crystal. Thus the mixing occurred with buildup on both the input frequencies. Other systems have been developed to generate cw ultraviolet light in this frequency range and beyond (Frölich *et al.*, 1976; Blit *et al.*, 1977, 1978; Couillaud, 1981; Couillaud *et al.*, 1982; Marshall *et al.*, 1980; Bergquist *et al.*, 1982; Hemmati *et al.*, 1983). The transparency of the nonlinear crystals limits cw UV generation to wavelengths longer than approximately 160 nm. That is not the case for UV generation in atomic vapors. Four-wave mixing has been used in strontium vapor to generate $\simeq$10 pW of broadband ($\simeq$6 GHz) cw UV radiation at 170 nm (Freeman *et al.*, 1978).

Noteworthy among these efforts to generate single-frequency cw UV radiation is the work of Bergquist and collaborators(Bergquist *et al.*, 1982; Hemmati *et al.*, 1983), who have combined the second harmonic of argon-ion laser radiation (fundamental = 514 nm) with tunable dye-laser radiation ($\simeq$792 nm) to generate UV radiation at 194 nm. Other colors could certainly be generated by these methods but the 194-nm radiation is of particular interest for high-resolution spectroscopy of Hg ions. Their system is diagrammed in Fig. 5.8. It serves as good example of the techniques and the 1980s state of the art for synthesizing single-frequency cw, UV radiation. Their system uses the nonlinear crystal KB5 (potassium pentaborate) to generate the sum frequency of 792- and 257-nm radiation. The 792 nm radiation comes from a cw ring dye laser (LD700 dye), and the 257 nm radiation is from the frequency-doubled 514 nm argon laser. One of the techniques they use to advantage is putting the nonlinear crystals in external ring buildup cavities. This enhances the optical power in the nonlinear crystals and avoids the problem of the crystal's loss in the laser resonator. A ring cavity containing a Brewster-cut ADP (ammonium dihydrogen phosphate) crystal is locked to and enhances the power of the 514 nm radiation by a factor of $\simeq$33. Thus 700 mW of 514 nm radiation generates about 30–40 mW of 257 nm radiation. Then, as shown in Fig. 5.8, they use two buildup cavities around the KB5 summing crystal; one ring cavity builds up the 792-nm radiation and the second ring resonator builds up the 257-nm radiation. Both these cavities have buildup factors of $\simeq$15. With input powers of $\simeq$120 mW at 792 nm and $\simeq$30 mW at 257 nm they obtain $\simeq$20 μW of cw single-frequency radiation at 194 nm. The output power at 194 nm is not limited by the availability of input powers

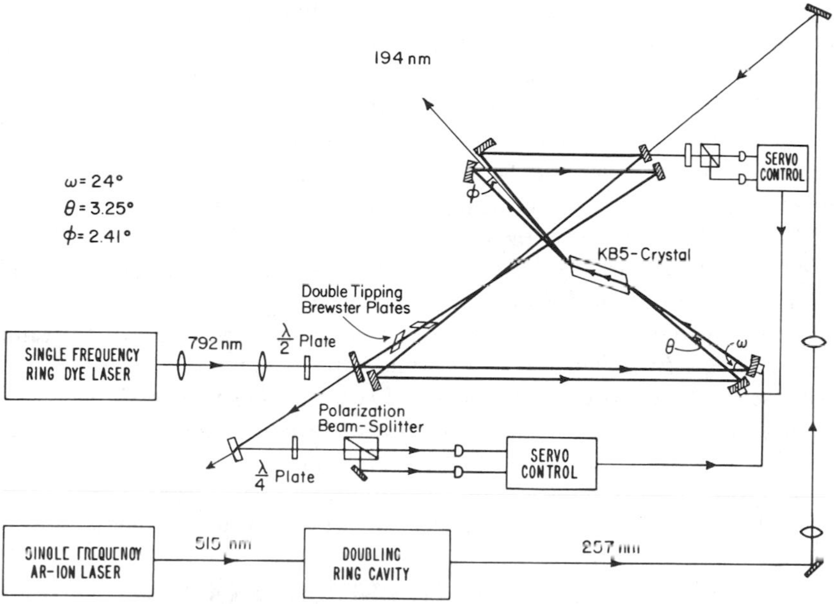

Fig. 5.8 Sum frequency generation of 194-nm radiation. This system, developed by Hemmati *et al.* (1983), and Bergquist (1989) uses the nonlinearity of the KB5 crystal to generate the sum frequency of 257- and 792-nm radiation. The power in both of these input beams is enhanced in ring buildup cavities whose waists intersect in the KB5 crystal. Servo-control loops (using the polarization lock method) keep the two buildup cavities locked to their respective input beams. The 194-nm radiation leaves the KB5 crystal at a slightly different angle and thus escapes from the two buildup cavities. The 792-nm input radiation comes from a frequency-stabilized ring dye laser, while the 257-nm is generated by frequency doubling the 514-nm radiation from a single-frequency argon-ion laser. The frequency doubling of the argon-ion laser radiation is done in a ring buildup cavity that is similar to the two used around the KB5 crystal shown here. With input powers of $\simeq$30 mW at 257 nm and 120 mW at 792 nm, this system generates up to 20 μW of single-frequency radiation at 194 nm. Unpublished figure reprinted with permission from Bergquist (1989).

but rather by the effects of thermal blooming due to nonuniform optical heating of the KB5 crystal. Another important technique of nonlinear optics that is used in this system is electronic servo-control systems to maintain the resonance condition for the buildup cavities (see section 4 on laser frequency control).

3.2. IR Generation by Difference-Frequency Mixing

The useful tuning range of dye lasers can be similarly extended into the infrared by difference-frequency mixing (Pine, 1974, 1976; Oka, 1988). The technique is analogous to the sum-frequency generation, but now the crystal's nonlinearity is used to generate the difference frequency between an ion-laser beam and a tunable cw dye laser. Using $LiNbO_3$ as the nonlinear crystal Pine (1974, 1976) has demonstrated that one can produce narrowband ($\simeq$15-MHz linewidth) radiation with wavelengths in the range of 2.2 to 4.2 microns and with power levels of a few microwatts. The tuning range can be extended by using other nonlinear crystals such as lithium iodate (Oka, 1988). Somewhat higher power levels can also be achieved.

One problem with describing the late 1980s state of the art is that there is a tendency to report the best performances rather than what is typically achieved. The output-power curves shown in Fig. 5.3 are an example of what is possible with an optimized dye laser with very powerful pump lasers. The performance that you should expect from a typical commercial dye laser system is probably a factor of two less power than shown there. Nevertheless the record performances are always interesting, because research continues to expand our possibilities and the best results today are indicative of the tools of the future. In addition, the extreme limits are often dominated by new and interesting science waiting to be explored.

4. LASER-FREQUENCY CONTROL

High-resolution dye lasers are common laboratory tools today because of early research efforts in laser frequency control. The broad bandwidth of the gain medium, combined with the mechanical instabilities of the rapidly flowing dye jet, conspire to make frequency selection and stabilization a challenge. Nonetheless, the technology required to control the frequency of dye lasers has been developed. Many years ago, Soffer and McFarland (1967) demonstrated that the wide bandwidth of pulsed dye lasers could be compressed to a relatively narrow bandwidth by inserting an optical grating inside the cavity. In addition to reducing the laser's linewidth, the frequency selectivity of the grating allowed the laser's wavelength to be tuned. This demonstration of spectral compression, as well as more recent techniques, takes advantage of the homogeneous broadening of the dye laser gain to produce a narrower spectral linewidth without sacrificing output power.

4.1. Frequency-Selective Elements

The laser's oscillation frequency can also be controlled by providing frequency-selective loss inside the resonator. To achieve this frequency-selective loss, we have freedom to control both the spatial and polarization boundary conditions of the resonator. Dispersive optical elements that create boundary conditions for the oscillating mode provide the appropriate frequency-selective loss. The dispersion can come from interferometry, which is solely wavelength-selective, or it can be provided by polarization-sensitive elements which act on the vector nature of the optical field. Some of the dispersive elements that have been used to control the frequency of cw dye lasers include gratings, prisms, birefringent filters, and Fabry-Perot etalons, as well as Michelson, Fox-Smith, and Mach-Zehnder interferometers (Smith, 1972).

Usually the addition of one frequency-selective element into a dye laser's cavity is not sufficient to uniquely determine the laser's oscillation frequency. The conventional technique to achieve mode-stable single-frequency operation is to use a hierarchy of frequency-selective elements inside the laser's cavity. Most single-mode lasers use three separate levels of frequency selectivity to force the laser to oscillate on a specific longitudinal mode of the laser's cavity. Lacking this hierarchy of frequency selectivity, the laser may oscillate with many modes, or it may oscillate with a single mode that is unstable to small perturbations and jump from mode to mode as a function of time.

Due to the homogeneous broadening of the gain, once the laser starts to oscillate on a specific mode, this mode quickly depletes the gain that is otherwise available for the other longitudinal modes. In a traveling-wave ring dye laser the homogeneous broadening causes strong mode competition, which is usually enough to ensure that the laser will oscillate on a single longitudinal mode. Unfortunately, without frequency selectivity in the cavity, the oscillating mode is unstable to small perturbations and will change in time. Thus even though the laser oscillates on a single mode, additional frequency-selective elements inside the cavity are necessary to select a particular frequency.

A typical hierarchy of frequency-selective elements might start on the least selective end with a birefringent filter, which typically has a half-power bandwidth in the laser resonator of about 1800 GHz (60 cm^{-1}). This is usually followed by two etalons, with progressively higher resolution; for example, a thin etalon with a free spectral range (FSR) of 200 GHz and a finesse of $\simeq 3$, then a thick etalon with a FSR of 10 GHz and again a finesse of $\simeq 3$. The overlap of the transmittance peaks of the three frequency-selective elements provides a net transmittance bandwidth that is narrow enough to uniquely determine a single mode for laser oscillation. The role of the various frequency-selective elements can be seen in Fig. 5.9, which shows the transmission functions of these elements overlaid with the laser longitudinal mode structure. The resolving powers of the three levels of frequency-selectivity are chosen so that the resolution of one level is sufficient to select a unique resolution element of the next higher level of selectivity. Thus, the birefringent filter has enough resolution to select a

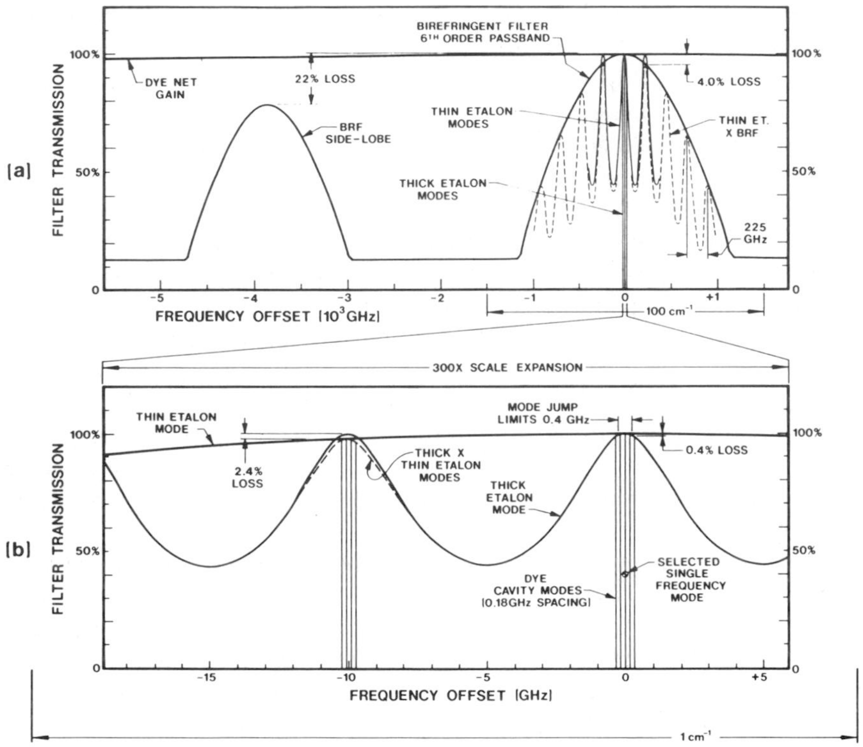

Fig. 5.9 Filter transmission functions of the various frequency-selective elements of a single-frequency dye laser. The traces show the bandpasses of the frequency-selective elements relative to the dye gain curve and the laser's cavity-mode structure. The upper half of the figure has a broad spectral range and shows the birefringent filter and thin-etalon transmission functions. The lower half of the figure shows the thick etalon's transmission function relative to the longitudinal mode spacing (c/L with L = cavity round-trip length) of the laser cavity. Also indicated are the losses created for radiation away from the central transmission peaks. This figure is reprinted with permission from Coherent Inc. and Johnston, T. F. (1987).

specific thin-etalon mode, the thin-etalon selects a specific thick-etalon mode, and the thick etalon has enough resolution to select a specific longitudinal mode of the laser cavity. For the system to work in an optimum way, the bandpasses of these filters need to be at least roughly centered on a specific cavity mode. Figure 5.6 is an example of a single-frequency dye laser with a slightly different hierarchy of frequency-selective elements, in this case a prism and two etalons.

The frequency stability of such single-frequency lasers is now determined by the mechanical stability of the optical elements, the gain medium, and the resonator's superstructure. We shall see in the spectral characteristics section (5.2) that "single frequency" is a bit optimistic, in that the spectral linewidth of single-mode lasers can still be substantial.

The birefringent filter is the most common coarse tuning element and is composed of quartz plates placed in the laser cavity inclined at Brewster's angle relative to the direction of propagation (Bloom, 1974; Bonarev and Kobtsev, 1986). The thickness of each plate in the filter is integrally related to the others (for example, 1:4:16 for a common three-plate birefringent filter). The quartz plates are cut with the optic axis parallel to the surface of the plates. The design of the birefringent filter for a laser follows the traditional Lyot filter design (Jenkins and White, 1957), with the important difference that the polarizers between the birefringent elements have been removed. The role of the polarizers is played by the Brewster-angle surfaces of the plates themselves and other Brewster surfaces in the cavity. The optical frequency selectivity of this filter results from the birefringence of the quartz waveplates combined with the polarization-dependent reflection loss from the Brewster surfaces of the plates. The plates are mounted together in the laser cavity so that they can be rotated about the normal to the plates, thereby changing the angle of the crystal axis with respect to the laser polarization (parallel to the plane of incidence) while maintaining the Brewster-angle orientation of the quartz surface relative to the laser mode propagation direction. For a given plate thickness and orientation of the optic axis, the plate will act as a wavelength-dependent waveplate, which in general changes the polarization character of the input laser mode. At some wavelength the polarization of the light that exits each plate will be the same linear polarization (p-polarization) that entered and, thus, will pass with no loss through the subsequent Brewster surfaces. In general the polarization of the beam will elliptical upon exiting each of the plates and will therefore experience reflection loss at the Brewster surfaces that it encounters in traversing the rest of the laser cavity. Birefringent plates with integrally related thicknesses can provide enough selectivity that lasing will be restricted to one color (a few cavity modes) within the dye gain curve. It is interesting to note the frequency selectivity of the birefringent filter, in the laser cavity, is higher than that of a traditional Lyot filter with polarizers. The net spectral width of a cw dye laser with only a three-plate birefringent filter can vary significantly depending on alignment; widths can vary from 1 to more than 100 GHz (Nieuwesteeg *et al.*, 1986). The loss provided by a single Brewster surface for a beam incident with a polarization that is orthogonal to Brewster's polarization is

$$\text{loss} = 1 - \left(\frac{2n}{n^2 + 1}\right)^2. \tag{5.6}$$

Here n represents the index of refraction of the Brewster plate, and the loss is the fractional reduction in power of the transmitted beam. A net gain reduction of less than 1% is usually sufficient to suppress other modes from lasing. A typical birefringent filter will have a free spectral range of $\simeq$100 nm. A dye laser with a three-plate birefringent filter has at least eight Brewster surfaces (including the jet surfaces) that can provide a net loss of $\simeq$4% for a 200-GHz detuning (this corresponds to the FSR of the thin etalon, see Fig. 5.9).

Usually etalons provide the next two higher levels of frequency selectivity. These parallel plate interferometers have the standard Fabry-Perot transmittance (Born and Wolf, 1964),

$$\mathrm{I}(\delta) = \frac{I_0}{1 + F\sin^2\left(\dfrac{\delta}{2}\right)} \tag{5.7}$$

where $F = \dfrac{4R}{(1-R)^2}$ and $\delta = \dfrac{4\pi nL}{\lambda}\cos\theta$.

Here R represents the power reflectance of the etalon's surfaces (assumed equal), n its index of refraction, L its thickness, and θ the angle between the etalon normal and the optical propagation direction inside the etalon. λ is the vacuum wavelength. A useful measure of the etalon's resolving power is its finesse, $\mathscr{F}$, which is given by,

$$\mathscr{F} = \pi\frac{\sqrt{F}}{2}. \tag{5.8}$$

A cw ring dye laser might have one thin etalon ($L = 200\ \mu$m with $R = 30\%$) and one thick etalon ($L = 1$ cm and $R = 30\%$), which, when combined with the birefringent filter, provide sufficient frequency-selectivity to force the laser to oscillate on a single longitudinal mode.

The frequency-selective components that we consider here are typical of single-frequency dye lasers used in 1989, but other frequency-control systems have also been developed. As noted earlier, the high gain of the dye laser allows some experimental flexibility in testing ideas about laser design. The system with a birefringent filter and two etalons is the most popular because of its dependability and relative ease of construction. That is not to imply that this system is optimum. The tuning elements that are alternatives to the birefringent filter are prisms and the tuning wedge that is now rarely used. A variety of interferometric methods have also been used to control the frequency of dye lasers. These include the Michelson, Fox-Smith, and Mach-Zehnder interferometers and multiple implementations of these. Interferometers generally compete for the higher selectivity roles normally filled by the etalons.

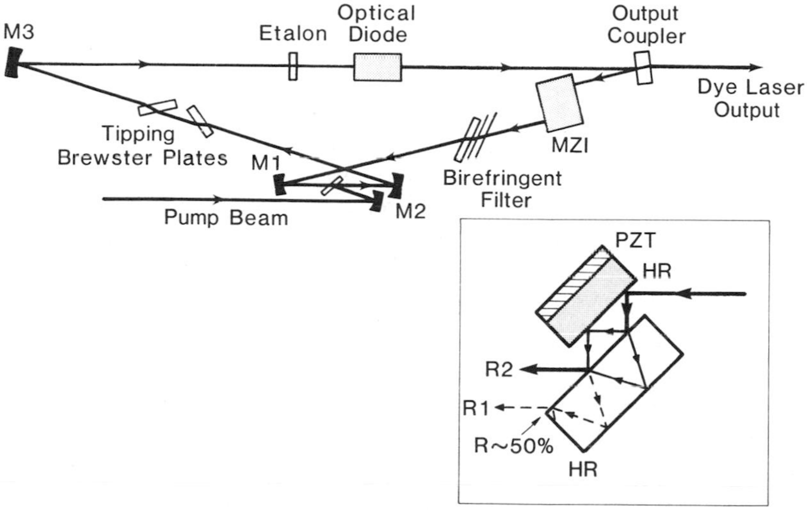

Fig. 5.10 Diagram of a single-frequency ring dye laser that uses a Mach-Zehnder interferometer for longitudinal mode control. This laser also uses a birefringent filter and a thin etalon for frequency selectivity. An optical diode is used for unidirectional operation and tipping Brewster plates are used for frequency scanning. The inset shows the structure of one type of intracavity Mach-Zennder. This interferometer is constructed from two high-quality optical flats. One piece has a 50% reflectivity coating on its front surface and a high reflectivity coating (HR) on the back surface. The partial reflecting surface splits the input beam into two paths through the interferometer. The other optical flat is a high-reflectivity mirror and is mounted on a PZT translator so that the passband of the interferometer can be changed. The Mach-Zehnder has two output ports (R1 and R2). If the reflectivity of the partial reflector is 50%, and the losses in the two paths through the interferometer are matched, the contrast in the output ports can be 100%. With good low-loss coatings, the net insertion loss of this device in the laser cavity can be very small. Figure reprinted with permission from Bergquist and Burkins (1984).

One of the most intriguing of the interferometric frequency selectors for use with dye lasers is the Mach-Zehnder system that was proposed and developed by Bergquist and Burkins (1984). An example of one of the several possible implementations is diagrammed in Fig. 5.10. The advantage of the Mach-Zehnder interferometer is that, in principle, it can provide the necessary frequency-selectivity with substantially less excess loss than that of a comparable etalon. The reason for the unavoidable loss with intracavity etalons is that it is necessary to tilt the etalons away from normal incidence to avoid the disruptive effects of optical feedback on the oscillating mode of the laser. Tilting the etalon eliminates the feedback but we pay the penalty that power is lost from the laser mode. In laser jargon, this "etalon walkoff loss" results from the multiple reflections within the tilted etalon that bounce laser power out of the spatial mode of the laser cavity. A properly balanced Mach-Zehnder system does not have this limitation because the only additional optical output port (beam R1 in Fig. 5.10) has zero output under proper operating conditions. This occurs when the Mach-Zehnder interferometer is set for a interference maximum on beam R2 (which corresponds to a minimum for beam R1) and when the interference maximum is tuned to a longitudinal mode of the laser cavity. In this case the only excess loss introduced by the Mach-Zehnder to the laser is that due to the absorption and scattering from the optical coatings, which can be made negligibly small. In practice the Mach-Zehnder system has seen only limited application but appears to have great potential.

The most important attribute of dye lasers is their tunability. In addition to reaching a particular wavelength, we frequently want to scan the frequency of the laser over some region near this wavelength. Depending on the spectral resolution that is required, it may not be necessary for the laser to oscillate on a single longitudinal mode. For example, the spectral features observed in solid-state spectroscopy are relatively broad, which often means that the resolution provided by multimode dye lasers is adequate. In this case a dye laser with only a birefringent filter provides enough resolution, and the laser can be scanned simply by tuning the bandpass of the birefringent filter. On the other hand, the technology required to tune and continuously scan the frequency of single-mode (single-frequency) dye lasers is complicated by the many frequency-selective elements. It is necessary to individually center and then scan all of these elements synchronously. Such scanning systems have been developed and some even actively optimize the centering of the bandpasses of the frequency-selective elements. The actively controlled systems provide long-term mode stability and repeatable scans. More sophisticated dye-laser control systems incorporate computer-controlled wavelength calibration, frequency scanning, and data acquisition (Marshall *et al.*, 1980; Williams *et al.*, 1983; Clark *et al.*, 1986).

For a deeper understanding of the methods of laser-frequency control we focus attention on a typical tuning system for a single-mode ring dye laser (such as in Fig. 5.6). We assume that the frequency-selective elements are able to force the dye laser to operate on a single longitudinal mode of the laser cavity. The optical length of the cavity then determines the fine tuning of the laser frequency. The laser frequency can be adjusted over a small range by using the piezoelectrically translated mirror, PZT (see Fig. 5.6), and larger-frequency changes are induced by rotation of the Brewster plate. Changing the optical length of the cavity will scan the laser but only within the bandpass of the thick etalon. Since the thick etalon has enough resolution to select a single mode it also limits the scan to less than the laser cavity FSR. In order to achieve longer scans, the thick etalon must be scanned synchronously and likewise eventually the thin etalon must also be scanned. Usually these etalons are scanned with a piezoelectric translator to change the spacing between the plates of the thick etalon and by using a galvo-motor to rotate the thin etalon.

The alignment of the thick-etalon transmission peak to the cavity mode is critical and usually requires an automatic electronic servo to keep its bandpass properly aligned with the laser-cavity mode. This can be accomplished by using a conventional "modulation lock" which will be discussed shortly. For long scans it is necessary to tilt the thin etalon to track the thick etalon, but because of its low resolving power it is usually not necessary to have automatic feedback control—although such systems have been developed (Biraben and Labastie 1982). Tracking of the thin-etalon bandpass to a scanning laser frequency is also complicated by the fact that the bandpass of a tilted etalon scans approximately quadratically as a function of the tilt angle (see Eq. 5.2). This nonlinearity can be electronically compensated by using an electronic square-root–function module. Centering the passband of the birefringent filter is usually done manually by finding the maximum laser output power on a given mode. Automatic control systems for the birefringent filter can also be made by monitoring the polarization of the laser's output, which becomes slightly elliptical when the birefringent filter's bandpass is not properly centered on the cavity mode (Biraben, 1989).

A typical dye-laser scanning system thus consists of a Brewster plate rotated by a galvo-motor, a piezoelectrically driven thick etalon to track the specific cavity mode, and a thin etalon that is tilted by a galvo-motor to track the thick-etalon scan. All of these scanned elements are synchronized by a master control circuit that thus controls the laser's frequency. Commercial single-mode dye-laser systems can be scanned continuously with direct electromechanical control over ranges of about 30 GHz. Longer scans are certainly possible but they would put more stringent requirements on the electromechanical stability of the tuning elements. For example, with 1-MHz resolution, a 30-GHz scan already provides 3×10^4 resolution elements, which is usually adequate for most applications. There are some limits to the mechanical and electrical stability, linearity, and reproducibility of the electromechanical transducers used to position the tuning elements. Instability in these systems translates directly to laser-frequency fluctuations. In practice it is difficult to achieve 10 parts per million resolution in the electromechanical positioning. One can reconstruct a broader spectrum by computer-driven piecewise scans of about 30 GHz each. The total scan is then limited only by the bandwidth of the dye, but care must be taken to maintain some frequency reference in putting all of the separate scans together.

Pressure scanning of dye lasers is an alternative approach to mechanical scanning. As of the late 1980s, pressure scanning is not very popular, but it has some advantages. For example, it does not require moving parts in the laser. The idea is to enclose the entire laser inside a pressure-tight vessel so that when the pressure is changed the optical path length changes (due to changes in the index of refraction of the gas) and this causes the laser's frequency to change. If the etalons are designed with an open air space between the etalon surfaces, then when the pressure is changed, the etalons scan synchronously with the laser frequency.

4.2. Frequency-Stabilization Methods

The nature of the frequency fluctuations of lasers (as with other oscillators) is such that the measured frequency-stability depends on the time scale over which the measurement is made. The spectrum of dye laser frequency fluctuations is strongly peaked at low Fourier frequencies and then tapers down to a white noise level at high frequencies. In addition strong resonant peaks in the spectrum of the frequency noise result from a variety of technical problems (for example, noise at the AC line frequency). It is common practice to specify a laser's frequency-stability by a short-term "linewidth" and a longer-term center-frequency stability or drift rate. The implicit assumption here is that short-term fluctuations of the laser's phase determine the "linewidth," while slower fluctuations and drift dominate the stability the laser's center frequency. This model of the lasers frequency fluctuations has enough validity to be useful but it is far from the complete picture.

The frequency of the single-mode dye laser is determined by the boundary conditions imposed by the optical length of its resonator, which are usually arbitrary and generally lack long-term stability. The solution to the problem of frequency instability and inevitable drift is to lock the laser to a frequency reference such as a stable reference cavity or a molecular resonance. In principle the ultimate stability of the laser frequency can match that of the reference cavity if the control system is properly designed. Usually systematic errors degrade this performance considerably. Standard stabilized dye-laser systems have linewidths of $\simeq 1$ MHz with a center frequency of $\simeq 5 \times 10^{14}$ Hz. Typical drift rates are about 10 kHz/sec, which is adequate for most experiments. The best cavity-stabilized laser systems have laser-frequency drift rates of about 1 Hz/s (Salomon *et al.*, 1988; Hils and Hall, 1989; Helmcke *et al.*, 1987). This outstanding performance exceeds the precision that is required for all but the most demanding metrology and physics experiments.

The other compelling reason to have a stable reference cavity is that it can act as a fast frequency discriminator with a very large signal-to-noise ratio. The error signal derived from the cavity is then used in an electronic servo system to narrow the laser's linewidth and hence improve its resolution. The resulting linewidth of a laser stabilized to a reference cavity will depend on the laser's intrinsic frequency-fluctuation spectrum, the reference cavity's resolution, the servo system's bandwidth, and even the observation time. Some knowledge of the methodology of frequency-control systems is useful in developing an understanding of the operational characteristics of high-resolution dye lasers (see also Balykin *et al.* (1987) and Helmcke *et al.* (1982, 1987). Some of the common frequency-stabilization techniques are outlined in the following section.

4.2.1. *Cavity-Side-Lock*

One of the easiest and most generally applicable frequency-control methods is the cavity-side-lock system. It has been with us at least since the early days of the dye laser. This system uses a spectrally sharp Fabry-Perot transmittance peak to derive an electronic error signal that can be used to lock the laser to the side of the cavity resonance (see Fig. 5.11). The error signal is the difference between the photocurrents of the cavity transmittance peak (signal channel detected by D1) and the laser power (reference channel detected by D2). The light levels are adjusted so that the output from the difference amplifier is 0 when the laser frequency is tuned to the side of the cavity resonance. This provides a smoothly varying monotonic frequency discriminator for laser frequency excursions less than 1/2 of the

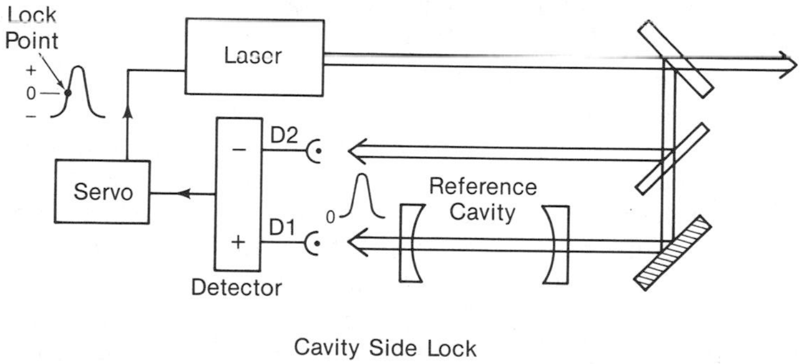

Fig. 5.11 Laser frequency control by the cavity side lock method. Part of the laser's output beam is directed into a Fabry-Perot cavity and onto a photodetector D1. A reference beam is directed onto detector D2 which measures the laser's power. The transmission function of the Fabry-Perot cavity is shown schematically near D1. The cavity acts as a frequency discriminator for the laser. The error signal is generated from the difference between the signals from D1 and D2, and is shown with the lock point indicated at zero voltage. A servo amplifier uses the error signal to control the laser's frequency.

cavity-resonance width. Depending on the overall sign of the correction signal, the laser frequency can be locked to one side of the fringe or the other. The zero crossing of the error signal is set at the point of fringe half maximum by adjusting a variable attenuator in the reference channel. The DC offset of the reference channel is derived from the laser power rather than from a voltage reference for a good reason. This reference signal generated by the laser power cancels laser-amplitude fluctuations that would otherwise cause an error in the lock signal. This is an important correction because, as we shall see, the amplitude fluctuations on dye lasers are often relatively large. Details of an advanced side-lock frequency-control systems can be found in Helmcke *et al.* (1982).

4.2.2. Modulation Lock

A second and very useful method of laser-frequency stabilization is the modulation lock. This system incorporates frequency modulation and synchronous detection to generate an error signal that is then used to lock the laser to the peak of a resonance. The resonance can be a Fabry-Perot cavity transmission signal but it can equally well be an atomic or molecular resonance. This method is very general and it allows us to stabilize any variable parameter to the maximum or a minimum of a response function. Because of the simplicity of the method and because we often want to maximize (or minimize) some response, the technique is widely used in servo-control systems. We described this method in terms of locking the oscillation frequency of a laser to the peak of a Fabry-Perot transmission fringe. The important features of the modulation lock are illustrated in Fig. 5.12a, 5.12b. The various electronic signals are shown in Fig. 5.12b where the Fabry-Perot transmission fringe is displayed as a function of the laser frequency. If the laser frequency is sinusoidally modulated about the peak of the cavity resonance, the transmitted light intensity varies as the second harmonic of the modulation frequency. However if it is modulated about the side of the cavity resonance, the transmitted power will vary at the modulation frequency with a phase that depends on which side of the fringe the laser is tuned. There is a 180° relative phase shift of the modulation response on the two sides of the fringe. The power of the laser light transmitted through the cavity is then detected and demodulated using a lock-in amplifier. The lock-in output is then lowpass filtered to produce the discriminator-shaped output (see Fig. 5.12b). This discriminator signal is used as the error signal for the servo-control loop. The phase reversal of the error signal across the fringe changes the sign of the output from the lock-in and produces an error signal that crosses 0 at the peak of the resonance. For small-modulation amplitudes, the shape of the discriminator signal is approximately the first derivative (with respect to frequency)

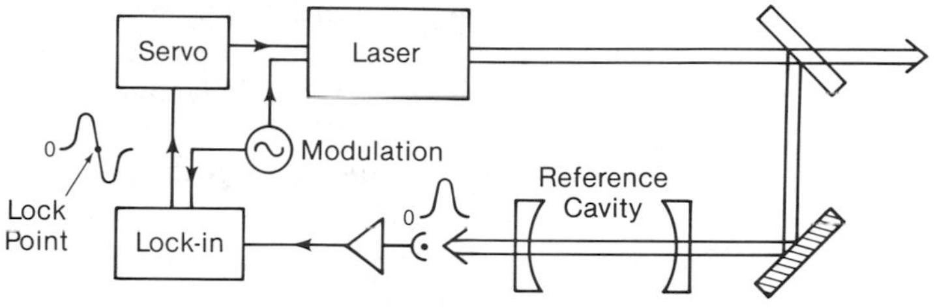

Modulation Lock

Fig. 5.12a Laser frequency control by the modulation lock method. A Fabry-Perot reference cavity is used as a frequency discriminator for the laser. The cavity's transmission function is indicated schematically near the photodetector and amplifier. The signal from the detector goes to the lock-in amplifier where it is compared with the modulation signal that is used to modulate the frequency of the laser. The lock point is indicated on the error signal that is shown schematically as the output of the lock-in. A servo amplifier uses the error signal to control the laser's frequency.

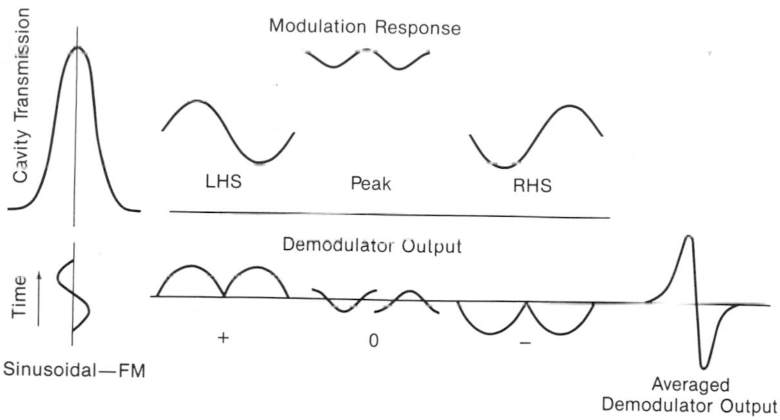

Modulation Lock Signals

Fig. 5.12b Signals used in the modulation locking method. The cavity transmission function and the modulation of the laser's center frequency are displayed on the left-hand edge of the figure. The center of the figure shows the variation of the cavity's transmission due to the modulation and is displayed as a function of the laser's center frequency relative to the cavity resonance. The demodulator (lock-in) inverts the modulation response synchronously every half-cycle of the modulation. Its output is shown at the bottom of the figure. When the demodulator's output is averaged for times that are long compared with the modulation frequency, the result is an output signal that is approximately the derivative of the cavity transmission function. This output is shown at the bottom right of the figure and is displayed as a function of the laser's center frequency.

of the transmission function. Similarly, high-order approximate derivative signals are obtained by demodulating the transmitted power at harmonics of the modulation frequency. For example, a third derivative locking signal is obtained by demodulating at the third harmonic of the modulation frequency and it has some advantages with respect to systematic errors. These frequency modulation–demodulation ideas are also useful for reducing noise and background problems in spectroscopy (e.g., laser derivative or FM spectroscopy).

4.2.3. rf-Optical Heterodyne Lock

Another laser stabilization system has been developed using rf heterodyne techniques (Drever *et al.*, 1983; Hough *et al.*, 1984; Helmcke *et al.*, 1987). Of all of the methods, this "rf-optical heterodyne" locking system has demonstrated the best performance, in terms of both center frequency control and the narrowest linewidths. The method is similar to the modulation lock described previously, with the distinction that it uses radio- or microwave phase modulation on the laser's output to generate laser sidebands that lie outside the resolution width of the resonance of interest. The high modulation frequencies are advantageous in suppressing noise, in achieving high-servo-control bandwidths, and in achieving good transient response characteristics. An example of an optical heterodyne system used for laser-frequency stabilization is diagrammed in Fig. 5.13. Here the system consists of a laser, an electro-optic phase modulator (ϕ mod, made from a crystal of ADP, ammonium dihydrogen phosphate), an optical directional coupler (made from two polarizers and a Faraday rotator), a precision reference cavity, a photodetector (det), some filters, a balanced mixer, and a servo-control amplifier. The cavity resonance is detected in reflection by this heterodyne technique and can provide two types of signals. These signals are derived from the two detection quadratures of the balanced mixer (determined by the phase of the reference signal) and correspond to the cavity transmission and dispersion functions. The cavity resonance acts to alter the pure phase modulation that was imposed on the laser light by the electro-optic modulator. This alteration of the phase modulation comes from the transmittance and phase shift of the laser light by the cavity. The cavity thus converts some of the phase modulation into amplitude modulation, which is then detected by the photodetector and subsequently demodulated by the mixer to generate a baseband (DC) response. The phase of the reference signal (derived from the original modulation signal) that is applied to the mixer determines the detection quadrature and, thus, whether we detect the effect of cavity transmittance or dispersion. Typical response signals for the two detection quadratures are shown in Fig. 5.14. The discriminator shape of the dispersion signal provides an excellent error

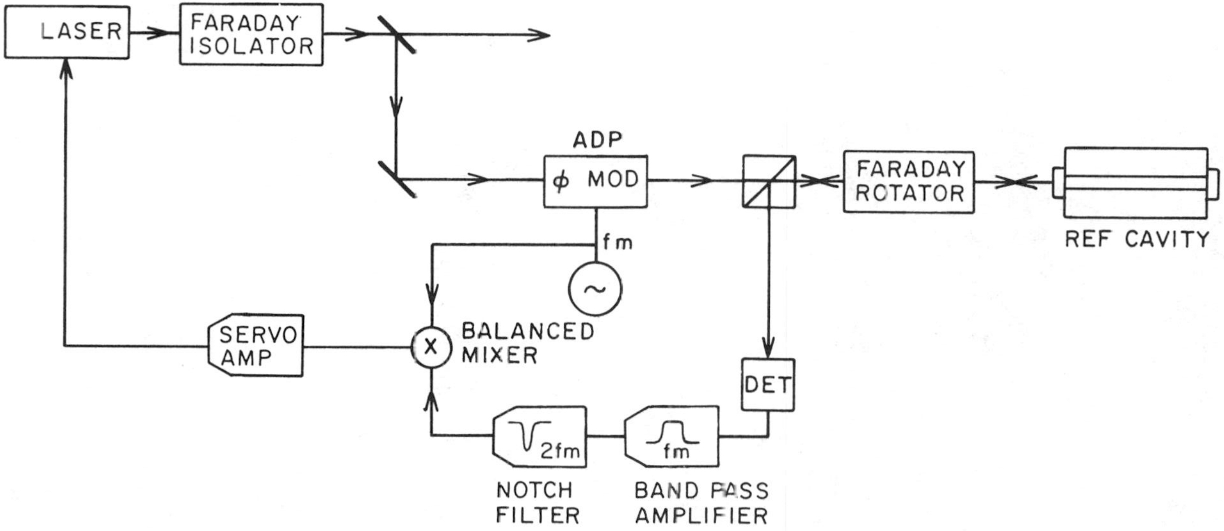

Fig. 5.13 Laser-frequency stabilization system using rf-optical heterodyne locking method. Part of the laser's output is phase-modulated using an electro-optic modulator (ϕ–mod). The phase-modulated laser beam then probes the reference cavity and is detected in reflection by the photodetector DET. The resulting signal at the modulation frequency passes through a set of filters to the balanced mixer. There it is mixed with a reference signal at the modulation frequency to produce the error signal (see Fig. 5.14) that is used by the servo-amp to control the laser's frequency. This figure is reproduced with permission from Drever *et al.*, (1983).

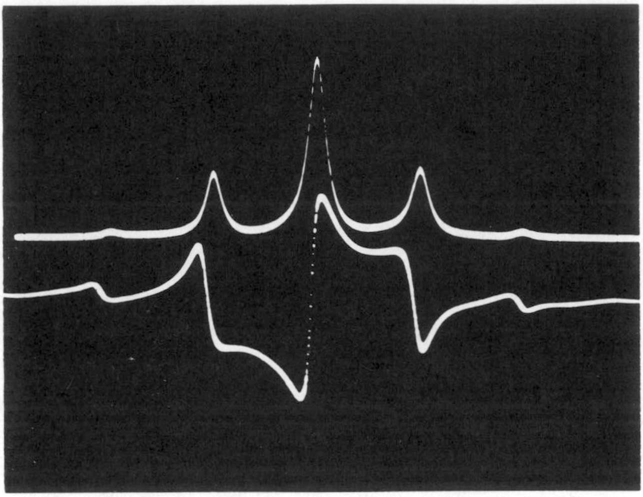

Fig. 5.14 Cavity signals generated by the rf-optical heterodyne locking method. The output of the mixer (from Fig. 5.13) is shown as a function of the laser's frequency which is scanned. The upper trace shows the output when the phase of the reference signal (applied to the mixer) is set to detect the cavity's transmission. The multiple-peaked structure results from the carrier and sidebands being transmitted into the cavity when they come into resonance. The carrier (large peak in the center) and the two adjacent sidebands are separated by the modulation frequency. Small peaks due to the second-order sidebands are also visible at a spacing that is twice the modulation frequency. The lower trace shows the output when the phase of the reference signal is shifted by $\pi/2$. It represents the effect of cavity dispersion as seen by the phase-modulated laser beam. The dispersion signal has a steep discriminator shape that crosses 0 when the carrier is tuned to the peak of the cavity resonance. This discriminator is used as the error signal in the frequency-control loop. This figure is reprinted with permission from Drever *et al.* (1983).

signal for laser-frequency control. Some of the components in this system can be compared directly to those of the modulation lock; the phase modulator serves the role of the laser FM, and the mixer replaces the lock-in amplifier. One important advantage of the heterodyne system is that the high modulation frequencies and the detection of the cavity in reflection allow the time scales of the detection process to be very short. This allows the response time of a servo system to be shorter than the characteristic response time of the cavity (Drever *et al.*, 1983) while maintaining a good signal-to-noise ratio. This is an important consideration for the high-speed servo systems that are required for very narrow laser linewidths. This method has been used by Bergquist and collaborators to observe $\simeq$ 50-Hz–wide mercury-ion resonances; this indicates that the dye laser's linewidth and center frequency stability are at least that good (Wineland *et al.*, 1989).

4.2.4. Polarization Lock

Noteworthy because of its good performance and simplicity is a related laser-frequency–stabilization method that was developed by Hänsch and Couillaud (1980). Analysis of the polarization state of the light reflected from the cavity is used to detect the dispersive nature of the cavity resonance. This method has the advantages that it produces a discriminator-shaped signal without modulation and it has a very large capture range. This results in a robust laser-locking method that is useful for laser-frequency stabilization and optical buildup cavities.

4.2.5. Postlaser Stabilization

An exciting and relatively new development in laser-frequency control is the implementation of a postlaser-frequency–correction system (Hall and Hañsch, 1984; Hall *et al.* 1988). The concept is that with the appropriate acousto-optic and electro-optic transducers, we can correct the frequency errors on the laser beam after the beam exits the laser. A system of this type is diagrammed in Fig. 5.15. The external frequency correction is implemented with an acousto-optic frequency shifter (driven by a voltage-controlled oscillator) and an electro-optic modulator (EOM). The EOM acts as frequency transducer through its ability to induce a time-dependent optical phase shift via its electric-field–dependent index of refraction. In this case the change in the laser frequency is proportional to the derivative with respect to time of the applied electric field. Thus the circuitry for the electronic feedback to the EOM is complicated by the fact that adding a fixed frequency shift to the laser requires that a voltage ramp be applied to the EOM. The EOM used in this way operates as an optical phase shifter and hence only works well for removing high-frequency fluctuations. The acousto-optic transducer is required for the low frequency fluctuations and fixed frequency offsets. Another challenge is the fact that the time scales for the response of the acousto-optic and electro-optic transducers are very different and appropriate crossover compensation must be designed. Clever solutions to these challenges have been developed by Hall and Hänsch (1984). The error signal for this system can be derived from any of the methods described previously, but perhaps the most appropriate is the rf-optical heterodyne method. Impressive performance has been achieved with these postlaser stabilization systems but, unfortunately, they are not available commercially as of 1989. Starting with a single-mode dye laser that had an intrinsic linewidth of $\simeq$1 MHz, the postlaser stabilization methods have produced linewidths on the order of 1 kHz. Even better performance can be anticipated. Another application of this method has been to make an optical phase lock between a cw dye laser and a HeNe laser

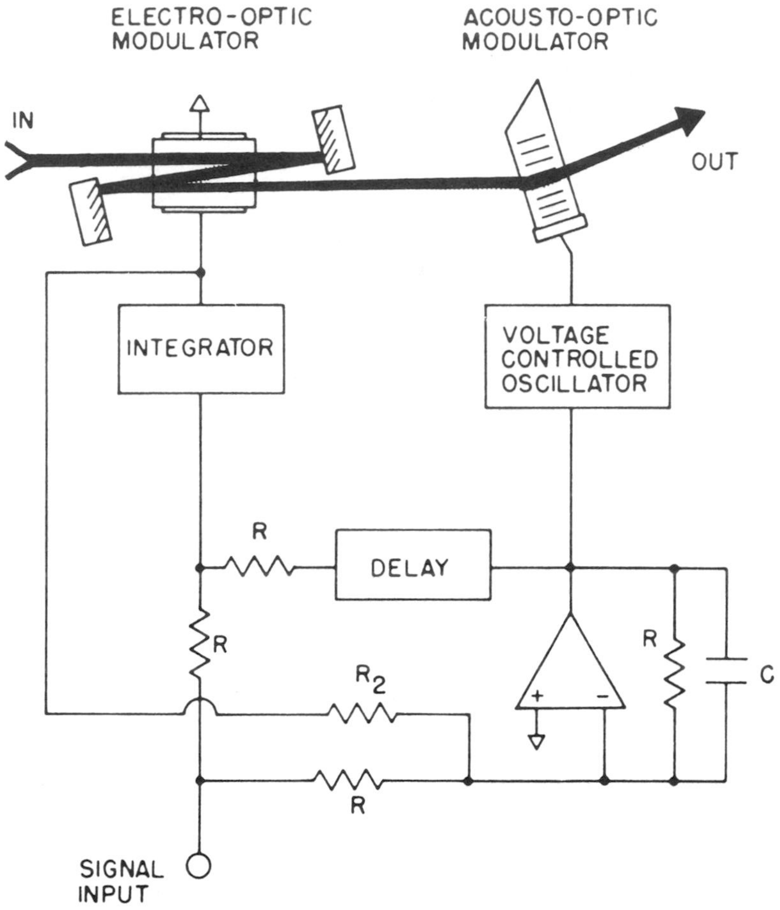

Fig. 5.15 Laser-frequency control by the post–laser-stabilization method. In this system the frequency of the laser's output is changed outside the laser by an electro-optic modulator and an acousto-optic modulator. A time rate of change of voltage applied to the electro-optic crystal will change the frequency of the laser beam by a small amount. Larger frequency shifts are generated by the acousto-optic modulator which is driven by a voltage-controlled oscillator. To optimally use the two transducers the input error signal is first processed electronically as indicated. Figure reprinted with permission from Hall and Hänsch (1984).

that was used as the reference oscillator (Hils and Hall, 1989; Salomon *et al.*, 1988).

It is humbling for laser scientists to learn that many of the basic principles in these methods had been developed in the prelaser days by scientists working with rf and microwave systems. The techniques were subsequently "reinvented" with laser technology.

Up to this point I have outlined various general techniques for deriving an error signal from a resonance, but have not discussed the requirements for a feedback loop that can use this error signal to control the laser frequency. The loop needs to process the error signal and drive the appropriate transducers to correct the laser's frequency fluctuations. The loop will thereby null (or attempt to null) the error signal. Common dye laser frequency transducers are galvo-motor–driven Brewster plates, piezoelectrically mounted mirrors, and electro-optic crystals (see Fig. 5.6). Each of these transducers can change the optical path length of the laser cavity and hence the laser frequency. Typical commercial frequency-stabilized dye-laser systems use a galvo-driven Brewster plate and a piezo-mounted mirror for their frequency-control transducers. These systems routinely achieve laser linewidths on the order of one MHz. With faster feedback control and better reference cavities it is possible to achieve laser linewidths of about 20–100 kHz with a piezo-mounted mirror and a galvo-driven Brewster plate. For very high resolution applications an electro-optic modulator (EOM) can be included in the laser cavity. With the EOM, dye-laser linewidths on the order of 50 hertz have been demonstrated and linewidths of one hertz or less are anticipated. These are truly impressive high-resolution oscillators when we recall that the oscillation frequency is $\approx 5 \times 10^{14}$ Hz. The narrow linewidths achieved with the intracavity EOMs are due to the high-speed response of the crystal and the high servo bandwidths that are possible. In general the higher the servo bandwidth, the narrower the resulting laser linewidth. The actual performance of the frequency-stabilization system depends on the spectral density of the frequency noise on the laser and the characteristics of the servo loop. In principle, and in some cases in practice, laser linewidths as narrow as you desire can be achieved with late 1980s technology (Hils and Hall, 1989). Often the limitation to the resolution is not the laser at all but rather the spectroscopic sample of interest.

Optimization of the filter characteristics of the feedback loop can be a difficult task given the complexity of the system with the multiplicity of transducers. In addition we need to take into account the spectrum of the dye laser's frequency noise and the laser's dynamics. The simplest control loops (termed proportional locks) use purely proportional control, where the feedback is linearly related (by the gain) to the error signal. These control systems are very stable but are lacking in both accuracy and dynamic response. A significant improvement in performance is obtained by integrating the error signal with respect to time before feeding back to the transducers. An integrating loop filter has gain that rolls off toward higher frequencies at a rate of 6 dB per octave for each stage of integration. Even a single-stage integrating-loop filter has the advantage that is guarantees zero error at DC and only a fixed, finite error for a ramping error signal. In contrast, the proportional control loop will in general always have a finite error. Adding some derivative feedback in the loop can often improve the dynamic response. It is common practice to use some combination of proportional, integral, and derivative feedback. These systems are best optimized with careful system analysis, but empirical adjustment of the gains and time constants is often productive.

As we have discussed, most of these laser-locking methods can be applied equally well to stabilizing the frequency of lasers to a cavity or to atomic or molecular resonances. In the latter case the most applicable techniques are the modulation lock and the rf heterodyne lock. The ability to stabilize the laser to an atomic resonance is important for some spectroscopic applications and is crucial for high-accuracy metrology and laser-frequency standards. The atomic and molecular resonances cannot provide the same signal-to-noise ratio that is available from a cavity, but their long-term (say, for times longer than a few seconds) stability is generally much better. For dye lasers to have high resolution and high accuracy, it is customary to lock the laser's frequency to a high-Q reference cavity (to achieve the narrow linewidths) and then to lock the cavity to an atomic or molecular resonance (which then provides the long-term stability).

The transformation of a broadband dye laser into a frequency-stabilized narrow-linewidth laser adds considerably to the cost and complexity of the laser system. There is also a penalty to be paid in the available output power when we add intracavity optical elements (this typically amounts to about a factor of 2). If we use an electro-optic crystal, the upper power limit may also be reduced due to the limited optical-power density that the crystals can endure.

In summary, commercial stabilized dye-laser systems can provide high resolution ($\simeq$1 MHz) with tunability and good power levels. Their tuning range extends throughout the visible and near-visible part of the spectrum. These high-resolution systems are more than adequate for most spectroscopic applications. For applications that demand higher resolution, very narrow-linewidth dye-laser systems have been developed in a number of research laboratories. The systems with the best reported resolution generally use optical heterodyne techniques and EOMs to achieve laser linewidths below 100 Hz. I see no reason that the linewidths could not be reduced further to below one Hz in the future. So while the performance levels achieved in 1989 are truly impressive, they are merely a glimpse of the present state of the art and they will certainly be surpassed.

5. SPECTRAL CHARACTERISTICS

The complexity of dye lasers encourages the user to have some diagnostic instruments to monitor the laser's output characteristics. Useful diagnostic tools include power meters as well as wavelength- and linewidth-measuring devices.

With dye tuning ranges of about 100 nm and linewidths of about 10^{-6} nm it is not a trivial task to determine, with precision, the lasing wavelength of cw dye lasers. Many interferometric instruments have been developed to measure the laser's wavelength relative to known standards; these include the lambda or wave meter, the sigma meter, the Fizeau wavemeter, calibrated Fabry-Perot interferometers, and dispersive birefringent filters (Juncar and Pinard, 1975; Kowalski *et al.*, 1976, 1978; Hall and Lee, 1976a; Woods *et al.*, 1978; Snyder, 1980; Williams *et al.*, 1983; Licthten 1985, 1986; DeVoe *et al.*, 1988). The accuracy of these interferometers varies from 5 to 8 digits. Special standards-laboratory instruments are capable of 10 digits of accuracy in wavelength measurement. Also extremely useful for general laboratory spectroscopy are the wavelength tabulations that exist for molecular spectra of some simple molecules, such as those for I_2 (in the red) (Gerstenkorn and Luc, 1978) and Te_2 (in the blue) (Cariou and Luc, 1980). These tabulations can be used to determine a dye laser's wavelength to about 3×10^{-7}.

We need certain mathematical and experimental tools to characterize the spectral properties of lasers. These tools certainly include the frequency and amplitude-noise spectral densities and maybe the Allan variance. The Allan variance was developed to quantify the frequency stability of oscillators and can be applied directly to lasers. It is discussed here because of its usefulness and its popularity in the literature and also because it is not found in the usual textbooks.

5.1. Amplitude Fluctuations

The spectral distribution of amplitude noise in the output of a single-frequency dye laser contains a great deal of useful information. It can be measured directly by monitoring the laser light with a fast photodetector and a rf-spectrum analyzer. The quantum theory of the laser predicts that the fluctuations in the power from an ideal laser source will be determined by the statistics of the quantum mechanical generation of the photons (Sargent *et al.*, 1974; London, 1983). In this model the ideal laser would produce light with a time distribution of photons that is Poissonian. Upon

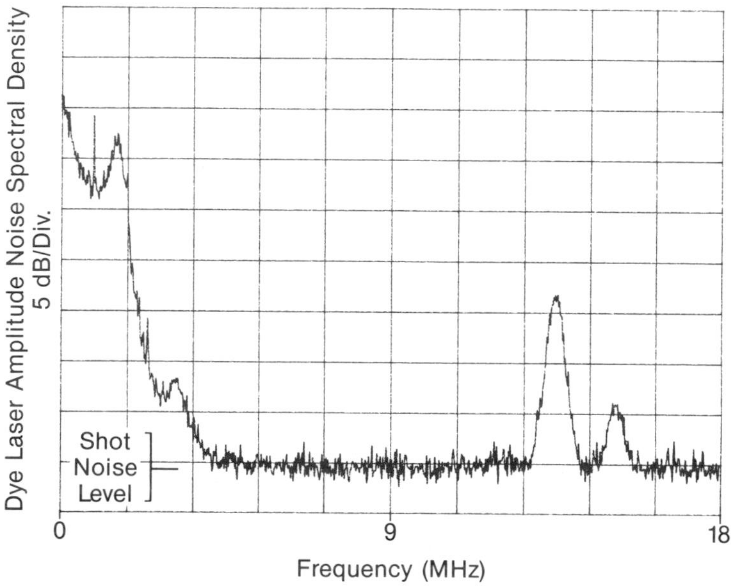

Fig. 5.16a Spectral density of the amplitude noise of a single-mode ring dye laser. The noise on the laser's output power is strongly peaked at the lower frequencies and finally reaches the shot-noise level at a frequency of about 4.5 MHz. The broad peaks in the noise come from small amounts of power oscillating in transverse modes in the argon pump laser. These pump-laser–related noise peaks extend out to hundreds of megahertzes but spectral regions can be found where the dye laser's amplitude noise is shot-noise–limited.

detection this photon distribution generates a Poissonian photoelectron distribution and hence produces the usual "shot-noise" fluctuations in the photocurrent,

$$i_{SN} = (2eiBW)^{1/2}. \tag{5.9}$$

Here i is the detected photocurrent, e is the electron charge, and BW is the detection bandwidth. The shot-noise distribution is spectrally flat. Figure 5.16a shows the measured spectral density of the amplitude noise of a good single-mode dye laser. The figure also indicates the expected shot-noise level for the same average optical power on the detector. Clearly, there is more going on in real lasers than is predicted by the quantum mechanical model of the ideal laser. Characteristically, the amplitude noise is very large at low Fourier frequencies and decreases progressively to higher frequencies until it reaches the shot-noise level at a frequency of approximately 4.5 MHz. We also see that there are many large resonant noise peaks throughout the spectrum. These are due to various technical noise

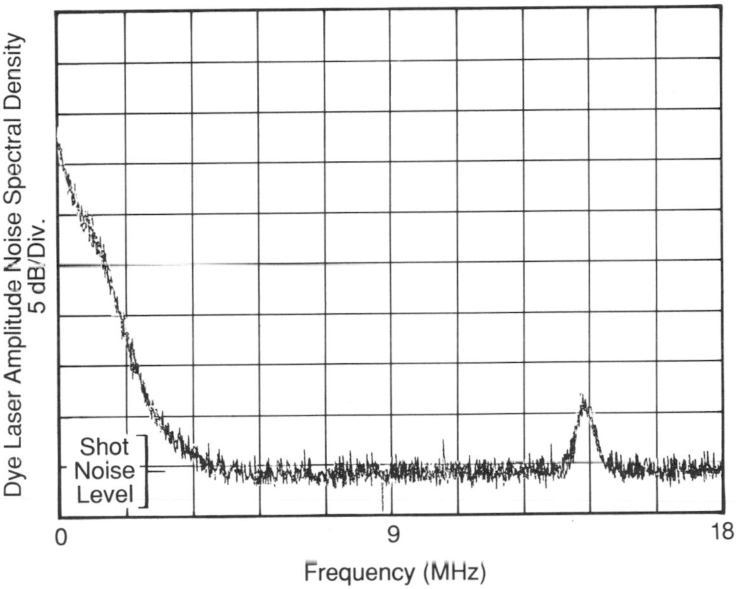

Fig. 5.16b Spectral density of the amplitude noise of the same dye laser as shown in Fig. 5.16a, but now with the aperture in the argon laser carefully adjusted to elliminate transverse modes. The dye laser's amplitude noise is improved, but the noise is still strongly peaked at low frequencies.

sources, the most important of which are the argon laser's intermode beats that are transferred through the gain of the dye jet to the dye laser's power (these intermode beats are observed out to several hundreds of megahertz). In Fig. 5.16b the spectral density of the amplitude noise of the same dye laser is shown but now with the argon laser resonator apertured to eliminate transverse modes. This removes most of the resonant peaks in the noise (in this frequency range) and significantly improves the dye laser's amplitude noise spectrum.

One of the important things we learn from the spectrum of amplitude noise is that any measurements that are made with a detection bandwidth centered at low frequencies (or near one of the resonant noise peaks) will be contaminated by excess noise on the laser. As seen in Fig. 5.16 this excess noise is not negligible. In this example there is $\approx$35 dB of excess noise at a detection frequency near 100 kHz (resolution bandwidth was 3 kHz). The noise is even greater at lower frequencies. Not surprisingly the signal-to-noise ratio can be improved by moving the center of the detection bandwidth to higher frequencies by using high-frequency modulation methods. If we look to high enough frequencies it is possible to reach the shot-noise level in some limited-detection bandwidth. In fact, it is even, possible to achieve noise levels below this shot-noise level by using nonlinear optical methods to generate nonclassical forms of light such as squeezed states. This regime of sub–shot-noise limited detection is a hot topic in modern optics with two compilations of papers on this and related subjects ("Squeezed States," 1987).

5.2. Frequency Fluctuations

One method for determining the frequency stability and linewidth of a dye laser is to monitor the laser's output with a high-resolution Fabry-Perot cavity (optical spectrum analyzer). These instruments are available commercially. Using a scanning Fabry-Perot cavity to measure the spectral distribution of light from a multimode standing-wave dye laser (with only a birefringent filter as the tuning element) shows the optical power distributed in a large number of discrete longitudinal modes. The spectral width of this array of modes can vary from 1 to 100 GHz. In this multimode case the distribution of modes is often unstable.

If the dye laser is made to run on a single mode, a high-finesse cavity can be used to determine the laser's linewidth and the spectral distribution of the residual frequency fluctuations. Another method for measuring the frequency stability of a laser is to use a fast photodiode to observe the beat note between the unknown laser and a narrow-linewidth reference laser. There is also the "self-heterodyne" method, which uses a fiber delay-line to measure the laser's linewidth.

The frequency noise spectrum of a single-mode laser can be measured best by detecting the rf spectrum of the heterodyne beat note between two lasers: the unknown laser and a reference laser that has a narrower spectral width (Hough *et al.*, 1984). This heterodyne method is generally preferred for the very high resolution systems but it requires an additional very stable reference laser to act as the local oscillator. A limitation here is that the linewidth of the very best lasers must be determined by deconvolution of two equally good lasers or by other methods (cavities or special narrow spectral lines).

The Fabry-Perot cavity method of measuring a laser's frequency-noise spectrum relies on using the cavity's transmittance or reflectance function as a frequency discriminator. By tuning the laser's center frequency to the side of the cavity resonance and measuring transmitted power, the cavity then acts to convert frequency fluctuations directly into amplitude fluctuations at the detector. A cavity (or a molecular resonance) can be an excellent frequency discriminator but care must be taken to avoid systematic errors (for example, detection bandwidth limits, cavity response time, and laser amplitude noise). The literature is full of "outstanding" results based on improper measurements. A good example of a very popular mistake is to lock a laser's frequency to some cavity or atomic resonance (by the methods described previously) and then monitor the error signal within the servo-loop to provide a measure of the residual frequency noise. This appears reasonable at first glance, but usually the error signal is contaminated at some level by systematic errors (such as a ground loop). These systematic error signals are then corrected by the servo system as if they were laser-frequency fluctuations. This puts these artificial errors directly on the laser frequency. Any measurement of the error signal would indicate that the laser's frequency fluctuations are small, because the servo loop is acting to cancel the error signal. This gives a misleading result when the measured error signal is then used to calculate the laser's frequency fluctuations. However, it is true that monitoring the residual error signal is useful in providing information about how well the servo electronics are working. It does *not* answer the question of how well the error signal represents the laser's frequency fluctuations. To make a definitive statement about the frequency fluctuations of a stabilized laser it is imperative to have a measure of the laser's frequency that is independent of the stabilization system.

The self-heterodyne method of measuring a laser's linewidth uses a time delay and a heterodyne detector to produce the autocorrelation of the laser's frequency noise at a given time delay (Okoshi *et al.*, 1980). This method compares the oscillation frequency of a laser at an earlier time with its oscillation frequency now. Self-heterodyne systems are very easily implemented by sending some of the laser's output through a long fiber (for the delay) and then combining this fiber-delayed output with the direct output on a fast photodetector to observe the beat note. In these systems one of the two beams is usually frequency shifted by an acousto-optic modulator (for example, by 80 MHz) to avoid technical noise at zero frequency difference upon detection. It is difficult to make accurate measurements with this method if the laser has a very narrow linewidth, and in general care must be used to avoid misleading results (Kikuchi and Okoshi, 1985).

A spectrum of frequency fluctuations can be used to provide an effective laser linewidth (Elliot *et al.*, 1982). In two limiting cases, which depend on the analytical character of the frequency noise, the linewidth can be calculated trivially. The important factor here is whether the Fourier frequency of the dominant frequency fluctuations is high or low, compared with the rms frequency fluctuations of the laser. That is, are the main frequency excursions caused by low frequencies with a high modulation index, or high frequencies with a low modulation index, where high and low are measured relative to the laser's rms frequency fluctuations. For example, if we make the assumption that the frequency-noise spectrum is rectangular with a bandwidth B, we can calculate the linewidth in these two limiting cases. If the laser's rms frequency fluctuations are large compared to the bandwidth of the frequency fluctuations, the lasers lineshape is Gaussian and the linewidth is given by

$$\Delta_{\text{FHM}} = 2(2\,\ell\text{n}\,2)^{1/2}\,\delta_{\text{rms}}. \tag{5.10}$$

Here δ^2_{rms} is the rms frequency deviation. On the other hand, when the bandwidth of frequency fluctuations is larger than the laser's rms frequency fluctuations, the laser's lineshape will be Lorentzian and the linewidth will be given by

$$\Delta_{\text{FWHM}} = \pi\frac{(\delta_{\text{rms}})^2}{B}. \tag{5.11}$$

Here we have again simplified the problem by assuming that the spectrum of frequency fluctuations is rectangular in shape with a bandwidth B. In the more general case when the frequency fluctuations are not spectrally flat, with important contributions coming from both high and low Fourier frequencies, the lineshape and linewidth could be determined from the spectrum of frequency fluctuations by numerical integration of the equations given by Elliot *et al.* (1982).

As with the amplitude noise, the spectrum of frequency fluctuations for dye lasers is large at low frequencies and then falls off at higher frequencies. There are often numerous large resonances throughout the spectrum. Figure 5.17 shows the spectrum of frequency fluctuations of a single-mode ring dye laser that was measured with a high finesse cavity and a rf-optical heterodyne detection system. Also shown is the reduction in the frequency noise that is achieved when the frequency-control loop is turned on. In this case, the servo is able to reduce the frequency noise (as measured by the servo-loop error signal) by as much as 50 dB. An independent frequency discriminator would be required for an accurate determination of the laser's linewidth.

A mathematical tool that is genuinely useful in describing the frequency stability of lasers (and other oscillators) is the Allan variance (Allan, 1966; Allan *et al.*, 1974). It is the variance of the second difference of a series of consecutive frequency measurements. The Allan variance is defined

$$\sigma(\tau) = \frac{1}{\nu_0}\left(\frac{1}{2(M-1)}\sum_{i=1}^{M-1}(\nu_{i+1}-\nu_i)^2\right)^{1/2} \tag{5.12}$$

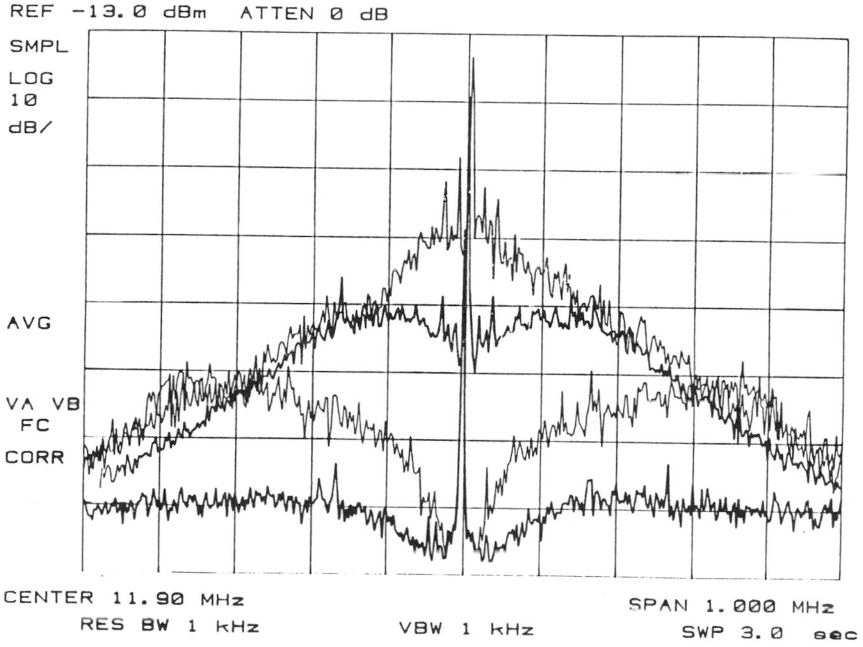

Hall et al Summary for IQEC - 88 Tokyo, Japan July 18-22, 1988

Fig. 5.17 Spectral density of the frequency noise of a single-mode ring dye laser with and without fast frequency control. The frequency fluctuations are measured using the rf-optical heterodyne locking method (modulation frequency ≃12 MHz) diagrammed in Fig. 5.13. The vertical scale displays the dye laser's frequency noise with a sensitivity of 10 dB/div. The horizontal scale is 100 kHz/div and the data were taken with a resolution bandwidth of one kHz and a video bandwidth of one kHz. The largest peak shows the frequency noise on the laser without fast frequency control. When a fast postlaser frequency control loop (of the type diagrammed in Fig. 5.15) is implemented, the frequency fluctuations are reduced. The three lower curves show the effect of incrementing the control loop gain by about 10 dB per step. The frequency noise remaining in the lowest trace is near the noise level of the measurement system. It would correspond to a dye-laser linewidth of less than one Hz if the measurement of the error signal was an accurate measure of the laser's frequency fluctuations. Figure is reproduced with permission from Hall *et al.* (1988).

Here M is the number of successive frequency measurements (assuming no dead time between measurements), ν_0 is the mean laser oscillation frequency, and ν_i is the ith frequency measurement made during the sample interval time τ. The Allan variance can also be calculated from the power spectral density of the frequency noise $S(f)$ as follows:

$$\sigma^2(\tau) = 2\int_0^\infty df\,\frac{S(f)}{\nu_0^2}\left(\frac{\sin^2 \pi f\tau}{\pi f\tau}\right)^2. \tag{5.13}$$

One of the wonderful properties of this function is that it is well behaved for common types of frequency noise, such as $1/f$ noise. This is not true for the usual standard deviation which is often divergent. The Allan variance serves as a one-parameter figure of merit for the frequency stability of oscillators. Without measuring the spectrum, it can provide a diagnostic of the type of frequency fluctuations that are present. For example, an Allan variance that decreases as $1/\tau$, where τ is the averaging time, indicates that the frequency noise spectral density is white.

6. SUMMARY

Cw dye lasers are a unique source of laser radiation with broad tunability, high spectral purity and good output-power levels. We have seen that these lasers can provide complete spectral coverage from 365 to 1000 nm with typical power levels that range from 0.1 to more than 1 W. Nonlinear optical techniques have been used to extend this spectral coverage into the UV (260–400 nm) by harmonic generation and sum frequency mixing and into the near-IR (2.2–5 microns) by difference frequency mixing. Other specific wavelengths not included here have been generated by nonlinear methods for special applications. Continued improvements in the pump lasers, the dyes, and the nonlinear optical materials will undoubtedly provide higher output powers and extended spectral coverage.

The spectral characteristics of cw dye lasers are both good and bad: good with respect to their linewidths and, hence, spectral-resolving capabilities, but generally poor with respect to their amplitude noise properties. The spectral density of amplitude noise on cw dye lasers is large for Fourier frequencies below a few megahertz. Because of necessity, techniques have been developed to alleviate the problems associated with this amplitude noise. Typical linewidths of cw frequency-stabilized dye lasers are $\simeq$1 MHz, which is more than adequate for most applications. The technology is available to reduce the dye laser's linewidth to $\simeq$1 Hz for those special applications that require it.

PROBLEMS

1. By making simple assumptions about the dye gain and available pump power, design a three-mirror folded linear dye laser cavity of the type shown in Fig. 5.2. For this cavity calculate the expected difference in frequency between the main longitudinal mode and the spatial hole-burning mode that could oscillate simultaneously.
2. Calculate the expected temperature rise of a laser dye (such as R6G in ethylene glycol) when it is pumped with 10 W of cw ion-laser radiation

at 514 nm. Make reasonable assumptions about the optical and physical properties of the dye and the dye-circulation system.

3. Estimate the feasibility of pumping a cw dye laser with solar radiation. For this estimate propose some simple and realistic design for the laser.
4. Assuming the spectrum of frequency noise shown in Fig. 5.17 is an accurate representation of the dye laser's frequency fluctuations, make an estimate of the contributions to the laser's linewidth due to the various spectral regions of the noise. To calibrate the vertical scale, use the fact that a perfectly coherent laser would produce a signal of −3 dBm (at 12 MHz) when the laser's center frequency was tuned to the half-power transmission point of the reference cavity. Note that the rf-heterodyne detection method transfers the frequency-fluctuation information from zero frequency to the modulation frequency (12 MHz in this case).
5. Assume that the Fourier spectrum of a laser's frequency fluctuations is described by a simple power law dependence (that is, $S(f) = H_a \cdot f^{-a}$; $a = 2, 3/2, 1, 1/2, 0, -1/2$, and -1, and H_a is a constant).
   (a) Find the functional dependence of the Allan variance on the averaging time τ.
   (b) Determine which simple power laws also give a standard deviation (using the usual definition) that is well behaved.
6. What sort of length stability is required for a Fabry-Perot reference cavity that is used to stabilize the frequency of a visible laser to one Hz? Can length stabilities with this performance be achieved given the physical properties of realistic materials?

REFERENCES

Allan, D.W. (1966). Statistics of atomic frequency standards, *Proc. IEEE* **54,** 221–230.

Allan, D.W., Shoaf, J.H., and Halford, D. (1974). *Time and Frequency: Theory and Fundamentals* (Blair, B.E., ed.) NBS (National Bureau of Standards), Monograph 140. U.S Government Printing Office, Washington, D.C., pp. 151–166.

Anliker, P., Luthi, H.R., Seelig, W., Steinger, J., Weber, H.P., Leutwyler, S., Schumacher, E., and Woste, L. (1977). 33-W cw dye laser. *IEEE J. Quant. Elect.* **QE-13,** 547–548.

Balykin, V.I., Ovchinnikov, Yu. B., and Sidorov, A.I. (1987). Single- and double-frequency cw dye laser with active frequency stabilization and electronic spectral scanning. *Sov. J. Quant. Elect.* **17,** 1535–1539.

Barger, R.L., Sorem, M.S., and Hall, J.L. (1973). Frequency stabilization of a cw dye laser. *Appl. Phys. Lett.* **22,** 573–575.

Barger, R.L., West, J.B., and English, T.C. (1975). Fast frequency stabilization of a cw dye laser. *Appl. Phys. Lett.* **27,** 31–33.

Bergquist, J.C., (1989). Private communication.

Bergquist, J.C., and Burkins, L. (1984). Efficient single mode operation of a cw ring dye laser with a Mach-Zehnder interferometer. *Opt. Comm.* **50,** 379–385.

Bergquist, J.C., Hemmati, H., and Itano, W.M. (1982). High power second harmonic generation of 257 nm radiation in an external ring cavity. *Opt. Comm.* **43,** 437–442.

Biraben, F. (1979). Efficacité des systemes unidirectionnels utilisables dans les lasers en anneau. *Opt. Comm.* **29,** 353–356.

Biraben, F. (1989). Private communication.

Biraben, F., and Labastie, P. (1982). Balayage d'un laser a colorant continu monomode sur 150 GHZ. *Opt. Comm.* **41,** 49–51.

Blit, S., Weaver, E.G., Dunning, F.B., and Tittel, F.K. (1977). Generation of tunable continuous-wave ultraviolet radiation from 257 to 320 nm. *Opt. Lett.* **1,** 58–60.

Blit, S., Weaver, E.G., Rabson,, T.A., and Tittel, F.K. (1978). Continuous wave UV radiation tunable from 285 nm to 400 nm by harmonic and sum frequency generation. *Appl. Optics*, **5,** 721–723.

Bloom, A.L. (1974). Modes of a laser resonator containing tilted birefringent plates. *J. Opt. Soc. Am.* **64,** 447–452.

Bonarev, B.V., and Kobtsev, S.M. (1986). Calculation and optimization of a birefringent filter for a cw dye laser. *Opt. Spectrosc.* **60,** 501–504.

Born, M. and Wolf, E. (1964). *Principles of Optics*, 2nd ed. Macmillan, New York, pp. 323–341.

Cariou, J., and Luc, P. (1980). *Atlas du Spectre d'Absorption de la Molecule de Tellure.* Laboratoire Aime-Cotton C.N.R.S. II, Orsay, France.

Clark, D.L., Martin, A.G., Dutta, S.B., and Rogers, W.F. (1986). A high resolution computer controlled dye laser system. *Advances in Laser Science-1*, AIP Conf. series 146, Optical Sci. and Eng. subseries 6, (Stwalley, W.C., and Lapp, M., eds.) American Institute of Physics, New York.

Couillaud, B. (1981). Generation of cw coherent radiation in the near U.V. *J. de Physique*, colloque C8, supplement 12, tome 42, 115–125.

Couillaud, B., Dabkiewicz, Ph., Bloomfield, L.A., and Hänsch, T.W. (1982). Generation of continuous-wave ultraviolet radiation by sum-frequency mixing in an external ring cavity. *Opt. Lett.* **7,** 265–267.

Danielmeyer, H.G. (1971). Effects of drift and diffusion of excited states on spatial hole burning and laser oscillations. *J. Appl. Phys.* **42,** 3125–3132.

DeVoe, R.G., Fabre, C., Jungmann, K., Hoffnagle, J., and Brewer, R.G. (1988). Precision optical-frequency difference measurements. *Phys. Rev. A* **37,** 1802–1805.

Drever, R.W.P., Hall, J.L., Kowalski, F.V., Hough, J., Ford, G.M., Munley, A.J., and Ward, H. (1983). Laser phase and frequency stabilization using an optical resonator. *Appl. Phys. B* **31,** 97–105.

Dunn, M.H., and Ferguson, A.I. (1977). Coma compensation in off-axis laser resonators. *Opt. Comm.* **20,** 214–219.

Elliot, D.S., Roy, R., and Smith, S.J. (1982). Extracavity laser band-shape and bandwidth modification. *Phys. Rev. A* **26,** 12–18.

Freeman, R.R., Bjorklund, G.C., Economou, N.P., Liao. P.F., and Bjorkholm, J.E. (1978). Generation of cw VUV coherent radiation by four-wave sum frequency mixing in Sr vapor. *Appl. Phys. Lett.* **33,** 739–742.

Frölich, D., Stein, L., Schröder, H.W., and Welling, H. (1976). Efficient frequency doubling of cw dye laser radiation. *Appl. Phys.* **11,** 97–101.

Gerstenkorn, S., and Luc, P. (1978). *Atlas du Spectre d'Absorption de la Molecule d'Iode.* Laboratoire Aime Cotton, C.N.R.S. II, Orsay, France.

Green, J.M., Hohimer, J.P., and Tittel F.K. (1973). Traveling-wave operation of a tunable cw dye laser. *Opt. Comm.* **7,** 349–350.

Hall, J.L. (1978). Stabilized lasers and precision measurements. *Science* **202,** 147–156.

Hall, J.L., and Hänsch, T.W. (1984). External dye-laser frequency stabilizer. *Opt. Lett.* **9,** 502–504.

Hall, J.L., and Lee, S.A. (1976a). Interferometric real-time display of cw dye laser wavelength with sub-doppler accuracy. *Appl. Phys. Lett.* **29,** 367–369.

Hall, J.L., and Lee, S.A. (1976b). Control techniques for cw dye lasers. In *Tunable Lasers and Applications* (Mooradian, A., Jaeger, T., and Stokseth, P., eds.). Springer-Verlag, Berlin, pp. 361–366.

Hall, J.L., Zhu, M., Shimizu, F.J., and Shimizu, K. (1988). External frequency stabilization of a commercial dye laser at the Hertz level. *XVI International Quantum Electronics Digest* (Japan Society of Applied Physics, Tokyo, 1988), pp. 4–5.

Hänsch, T.W. (1976). Spectroscopic applications of tunable lasers. In *Tunable Lasers and Applications.* (Mooradian, A., Jaeger, T., and Stokseth, P., eds.), Springer-Verlag, Berlin, pp. 326–339.

Hänsch, T.W., and Couillaud, B. (1980). Laser frequency stabilization by polarization spectroscopy of a reflecting reference cavity. *Opt. Comm.* **35,** 441–444.

Harri, H.-P., Leutwyler, S., and Schumacher, E. (1982). Nozzle design yielding interferometrically flat fluid jets for use in single-mode dye lasers. *Rev. Sci. Instrum.* **53,** 1855–1858.

Helmcke, J., Lee, S.A., and Hall, J.L. (1982). Dye laser spectrometer for ultrahigh spectral resolution: Design and performance. *Appl. Opt.* **21,** 1686–1694.

Helmcke, J., Snyder, J.J., Morinaga, A., Mensing, F., and Glaser, M. (1987). New ultra-high resolution dye laser spectrometer utilizing a non-tunable reference resonator. *Appl. Phys. B* **43,** 85–91.

Hemmati, H., Bergquist, J.C., and Itano, W.M. (1983). Generation of continuous-wave 194-nm radiation by sum-frequency mixing in an external ring cavity. *Opts. Lett.* **8,** 73–75, and Hemmati, H., and Bergquist, J.C. (1983). Generation of continuous-wave 243-nm radiation by sum-frequency mixing. *Opt. Comm.* **47,** 157–160, and Wineland, D.J., Itano, W.M., Bergquist, J.C., Bollinger, J.J., and Prestage, J.D. (1985). In Spectroscopy of stored atomic ions. *Atomic Physics* **9,** (Vandyck Jr., J.S., and Fortson, E.N., eds.). World Scientific, Singapore.

Hills, D., and Hall, J.L. (1989). Ultra stable cavity-stabilized lasers with subhertz linewidth. In *4th Symposium on Frequency Standards and Metrology*, Ancona, Italy, Sept. 1988) De Marchi, A. ed.). Springer-Verlag, Heidelberg.

Holberg, L.W. (1984). Measurement of the Atomic Energy Level Shift Induced by Blackbody Radiation. Thesis. U. of Colorado, Boulder.

Hough, J., Hils, D., Rayman, M.D., Ma, L.s., Hollberg, L., and Hall, J.L. (1984). Dye-laser frequency stabilization using optical resonators. *Appl. Phys. B* **33,** 179–185.

Jarrett, S.M., and Young, J.F. (1979). High-efficiency single-frequency cw ring dye laser. *Opt. Lett.* **4,** 176–178.

Jenkins, F.A., and White, H.E. (1957). *Fundamentals of Optics*, 3rd ed. McGraw-Hill, New York, p. 565.

Johnston, T. (1988). High power single frequency operation of dyes over the spectrum from 364 nm to 524 nm pumped by an ultraviolet argon ion laser. *Opt. Comm.* **69,** 147–152.

Johnston, T.F., Jr. (1987). Tunable dye lasers. In *Encyclopedia of Physical Science and Technology* **14,** Academic Press, Orlando, Fla., pp. 96–141.

Johnston, T.F., JR., and Proffitt, W. (1980). Design and performance of a broad-band optical diode to enforce one-direction traveling-wave operation of a ring laser. *IEEE J. Quant. Elect.* **QE-16,** 483–488.

Johnston, T.F. Jr., Brady, R.H., and Proffitt, W. (1982). Powerful single-frequency ring dye laser spanning the visible spectrum. *Appl. Phys.* **21,** 2307–2316.

Johnston, W.D., Jr., and Runge, P.K. (1972). An improved astigmatically compensated resonator for cw dye laser. *IEEE J. Quant. Elect.* **QE-8** (Corresp.), 724–725.

Juncar, P., and Pinard, J. (1975). A new method for frequency calibration and control of a laser. *Opt. Comm.* **14,** 438-441.

Kikuchi, K., and Okoshi, T. (1985). Dependence of semiconductor laser linewidth on measurement time: Evidence of predominance of 1/f noise. *Electron. Lett.* **21,** 1011–1012.

Kogelnik, H.W., Ippen, E.P., Dienes, A., and Shank, C.V. (1972). Astigmatically compensated cavities for cw dye lasers. *IEEE J. Quant. Elect.* **QE-8,** 373-379.

Kowalski, F.V., Hawkins, R.T., and Schawlow, A.L. (1976). Digital wavemeter for cw lasers. *J. Opt. Soc. Am.* **66,** 965–966.

Kowalski, F.V., Teets, R.E., Demtroder, W., and Schawlow, A.L. (1978). An improved wavemeter for cw lasers. *J. Opt. Soc. Am.* **68,** 1611–1613.

Kuhlke, D., Dietel, W. (1977). Mode selection in cw laser with homogeneously broadened gain. *Opt. Quant. Elect.* **9,** 305–313.

Lee, J.H., Kim, K.C., and Kim K.H. (1988). Threshold pump power of a solar-pumped dye laser. *Appl. Phys. Lett.* **53,** 2021–2022.

Liberman, S., and Pinard, J. (1973). Single-mode cw dye laser with large frequency range tunability. *Appl. Phys. Lett.* **24,** 142–144.

Lichten, W. (1985). Precise wavelength measurements and optical phase shifts. I. general theory. *J. Opt. Soc. Am. A* **2,** 1869–1876.

Lichten, W. (1986). Precise wavelength measurements and optical phase shifts II. Applications. *J. Opt. Soc. Am. A* **3,** 909–915.

Loudon, R. (1983). *The Quantum Theory of Light*, 2nd ed. Clarendon Press, Oxford, England.

Marowsky, G., and Zaraga, F. (1974). A comparative study of dye prism ring lasers. *IEEE J. Quant. Elect.* **QE-10,** 832–837.

Marshall, C.M., Stickel, R.E., Dunniway, F.B., and Tittel, F.K. (1980). Computer controlled intracavity SHG in cw ring dye laser. *Appl. Optics* **19,** 1980–1983.

Mollenauer, L.F. (1985). Color center lasers. In *Laser Handbook*, **vol. 4** (Stitch, M.L., and Bass, M., eds.). North-Holland, Amsterdam, pp. 143–228.

Mollenauer, L.F., and White, J.C. (1987). General principles and some common features. In *Tunable Lasers* (Mollenauer, L.F., and White, J.C., eds.). Springer-Verlag, Berlin, Heidelberg, pp. 1–18.

Nieuwesteeg, K.J.B.M., Hollander, Tj., and Alkemade, C.Th.J. (1986). Spectral properties of an Ar-ion laser-pumped tunable cw dye laser. *Opt. Comm.* **59,** 285–289.

Oka, T. (1988). Report of IR difference frequency generation in the 2 to 5 μm range using the crystal lithium iodate and the method developed by Pine (1974). Presentation at the conference, Coherent Sources for Frontier Spectroscopy, International Center for Theoretical Physics, Aug. 1988, Triesta, Italy.

Okoshi, T., Kikuchi, K., and Nakayama, A. (1980). Novel method for high resolution measurement of laser output spectrum. *Electron. Lett.* **16,** 630–631.

Peterson, O.G. (1979). Dye lasers. In *Quantum Electronics, part A. Methods of Experimental Physics*, **15,** (Tang, C.L., ed.). Academic Press, New York, pp. 251–359.

Peterson, O.B., Tuccio, S.A., and Snavely, B.B. (1970). Cw operation of an organic dye solution laser. *Appl. Phys. Lett.* **17,** 245–247.

Peterson, O.G., Webb, J.P., McColgin, W.C., and Eberly, J.H. (1971). Organic dye laser threshold. *J. Appl. Phys.* **42,** 1917–1928.

Pike, C.T. (1974). Spatial hole burning in cw dye lasers. *Opt. Comm.* **10,** 14–17.

Pine, A.S. (1974). Doppler-limited molecular spectroscopy by difference-frequency mixing. *J. Opt. Soc. Am.* **64,** 1683-1690.

Pine, A.S. (1976). High-resolution methane ν_3-band spectra using a stabilized tunable difference-frequency laser system. *J. Opt. Soc. Am.* **66,** 97–108.

Runge, P.K., and Rosenberg, R. (1972). Unconfined flowing-dye films for cw dye lasers. *IEEE J. Quant. Elect.* **QE-8,** 910–911.

Salomon, Ch., Hils, D., and Hall, J.L. (1988). Laser stabilization at the millihertz level. *J. Opt. Soc. Am. B* **5,** 1576–1587.

Sargent, M., Scully, M.O., and Lamb, W.E. (1974). *Laser Physics.* Addison-Wesley, London.

Schäfer, F.P. (1977). *Dye Lasers, 2nd ed.* **Vol. 1** of Topics in Applied Physics. Springer-Verlag, Berlin, Heidelberg. (Good summary of dye lasers as of 1977.)

Schäfer, F.P., Schmidt, W., and Volze, J. (1966). Organic dye solution laser. *Appl. Phys. Lett.* **9,** 306–309.

Schearer, L.D., Leduc, M., Vivien, D., Lejus, A.M., and Thery, J. (1986). LNA: A new cw Nd laser tunable around 1.05 and 1.08 μm. *IEEE J. Quant. Elect.* **QE-22,** 713–717.

Schmidt, W., and Schäfer, F.P. (1967). Blitzlampengepumpte farbstofflaser. *Z. Naturforschg.* **22a,** 1563–1566.

Schröder, H.W., Welling, H., and Wellegehausen, B. (1973). A narrowband single-mode dye laser. *Appl. Phys.* **1,** 343–348.

Schröder, H.W., Dux, H., and Welling, H. (1975). Single mode operation of cw dye lasers. *Appl. Phys.* **7,** 21–28.

Schröder, H.W., Stein, L., Frölich, D., Fugger, B., and Welling, H. (1977). A high-power single-mode cw dye ring laser. *Appl. Phys.*, **14,** 377–380.

Schulz, P.A. (1988). Single-frequency Ti:Al_2O_3 ring laser. *IEEE J. Quant. Elect.* **24,** 1039–1044.

Shank, C.V. (1975). Physics of dye lasers. *Rev. Mod. Phys.* **47,** 649–657.

Siegman, A.E. (1986). *Lasers.* University Science Books, Mill Valley, calif.

Smith, P.W. (1972). Mode Selection in Lasers. *Proc. IEEE* **60,** 422–440.

Snavely, B.B. (1977). Continuous-wave dye lasers. In *Dye Lasers*, **vol. 1**, 2nd ed. Topics in Applied Physics (Schäfer, F.P., ed.). Springer-Verlag, Berlin, Heidelberg, pp. 86–120.

Snavely, B.B., and Schäfer, F.P. (1969). Feasibility of CW Operation of dye-lasers. *Phys. Lett.* **28A,** 728–729.

Snyder, J.J. (1980). Algorithm for fast digital analysis of interference fringes. *Appl. Opts.* **19,** 1223–1225.

Soffer, B.H., and McFarland, B.B. (1967). Continuously tunable, narrow-band organic dye lasers. *Appl. Phys Lett.* **10,** 266–267.

Sorokin, P.P., and Lankard, J.R. (1966). Stimulated emission observed from an organic dye, chloro-aluminum phthalocyanine. *IBM J. Res. Dev.* **10,** 162–163.

Sorokin, P.P., and Lankard, J.R. (1967). Flashlamp excitation of organic dye lasers: A short communication. *IBM J. Res. Dev.* **11,** 148.

Spaeth, M., and Bortfeld, D.P. (1966). Stimulated emission from polymethine dyes. *Appl. Phys. Lett.* **9,** 179–181.

"Squeezed States" (1987). *J. Opt. Am. B* **4** (10), whole issue, and *J. Mod. Optics* **34** (6, 7), whole issue.

Stepanov, B.I., Rubinov, A.N., and Mostovnikov, V.A. (1967). Optic generation in solutions of complex molecules. *JETP Lett.* **5,** 117–119.

Tang, C.L., Statz, H., and deMars, G. (1963). Regular spiking and single-mode operation of ruby laser. *Appl. Phys. Lett.* **2,** 222–224. And Tang, C.L., Statz, H., and de Mars, G. (1963). Spectral output and spiking behavior of solid state lasers. *J. Appl. Phys.* **34,** 2289–2295.

Thiel, E., Zander, C., and Drexhage, K.H. (1986). Continuous wave dye laser pumped by a HeNe laser. *Opt. Comm.* **60,** 396–398.

Theil, E., Zander, C., and Drexhage, K.H. (1987). Incoherently pumped continuous wave dye laser. *Opt. Comm.* **62,** 171–173.

Tuccio, S.A., and Strome, F.C., Jr. (1972). Design and operation of a tunable continuous dye laser. *Appl. Optics* **11,** 64–73.

Williams, G.H., Hobart, J.L., and Johnston, T.F., Jr. (1983). A 10-THz scan range dye laser, with 0.5-MHz resolution and integral wavelength readout. In Laser Spectroscopy VI (Weber, H.P., and Luthy, W., eds.). Springer-Verlag, Heidelberg, pp. 422–423.

Wineland, D.J., Berquist, J.C., Bolinger, J.J., Itano, W.M., Gilbert, S.L., and Hulet, R.G. (1989). Quantum jumps, ion crystals and solid plasmas. CLEO/QELS conference plenary session JA2, Baltimore, Md., unpublished.

Woods, P.T., Shotton, K.C., and Rowley, W.R.C. (1978). Frequency determination of visible laser light by interferometric comparison with upconverted CO_2 laser radiation. *Appl. Opt.* **17,** 1048–1054.

Wu, F.Y., Grove, R.E., and Ezekial, S. (1974). Cw dye laser for ultrahigh-resolution spectroscopy. *Appl. Phys. Lett.* **25,** 73–75.

Yariv, A. (1975). *Quantum Electronics*, 2nd ed. Wiley, New York.

Chapter 6

TECHNOLOGY OF PULSED DYE LASERS

F.J. Duarte

Photographic Research Laboratories
Photographic Products Group
Eastman Kodak Company
Rochester, New York

1. INTRODUCTION

In this chapter various aspects of the technology and design of narrow-linewidth dye-laser oscillators are considered. The discussion starts with a brief description of available pump sources and their spectral and energetic characteristics. This is followed by an outline of pumping geometries, dye cells, and dye flow methods. The discussion then is focused on amplified spontaneous emission, polarization matching, tuning methods, and oscillator-amplifier arrangements.

ISBN 0-12-222700-X

2. EXCITATION SOURCES

Photon excitation of the active medium in dye lasers, dye molecules, is accomplished by utilizing either laser sources or flashlamp radiation. In this section a brief survey of spectral and energetic characteristics of available sources is provided. First, we consider laser sources and then flashlamp systems are described. In the case of laser-pump sources, the approximate bandwidth of the emission is provided (see Table 6.1). This information is given to highlight the availability of direct fine-tuning of narrow-linewidth radiation centered at the main emission wavelength utilizing grazing-incidence, MPL, HMPGI, and other oscillator configurations (see Section 8, Chapter 4). This alternative is particularly relevant in the case of excimer lasers, which emit via molecular transitions in the ultraviolet.

Table 6.1

Available Lasers for Excitation of Pulsed Dye Lasers

Lasers	Transition	λ (nm)	prf[a]	~Bandwidth (GHz)	References[b]
KrF	$B^2\Sigma^+_{1/2}$–$X^2\Sigma^+_{1/2}$	248	200 Hz	10500[c]	Loree *et al.* (1978)
				2583[c]	Caro *et al.* (1982)
XeCl		308	500 Hz	374	Lyutskanov *et al.* (1980)
				204	McKee (1985)
		308.2		397	Lyutskanov *et al.* (1980)
				223	McKee (1985)
XeF		351	200 Hz	187	Yang *et al.* (1988)
		353		330	Yang *et al.* (1988)
N_2	$C^3\Pi_u$–$B^3\Pi_g$	337.1	100 Hz	203	Woodward *et al.* (1973)
HgBr	$B^2\Sigma^+_{1/2}$–$X^2\Sigma^+_{1/2}$	502	<100 Hz	918	Shay *et al.* (1981)
		504		1012	Shay *et al.* (1981)
Ca^+	$5^2S_{1/2}$–$4^2P_{3/2}$	373.7	—	—	—
Sr^+	$6^2S_{1/2}$–$5^2P_{3/2}$	430.5	0.5–15 kHz	2–12[c]	Bukshpun *et al.* (1981)
Cd^+	$4^2F_{5/2}$–$5^2D_{3/2}$	533.7	—	—	—
Cu	$^2P_{3/2}$–$^2D_{5/2}$	510.5	5–20 kHz	7	Tenenbaum *et al.* (1980)
	$^2P_{1/2}$–$^2D_{3/2}$	578.2	5–20 kHz	11	Tenenbaum *et al.* (1980)
Au	$^2P_{1/2}$–$^2D_{3/2}$	627.8	5–20 kHz	—	—
$Al_2O_3:Cr^{3+}$	$\bar{E}(^2E) - {}^4A_2$	694.3	<1 Hz	7	D'Haenens and Asawa (1962)
Nd:YAG	$^4F_{3/2}$–$^4I_{11/2}$	1064	10–50 Hz	15–30	—
	Second harmonic	532			—
	Third harmonic	355			—
	Fourth harmonic	266			—

[a] prf figures do not represent absolute limits.
[b] References relate to the full-width bandwidth exclusively.
[c] Tuning range.

In selecting an appropriate pump for a particular dye laser, several parameters have to be considered. These variables include photon-conversion efficiency, spectral region, dye lifetime, peak power, pulse length, pulse repetition frequency (prf), and cost.

A brief example related to a particular dye can be used to illustrate the efficiency alternative. In essence, dyes are large molecules (with molecular weights ranging from about 175 to 1000) that can absorb light from a wide spectral region and fluoresce at longer wavelengths. In these molecules the largest absorption cross section corresponds to the $S_0 \rightarrow S_1$ electronic transition. In addition, lasing occurs from the first excited-singlet electronic state to the ground state in the $S_1 \rightarrow S_0$ transition. Thus, excitation of the dye molecules at wavelengths exceeding the energy of the S_1 level lead to the population of higher-lying electronic singlet states (such as S_2 and S_3) and is followed by rapid (subpicosecond) nonradiative transitions to the first-lying electronic singlet S_1. Excitation of higher singlet states reduces conversion efficiency and leads to inferior photochemical stability of the dye molecules. Therefore, ultraviolet excitation of a red dye such as rhodamine 590 ($MW = 479$, and maximum absorption at ~528 nm) leads to population of higher-lying electronic singlets and is not very efficient. On the other hand, for the same dye, efficiencies ranging from 29 to 55% have been reported for copper-laser excitation (at 510.5 nm) (Morey, 1980; Hargrove and Kan, 1980). As expected, the conversion efficiency resulting from copper-laser pumping decreases substantially (to <10%) in the case of IR 144 emitting at 900 nm (Morey, 1980). The concepts outlined here are neatly summarized in the ideal optimum efficiency relation discussed by Shank (1975), $\eta \approx (\lambda_p/\lambda_e)$, where λ_p is the wavelength of the pump source and λ_e is the dye-emission wavelength.

Therefore, if conversion efficiency is an important parameter, then one should use excimer or nitrogen lasers in the excitation of near-UV and blue-green dye lasers, copper or frequency-doubled Nd:YAG lasers for the excitation of orange-red dyes, ruby or gold vapor lasers for near-infrared dyes, and Nd:YAG lasers for the excitation of infrared dyes emitting in the 1.1 to ~1.3-μm region.

For high-prf excitation the alternatives available are recombination lasers for dyes emitting in the blue-green and certainly copper-vapor lasers for dyes lasing in the orange-red. On the other hand, if cost is the main factor to consider, then the best coherent source alternative is the nitrogen laser.

Flashlamp excitation offers the advantage of a single-device system and long pulse lasing well into the μs regime.

In summary, it is difficult to point to the best or most appropriate dye-laser pump without specifying application requirements, emission characteristics, and cost.

2.1. Excimer Lasers

Excimer lasers utilized in dye-laser pumping are mainly excited by electrical discharge. These lasers offer excitation at various wavelengths starting with F_2 and ArF at 157 and 193 nm, respectively. Typically, excimer lasers provide pulse lengths in the 10–30-ns range although recently pulse lengths >200 ns have become available. Pulse repetition frequencies in excimer lasers vary from less than one Hz to a few hundred Hz. However, some research devices can operate at a prf of up to one kHz (Sze and Seegmiller, 1981; Baranov *et al.*, 1984). On a comparative basis it should be mentioned that employing the same laser head, the KrF medium can provide up to twice the energy of ArF or XeCl. Conversion efficiencies are in the 3–4% range for KrF and 2–3% for XeCl. Single-pulse output energies of commercial excimer lasers are in the 10–1000-mJ range.

In the late 1980s, the most widely used excimer dye-laser pump is the XeCl laser. The KrF system is utilized mainly in the excitation of dyes emitting in the near UV.

As mentioned in Chapter 4 the molecular transitions involved in these lasers offer a limited tunability range. The tunability ranges and corresponding electronic transitions are included in Table 6.1. It should be clear that these transitions involve many vibrational-rotational levels and that the bandwidth quoted does not offer a uniform intensity profile but exhibits significant structure. This is highlighted in the case of the XeCl and the XeF lasers where two main frequencies are quoted. In the XeF laser the emission near 351 nm involves rotational levels associated with the $(B, v' = 1) \rightarrow (X, v'' = 4)$ and $(B, v' = 0) \rightarrow (X, v'' = 2)$ transitions, whereas emission at 353 nm involves primarily J levels from the $(B, v' = 0) \rightarrow (X, v'' = 3)$ transition (Yang *et al.*, 1988). Another electronic transition of the XeF laser $(C \rightarrow A)$ may offer further prospects for excitation in the blue-green region of the spectrum (see, for example, Tittel *et al.* (1986)). Observed spectral bandwidth depends on discharge parameters, such as pressure, and optical cavity configurations. Thus, quoted bandwidth may vary for the same type of laser.

The approximate bandwidth information provided in Table 6.1 contains two types of emission. In the case of the KrF laser, intracavity tuning of narrow-bandwidth emission was reported for about 10,500 GHz; this should be compared with a bandwidth of ~2853 GHz (at full width) for the untuned (or broadband) lasing (Loree *et al.*, 1978). The tuning range (2583 GHz) reported for the narrow-linewidth master-oscillator forced-oscillator system of Caro *et al.* (1982) is close to the full width of the emission in the absence of injection. For the case of the XeCl and XeF lasers the quoted bandwidth corresponds to broadband stimulated emission.

It is interesting to note that perhaps the largest energy per pulse reported for a dye laser as of 1990 has been obtained utilizing an e-beam-excited XeCl laser pump (Tang *et al.*, 1987). These authors report 600 J per pulse for rhodamine 6G and ~800 J per pulse for coumarin 480, at a pulse width of about 500 ns. The dye laser was excited in a transverse configuration.

2.2. Other Molecular Lasers

One of the most widely used dye-laser excitation sources is the nitrogen laser emitting in its second positive system ($C^3\Pi_u - B^3\Pi_g$) at 337.1 nm. This electronic transition is compatible with the excitation of a large number of dye species. In addition, the nitrogen laser is inexpensive and simple to construct and operate. A convenient transverse discharge design for dye-laser pumping is described by Schenck and Metcalf (1973).

In general, nitrogen lasers utilized in dye-laser pumping yield 1–10 mJ per shot at pulse lengths in the 1–10-ns range. Pulse-repetition frequencies range from 10 to 100 Hz. Typical efficiencies are in the 0.06–0.1% range, although higher efficiencies have been reported (Godard, 1974; Oliveira dos Santos *et al.*, 1986).

The $C^3\Pi_u - B^3\Pi_g$ system in N_2 at 337.1 nm exhibits clearly defined structure and it involves numerous rotational levels in the $v' = 0 \rightarrow v'' = 0$ transition (see, for example, Parks *et al.* (1968) and Woodward *et al.* (1973)). This system also gives rise to additional vibrational transitions $v' = 0 \rightarrow v'' = 1$, $v' = 1 \rightarrow v'' = 0$, at around 357.7 and 315.9 nm respectively; further the $B^3\Pi_g - A^3\Sigma_u^+$ first positive system yields transitions in the near IR (for more information and references on the N_2 laser, see Willett (1974)).

The HgBr laser emits in the $B^2\Sigma_{1/2}^+ - X^2\Sigma_{1/2}^+$ electronic transition in the blue-green region of the spectrum at wavelengths centered at 502 and 504 nm. Laser energy can be in the 10–100-mJ range in pulses 40–80 ns long. However, using a cavity-dumped configuration, Leslie *et al.* (1984) reported pulses of about 8-ns duration. Pulse-repetition frequencies vary from a few Hz up to 100 Hz (Burnham and Schimitschek, 1981). Overall efficiencies are in the 1–2% range (Shay *et al.*, 1985).

It should be noted that the bandwidth quoted in Table 6.1 refers to the broadband emission output spectrum of the HgBr laser. Shay *et al.* (1981) have demonstrated continuous tuning in the 501–505-nm region (at $\Delta\nu \sim 59$ GHz) using a master-oscillator forced-oscillator configuration. In these injection-locking experiments the master oscillator was a flashlamp-pumped dye laser yielding 200-ns pulses at a linewidth of ~24 GHz.

Tellinghuisen and Ashmore (1982) provide a vibrational assignment of transitions in the $B \rightarrow X$ system of HgBr that includes v' levels 0–5 and v''

levels 22–27. The rich output spectrum observed in the HgBr laser indicates that this blue-green source is an excellent candidate for fine frequency tuning of narrow-linewidth emission utilizing the multiple-prism grating techniques described in Chapter 4.

Since mercury halide lasers (including HgCl) have not been developed commercially, their use as dye-laser pumps has yet to be fully exploited.

2.3. Recombination Lasers

Recombination lasers provide an additional alternative for dye-laser pumping. In this regard, the Ca^+ and Sr^+ systems emitting in the near UV (373.7-nm) and blue (430.5-nm) regions of the spectrum are particularly interesting since they can offer high prf operation.

In an excellent paper Latush *et al.* (1982) list a number of recombination laser systems and corresponding transitions. In the same paper the operation of Sr^+ and Ca^+ lasers in a self-heated discharge tube at a prf of 6.5 kHz was reported. For Sr^+ an average power of 2 W was obtained in 200-ns pulses at an efficiency of 0.15% (Latush *et al.*, 1982). Bukshpun *et al.* (1981) report on pulse-repetition frequencies in the 5–15-kHz range for the Sr^+ laser.

Operation at lower repetition rates in the Sr^+ laser has been reported to yield up to 6 mJ per pulse at a pulse duration of ~200 ns (Butler and Piper, 1985). Brandt (1984) has reported on a transversely excited Sr^+ laser operating in the 200–500 Hz regime.

For the Sr^+ laser Bukshpun *et al.* (1981) reported an intrinsic linewidth of 1.5–1.8 GHz. Using an intracavity etalon they achieved continuous frequency tuning in the 2–12 GHz range. The tuning range was a function of the He pressure and was approximately equal to the width of the spontaneous emission line. At a helium pressure of 700 Torr, the tuning range was 12 GHz (Bukshpun *et al.*, 1981).

2.4. Metal-Vapor Lasers

Copper lasers provide probably the best alternative for the excitation of dyes emitting in the orange-red region of the spectrum. Among the unique properties of copper lasers are high prf operation (up to 30 kHz) and high overall conversion efficiency (1–3% in transversely excited devices) (Artemev *et al.*, 1982). The most well known utilization of copper-vapor lasers as excitation sources of dye lasers is in the large-scale atomic vapor laser isotope separation (LIS) program at Lawrence Livermore National Laboratory (see, for example, Paisner (1988)).

For a transverse-discharge copper-vapor laser, Artemev *et al.* (1982) reported on an average power of ~100 W at a prf of 3 kHz. The corresponding pulse energy was 33 mJ and the peak power exceeded 1 MW. Copper-vapor lasers are produced commercially in a variety of models offering average powers in the 2–100-W range at pulse-repetition frequencies in the 4–20-kHz region. For this type of laser, pulse length is in the 10–60-ns range. From a different perspective, single pulse energies are available in the 0.2 to ~20-mJ range. Efficiency in longitudinal discharge copper-vapor lasers is ~1.2% (Webb, 1988).

As mentioned in Chapter 4, in addition to copper-vapor lasers operating in the high-prf mode, low-repetition transverse-excitation devices utilizing CuBr have been employed in the pumping of dye lasers (Duarte and Piper, 1982). These lasers, which operate at a temperature range of 420–520°C (substantially lower than required temperatures for the elemental metal devices (~1500°C)), have been shown to yield 2.5 mJ per pulse at pulse lengths of about 25 ns (Piper, 1978; Brandt and Piper, 1981).

In copper lasers the output energy ratio of the 510.5 and 578.2 nm lines varies according to design and discharge characteristics. For instance, one can find in the literature green-to-yellow power ratios in the range of 1.2:1 (Webb, 1988) to 3:1 (Zenchenko *et al.*, 1985).

Copper occurs in two isotopes (^{63}Cu and ^{65}Cu) and each of these isotopes contributes a set of hyperfine components. Certainly, each of the hyperfine transitions is Doppler-broadened and the resulting overall linewidth is at least several GHz wide. Linewidth measurements for an elemental metal device (at a buffer gas pressure of 25 Torr) are given by Tenenbaum *et al.* (1980). Here, $\Delta\nu \sim 7$ GHz at 510.5 nm and $\Delta\nu \sim 11$ GHz at 578.2 nm. Laser linewidth measurements as a function of buffer gas pressure and temperature in a CuBr laser are provided by Wang *et al.* (1988).

Gold-vapor lasers ($\lambda = 627.8$ nm), utilizing similar technology to the pure metal commercial copper-vapor laser designs mentioned previously yield 0.2–1.5 mJ per pulse at similar pulse lengths. These lasers have been shown to exhibit a constant 0.2% efficiency as the repetition rate is increased from 5 to 20 kHz (Webb, 1988). The lower efficiency quoted for the gold laser is partly due to the fact that gold is less volatile than copper and, for technological reasons, the discharge must be maintained at temperatures slightly below the optimum operating temperature (1600°C) at 627.8 nm (Webb, 1988). An additional metal-vapor laser system of interest in the red is the lead laser (722.9 nm) (Bricks and Karras, 1980).

2.5. Solid-State Lasers

The most widely used solid state-laser in dye-laser pumping is the Nd:YAG laser ($\lambda = 1064$ nm). Commercial lasers of this type deliver

100–1000 mJ per pulse, in the fundamental, at pulse repetition frequencies in the 10–50-Hz range. Available pulse lengths vary from 5 to 30 ns.

The efficiency of a high beam quality Nd:YAG laser utilized in dye-laser excitation can be ~1% (at 1064 nm). However, efficiency depends on system design and can be considerably lower in certain cases. Conversion to second, third, and fourth harmonics is accomplished using KDP or KD*P crystals. Conversion efficiency of harmonic generation depends on several factors, including the properties of the crystal, laser wavelength, spectral purity at the fundamental frequency, and beam quality. For instance, second-harmonic (532-nm) generation can be accomplished in the 35–50% efficiency range. On the other hand, generation of the fourth harmonic (266 nm) offers much lower efficiencies (5–10%). Thus, a particular Nd:YAG laser system yielding X mJ at 1064 nm upon harmonic conversion may produce $(X/2.2)$ mJ at 532 nm, $(X/5.5)$ mJ at 355 nm, and $(X/13.5)$ mJ at 266 nm.

In addition to the versatile wavelength options available with Nd:YAG lasers, further characteristics of these systems include high peak powers and excellent beam quality (close to TM_{00} mode). Fountain and Bass (1982) report that a Nd:YAG oscillator with an intrinsic bandwidth of 15 GHz can easily yield a linewidth of 1.5 GHz with the introduction of an intracavity etalon. These authors describe the design of an oscillator yielding single–longitudinal-mode operation at an energy output of 100 mJ at the fundamental.

The ruby laser ($\lambda = 694.3$) was the first excitation source to be utilized to pump dye lasers (Sorokin and Lankard, 1966; Schäfer *et al.*, 1966; Spaeth and Bortfeld, 1966). The early dye lasers were excited directly at 694.3 nm and emitted in the 731–835-nm region of the spectrum (Schäfer *et al.*, 1966). Second-harmonic generation (at 347 nm) has also been used to excite dye lasers emitting in the visible region of the spectrum (see, for example, Soffer and McFarland (1967)). Ruby lasers utilized in dye-laser pumping emit high peak powers (in the 5–100-MW range) in pulses 10–30 ns long at low pulse repetition frequencies. The bandwidth quoted in Table 6.1 refers to the stimulated emission linewidth of a laser working at a temperature of 300°K (D'Haenens and Asawa, 1962).

2.6. Flashlamp-Pumped Dye Lasers

A different form of excitation of dye molecules involves the use of broadband UV-VIS emission from flashlamps (Sorokin and Lankard, 1967; Schmidt and Schäfer, 1967; Snavely *et al.*, 1967). This broadband emission is usually the result of a xenon discharge although other gases such as air, argon, and krypton are also used. The first flashlamp-pumped dye laser utilized a coaxial lamp filled with a 4:1 air-argon mixture (Sorokin and Lankard, 1967). This alternative form of optical pumping offers inherent symmetric excitation and allows the attainment of large laser output energies per pulse in a single device. In addition, the availability of relatively long pulses in the microsecond regime makes these lasers very attractive for a wide range of applications. Table 6.2 gives the output characteristics of several flashlamp-pumped dye lasers. Parameters tabulated are type of excitation configuration, dye (and concentration), single-pulse laser energy, pulse duration, and efficiency.

In this section we consider several parameters and characteristics unique to flashlamp-pumped dye lasers. It is apparent from Table 6.2 that there are two methods of flashlamp excitation: one that involves the use of linear flashlamps, whereas the other utilizes coaxial flashlamps. Briefly, linear excitation usually involves the use of two or more lamps arranged symmetrically around the active region. For example, Jethwa and Schäfer (1974) and Jethwa *et al.* (1982) describe single- and double-elliptical reflectors used to configure double and quadruple linear flashlamp excitation, respectively. Pumping arrangements involving the use of 8 and 12 linear flashlamps placed around the dye cell, configured to provide cylindrical excitation, are described by Hirth *et al.* (1981) and Fort and Moulin (1987), respectively. Features common to the linear systems quoted here involve the use of outer boundary reflectors and cooling fluid around the flashlamps (see Fig. 6.1a).

Coaxial flashlamp-pumping (see, for example, Sorokin *et al.* (1968) and Furomoto and Ceccon (1969)) provides inherent concentric excitation that offers better optical coupling. In addition, this type of excitation has been shown to be suitable to provide large output energies, in the range of 110–400 J per pulse (Baltakov *et al.*, 1973; Baltakov *et al.*, 1974) and attractive efficiencies (up to 1.6% (Neister, 1977a)). Coaxial excitation is depicted in Fig. 6.1b. A disadvantage associated with coaxial pumping is the presence of shock waves and thermal effects which can induce changes in the refractive index of the active medium. These effects can cause early termination in long-pulse lasers (Blit *et al.*, 1974; Ewanizky *et al.*, 1973).

Flashlamp studies conducted by Everett *et al.* (1986) indicate that a substantial portion of the UV-VIS flashlamp photon flux originates around the 200-nm region. As discussed earlier, such emission leads to the excitation of higher singlet states (such as S_3) and to photodegradation of visible-emitting dyes. To avoid hard UV pumping, authors have employed a number of UV filters such as glass, Pyrex, and caffeine (Calkins *et al.*, 1982). It should be clear that additives such as caffeine are used in conjunction with the flashlamp cooling fluid. Authors have reported the use of a variety of such ultraviolet absorbers including $NaCOOCH_3$ (Baltakov

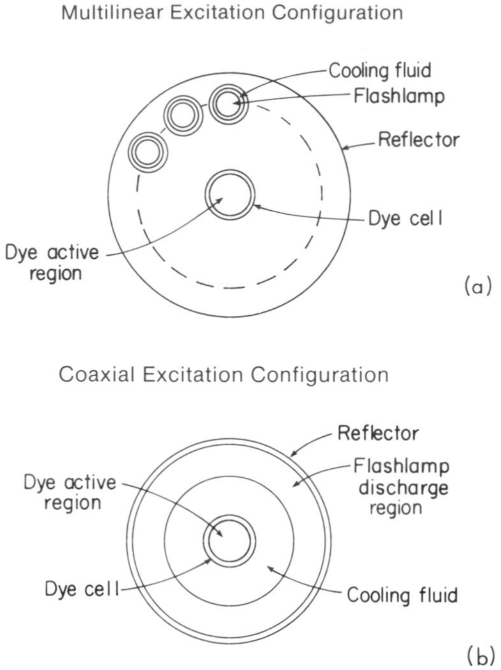

Fig. 6.1 (a) Cross-sectional view of a multiple-lamp linear-excitation configuration. Up to 12 lamps are placed symmetrically around the active region (see text). (b) Cross section of a coaxial flashlamp dye laser.

et al., 1973), ethanol (Jethwa *et al.*, 1982), and $CuSO_4$ (or $NaNO_3$) (Knyazev *et al.*, 1983). Jethwa and Schäfer (1974) indicate that the use of $CuSO_4$, in addition to its function as filter, prevents shock-formation in their two-lamp linear configuration.

In addition to the use of UV absorbers, some authors have employed UV radiation utilizing transfer dyes around the flashlamps to improve the excitation coupling between flashlamp and the dye molecules (see, for example, Burlamacchi and Cutter (1977), Mazzinghi *et al.* (1981), and Everett *et al.* (1986)). Of particular interest are the results of Everett *et al.* (1986), who reported a 75% enhancement in their coumarin 504 laser output utilizing the transfer dye stilbene 420 at a concentration of 2.5×10^{-4} M in EG/H_2O.

A feature common to both methods of excitation is the use of relatively low dye concentrations (see Table 6.2) and relatively large active volumes. Typically, the length of an active region is in the 15–75-cm range and the diameter of active regions vary from a few millimeters to 1–2 cm or more

Table 6.2

Output Characteristics of Flashlamp-Pumped Dye Lasers

System	Dye	Output energy	Pulse duration	% Eff.	Reference
Linear					
	rhodamine 6G				
2	1.15×10^{-4} M	2.54 J[b]	—	0.8	Jethwa *et al.* (1982)
4		5.4 J	—	0.34	Jethwa *et al.* (1982)
2	8×10^{-4} M[a]	5 J	15 μs	1.8	Mazzinghi *et al.* (1983)
	coumarin 504				
2	1.9×10^{-4} M	7 J	~3 μs	1.0	Everett *et al.* (1986)
	rhodamine 6G				
6	8×10^{-4} M[a]	~5 J[c]	~12 μs	0.6	Mazzinghi *et al.* (1981)
8	4×10^{-4} M	12 J	5 μs	1.2	Anliker *et al.* (1972)
8	1×10^{-4} M	5.4 J[d]	2.5 μs	0.6	Hirth *et al.* (1981)
12	8×10^{-5} M	40 J	7 μs	0.4	Fort and Moulin (1987)
Coaxial	rhodamine 6G				
	1.5×10^{-4} M	6.82 J	600 ns	1.1	Bunkenburg (1972)
	2.3×10^{-4} M	8 J	5 μs	0.3	Knyazev *et al.* (1983)
	2.2×10^{-5} M	400 J	10 μs	0.8	Baltakov *et al.* (1974)

[a] Slab geometry.
[b] Provides an average power of 50 W at a prf of 30 Hz.
[c] Provides an average power of 200 W at a prf of 50 Hz.
[d] Can be operated at a prf of up to 10 Hz.

(Baltakov *et al.* (1974) reports on a dye cell inner diameter as large as 6 cm). Hence, lower dye concentrations are necessary to avoid nonuniform absorption along the radius. An exception in this regard is the slab geometry utilized by Mazzinghi *et al.* (1981, 1983), which allows the use of relatively high dye concentrations (see Table 6.2).

The long-pulse characteristics of flashlamp-pumped dye lasers make the excitation process of these devices susceptible to triplet effects. Briefly, triplet states are long-lived metastables, which contribute to losses in the laser excitation cycle. Besides contributing to the depletion of the ground-level population, the presence of triplet states can lead to triplet–triplet absorption, which may overlap the laser emission spectrum. These effects can be reduced via collisional deexcitation of the triplet level. Thus, it is a common practice to use triplet-state quenchers in the dye solution. A brief survey indicates the use of various triplet quenchers including C_8H_8 (COT) (Pappalardo *et al.*, 1970; Anliker *et al.*, 1972; Bunkenburg, 1972; Jethwa and Schäfer, 1974; Mazzinghi *et al.*, 1981; Fort and Moulin, 1987),

C_7H_8 (CHT) (Pappalardo *et al.*, 1970), oxygen (Marling *et al.*, 1970a; Schäfer and Ringwelski, 1973; Hirth *et al.*, 1981), and diacetam-5 (Denisov *et al.*, 1985). Marling *et al.* (1970b) measured the effect, on output energy and pulse length, caused by numerous triplet-state quenchers including oxygen, NO_2^-, and C_8H_8. These authors report considerable improvement in laser characteristics for some dyes including coumarin 1 (or coumarin 460), rhodamine 6G, and kiton red s. A detailed study on the effect of oxygen as a triplet-state quencher in numerous dyes has been provided by Marling *et al.* (1970a). These experiments indicated that the addition of oxygen yields good results in dyes such as rhodamine 6G, rhodamine B, 7-hydroxycoumarin, and coumarin 1. However, a different behavior was observed with other dyes (such as 3-aminofluoranthene) in which the addition of oxygen caused lasing to cease (Marling *et al.*, 1970a). These authors also report that the output for a third group of dyes was optimized at a given oxygen concentration. These results indicate that in addition to being an effective triplet-state quencher, oxygen has the ability to deexcite the first singlet state (S_1) by augmenting intersystem crossing to the triplet state (Marling *et al.*, 1970a). Further results on this topic are discussed by Schäfer and Ringwelski (1973).

Flashlamp-pumped dye lasers have been successfully operated at relatively high prf's. For instance, Jethwa and Schäfer (1974) report on a 6.6-W (average power) rhodamine 6G dye laser excited by two linear flashlamps at a prf of 100 Hz. An average power of 50 W was obtained by Jethwa *et al.* (1982) at a prf of 30 Hz and an average power of 200 W was reported by Mazzinghi *et al.* (1981) for a prf of 50 Hz. At this stage it should be indicated that the single-pulse energies listed in Table 6.2 are higher than the single-pulse energies achieved at high-prf operation. Notice that the system utilized by Mazzinghi *et al.* (1981) employed a planar waveguide dye laser with a transverse dye flow configuration and transverse illumination by six flashlamps (three on each side). The flashlamp system described by Everett *et al.* (1986) also utilizes transverse dye flow and was developed to meet high-prf operation (up to 500-Hz) requirements for LIS applications.

Everett and Zollars (1988) and Zollars and Everett (1988) discuss the engineering of a module of a multibeam flashlamp-pumped dye-laser system designed to yield an average of 300 W. The module tested gives 7 J per pulse at an efficiency of 0.35%. An interesting feature of this project is the use of an automated dye-cleaning system, incorporating carbon filters, which provides the ability to operate the lasers for prolonged periods of time without the detrimental effects of dye photodegradation. Zollars and Everett (1988) also discuss the use of several solvents including an acetamine and water mixture, which was found to provide low risk of fire hazard and toxicity. Another development of interest in the field of flashlamp excitation of dye molecules is the use of zigzag slab geometry to attempt to reduce thermal distortions (Dearth *et al.*, 1988).

Soffer and McFarland (1967) demonstrated stimulated emission using laser excitation of polymethyl methacrylate (PMMA) doped with rhodamine 6G. Peterson and Snavely (1968) achieved laser action utilizing flashlamp-excitation of rods of PMMA–rhodamine 6G (at 1.4×10^{-4} M) and PMMA–rhodamine B (at 6×10^{-5} M). In an interesting comparison with lasing in the liquid state, Peterson and Snavely (1968) indicate that the use of a solid-state medium leads to a reduction of laser-emission quenching by the triplet state. The use of transfer dyes to improve conversion efficiency in dye-doped PMMA rods was reported by Drake *et al.* (1972). Studies in solid-state flashlamp-pumped dye lasers employing polymer hosts have been reported by Pacheco *et al.* (1988) and Erickson (1988). Pacheco *et al.* (1988) describe results obtained with coaxial-lamp excitation of PMMA rods doped with rhodamine 590 (at 1.1×10^{-4} M) and coumarin 540 (at 1.4×10^{-4} M). In the case of rhodamine 590, more than 50 mJ, in a 300-ns pulse, was obtained for a rod 0.95 cm in diameter and 18 cm long. For coumarin 540 the output energy was 50 mJ, in a 475-ns pulse, for a rod of the same diameter and 42 cm long.

On the other hand, a substantial amount of work has been devoted to dye lasers in the vapor phase. A good review on this subject, which includes a comprehensive reference list, is given by Nair (1982). Studies on discharge excitation of dye molecules in the vapor phase have been conducted by Eden (1986). In this regard, it should be noted that Basov *et al.* (1981) obtained laser action from coumarin 6 in the vapor phase using flashlamp excitation. In these experiments coumarin 6 was stabilized with ether vapor and the operating temperature was in the 200–240°C range. A study on the fluorescence dynamics of POPOP under broadband flashlamp excitation is given by Trusov (1981).

3. PULSED LASER EXCITATION GEOMETRIES

The laser-excited narrow-linewidth oscillators described in Chapter 4 assumed transverse pumping. In this section a brief description of the parameters involved in transverse excitation and alternative pump schemes are provided.

Transverse excitation is perhaps the most general choice for laser pumping of narrow-linewidth oscillators. As illustrated in the various arrangements of Chapter 4, this form of excitation is quite simple, avoids

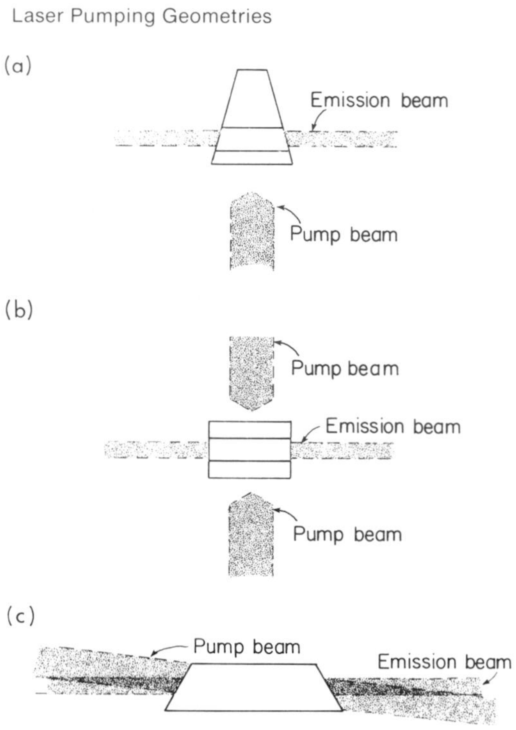

Fig. 6.2 Laser pumping geometries. (a) Typical transverse laser excitation geometry. (b) Colinear two-sided transverse excitation arrangement. (c) Semilongitudinal excitation.

interaction of the pump beam with mirrors and other intracavity components of the oscillator, and allows a compact geometry (Fig. 6.2a). A limitation of transverse laser pumping is an uneven spatial excitation of the active volume. This lack of uniformity occurs in the direction of the path of the excitation beam owing to absorption, and along the axis of propagation (orthogonal to the exitation beam) in the active region due to the inherent beam profile of the pump laser.

Focusing of the entire pump-laser energy into a central spot at the dye cell may lead to optical damage of the cell and to side regions of unexcited dye solution, which may reabsorb and provide favorable conditions for thermal lensing effects (Duarte, 1983). Thus, in the oscillator designs outlined in Chapter 4, the entire width (~10 mm) of the dye cell is illuminated.

In the case of excitation by lasers with a rectangular beam cross section, an appropriate convex lens (silica) can be used to provide a very thin and wide beam (usually ~10 mm wide and 0.1–0.5 mm high) at the focal plane. This approach is applicable to nitrogen and excimer laser pumps (see, for example, Hänsch (1972), Duarte and Piper (1980), and Duarte (1987)). On the other hand, copper-vapor lasers offer a beam with a circular cross section. Hence, for the case of a 13 mm diameter beam a cylindrical lens was utilized to illuminate the whole width of the dye cell by focusing to a thin horizontal stripe ~0.15 mm high (Duarte and Piper, 1984). Ideally, in this type of excitation the dye concentration should be adjusted to ensure that the resulting width of the active region is close to the height of the focused excitation beam (for example, ~150 μm, see Hänsch (1972)). Since the active volume is often formed close to the front window of the dye cell, the required depth of the cell is less than 1 mm. This transverse geometry leads to the excitation of long dye volumes with small cross-sectional areas yielding beam waists in the 100–200 μm range, which provide convenient Rayleigh lengths, and conditions favorable to yield near diffraction-limited beams.

This simple type of transverse pumping is also widely used in the excitation of amplifier stages. In the case of amplifiers, improved excitation uniformity has been demonstrated by pumping the active region with a transverse dual-beam approach that employs two colinear beams from opposite directions (see Fig. 6.2b). This type of geometry has been employed with copper lasers (Hargrove and Kan, 1980) and high-average-power excimer lasers (Klick *et al.*, 1986).

Longitudinal pumping was frequently utilized in early laser-pumped dye lasers (see, for example, Sorokin *et al.* (1966) and Sorokin *et al.* (1967)). Using an appropriate dye concentration and absorption length this approach can provide better spatial uniformity of the active dye volume. In order to avoid interaction of the high-power pump beam with the optical elements of the dye laser, Bradley *et al.* (1968) introduced a variation of longitudinal pumping where the excitation beam and the optical axis of the dye laser are at a small angle in a semilongitudinal configuration (Fig. 6.2c). A number of publications discuss the use of this pumping scheme in the excitation of dye-laser oscillators (Morey, 1980; Soldatov and Sukhanov, 1983; Littman, 1984). Using semilongitudinal bilateral laser pumping of a MPL oscillator, Nechaev (1983) reported improved conversion efficiency. Further, this pumping configuration has been used successfully in the excitation of amplifier stages (see, for example, Hargrove and Kan (1980)) and is commonly used in the pumping of cw dye lasers.

The excitation of dye solutions in quasi-waveguide geometries has been discussed by Arutunyan *et al.* (1984).

4. DYE CELLS

In this section the geometry of various dye-cell designs is described. The discussion is initiated with geometric requirements for dye cells designed primarily to emit low ASE levels. Then we consider several options to extend the applicability of these cells to a high-prf mode of operation.

A primitive transversely excited dye laser only requires a square or rectangular cell with a reflective window and a partially transmitting window. Alternatively, a simple broadband device can incorporate a square or rectangular dye cell and two external mirrors. However, in the case of transversely excited narrow-linewidth dye lasers, it was clear from the early stages that it is advantageous to utilize cell designs with the emission windows tilted at an angle relative to an axis, on the plane of propagation, perpendicular to the cavity axis (see, for example, Hänsch (1972)). This simple wedge-type geometry is quite effective in reducing parasitic internal reflections. Antireflection coatings are used to provide further improvements. Roncin and Damany (1981) have designed demountable trapezoidal dye cells.

Alternatives to the trapezoidal designs include dye cells of the parallelogram type (see, for example, Littman and Metcalf (1978)) and rectangular cells oriented at an angle on the plane formed by the cavity axis and the axis perpendicular to the plane of propagation (see, for example, Itzkan and Cunningham (1972) and Duarte (1987)). These alternatives are outlined in Fig. 6.3.

For low-prf operation the dye either is not flowed or is flowed at a relatively low speed in a direction perpendicular to the plane of propagation. Other arrangements include a small magnetic stirrer. In most cases the cross-sectional area of dye cells, utilized in low-prf operation, is of the order of $\sim 1\ cm^2$.

High-prf operation demands flowing the dye solution at relatively high speeds. A well-known consequence of Bernoulli's equation is that, for a given pressure, reduction in the cross-sectional area results in higher flow velocity. Thus, in addition to the use of a fluid pressure pump it is often necessary to reduce the cross-sectional area of the dye cell.

A simple versatile trapezoidal dye-laser cell, applicable to low- and high-prf laser excitation, is that utilized by Duarte and Piper (1980, 1984). The cross-sectional area of this cell is 11 mm wide by $\leqslant 1$ mm deep, the height is about 20 mm, and the side windows are tilted to $\sim 12°$. For low-prf operation this cell was employed with a static dye fill. For high-prf (8–10 kHz) copper-vapor laser pumping, the dye solution in ethanediol flowed through the narrow passage at a speed of $\sim 5\ ms^{-1}$. This flow velocity allows several active volume clearances between pump pulses." The dye-

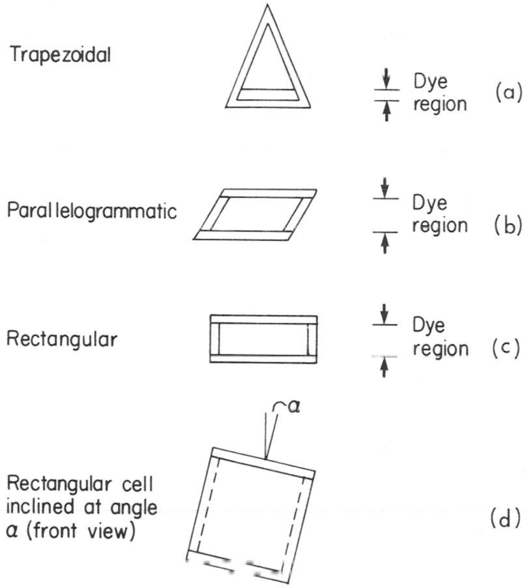

Fig. 6.3 Various dye-cell geometries. (a) Trapezoidal, (b) parallelogrammatic, and (c) rectangular. In dye cell geometries (a)–(c), top view is shown (here the dye flow is perpendicular to the page). (d) Front view of inclined rectangular cell.

circulation system incorporated a heat exchanger to maintain the dye temperature at 22 ± 2°C and an accumulator to reduce dye flow-rate variations" (Duarte and Piper, 1984). Tight control of the flow rate was important to eliminate turbulence and ensure a high optical-quality laminar flow in the region near the center of the dye cell along the propagation axis (the influence of this effect on narrow-linewidth stability is outlined in Section 9, Chapter 4).

Other authors utilizing high-prf copper-vapor laser pumping of dye lasers have used cross-sectional areas 15 mm × 0.5 mm (Zherikin *et al.*, 1981) and 13.5 mm × 0.4 mm (Broyer *et al.*, 1984). Hargrove and Kan (1980) utilized a 10 mm × 0.3 mm cross-sectional area in a cell designed for two-sided excitation. Certainly, as indicated by Bernhardt and Rasmussen (1981), the required flow velocity depends on the prf. In their case a flow velocity of ~3 ms^{-1} was required for operation at 6 kHz. Copper-vapor lasers have also been used to excite jet arrangements at a prf of 10 kHz (Zemskov *et al.*, 1976).

Alternative dye cells designed to provide a uniform mode of excitation include prismatic dye cells (Bethune, 1981; Wright and Falconer, 1988) and conic dye cells (Simon and Juhasz, 1985). These cells offer better spatial distribution of the pump energy utilizing total internal reflection to provide uniform illumination of the dye solution.

5. AMPLIFIED SPONTANEOUS EMISSION

The subject of amplified spontaneous emission (ASE) has been discussed in numerous publications. In this section we focus on aspects of ASE relevant to pulsed dye lasers including methods of ASE measurement. Further, we describe techniques to suppress the level of ASE in practical dye-laser oscillators.

First, it should be indicated that the theory of ASE in dye lasers has been discussed by several authors including Ganiel *et al.*, (1975), Dujardin and Flamant (1978a,b), Penzkofer and Falkenstein (1978), Haag *et al.* (1983), Nair and Dasgupta (1985), and Hnilo and Martinez, (1987).

In practical dye-laser oscillators ASE can be easily differentiated from laser narrow-linewidth emission given the broadband and low-intensity nature of the former. Using this observation a simple illustration of the emission process in transversely excited narrow-linewidth dye lasers can be used to provide a qualitative description of ASE. As discussed in Chapter 4, closed-cavity, transversely excited, narrow-linewidth oscillators are asymmetric resonators terminated at one extreme by a broadband output coupler (R_1). At the other side of the gain region we usually have a multiple-prism grating assembly that can be represented by a reflector $R_2(\lambda)$ whose reflectivity obeys a Gaussian function that depends on the dispersive $\Delta\lambda$. Using the notation of Munz *et al.* (1980), the photon flux is considered to have two components propagating in the positive and negative x direction (where x is the optical axis of the cavity)

$$I_\ell = I_\ell^+(x, t, \lambda) + I_\ell^-(x, t, \lambda). \tag{6.1}$$

Now, early in the excitation cycle (prior to laser threshold), one of the components of the broadband photon flux first interacts with the output coupler R_1. Certainly, only part of this photon flux component returns to the gain region, for amplification and then interaction with the frequency-selective elements of $R_2(\lambda)$. The other part of this component contributes directly to the generation of ASE throughout the duration of the buildup time. On the other hand, the photon flux component propagating in the opposite direction interacts first with $R_2(\lambda)$, thus undergoing frequency selectivity prior to returning to the gain region for further amplification. The process outlined here continues to amplify the narrow-linewidth emission depending on the number of return passes allowed by the cavity length and the characteristics of the excitation pulse.

In addition to the ASE outcoupled during the build up time, prior to laser threshold, other contributing sources include broadband reflections from the dye-cell windows and back reflections from the prism output faces. In a worst case these reflections can amount to ~3.6% per surface depending on the refractive index of the material.

Considerably higher ASE levels can be created in a transversely excited open-cavity configuration where the laser output is coupled via the losses of the frequency-selectivity elements such as prisms and gratings. For a comparison of the ASE performance of these two configurations, see Fig. 4.8 in Chapter 4.

The measurement convention of the ASE background is rather important. Methods to determine the ASE level exploit the broadband and highly divergent characteristics of this radiation. One of the techniques employed takes advantage of the difference in divergence between ASE and the narrow-linewidth laser emission. Hence, employing geometric means, the overall output energy is measured with and without the central laser beam. This technique has been discussed by McKee *et al.* (1982). This type of measurement is quite useful in determining the percentage of ASE but does not provide information about the spectral distribution of ASE.

An alternative method, used by several authors (Duarte and Piper, 1980; Berik *et al.*, 1985; Nair and Dasgupta, 1985), consists in using a spectrometer to determine the intensity and frequency distribution of the overall emission. In this method the output-intensity profile of ASE is recorded as a function of wavelength using small incremental steps. For the intense laser emission the linewidth is determined interferometrically.

It should be emphasized that all ASE measurements should be performed under normal laser-operating conditions. In other words, ASE should not be determined under single-pass conditions.

The measuring methods just described relate quite well to the percentage and spectral-density definitions of ASE. First, the ASE percentage can be defined as

$$\text{ASE\%} = \frac{\text{spectrally integrated energy in broadband ASE}}{\text{energy in narrow-linewidth laser emission}} = \frac{\int_{\Lambda_1}^{\Lambda_2} W(\Lambda)\, d\Lambda}{\int_{\lambda_1}^{\lambda_2} E(\lambda)\, d\lambda} \tag{6.2}$$

where $W(\Lambda)$ is the broadband ASE energy contained in the wavelength region, defined by the gain characteristics of the dye, from Λ_1 to Λ_2. For the narrow-linewidth emission energy $E(\lambda)$, λ_1 and λ_2 represent the boundaries of the linewidth $(\Delta\lambda)$ at its full width. For the case of single–longitudinal-mode oscillation, $E(\lambda)$ may be considered as a Gaussian.

The percentage assessment of ASE does not provide information about spectral brightness, which is a very important parameter for many applications. For instance, it is easy to see that two lasers yielding identical ASE percentages could have very different absolute ratios of maximum ASE intensity to maximum laser intensity if the respective laser linewidths differ.

Hence, it is useful to define the level of ASE in terms of energy spectral density (Duarte, 1987). For this definition the spectral purity figure of merit becomes

$$\left(\frac{\rho_{\mathrm{ASE}}}{\rho_\ell}\right) = \frac{(\Delta\Lambda)^{-1} \sum_{n=1}^{r} W(\Lambda)_n \Delta\Lambda_n}{(\Delta\lambda)^{-1} \int_{\lambda_1}^{\lambda_2} E(\lambda)\, d\lambda} \approx \frac{(\Delta\Lambda)^{-1} \int_{\Lambda_1}^{\Lambda_2} W(\Lambda)\, d\Lambda}{(\Delta\lambda)^{-1} \int_{\lambda_1}^{\lambda_2} E(\lambda)\, d\lambda} \tag{6.3}$$

for very small $(\Delta\Lambda)_n$ segments. Here, $\Delta\Lambda = |\Lambda_2 - \Lambda_1|$ is the total bandwidth of the broadband emission and $\Delta\lambda = |\lambda_2 - \lambda_1|$ is the laser linewidth at full width. As in the case of the percentage definition. $E(\lambda)$ can assume a Gaussian profile for the case of single–longitudinal-mode lasing.

For lasing at the center of the gain curve, Duarte (1987) reports about 0.3% ASE for an excimer-laser-pumped MPL oscillator ($\Delta\nu \sim 714$ MHz) corresponding to a $\rho_{\mathrm{ASE}}/\rho_\ell$ factor $\sim 7.6 \times 10^{-8}$ and an absolute ratio of maximum ASE intensity to maximum laser intensity of $\sim 6.4 \times 10^{-8}$. Thus, in practice, the simple ASE:laser-intensity ratio provides the same order of magnitude information as the more involved $(\rho_{\mathrm{ASE}}/\rho_\ell)$ factor and appears to offer a fair assessment of the brightness and purity of the emission. The $\sim 10^{-7}$ ASE:laser ratio quoted here should be compared with a $\sim 7.5 \times 10^{-6}$ ratio given for the early nitrogen-pumped MPL oscillator, at $\Delta\nu \sim 1.61$ GHz (Duarte and Piper, 1980). For copper-laser-pumped, narrow-linewidth, multiple-prism-grating oscillators, reported ASE:laser ratios include 2×10^{-7} for $\Delta\nu = 60$ MHz (Bernhardt and Rasmussen, 1981) and 5×10^{-7} for $\Delta\nu \approx 400$ MHz (Duarte and Piper, 1984; Duarte, 1987).

It is well known from experimental data and theoretical studies that the level of ASE in dye lasers is a function of several interdependent parameters including number of dye molecules per unit volume, pump photon flux, and output coupler reflectivity. In addition, as discussed earlier, characteristics of cavity components and cavity configuration also influence the outcome.

It has already been mentioned that a closed-cavity configuration provides much lower ASE levels than a comparable open-cavity laser. For instance, closed-cavity, narrow-linewidth MPL oscillators have been shown to provide about half the energy of the open-cavity configuration at ASE levels reduced more than two orders of magnitude (Duarte and Piper, 1980). The advantage of this trade-off was highlighted by Nair (1982) in his review of dye lasers. Trebino *et al.* (1982) and Duarte and Piper (1982) provide further ASE comparisons for alternative cavity configurations.

Measures to reduce the ASE level in MPL and HMPGI oscillators include the use of trapezoidal or parallelogram dye cells (see Section 4) with antireflection coatings on the windows. Although prisms are usually designed for orthogonal beam exit, in practice only near-orthogonality is achieved. This helps to reduce back reflections at the output face of the prisms and hence reduce ASE. However, best results are achieved with the use of antireflection coatings that reduce reflectivity down to ~0.2% per surface. Indeed, two of the lowest ASE:laser-intensity ratios have been achieved with multiple-prism expanders utilizing antireflection coatings (Bernhardt and Rasmussen, 1981; Duarte, 1987). Additional measures may include a reduction in dye cencentration and optimization of the output coupler reflectivity. Also, Berik *et al.*, (1985) have reported on the use of efficient single-pass stimulated Raman scattering to suppress the ASE further.

The level of ASE has been measured to be a minimum at the maximum of the dye gain and to increase as the laser is tuned toward the wings of the gain profile (Berik *et al.*, 1985; Nair and Dasgupta, 1985). Distributed-feedback dye lasers are reported to have inherent lower ASE levels than conventional oscillators (Bor, 1981; McIntyre and Dunn, 1984) since the required buildup times to reach laser threshold are shorter.

The issue of ASE in oscillator-amplifier systems has been considered by several authors (see, for example, Ganiel *et al.* (1975), McKee *et al.* (1982), and Haag *et al.* (1983)). As outlined previously, most of the ASE is produced in the buildup time prior to laser threshold. This characteristic can be exploited in oscillator-amplifier systems to limit the overall ASE level (Bos, 1981; McKee *et al.*, 1982). In this regard, McKee *et al.* (1982) utilized an excimer-laser-pumped oscillator-amplifier to demonstrate that delay of the amplifier excitation pulse does indeed reduce the ASE background. In this experiment the amplifier pump pulse was delayed by 4 ns in order to synchronize the arrival of the excitation pulse with the narrow-linewidth oscillator signal. Reported delay factors vary from 2–3 (Lavi *et al.* 1979) to ~9 ns (Dupre, 1987). An additional method employed in multistage oscillator-amplifier systems is the use of spectral and spatial filters between amplification stages to filter out ASE (Wallenstein and Hansch, 1975).

6. POLARIZATION

For many applications it is advantageous for dye lasers to provide strongly polarized narrow-linewidth radiation. In this regard, one further advantage of multiple-prism grating dye-laser oscillators is that they inherently produce output emission strongly polarized parallel to the plane of incidence (p-polarization) (see, for example, Duarte and Piper, 1982, 1984).

In this section, we discuss the influence on polarization by the various elements integrated in the architecture of narrow-linewidth dye-laser oscillators. These elements include the pump beam, dye solution, dye cell, prisms, and grating.

In the case of dye molecules, the polarization characteristics are related to the relative orientation of the transition moments for absorption and emission and the rotational diffusion-relaxation time (Schäfer, 1977). For the case involving excitation pulses shorter than the diffusion-relaxation times, Schäfer (1977) provides a table listing the possible emission polarization orientation, for pump beams of s- and p-polarization, in the two cases where the relative orientation of the transition moments of absorption and emission are either parallel or perpendicular to each other. Early experiments in this area were performed by Sorokin *et al.* (1967), McFarland (1967), and Sevchenko *et al.* (1968). Further information on relevant polarization phenomena can be obtained from Förster (1951) and Feofilov (1961).

In practice it is found that the optical elements in the cavity play an important role in determining the final polarization state of the laser emission. In this regard, the elements that influence the emission polarization include trapezoidal- (or wedge-) type dye cells, prisms, and gratings. The influence of the dye cell windows and the prisms can be quantified using the Fresnel formulae (Born and Wolf, 1975) for the loss factors for the s and p components of polarization,

$$\mathrm{l}_p = \tan^2(\phi - \psi)/\tan^2(\phi + \psi) \tag{6.4}$$

$$\mathrm{l}_s = \sin^2(\phi - \psi)/\sin^2(\phi + \psi) \tag{6.5}$$

where l_m represents the individual loss at mth prism for either p- or $\sin\phi = n\sin\psi$. As discussed in Chapter 4 (Section 4), the cumulative losses at the mth prism for either $p-$ or s-polarized radiation can be estimated utilizing the expression,

$$\mathrm{L}_m = \mathrm{L}_{m-1} + (1 - \mathrm{L}_{m-1})\mathrm{l}_m \tag{6.6}$$

where l_m represents the individual loss at mth prism for either p- or s-polarization as outlined in Eqs. (6.4) and (6.5).

Grating efficiency for the two components of polarization can be determined experimentally (see Chapter 4, Fig. 4.6) or by using the recent developments of the electromagnetic theory of gratings (Maystre, 1980). Given this information one can estimate the passive losses for both polarization components for a typical MPL or HMPGI oscillator.

For high-prf, copper-laser-pumped multiple-prism grating oscillators Duarte and Piper (1984) report outputs ~100% *p*-polarized. Conversion efficiencies achieved with the polarization mode of the copper laser matched to the preferred *p*-polarization of the multiple-prism grating combinations were ~15% higher than for pumping with an *s*-polarized beam (similar results were achieved in earlier low-prf experiments of Duarte and Piper (1982)).

For rhodamine 590, the transition moments of absorption and emission are parallel to each other for excitation in the green region of the spectrum since the same electronic transition is involved (Schäfer, 1977, 1983). Indeed, measurements show that the intrinsic broadband emission is ~90% *s*-polarized for transverse excitation with an *s*-polarized copper-laser beam and almost unpolarized for excitation with a *p*-polarized beam (Duarte and Piper, 1985) (see Fig. 6.4). Utilizing the polarization efficiency response of the grating in conjunction with Eqs. (6.4)–(6.6), it is not difficult to show that overall intracavity passive losses, in MPL or HMPGI configurations, are substantially larger for the case involving an

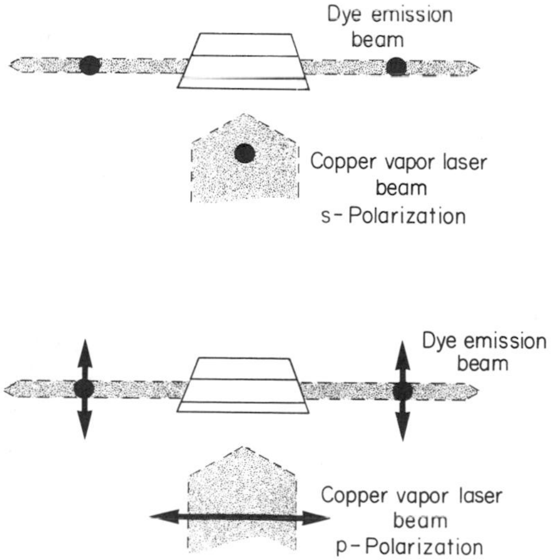

Fig. 6.4 Polarization of single-pass emission from rhodamine 590 when excited with a copper-vapor laser at 510.5 nm.

s-polarized pump beam than the losses incurred when using *p*-polarized excitation (Duarte and Piper, 1985).

It should be emphasized that even in the absense of a well-defined excitation polarization orientation, as in the case of N_2 laser pumping and flashlamp excitation, the dye-laser output is strongly *p*-polarized due to significant influence exerted by the anisotropy of the multiple-prism expander (Duarte and Piper, 1980, 1981; Duarte and Conrad, 1987). By contrast, resonators incorporating a telescopic beam expander necessitate the inclusion of an intracavity polarizer (Hänsch, 1972).

Additional literature on polarization in long-pulse flashlamp-pumped dye lasers includes experiments on the effects of cavity anisotropy (Dugan *et al.*, 1978) and studies on the dependence of polarization on solvent viscosity (Morgan and Dugan, 1979).

7. THERMAL EFFECTS

Thermal equilibrium and stability are two very important requirements necessary to control the emission characteristics from dye lasers. In general, thermal effects can be classified in two categories: those that affect the molecular dye medium and those that alter the physical outline of the cavity.

To some extent thermal effects have already been considered in Sections 4, 6, and 9 of Chapter 4. As indicated in Chapter 4, thermal gradients in the active medium of flashlamp-excited dye lasers can affect laser performance in more than one way. For instance, Fletcher *et al.* (1986) report a rather sensitive dependence of the laser output signal as a function of temperature difference between the dye solution and the coolant. In their study on several dyes including rhodamine 590 and coumarin 1, Fletcher *et al.* (1986) report that the optimum temperature difference may be bimodal and depends on dye, solvent, and, to a certain degree, pump energy. In addition, thermal gradients in conjunction with turbulence are identified with the mechanism causing linewidth instabilities (Duarte and Conrad, 1987; Duarte *et al.*, 1988). Further details on thermal effects in flashlamp-pumped dye lasers are given by Neister (1977b) and Fletcher *et al.* (1982).

In the case of transverse laser-excited dye lasers, temperature studies indicate that beam divergence increases and output power decreases as temperature increases (Peters and Mathews, 1980; Duarte, 1983). Measurements of single-pass broadband emission output (using a single mirror) as a function of temperature in several dyes (see Fig. 6.5) indicate that the rate of decrease in output yield is similar to that measured for a narrow-linewidth MPL oscillator (Duarte, 1983). This indicates that deterioration in the laser output is related to the thermal characteristics of the dye

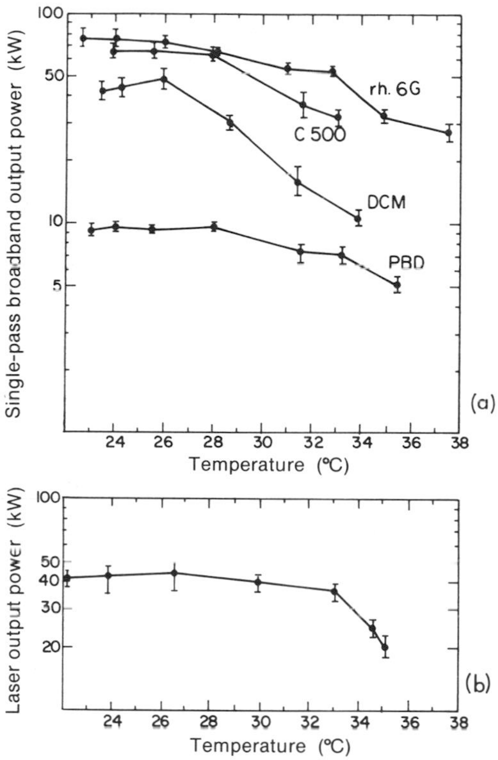

Fig. 6.5 (a) Output power of single-pass broadband emission as a function of temperature for several dyes. (b) Output power as a function of temperature in a narrow-linewidth MPL oscillator utilizing rhodamine 6G [from Duarte, 1983]. © 1983 IEEE.

solution itself. This decline in output energy, as temperature is increased, may be due to various effects including conformational changes in the molecular structure of the excited dye molecules (see Chapter 7, Section 2.2). As indicated by Duarte (1983) laser-power variations as a function of temperature may be reduced by utilizing dyes with rigidized structure (such as rhodamine 101, coumarin 102 (or 480), and coumarin 314 (or 504)) (see, for example, Drexhage (1977)). It is interesting to note that some of the dyes investigated (rhodamine 6G and PBD) exhibit good recovery characteristics when recooled.

Beam divergence as a function of temperature for the MPL oscillator is shown in Fig. 6.6. One of the contributing factors to this increment in $\Delta\theta$, as the temperature increases, is thermal lensing which results from thermal gradients in the index of refraction of the dye solution (Peters and

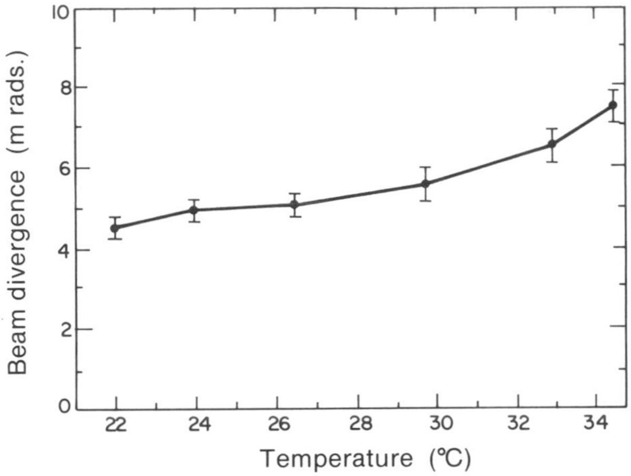

Fig. 6.6 Full-angle beam divergence in MPL oscillator as a function of temperature [from Duarte, 1983]. © 1983 IEEE.

Mathews, 1980). To illustrate the influence of refractive-index variations on the single-pass divergence we consider a hypothetical case of a dye laser where the dye solution is represented by a complex *ABCD* matrix. The single-pass beam divergence in a MPL or HMPGI oscillator, without considering the gain medium, can be described as

$$\Delta\theta = (w/L_R)\,[1 + (L_R/B')^2 + (L_R M/B')^2]^{1/2} \tag{6.7}$$

(see Section 6, Chapter 4). If the dye cell is represented as a Gaussian aperture in conjunction with a thin lens (see Siegman (1986)), then Eq. (6.7) may be modified to

$$\Delta\theta' = (w/L_R)\,\{1 + (L_R/B')^2 + [L_R(M + B'\chi)/B']^2\}^{1/2} \tag{6.8}$$

(Duarte, 1989) where the χ factor is given by the **C**-component of the **ABCD** matrix. If we consider just the real part of the **C**-component, then $\chi = -(1/f)$.

For a long focal length χ becomes small and Eq. (6.8) approaches the form of Eq. (6.7) so that $\Delta\theta' \approx \Delta\theta$. Thus, for this condition there is virtually no effect from thermal lensing. This result is valid for either a positive or a negative lens.

For large beam magnification factors, $B' \approx ML$, and $ML \gg L_R$ (for $L_R = L$). Under these special conditions it is found that for the case of $|f| \ll L_R$, Eq. (6.8) reduces to

$$\Delta\theta' \approx (\lambda/(\pi w))(L_R/f), \tag{6.9}$$

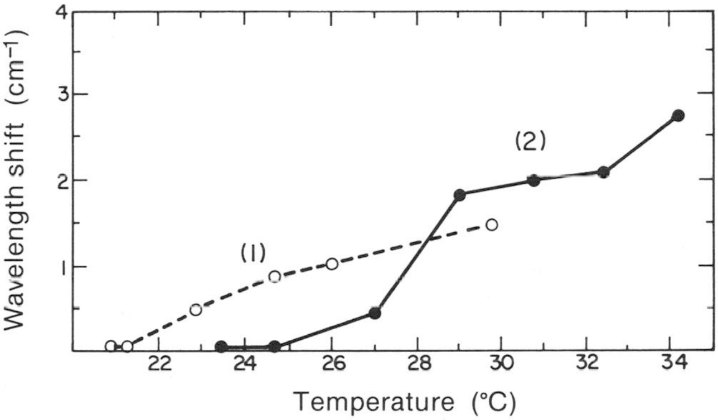

Fig. 6.7 Wavelength drift as a function of temperature for two different runs [from Duarte, 1983]. © 1983 IEEE.

which implies that $\Delta\theta'$ is now greater than $\Delta\theta$. Again, these results apply for either a positive or a negative lens. For the hypothetical case of $|f| \approx L_R$ then $\Delta\theta$ approaches $[\lambda/(\pi w)]$ for a positive lens. On the other hand, $\Delta\theta \approx (\lambda\sqrt{5})/(\pi w)$ in the case of a negative lens. For a discussion of thermal lensing effects generated in a transient regime, see Morey (1980).

One of the most serious consequences of changes in temperature in narrow-linewidth oscillators is the wavelength drift illustrated in Fig. 6.7. This effect can be caused by a number of processes, including the slight physical displacement of cavity components, such as prisms and gratings, due to thermally induced material expansion and/or contraction. In addition, as indicated in Eq. (4.46), the angular displacement of the beam emerging from the prism expander can deviate owing to the (dn/dT) factor. This effect can be reduced significantly by using achromatic expanders where $(d\phi_{2,r}/d\lambda) \approx 0$.

Thermal drifts in a pressure-tuned pulsed dye laser, utilized in atomic uranium spectroscopy, have been observed and characterized by Broglia *et al.* (1986).

The measurements discussed in this section strongly suggest that dye lasers ideally should be constructed of materials with a low thermal expansion coefficient, such as invar, and that the ambient temperature should be controlled at least within ±0.5°C.

8. TUNING METHODS

The most distinct characteristic of dye lasers is their ability to provide tunable coherent radiation. This inherent advantage is perhaps the principal reason that has made the dye laser one of the most successful and widely used lasers today. In this regard, it should be realized that tuning requirements and techniques may differ from application to application. Certainly, drastic, coarse broadband tuning can be accomplished by the simple use of different dye molecules in a basic mirror cavity. A more refined type of braodband tuning for a given dye molecule was demonstrated by changing the dye solution concentration (Schäfer *et al.* (1966); McFarland (1967)).

In this section we discuss tuning methods involving intracavity elements such as gratings, prisms, and etalons. Alternative techniques for pressure tuning are also considered.

Basic tunable dye lasers utilize dispersive elements such as gratings and prisms as means of tuning. In the type of mirror-grating cavity introduced by Soffer and McFarland (1967), tuning is governed by the Littrow grating equation,

$$m\lambda = 2a \sin \Theta \tag{6.10}$$

Thus, for a given order (m) and groove spacing (a), changing the angle Θ leads to wavelength tuning. For a grating mounted in the grazing-incidence configuration, the basic grating equation is utilized,

$$m\lambda = a(\sin \Theta + \sin \Theta'), \tag{6.11}$$

where Θ is the angle of incidence and Θ' is the angle between the normals of the grating and tuning mirror (see Fig. 4.9, Chapter 4). In this case, wavelength is varied by rotating the tuning mirror (Shoshan *et al.*, 1977; Littman and Metcalf, 1978). Using Eqs. (6.10) and (6.11) Littman (1978) has derived wavelength expressions for double-grating arrangements where the tuning mirror is replaced by a Littrow grating.

In the case of prismatic cavities (Strome and Webb, 1971; Schäfer and Muller, 1971) tuning is accomplished by dispersing the intracavity broadband photon flux and utilizing a mirror, at the exit of the prism tuner, to achieve resonance at various angular positions. Certainly, this is due to the fact that for a prism at a fixed position, the angle of emergence is a function of wavelength, as can be illustrated by the expression

$$\phi_{2,1} = \arcsin\left\{n(\lambda) \sin\left[\alpha_1 - \arcsin\left(\frac{\zeta}{n(\lambda)}\right)\right]\right\} \tag{6.12}$$

where $\zeta = \sin \phi_{1,1}$ is a constant for a stationary prism, α_1 is the apex angle, and $n(\lambda)$ is given by the appropriate material dispersion function. Similarly, the output angle at the mth prism can be expressed as

$$\phi_{2,m} = \arcsin\left\{n(\lambda) \sin\left[\alpha_m - \arcsin\left(\frac{\sin \phi_{1,m}}{n(\lambda)}\right)\right]\right\}. \tag{6.13}$$

Here the incidence angle of the mth prism ($\phi_{1,m}$) is related geometrically to the exit angle of the previous prism ($\phi_{2,(m-1)}$).

For an array of r identical prisms, arranged in a symmetric configuration, the cumulative angular spread achieved at the exit of the last prism is augmented considerably since the overall dispersion is given by

$$(d\phi_{2,r}/d\lambda) = r(d\phi_{2,1}/d\lambda). \tag{6.14}$$

Hence, in addition to linewidth refinement (see Section 4, Chapter 4) multiple-prism chains can be employed to improve significantly the wavelength-tuning resolution achieved by angular rotation of the mirror at the exit of the multiple-prism array. Strome and Webb (1971) report a tuning range from 571 to 615 nm (half-power points) utilizing a four-prism tuner and rhodamine 6G in a flashlamp-pumped, pulsed dye laser.

In conjunction with gratings and prisms, intracavity etalons provide a further avenue of fine wavelength tuning. In this case the etalon can be considered as a periodic filter that satisfies the condition

$$m_e\lambda = 2n\, d_e \cos\theta_e' \tag{6.15}$$

for transmission maxima (Born and Wolf, 1975). Here, m_e is an integer, d_e is the distance between the etalon surfaces, θ_e' is the refraction angle that is related to the tilt angle θ_e by $\sin\theta_e = n\sin\theta_e'$, and n is the refractive index. Wavelength tuning is accomplished by varying the tilt angle. The angular dispersion for an etalon can be expressed as

$$(d\theta_e/d\lambda) = (\sin\theta_e'/\cos\theta_e)(dn/d\lambda) + n(\cos\theta_e'/\cos\theta_e)(d\theta_e'/d\lambda) \tag{6.16}$$

where

$$(d\theta_e'/d\lambda) = [\tan\theta_e']^{-1}[(1/n)\, dn/d\lambda - (1/\lambda)]. \tag{6.17}$$

For $n = 1$ we have $(d\theta_e/d\lambda) = (d\theta_e'/d\lambda)$, and subsequently we obtain the result given by Schäfer (1977),

$$(d\theta_e/d\lambda) = -(\lambda\tan\theta_e)^{-1}. \tag{6.18}$$

According to Schäfer (1977) the wavelength shift resulting from a displacement in θ_e (from $\theta_e = 0$, at λ_0) can be written as

$$\delta\lambda_e = (1 - \cos\theta_e)\lambda_0 \tag{6.19}$$

for $n = 1$.

An alternative method of wavelength tuning in gas-spaced Fabry-Perots is pressure scanning. As discussed by Meaburn (1976) the refractive index of the gas can be varied according to

$$K\Delta P \approx (n - 1) \tag{6.20}$$

where K is a constant related to the intrinsic properties of the gas employed, and ΔP is the pressure gradient exerted over the Fabry-Perot plates. The wavelength shift accomplished with this method can be expressed as

$$\delta\lambda_e \approx K\,\Delta P\lambda \tag{6.21}$$

(see, for example, Meaburn (1976)). A listing of the approximate values for K for several gases including O_2, N_2, CO_2, and C_3H_8 is given by Meaburn (1976).

This principle of pressure tuning in etalons was adopted by Wallenstein and Hänsch (1974) to vary the frequency of a telescopic dye laser mounted in a pressure chamber. Frequency-selective elements in this oscillator included a tilted etalon and a grating in Littrow mounting. Wallenstein and Hänsch (1974) point out that the light frequency changes almost linearly with pressure according to

$$\delta\nu \approx -(c/\lambda)K\,\Delta P. \tag{6.22}$$

For a pressure change of 760 Torr (using nitrogen gas) these authors report a tuning range over 142.5 GHz with a nonlinearity factor of 3×10^{-4} in the equation for $\delta\nu$. Bollen *et al.* (1987) discuss the performance of an oscillator-amplifier system utilizing a pressure-tuned MPL oscillator incorporating an intracavity etalon and an external confocal Fabry-Perot etalon.

As discussed in Chapter 4, narrow-linewidth dye lasers can incorporate gratings, prisms, etalons, or integrated frequency-selective arrangements combining these elements. An example of a high-dispersion dye-laser design incorporating an integrated system is the telescopic cavity employing an intracavity etalon in addition to the Littrow grating. Other high-dispersion resonators incorporating integrated systems are the MPL and the HMPGI oscillators with and without intracavity etalons. As such, it should be absolutely clear that narrow-linewidth dye lasers are intrinsically tunable oscillators offering more than one option to shift the resonant frequency.

With all these options available, an obvious concern is to identify the most effective or the most appropriate form of wavelength tuning. The answer depends on the application. For instance, in many physics-type applications, dye-laser oscillators are utilized to provide narrow-linewidth emission at a specific wavelength that cannot be obtained from traditional lasers. Examples of this type of application include optical pumping of molecular lasers and multiphoton excitation requiring a high degree of spectral specificity. In this regard, all that may be necessary to obtain the required wavelength is to select an appropriate dye medium and then fine-tune the laser using an appropriate mechanism, such as rotation of the grating, until the specified frequency is obtained. In this type of application wavelength stability is very important and frequency jitter is often required to be within the boundaries provided by the linewidth (see, for example, Duarte (1988)). For further details see the discussion of Fig. 4.10 (Chapter 4).

A different type of application, which requires scanning over relatively large segments of the spectrum, is spectroscopy. First we consider the traditional grating tuning method. This is accomplished by precision rotation of the grating mount using a mechanical or piezoelectric approach. In Fig. 6.8 we include a portion of fluorescence spectrum of D_2CO excited at 364 nm with (a) commercial telescopic dye laser utilizing mechanical rotation of the grating and (b) a MPL oscillator ($\Delta\nu \approx 2$ GHz) employing a long-travel, high-resolution, sequential piezoelectric driver for the grating rotation. The spectral features obtained with the latter method are in good agreement with those obtained with a Fourier-transform spectrometer. For the spectrum shown in Fig. 6.8, the wavelength range covered by the laser was 0.227 nm (that is, 514 GHz or 17 cm^{-1}). Using a HMPGI oscillator Dupre (1987) investigated the absorption spectrum of $^{130}Te_2$ for a 18-cm^{-1} span and reported excellent agreement with the equivalent spectrum recorded using Fourier-transform spectrometry.

Using a telescopic cavity Olcay *et al.* (1985) synchronized the motion of the Littrow grating and the air-spaced etalon to resolve the iodine spectrum for over more than 30 nm. Suzuki *et al.* (1981) report on a synchronized, fully automatic, prismatic dye laser, incorporating a grating in Littrow mount and an intracavity etalon, utilized to obtain a Doppler-limited absorption spectrum over 1.3 nm at around 591 nm (that is ~1116 GHz or 37 cm^{-1}). In addition, Raymond *et al.* (1984) describe a linear grating drive for a grating in Littrow mounting, and McNicholl and Metcalf (1985) discuss synchronous wavelength scanning in cavities with gratings in Littrow or grazing-incidence configurations. Mory *et al.* (1979) discuss the use of a single-grating grazing-incidence arrangement for tuning a cw dye laser, and German (1981) compares the performance of a Littrow-grating-etalon combination with a grazing-incidence grating arrangement in a cw color center laser.

Wavelength scanning in short-cavity, single-mode, grazing-incidence dye lasers has been described by Liu and Littman (1981) and Littman (1984). In these cavities synchronous tuning is accomplished by rotating the tuning mirror about an axis defined approximately by the convergence point of the surface planes of the two mirrors and the grazing-incidence grating (Liu and Littman, 1981). This type of arrangement provides a simple alternative

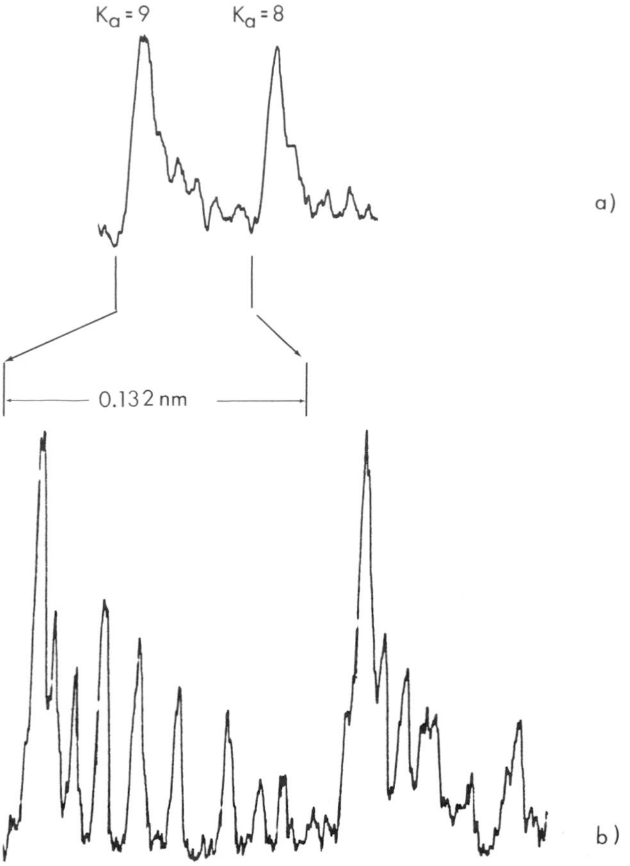

Fig. 6.8 Fluorescence spectra from D_2CO excited at ~364 nm. (a) Response obtained with a commercial dye laser. (b) Fluorescence spectrum obtained with a MPL oscillator. This spectrum involves the 4^0_1 band of the $\tilde{A}\ {}^1A_2$–$\tilde{X}\ {}^1A_1$ electronic system of D_2CO (this was recorded in 1982 when the author was working with Prof. B.J. Orr at the University of New South Wales) [from Duarte, 1988].

for adjusting the cavity length as the wavelength is varied and has enabled researchers to achieve a single-mode scan range of more than 15 cm^{-1} (Littman, 1984).

An additional method of wavelength tuning, demonstrated in a HMPGI oscillator, consists in rotating a fine-tuning wedge inserted between the grating and the tuning mirror (Greenhalgh and Sarkies, 1982).

9. NARROW-LINEWIDTH OSCILLATOR-AMPLIFIERS

It is not difficult to see that there are energetic limitations to the performance of pulsed narrow-linewidth oscillators. For instance, oscillators that produce narrow-linewidth emission utilizing coated intracavity etalons in conjunction with gratings are susceptible to optical damage by high intracavity energy fluxes. An alternative class of oscillators incorporating intracavity beam expansion, such as the MPL and HMPGI configurations, avoid optical damage on the frequency-selective elements but require a relatively small aperture to select a beam with near single transverse mode (TM_{00}) characteristics. In the case of laser-pumped designs, this condition imposes a limitation on the amount of excitation power deposited at the dye cell and in the case of flashlamp-pumped lasers, the insertion of an intracavity aperture reduces the available active volume. In addition, most frequency-selective elements utilized in dye-laser oscillators induce losses, which limit the conversion efficiency.

9.1. Laser-Pumped Narrow-Linewidth Oscillator-Amplifiers

An effective alternative for producing high-energy and/or high–peak-power, narrow-linewidth emission is to use dye-laser amplifiers in conjunction with the oscillator.

In this section we provide a description of narrow-linewidth oscillator-amplifier configurations in both laser- and flashlamp-pumped dye lasers. Emphasis is given to systems incorporating narrow-linewidth pulsed dye-laser oscillators.

The excitation mechanism in dye-laser amplifiers has been considered by several authors including Ganiel *et al.* (1975), Dujardin and Flamant (1978a), Haag *et al.* (1983), and Hnilo and Martinez (1987). Ganiel *et al.* (1975) report a fair agreement between calculated and measured gain factors, as a function of input signal, for a single-pass amplifier. In these experiments the input signal was provided by an oscillator incorporating a telescope and a Littrow grating at a linewidth of 42.2 GHz (0.05 nm). In this configuration, Ganiel *et al.* (1975) report 40-kW output, at the amplifier, for an oscillator signal of 6.7 kW. The pump power used for the amplifier stage is quoted at 70 kW.

Table 6.3 lists several laser-pumped oscillator-amplifier systems. Notice that one condition common to all the systems listed in Table 6.3 is the use of a single-pass configuration at the amplifier stage. In this regard, it is interesting that in at least three of those systems, specific reference is made to the use of an appropriate delay in the excitation of the amplifier stage in

Table 6.3

Laser-Pumped Oscillator-Amplifier Systems

Oscillator	Excitation method	Output energy	%Eff.	Reference
Littrow grating and two etalons				
$\Delta\nu \approx 2.4$ GHz	Transverse[a]	0.25 mJ	25 at 460 nm	Itzkan and Cunningham (1972)
Littrow grating				
$\Delta\nu \approx 24$ GHz	Longitudinal	150 mJ	12 at ~790 nm	Carlsten and McIlrath (1973)
Single-prism and Littrow grating				
$\Delta\nu = 12$ GHz	Longitudinal	35 mJ	35 at ~560 nm	Hartig (1978)
Telescope and Littrow grating plus etalon				
$\Delta\nu = 1.8$ GHz	Transverse	1.8 mJ	30 at ~590 nm	Lavi *et al.* (1979)
$\Delta\nu = 320$ MHz	Transverse[b]	165 mJ	55 at ~590 nm	Bos (1981)
HMPGI				
$\Delta\nu = 650$ MHz	Transverse[b]	3.5 mJ	~9 at 440 nm	Dupre (1987)

[a] Oscillator and amplifier pumped synchronously by two N_2 lasers.
[b] Incorporates preamplifier in addition to amplification stage(s).

order to minimize ASE (Lavi *et al.*, 1979; Bos, 1981; Dupre, 1987) (see Section 5).

Some specific gain factors for single-amplification stage systems include those of Carlsten and McIlrath (1973) (~21) and Itzkan and Cunningham (1972) (10–20 dB).

An oscillator-amplifier system employing double-stage amplification is described by Wallenstein and Hänsch (1975). These authors utilized a pressure-tuned telescopic oscillator, incorporating an intracavity etalon, yielding ~1 kW at a linewidth of ~1 GHz. Following the oscillator an external confocal interferometer was used to narrow the linewidth further to 50 or 10 MHz. Then, a two-stage amplifier was used to provide a peak power of 50 kW. In their report Wallenstein and Hänsch (1975) discuss the use of gain saturation (Curry *et al.*, 1973) to improve amplitude stabilization at the amplification stage.

Of particular interest in multistage amplification is the report of Bos (1981), in which specific information is given on the energy and efficiency at each particular amplification stage. Using rhodamine 590 as the active medium and excitation from the second harmonic of a Nd:YAG laser, Bos

(1981) reports 6% efficiency at the oscillator, 20% at the preamplifier, and 60% at the amplifiers. Thus, the overall gain from the oscillator to the final amplifier is of the order of 229. Notice that only 4% of the total 300-mJ pump energy was used to excite the oscillator (Bos, 1981). A similar multistage system is reported by Dupre (1987), using a frequency-tripled Nd:YAG laser (355 nm) to excite coumarin 120. Here 1.5% of the total pump energy was delivered to the oscillator and 4% was used at the preamplifier (Dupre, 1987). Thus, the overall system efficiency (~9%) depends largely on the efficiency of the main amplification stage.

In oscillator-amplifier systems employing a single pump laser, it is a relatively simple matter to synchronize, utilizing optical delay lines, the arrival of the pump pulse at the amplifier with the arrival of the input signal. In systems having more than one pump laser, precise electronic triggering is needed to synchronize the arrival of the excitation pulse and the input signal at the amplifier. Examples of laser-pumped oscillator-amplifier dye lasers utilizing electronic synchronous triggering include the N_2 laser-pumped system of Itzkan and Cunningham (1972) (operating at a prf of 500 Hz) and the copper-laser-pumped dye laser described by Hargrove and Kan (1980) (at a prf of 6 kHz). The latter laser used two-sided transverse pumping at the amplifier and provided efficiencies in the 18–30% range.

In addition to the dispersive oscillators incorporated in the systems listed in Table 6.3, some authors have reported the use of cw or quasi-cw lasers as the injection source. Erikson and Szabo (1971) report linewidths of about 190 MHz with the use of a pulsed Ar^+ laser (514.5 nm) as injection source of a simple mirror–mirror dye laser cavity excited transversely by a nitrogen laser. Vrehen and Breimer (1972) report on the use of a cw HeNe laser (632.8 nm) as injection source of a simple mirror–mirror dye laser resonator excited longitudinally by a frequency-doubled Nd:YAG laser. These authors report a 300-MHz linewidth. Clearly, the main disadvantage of this type of arrangement is their restricted tuning range.

9.2. Flashlamp-Pumped Narrow-Linewidth Multistage Systems

Table 6.4 lists the performances of various flashlamp-pumped multistage systems. The table covers systems incorporating narrow-linewidth, flashlamp-excited oscillators and cw dye-laser injection sources. Information on the configuration of the amplification stage is also included.

Energy-gain coefficients for systems listed in Table 6.4 vary substantially from system to system. For instance, an energy-gain factor of 6 is quoted for the single-pass, multistage amplifier system driven by a single-etalon oscillator (Flamant and Meyer, 1971). On the other hand, a factor of ~267

Table 6.4

Flashlamp-Excited Narrow-Linewidth Multistage Systems

Oscillator	Amplifier configuration	Output energy	Reference
Flashlamp-pumped oscillators	*Amplifier*		
One etalon $\Delta\nu \approx 435$ GHz	Single-pass, six stages	33 mJ at 587 nm	Flamant and Meyer (1971)
Flashlamp-pumped master-oscillator	*Forced-oscillator*		
Two etalons $\Delta\nu \approx 8.65$ GHz	Flat-mirror cavity	600 mJ at 589 nm	Magyar and Schneider-Muntau (1972)
Three etalons $\Delta\nu = 4$ GHz	Plano-concave resonator	4 J	Maeda *et al.* (1975)
Two etalons $\Delta\nu = 346$ MHz	Flat-mirror cavity	300 mJ at 589 nm	Flamant *et al.* (1984)
MPL $\Delta\nu \leqslant 375$ MHz	Positive-branch unstable resonator	600 mJ at 590 nm	Duarte and Conrad (1987)
cw dye laser oscillators			
$\Delta\nu = 30$ MHz	Ring configuration	50 mJ	Blit *et al.* (1977)
$\Delta\nu = 8$ MHz	Ring configuration	~1 mJ	Trehin *et al.* (1979)

was obtained by Maeda *et al.* (1975) for their forced-oscillator system, incorporating a plano-concave resonator, driven by a three-etalon master oscillator.

At this stage it should be indicated that a forced oscillator is an amplification stage consisting of a gain region within a resonator. Under these circumstances the driving oscillator is referred to as master oscillator. An example of such type of configuration is shown in Fig. 6.9.

Duarte and Conrad (1987) describe a master-oscillator forced-oscillator

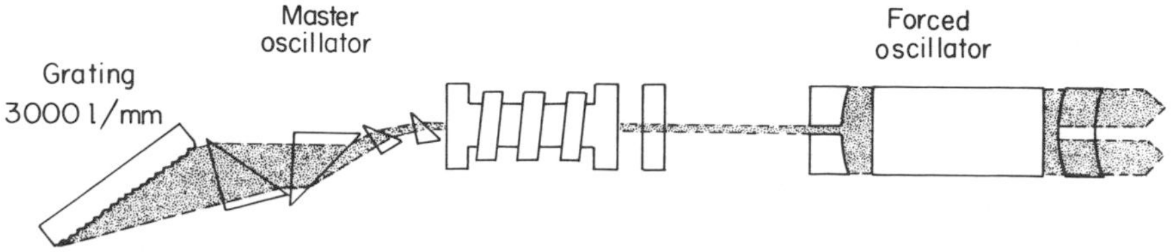

Fig. 6.9 Flashlamp-excited MPL master-oscillator forced-oscillator configuration [from Duarte and Conrad, 1987].

system yielding 600 mJ, at $\Delta\nu \leq 375$ MHz. The injection frequency in these experiments was provided by a MPL master oscillator delivering 5–11 mJ in a 99% *p*-polarized beam. The forced oscillator utilized a positive-branch unstable resonator with a magnification factor of 2 and rhodamine 590 dye at a concentration of 2.5×10^{-5} M.

Successful frequency locking in a flashlamp-excited, master-oscillator, forced-oscillator system demands some well-defined requirements. For instance, synchronized triggering is important and the master-oscillator signal must be timed to arrive at the outset of the forced-oscillator pulse. Also, as expected, optimum lasing is achieved when the master-oscillator emission wavelength is tuned to the central emission wavelength of the amplification stage. In addition, quality of emission depends rather critically on exact concentric alignment of the master-oscillator beam relative to the optical axis of the unstable resonator. Further information on these topics and comparisons with the performance of forced oscillators in the absence of injection can be found in Magyar and Schneider-Muntau (1972) and Duarte and Conrad (1987).

An alternative to the use of pulsed narrow-linewidth oscillators is the utilization of cw dye lasers as injection sources. Certainly, narrower linewidths are available with this method. However, the use of a cw dye laser as oscillator implies considerable additional complexity and the availability of significantly reduced injection peak powers. In addition to the references in Table 6.4 it should be mentioned that Bigio (1978) reported on the cw dye-laser injection of a coaxial flashlamp-pumped dye laser utilizing an unstable resonator. Also, Verkerk *et al.* (1986) report on an improved cw dye-laser-injected, ring-cavity, flashlamp-pumped dye laser yielding a linewidth of 3.5 MHz.

10. FREQUENCY CONVERSION OF TUNABLE DYE LASER RADIATION VIA STIMULATED RAMAN SCATTERING

A method to generate tunable radiation in spectral regions beyond the reach of traditional dye lasers is to employ stimulated Raman scattering (SRS) in gases and vapors. This method allows frequency conversion, via Stokes or anti-Stokes, of efficient narrow-linewidth tunable dye laser radiation. Here we provide a brief overview on this subject.

Frequency conversion of dye lasers via SRS was first demonstrated by Schmidt and Appt (1972) using room-temperature hydrogen at a pressure of 200 atmospheres. In principle, the design of the hydrogen Raman cells is quite simple. However, proper safety considerations must be observed, such as the use of stainless-steel tubing, given the high pressures involved.

Also the material of the cell windows must be suitable to allow efficient transmission at the desired wavelengths.

Using an oscillator-amplifier system, excited by a frequency-doubled Nd:YAG laser, Wilke and Schmidt (1978) generated SRS radiation in H_2 from the eighth anti-Stokes to the third Stokes (2064 nm), at an overall conversion efficiency of up to 50%. Using frequency doubling of the dye laser these authors report SRS emission from the fourth anti-Stokes (189 nm) to the fifth Stokes, at an overall conversion efficiency of up to 75%. In these experiments Wilke and Schmidt (1978) used a single-prism Littrow oscillator and rhodamine 6G dye. Using a similar dye-laser configuration Hartig and Schmidt (1979) employed a capillary waveguide H_2 cell (with a CaF_2 output window) to obtain a wavelength range from 0.7 to 7 μm (first, second, and third Stokes). The optimum conversion efficiency of the third Stokes was 1.3%. Mannik and Brown (1986) report on a conversion efficiency of 4% at the third Stokes using a waveguide Raman shifter and a dye laser emitting at 568 nm.

Schomburg *et al.* (1983) utilized a dye-laser system incorporating an MPL oscillator and two stages of amplification pumped by a Nd:YAG laser to obtain up to the thirteenth anti-Stokes at 138 nm. The laser wavelength was ~548 nm and $\Delta\nu \sim 24$ GHz (Schomburg *et al.*, 1983). Brink and Proch (1982) report on improvements in SRS conversion efficiency of up to 70% (at the seventh anti-Stokes) by lowering the hydrogen temperature to 78°K. Hanna *et al.* (1985) report on a 90% conversion efficiency to the first Stokes utilizing an oscillator-amplifier H_2 SRS configuration.

Raman conversion of tunable dye laser radiation in other media include Tl and Sn (see, for example, White and Henderson (1983) and Ludewigt *et al.* (1984)). Wyatt *et al.* (1982) report on the use of 6s–9s electronic Raman scattering in cesium vapor to obtain tunable emission at 16 μm.

PROBLEMS

1. Show that the wavelength shift in an etalon resulting from a displacement in θ_e can be written as Eq. (6.19). [Hint: use Eq. (6.15).]
2. Using a four-prism expander providing $M = 100$, calculate the single-pass passive losses for the *s*- and *p*-polarization components. Use Eqs. (6.4)–(6.6) and assume $k_{1,1} = k_{1,2} = k_{1,3} = k_{1,4}$.
3. Calculate the wavelength shift resulting from a 3°C change in temperature for a prism-grating assembly. Assume a double-prism expander in additive configuration ($M = 49$ with $k_{1,1} = k_{1,2}$) and a Littrow grating (2400 ℓ/mm) utilized in first order at 590 nm. Use $n = 1.5$, $(dn/d\lambda) \sim 7 \times 10^4\ \mathrm{m}^{-1}$, and $(dn/dT) \sim 1.5 \times 10^{-6\circ}$ C. [Hint: use Eq. (4.46) in Chapter 4.]

4. Use Eqs. (6.10) and (6.11) to find a wavelength expression for a double-grating arrangement utilizing a grazing-incidence grating and a Littrow grating. Assume that both gratings are identical.
5. (a) Discuss the design of an efficient blue-green flashlamp dye laser. Consider type of excitation geometry and triplet quencher to be used. Also select a UV filter or transfer dye to be employed.
   (b) Select laser pump sources for efficient excitation of rhodamine 6G for (i) high-prf operation and (ii) low-prf and high-peak powers.

REFERENCES

Anliker, P., Gassmann, M., and Weber, H. (1972). 12 Joules rhodamine 6G laser. *Opt. Commun.* **5,** 137–138.

Artemev, A.Y., Borovich, B.L., Vasilev, L.A., Gerts, V.E., Nalegach, E.P., Negashev, S.A., Radostin, E.G., Rybin, V.M., Ryazanskii, V.M., Tatarintsev, L.V., and Ulyanov, A.N. (1982). Multisection copper vapor laser. *Sov. J. Quantum Electron.* **12,** 456–458.

Arutunyan, V.M., Djotyan, G.P., Karmenyan, A.V., and Meliksetyan, T.E. (1984). Thin film dye laser amplifier. *Opt. Commun.*, **52,** 114–116.

Baltakov, F.N., Barikhin, B.A., Kornilov, V.G., Mikhnov, S.A., Rubinov, A.N., and Sukhanov, L.V. (1973). 110-J pulsed laser using a solution of rhodamine 6G in ethyl alcohol. *Sov. Phys.—Tech. Phys.* **17,** 1161–1163.

Baltakov, F.N., Barikhin, B.A., and Sukhanov, L.V. (1974). 400-J pulsed laser using a solution of rhodamine 6G in ethanol. *JETP Lett.* **19,** 174–175.

Baranov, V.Y., Borisov, V.M., Vinokhodov, A.Y., Vysikailo, F.I., and Kiryukhin, Y.B. (1984). Increase in the repetition frequency of XeCl laser pulses to 1 kHz. *Sov. J. Quantum Electron.* **14,** 558–560.

Basov, N.G., Logunov, O.A., Nurligareev, D.K., and Trusov, K.K. (1981). Lasing in dye vapor exposed to wide-band optical pumping. *Sov. J. Quantum Electron.* **11,** 1400–1401.

Berik, E., Davidenko, B., Mihkelsoo, V., Apanasevich, P., Grabchikov, A., and Orlovich V. (1985). Stimulated Raman scattering of dye laser radiation in hydrogen. Improvement of spectral purity. *Opt. Commun.* **56,** 283–287.

Bernhardt, A.F., and Rasmussen, P. (1981). Design criteria and operating characteristics of a single-mode pulsed dye laser. *Appl. Phys. B* **26,** 141–146.

Bethune, D.S. (1981). Dye cell design for high-power low-divergence excimer-pumped dye lasers. *Appl. Opt.* **20,** 1897–1899.

Bigio, I.J. (1978). Injection-locked, unstable resonator dye laser. In *High-Power Lasers and Applications* (Kompa, K.L., and Walther, H., eds.). Springer-Verlag, Berlin, pp. 116–118.

Blit, S., Fischer, A., and Ganiel, U. (1974). Early termination of flashlamp pumped dye laser pulses by shock wave formation. *Appl. Opt.* **13,** 335–340.

Blit, S., Ganiel, U., and Treves, D. (1977). A tunable, single-mode, injection-locked flashlamp pumped dye laser. *Appl. Phys.* **12,** 69–74.

Bollen, G., Kluge, H.J., Wallmeroth, K., Schaaf, W.H., and Moore, R.B. (1987). High-power pulsed dye laser with Fourier-limited bandwith. *J. Opt. Soc. Am. B* **4,** 329–336.

Bor, Z. (1981). Amplified spontaneous emission from N_2-laser pumped dye lasers. *Opt. Commun.* **39,** 383–386.

Born, M., and Wolf, E. (1975). *Principles of Optics*, 5th ed. Pergamon, New York.

Bos, F. (1981). Versatile high-power single–longitudinal-mode pulsed dye laser. *Appl. Opt.* **20,** 1886–1890.

Bradley, D.J., Gale, G.M., Moore, M., and Smith, P.D. (1968). Longitudinally pumped, narrow-band continuously tunable dye laser. *Phys. Lett.* **26A,** 378–379.

Brandt, M. (1984). Repetitively pulsed transversely excited Sr^+ recombination laser. *IEEE J. Quantum Electron.* **QE-20,** 1006–1007.

Brandt, M., and Piper, J.A. (1981). Operating characteristics of TE copper bromide lasers. *IEEE J. Quantum Electron.* **QE-17,** 1107–1115.

Bricks, B.G., and Karras, T.W. (1980). Power scaling experiments with a discharge-heated lead vapor laser. In *Proceedings of the International Conference on Lasers '79* (Corcoran, V.J., ed.). STS Press, McLean, Va. pp. 309–314.

Brink, D.J., and Proch, D. (1982). Efficient tunable ultraviolet source based on stimulated Raman scattering of an excimer-pumped dye laser. *Opt. Lett.* **7,** 494–496.

Broglia, M., Zampetti, P., and Benetti, P. (1986). Single longitudinal mode interaction between a uranium atomic beam and a modified Hansch-type pulsed dye laser. *Appl. Phys. B* **39,** 73–76.

Broyer, M., Chevaleyre, J., Delacretaz, G., and Woste, L. (1984). CVL-pumped dye laser for spectroscopic application. *Appl. Phys. B.* **35,** 31–36.

Bukshpun, L.M., Zhukov, V.V., Latush, E.L., and Sem, M.F. (1981). Frequency tuning and mode self-locking in He-Sr recombination laser. *Sov. J. Quantum Electron.* **11,** 804–805.

Bunkenburg, J. (1972). An 11 megawatt 6.8 Joule flashlamp pumped coaxial liquid dye laser. *Rev. Sci. Instrum.* **43,** 1611–1612.

Burlamacchi, P., and Cutter D. (1977). Energy transfer in flashlamp pumped organic dye lasers. *Opt. Commun.* **22,** 283–287.

Burnham, R., and Schimitschek, E. J. (1981). High-power blue-green lasers. *Laser Focus* **17**(6), 54–66.

Butler, M.S., and Piper, J.A. (1985). Pulse energy scaling characteristics of longitudinally excited Sr^+ discharge recombination lasers. *IEEE J. Quantum Electron.* **QE-21,** 1563–1566.

Calkins, J., Colley, E., and Hazle, J. (1982). The use of caffeine as a liquid filter in coaxial flashlamp pumped dye lasers. *Opt. Commun.* **42,** 275–277.

Carlsten, J.L., and McIlrath, T.J. (1973). An oscillator-amplifier dye laser: tuneable high powers without grating damage. *Opt. Commun.* **8,** 52–55.

Caro, R.G., Gower, M.C., and Webb, C.E. (1982). A simple tunable KrF laser system with narrow bandwidth and diffraction-limited divergence. *J. Phys. D: Appl. Phys.* **15,** 767–773.

Curry, S.M., Cubeddu, R., and Hänsch, T.W. (1973). Intensity stabilization of dye laser radiation by saturated amplification. *Appl. Phys.* **1,** 153–159.

Dearth, J.J., Vaughn, V.V., McGowan, R.B., Ehrlich, J., and Conrad, R.W. (1988). A flashlamp pumped zig-zag slab dye laser. In *Proceedings of the International Conference on Lasers '87* (Duarte, F.J., ed.). STS Press, McLean, Va. pp. 320–329.

Denisov, L.K., Krasnov, I.V., and Uzhinov, B.M. (1985). Effect of triplet state of unsubstituted rhodamine on its lasing efficiency with flashlamp pumping. *Opt. Spectrosc.* **59,** 241–243.

D'Haenens, I.J., and Asawa, C.K. (1962). Stimulated and fluorescent optical emission in ruby from 4.2° to 300°K: zero-field splitting and mode structure. *J. Appl. Phys.* **33,** 3201–3208.

Drake, J.M., Tam, E.M., and Morse, R.I. (1972). The use of light converters to increase the power of flashlamp-pumped dye lasers. *IEEE J. Quantum Electron.* **QE-8,** 92–94.

Drexhage, K.H. (1977). Structure and properties of laser dyes. In *Dye Lasers* (Schäfer, F.P., ed.). Springer-Verlag, New York, pp. 144–193.

Duarte, F.J. (1983). Thermal effects in double prism dye laser cavities. *IEEE J. Quantum Electron.* **QE-19,** 1345–1347.

Duarte, F.J. (1987). Critical assessment of ASE in laser-pumped pulsed dye lasers. In *Proceedings of the International Conference on Lasers '86* (McMillan, R.W., ed.). STS Press, McLean, Va., pp. 416–419.

Duarte, F.J. (1988). Narrow-linewidth, multiple-prism, pulsed dye lasers. *Lasers and Optronics* 7(2), 41–45. Gordon Publications Inc. © 1988. All rights reserved.

Duarte, F.J. (1989). Ray matrix analysis of multiple-prism dye laser oscillators. *Opt. Quantum Electron.* **21,** 47–54.

Duarte, F.J., and Conrad, R.W. (1987). Diffraction-limited single-longitudinal-mode multiple-prism flashlamp-pumped dye laser oscillator: linewidth analysis and injection of amplifier system. *Appl. Opt.* **26,** 2567–2571.

Duarte, F.J., Ehrlich, J.J., Patterson, S.P., Russell, S.D., and Adams, J.E. (1988). Linewidth instabilities in narrow-linewidth, flashlamp-pumped dye laser oscillators. *Appl. Opt.* **27,** 843–846.

Duarte, F.J., and Piper, J.A. (1980). A double-prism beam expander for pulsed dye lasers. *Opt. Commun.* **35,** 100–104.

Duarte, F.J., and Piper, J.A. (1981). Prism preexpanded grazing-incidence grating cavity for pulsed dye lasers. *Appl. Opt.* **20,** 2113–2116.

Duarte, F.J., and Piper, J.A. (1982). Comparison of prism-expander and grazing-incidence grating cavities for copper laser pumped dye lasers. *Appl. Opt.* **21,** 2782–2786.

Duarte, F.J., and Piper, J.A. (1984). Narrow linewidth, high prf copper laser-pumped dye-laser oscillators. *Appl. Opt.* **23,** 1391–1394.

Duarte, F.J., and Piper, J.A. (1985). Design and operation of efficient UV-VIS narrow linewidth pulsed dye lasers. In *Proceedings of the International Conference on Lasers '83* (Powell, R.C., ed.). STS Press, McLean, Va., pp. 677–681.

Dugan, C.H., Lee, A., and Morgan, F.J. (1978). Polarization of light from a pulsed dye laser. *Appl. Opt.* **17,** 1012–1014.

Dujardin, G., and Flamant, P. (1978a). Conversion d'energie dans les amplificateurs a colorants en presence de superfluorescence. *Opt. Acta* **25,** 273–283.

Dujardin, G., and Flamant, P. (1978b). Amplified spontaneous emission and spatial dependence of gain in dye amplifiers. *Opt. Commun.* **24,** 243–247.

Dupre, P. (1987). Quasiunimodal tunable pulsed dye laser at 440 nm: theoretical development for using a quad prism beam expander and one or two gratings in a pulsed dye laser oscillator cavity. *Appl. Opt.* **26,** 860–871.

Eden, J.G. (1986). Vapor phase dye lasers. Presented at the International Conference on Lasers '86, Orlando, Fla., paper WH.1.

Erickson, G.F. (1988). Solid hosts for dye laser rods—part 2. Some experimental results. In *Proceedings of the International Conference on Lasers '87* (Duarte, F.J., ed.). STS Press, McLean, Va., pp. 342–348.

Erickson, L.E., and Szabo, A. (1971). Spectral narrowing of dye laser output by injection of monochromatic radiation into the laser cavity. *Appl. Phys. Lett.* **18,** 433–435.

Everett, P.N., Aldag, H.R., Ehrlich, J.J., Janes, G.S., Klimek, D.E., Landers, F.M., and Pacheco, D.P. (1986). Efficient 7-J flashlamp-pumped dye laser at 500 nm wavelength. *Appl. Opt.* **25,** 2142–2147.

Everett, P.N., and Zollars B.G. (1988). Engineering of a multi-beam 300-watt flashlamp-pumped dye-laser system. In *Proceedings of the International Conference on Lasers '87* (Duarte, F.J., ed.). STS Press, McLean, Va., pp. 291–296.

Ewanizky, T.F., Wright, R.H., and Theissing H.H. (1973). Shock-wave termination of laser action in coaxial flash lamp dye lasers. *Appl. Phys. Lett.* **22,** 520–521.

Feofilov, P.P. (1961). *The Physical Basis of Polarized Emission*. Consultants Bureau, New York.

Flamant, P.H., Josse, D., and Maillard, M. (1984). Transient injection frequency-locking of a microsecond-pulsed dye laser for atmospheric measurements. *Opt. Quantum Electron.* **16,** 179–182.

Flamant, P., and Meyer, Y.H. (1971). Absolute gain measurements in a multistage dye amplifier. *Appl. Phys. Lett.* **19,** 491–493.

Fletcher, A.N., Bliss, D.E., Pietrak, M.E., and McManis, G.E. (1986). Improving the output and lifetime of flashlamp-pumped dye lasers. In *Proceedings of the International Conference on Lasers '85* (Wang, C.P., ed.). STS Press, McLean, Va., pp. 797–804.

Fletcher, A.N., Knipe, R.H., and Pietrak, M.E. (1982). Laser dye stability. Part 7. Effects of temperature, UV filter, and solvent purity. *Appl. Phys. B* **27,** 93–97.

Förster, T. (1951). *Fluoreszenz Organischer Verbindungen*. Vandenhoeck and Ruprecht, Göttingen, West Germany.

Fort, J., and Moulin, C. (1987). High power high-energy linear flashlamp-pumped dye laser. *Appl. Opt.* **26,** 1246–1249.

Fountain, W.D., and Bass, M. (1982). Single-axial-mode operation of a polarization-coupled stable/unstable-resonator Nd:YAG laser oscillator. *IEEE J. Quantum Electron.* **QE-18,** 432–437.

Furumoto, H.W., and Ceccon, H.L. (1969). Optical pumps for organic dye lasers. *Appl. Opt.* **8,** 1613–1623.

Ganiel, U., Hardy, A., Neumann, G., and Treves, D. (1975). Amplified spontaneous emission and signal amplification in dye-laser systems. *IEEE J. Quantum Electron.* **QE-11,** 881–891.

German, K.R. (1981). Grazing angle tuner for cw lasers. *Appl. Opt.* **20,** 3168–3171.

Godard, B. (1974). A simple high-power large-efficiency N_2 ultraviolet laser. *IEEE J. Quantum Electron.* **QE-10,** 147–153.

Greenhalgh, D.A., and Sarkies, P.H. (1982). Novel geometry for simple accurate tuning of lasers. *Appl. Opt.* **21,** 3234–3236.

Haag, G. Munz, M., and Marowsky, G. (1983). Amplified spontaneous emission (ASE) in laser oscillators and amplifiers. *IEEE J. Quantum Electron.* **QE-19,** 1149–1160.

Hanna, D.C., Pacheco, M.T.T., and Wong, K.H. (1985). High efficiency and high brightness Raman conversion of dye laser radiation. *Opt. Commun.* **55,** 188–192.

Hänsch, T.W. (1972). Repetitively pulsed tunable dye laser for high resolution spectroscopy. *Appl. Opt.* **11,** 895–898.

Hargrove, R.S., and Kan, T. (1980). High power efficient dye amplifier pumped by copper vapor lasers. *IEEE J. Quantum Electron.* **QE-16,** 1108–1113.

Hartig, W. (1978). A high power dye-laser pumped by the second harmonic of a Nd-YAG laser. *Opt. Commun.* **27,** 447–450.

Hartig. W., and Schmidt, W. (1979). A broadly tunable IR waveguide Raman laser pumped by a dye laser. *Appl. Phys.* **18,** 235–241.

Hirth, A., Meyer, R., and Schetter, K. (1981). A reliable high power repetitive pulsed dye laser. *Opt. Commun.* **40,** 63–67.

Hnilo, A.A., and Martinez, O.E. (1987). On the design of pulsed dye laser amplifiers. *IEEE J. Quantum Electron.* **QE-23,** 593–599.

Itzkan, I., and Cunningham, F.W. (1972). Oscillator-amplifier dye-laser system using N_2 laser pumping. *IEEE J. Quantum Electron.* **QE-8,** 101–105.

Jethwa, J., and Schäfer, F.P. (1974). A reliable high average power dye laser. *Appl. Phys.* **4,** 299–302.

Jethwa, J., Anufrik, S.S., and Docchio, F. (1982). High-efficiency high-energy flashlamp-pumped dye laser. *Appl. Opt.* **21,** 2778–2781.

Klick, D., Akerman, M.A., Tsuda, H., and Supurovic D. (1986). High-power broadband dye laser for manufacturing applications. In *Proceedings of the International Conference on Lasers '85* (Wang, C.P., ed.). STS Press, McLean, Va., pp. 531–538.

Knyazev, B.A., Lebedev, S.V., and Fokin, E.P. (1983). High-power rhodamine 6G laser with an extended service life. *Sov. J. Quantum Electron.* **13,** 146–150.

Latush, E.L., Zhukov, V.V., Mikhalevsky, V.S., and Sem, M.F. (1982). Metal vapor recombination laser research. In *Proceedings of the International Conference on Lasers '81* (Collins, C.B., ed.). STS Press, McLean, Va., pp. 1121–1128.

Lavi, S., Levin, L.A., Liran, J., and Miron, E. (1979). Efficient oscillator-amplifier dye laser pumped by a frequency-doubled Nd:YAG laser. *Appl. Opt.* **18,** 525–527.

Leslie, S.G., Liu, C.S., and Liberman, I. (1984). Efficient cavity dumped HgBr laser. *Appl. Opt.* **23,** 36–39.

Littman, M.G. (1978). Single-mode operation of grazing-incidence pulsed dye laser. *Opt. Lett.* **3,** 138–140.

Littman, M.G. (1984). Single-mode pulsed tunable dye laser. *Appl. Opt.* **23,** 4465–4468.

Littman, M.G., and Metcalf, H.J. (1978). Spectrally narrow pulsed dye laser without beam expander. *Appl. Opt.* **17,** 2224–2227.

Liu, K., and Littman, M.G. (1981). Novel geometry for single-mode scanning of tunable lasers. *Opt. Lett.* **6,** 117–118.

Loree, T.R., Butterfield, K.B., and Barker, D.L. (1978). Spectral tuning of ArF and KrF discharge lasers. *Appl. Phys. Lett.* **32,** 171–173.

Ludewigt, K., Birkmann, K., and Wellegehausen, B. (1984). Anti-Stokes Raman laser investigations on atomic Tl and Sn. *Appl. Phys. B* **33,** 133–139.

Lyutskanov, V.L., Khristov, K.G., and Tomov, I.V. (1980). Tuning of the emission frequency of a gas-discharge XeCl laser. *Sov. J. Quantum Electron.* **10,** 1456–1457.

Maeda, M., Uchino, O., Okada, T., and Miyazoe, Y. (1975). Powerful narrow-band dye laser forced oscillator. *Jap. J. Appl. Phys.* **14,** 1975–1980.

Magyar, G., and Schneider-Muntau, H.J. (1972). Dye laser forced oscillator. *Appl. Phys. Lett.* **20,** 406–408.

Mannik, L., and Brown, S.K. (1986). Tunable infrared generation using third Stokes output from a waveguide Raman shifter. *Opt. Commun.* **57,** 360–364.

Marling. J.B., Gregg, D.W., and Thomas, S.J. (1970a). Effect of oxygen on flashlamp-pumped organic-dye lasers. *IEEE J. Quantum Electron.* **QE-6,** 570–572.

Marling, J.B., Gregg, D.W., and Wood, L. (1970b). Chemical quenching of the triplet state in flashlamp-excited liquid organic lasers. *Appl. Phys. Lett.* **17,** 527–530.

Maystre, D. (1980). Integral methods. In *Electromagnetic Theory of Gratings* (Petit, R., ed.). Springer-Verlag, Berlin, pp. 63–100.

Mazzinghi, P., Burlamacchi, P., Matera, M., Ranea-Sandoval, H.F., Salimbeni, R., and Vanni, U. (1981). A 200 W average power, narrow bandwidth, tunable waveguide dye laser. *IEEE J. Quantum Electron.* **QE-17,** 2245–2249.

Mazzinghi, P., Rivano, V., and Burlamacchi, P. (1983). High efficiency high-energy slab dye laser for photobiological experiments. *Appl. Opt.* **22,** 3335–3337.

McFarland, B.B. (1967). Laser second-harmonic-induced stimulated emission of organic dyes. *Appl. Phys. Lett.* **10,** 208–209.

McIntyre, I.A., and Dunn, M.H. (1984). Amplified spontaneous emission in distributed feedback dye lasers. *Opt. Commun.* **50,** 169–172.

McKee, T.J. (1985). Spectral-narrowing techniques for excimer laser oscillators. *Can. J. Phys.* **63,** 214–219.
McKee, T.J., Lobin, J., and Young, W.A. (1982). Dye laser spectral purity. *Appl. Opt.* **21,** 725–728.
McNicholl, P., and Metcalf, H.J. (1985). Synchronous cavity mode and feedback wavelength scanning in dye laser oscillators with gratings. *Appl. Opt.* **24,** 2757–2761.
Meaburn, J. (1976). *Detection and Spectrometry of Faint Light.* Reidel, Boston.
Morey, W.W. (1980). Copper vapor laser pumped dye laser. In *Proceedings of the International Conference on Lasers '79* (Corcoran, V.J., ed.). STS Press, McLean, Va., pp. 365–373.
Morgan, F.J., and Dugan, H. (1979). Polarization of light from pulsed dye laser: effects of solvent viscosity. *Appl. Opt.* **18,** 4112–4115.
Mory, S., Patzold, H.J., Rosenfeld, A., Polze, S., Korn, G., and Tilch, J. (1979). Spectral narrowing and continuous tuning of a cw dye laser by grating beam-expansion. *Opt. Commun.* **29,** 201–203.
Munz, M., Haag, G., and Marowsky, G. (1980). Optimization of dye-laser output coupling by consideration of the spatial gain distribution. *Appl. Phys.* **22,** 175–184.
Nair, L.G. (1982). Dye lasers. *Prog. Quantum Electron.* **7,** 153–268.
Nair, L.G., and Dasgupta, K. (1985). Amplified spontaneous emission in narrow-band pulsed dye laser oscillators—theory and experiment. *IEEE J. Quantum Electron.* **QE-21,** 1782–1794.
Nechaev, S.Y. (1983). Investigation of a pulsed dye laser under various pumping conditions. *Sov. J. Quantum Electron.* **13,** 1012–1014.
Neister, S.E. (1977a). Flashlamp improvement. U.S. Air Force Report AFAL-TR-77–139.
Neister, S.E. (1977b). The dye laser dye update. *Opt. Spectra* **11**(2), 34–36.
Olcay, M.R., Pasqual, J.A., Lisboa, J.A., and Francke, R.E. (1985). Tuning of a narrow linewidth pulsed dye laser with a Fabry-Perot and diffraction grating over a large wavelength range. *Appl. Opt.* **24,** 3146–3150.
Oliveira dos Santos, B., Fellows, C.E., de Oliveira e Souza, J.B., and Massone C.A. (1986). A 3% efficiency N_2 laser. *Appl. Phys. B.*, **41,** 241–244.
Pacheco, D.P., Aldag, H.R., Itzkan, I., and Rostler, P.S. (1988). A solid-state flashlamp-pumped dye laser employing polymer hosts. In *Proceedings of the International Conference on Lasers '87* (Duarte, F.J., ed.). STS Press, McLean, Va., pp. 330–337.
Paisner, J.A. (1988). Atomic vapor laser isotope separation. *Appl. Phys. B.* **46,** 253–260.
Pappalardo, R., Samelson, H., and Lempicki, A. (1970). Long pulse laser emission from rhodamine 6 G. *IEEE J. Quantum Electron.* **QE-6,** 716–725.
Parks, J.H., Rao, D.R., and Javan, A. (1968). A high-resolution study of the C $^3\Pi_u \rightarrow$ B $^3\Pi_g$ (0, 0) stimulated transitions in N_2. *Appl. Phys. Lett.* **13,** 142–144.
Penzkofer, A., and Falkenstein, W. (1978). Theoretical investigation of amplified spontaneous emission with picosecond light pulses in dye solutions. *Opt. Quantum Electron.* **10,** 399–423.
Peters, D.W., and Mathews, C.W. (1980). Temperature dependence of the peak power of a Hansch-type dye laser. *Appl. Opt.* **19,** 4131–4132.
Peterson, O.G., and Snavely B.B. (1968). Stimulated emission from flashlamp-excited organic dyes in polymethyl methacrylate. *Appl. Phys. Lett.* **12,** 238–240.
Piper, J.A. (1978). A transversely excited copper halide laser with large active volume. *IEEE J. Quantum Electron.* **QE-14,** 405–407.
Raymond, T.D., Walsh, S.T., and Keto, J.W. (1984). Narrowband dye laser with a large scan range, *Appl. Opt.* **23,** 2062–2064.

Roncin, J.Y., and Damany, H. (1981). Improved cell design for pulsed dye lasers. *Rev. Sci. Instrum.* **52,** 1922–1923.

Schäfer, F.P. (1977). Principles of dye laser operation. In *Dye Lasers* (Schäfer, F.P., ed.). Springer-Verlag, Berlin, pp. 1–85.

Schäfer, F.P. (1983). Private communication.

Schäfer, F.P., and Müller, H. (1971). Tunable dye ring-laser. *Opt. Commun.* **2,** 407–409.

Schäfer, F.P., and Ringwelski, L. (1973). Triplet quenching by oxygen in a rhodamine 6G laser, *Z. Naturforsch.* **28a,** 792–793.

Schäfer, F.P., Schmidt, W., and Volze, J. (1966). Organic dye solution laser. *Appl. Phys. Lett.* **9,** 306–309.

Schenck, P., and Metcalf, H. (1973). Low cost nitrogen laser design for dye laser pumping. *Appl. Opt.* **12,** 183–186.

Schmidt, W., and Appt, W. (1972). Tunable stimulated Raman emission generated by a dye laser, *Z. Naturforsch.* **27a,** 1373–1375.

Schmidt, W., and Schäfer, F.P. (1967). Blitzlampengepumpte farbstofflaser. *Z. Naturforschg.* **22a,** 1563–1566.

Schomburg, H., Döbele, H.F., and Rückle, B. (1983). Generation of tunable narrow-bandwidth VUV radiation by anti-Stokes SRS in H_2. *Appl. Phys. B* **30,** 131–134.

Sevchenko, A.N., Kovalev, A.A., Pilipovich, V.A., and Razvin, Y.V. (1968). Polarized laser radiation from solutions of organic dyes. *Sov. Phys.—Doklady* **13,** 226–228.

Shank, C.V. (1975). Physics of dye lasers. *Rev. Mod. Phys.* **47,** 649–657.

Shay, T.M., Gookin, D., Jordan, M.C., Hanson, F.E., and Schimitschek, E.J. (1985). Experimental diagnostics of an avalanche discharge excited HgBr/$HgBr_2$ dissociation laser. *IEEE J. Quantum Electron.* **QE-21,** 1271–1277.

Shay, T.M., Hanson, F.E., Gookin, D., and Schimitschek, E.J. (1981). Line narrowing and enhanced efficiency of an HgBr laser by injection locking. *Appl. Phys. Lett.* **39,** 783–785.

Shoshan, I., Danon, N.N., and Oppenheim, U.P. (1977). Narrowband operation of a pulsed dye laser without intracavity beam expander. *J. Appl. Phys.* **48,** 4495–4497.

Siegman, A.E. (1986). *Lasers.* University Science Books, Mill Valley, Ca.

Simon, P., and Juhasz, N. (1985). Dye cell design for high-efficiency amplifier of high beam quality. *J. Phys. E: Sci. Instrum.* **18,** 829–830.

Snavely, B.B., Peterson, O.G., and Reithel, R.F. (1967). Blue laser emission from a flashlamp-excited organic dye solution. *Appl. Phys. Lett.* **11,** 275–276.

Soffer, B.H., and McFarland, B.B. (1967). Continuously tunable, narrowband, organic dye lasers. *Appl. Phys. Lett.* **10,** 266–267.

Soldatov, A.N., and Sukhanov, V.B. (1983). Metal vapor laser-pumped dye laser. In *Proceedings of the International Conference on Lasers '82* (Powell, R.C., ed.). STS Press, McLean, Va., pp. 493–500.

Sorokin, P.P., Culver, W.H., Hammond, E.C., and Lankard, J.R. (1966). End-pumped stimulated emission from a thiacarbocyanine dye. *IBM J. Res. Develop.* **10,** 401.

Sorokin, P.P., and Lankard, J.R. (1966). Stimulated emission observed from an organic dye, chloro-aluminum phthalocyanine. *IBM J. Res. Develop.* **10,** 162–163.

Sorokin, P.P., and Lankard, J.R. (1967). Flashlamp excitation of organic dye lasers: a short communication. *IBM J. Res Develop.* **11,** 148.

Sorokin, P.P., Lankard, J.R., Hammond, E.C., and Moruzzi, V.L. (1967). Laser-pumped stimulated emission from organic dyes: experimental studies and analytical comparisons. *IBM J. Res. Develop.* **11,** 130–147.

Sorokin, P.P., Lankard, J.R., Moruzzi, V.L., and Hammond, E.C. (1968). Flashlamp-pumped organic-dye lasers. *J. Chem. Phys.* **48,** 4726–4741.

Spaeth, M.L., and Bortfeld, D.P. (1966). Stimulated emission from polymethine dyes. *Appl. Phys. Lett.* **9,** 179–181.
Strome, F.C., and Webb, J.P. (1971). Flashtube-pumped dye laser with multiple-prism tuning. *Appl. Opt.* **10,** 1348–1353.
Suzuki, T., Kato, H., Taira, Y., Adachi, Y., Konishi, N., and Kasuya, T. (1981). Full-automatic broad-band wavelength control of a pulsed dye laser. *Appl. Phys.* **24,** 331–340.
Sze, R.C., and Seegmiller, E. (1981). Operating characteristics of a high repetition rate miniature rare-gas halide laser. *IEEE J. Quantum Electron.* **QE-17,** 81–91.
Tang, K.Y., O'Keefe, T., Treacy, B., Rottler, L., and White, C. (1987). Kilojoule output XeCl dye laser: optimization and analysis. In *Proceedings: Dye Laser/Laser Dye Technical Exchange Meeting, 1987* (Bentley, J.H., ed.). U.S. Army Missile Command, Redstone Arsenal, Al., pp. 490–502.
Tellinghuisen, J., and Ashmore, J.G. (1982). The B→X transition in ^{200}Hg ^{79}Br. *Appl. Phys. Lett.* **40,** 867–869.
Tenenbaum, J., Smilanski, I., Gabay, S., Levin, L.A., Erez, G., and Lavi, S. (1980). Structure of 510.6 and 578.2 nm copper laser lines. *Opt. Commun.* **32,** 473–477.
Tittel, F.K., Marowsky, G., Nighan, W.L., Zhu, Y., Sauerbrey, R.A., and Wilson, W.L. (1986). Injection-controlled tuning of an electron-beam excited XeF(C→A) laser. *IEEE J. Quantum Electron.* **QE-22,** 2168–2173.
Trebino, R., Roller, J.P., and Siegman, A. (1982). A comparison of the Cassegrain and other beam expanders in high power pulsed dye lasers. *IEEE J. Quantum Electron.* **QE-18,** 1208–1213.
Trehin, F., Biraben, F., Cagnac, B., and Grynberg, G. (1979). Flashlamp pumped tunable dye laser of ultra-narrow bandwidth. *Opt. Commun.* **31,** 76–80.
Trusov, K.K. (1981). Efficiency of excitation of POPOP vapor by wide-band flashlamp radiation. *Sov. J. Quantum Electron.* **11,** 1288–1292.
Verkerk, P., Grand-Clement, D., Trehin, F., and Grynberg, G. (1986). Spectral analysis of an injection-locked flash-lamp pumped dye-laser of ultranarrow linewidth. *Opt. Commun.* **58,** 413–416.
Vrehen, Q.H.F., and Breimer, A.J. (1972). Spectral properties of a pulsed dye laser with monochromatic injection. *Opt. Commun.* **4,** 416–420.
Wallenstein, R., and Hänsch, T.W. (1974). Linear pressure tuning of a multielement dye laser spectrometer. *Appl. Opt.* **13,** 1625–1628.
Wallenstein, R., and Hänsch, T.W. (1975). Powerful dye laser oscillator-amplifier system for high resolution spectroscopy. *Opt. Commun.* **14,** 353–357.
Wang, Y., Shen, S., Xia, T., and Wu, Z. (1988). Spectral structure of CuBr vapor laser lines. *Appl. Phys. B.* **46,** 191–195.
Webb, C.E. (1988). Copper and gold vapor lasers: recent advances and applications. In *Proceedings of the International Conference on Lasers '87* (Duarte, F.J., ed.). STS Press, McLean, Va., pp. 276–284.
White, J.C., and Henderson, D. (1983). Tuning and saturation behavior of the anti-Stokes Raman laser. *Opt. Lett.* **8,** 15–17.
Wilke, V., and Schmidt, W. (1978). Tunable UV-radiation by stimulated Raman scattering in hydrogen. *Appl. Phys.* **16,** 151–154.
Willett, C.S. (1974). *An Introduction to Gas Lasers: Population Inversion Mechanisms.* Pergamon, New York.
Woodward, B.W., Ehlers, V.J., and Lineberger, W.C. (1973). A reliable, repetitively pulsed, high-power nitrogen laser. *Rev. Sci. Intrum.* **44,** 882–887.
Wright, W., and Falconer, I.S. (1988). A transversely pumped prismatic dye cell for high power dye lasers. *Opt. Commun.* **67,** 221–224.

Wyatt, R., Ernsting, N.P., and Wrobel, W.G. (1982). Tunable electronic Raman laser at 16 μm. *Appl. Phys. B* **27,** 175–176.

Yang, T.T., Burde, D.H., Merry, G.A., Harris, D.G., Pugh, L.A., Tillotson, J.H., Turner, C.E., and Copeland, D.A. (1988). Spectra of electron-beam pumped XeF lasers. *Appl. Opt.* **27,** 49–57.

Zemskov, K.I., Isaev, A.A., Kazaryan, M.A., Petrash, G.G., Adamushko, A.V., Belokon, M.V., Rubinov, A.N., and Evtukhovich, P.G. (1976). Jet dye laser pumped by copper vapor laser. *Sov. J. Quantum Electron.* **6,** 727–728.

Zenchenko, S.A., Ivanov, V.I., Malevich, I.A., and Shulekin, S.F. (1985). Switching of the emission wavelengths of a copper vapor laser. *Sov. J. Quantum Electron.* **15,** 124–125.

Zherikin, A.N., Letokhov, V.S., Mishin, V.I., Belyaev, V.P., Evtyunin, A.N., and Lesnor, M.A. (1981). High-repetition-rate tunable dye lasers pumped by copper vapor laser. *Sov. J. Quantum Electron.* **11,** 806–808.

Zollars, B.G., and Everett, P.N. (1988). Performance of a prototype module of a 300-watt flashlamp-pumped dye laser. In *Proceedings of the International Conference on Lasers '87* (Duarte, F.J., ed.). STS Press, McLean, Va., pp. 297–303.

Chapter 7

PHOTOCHEMISTRY OF LASER DYES*

Guilford Jones II

Department of Chemistry
Boston University
Boston, Massachusetts

1. INTRODUCTION

The development of the dye laser has been intimately tied to the systematic search for organic dye structures and dye media in which stimulated emission is observed. An array of photophysical properties such as absorptivity, emission (fluorescence) yield, Stokes shift, and triplet formation influence threshold and peak power lasing characteristics. The tailoring of dyes must also contend with variables having to do with

* The author wishes to thank the U.S. Office of Naval Research for support of work on laser-dye photochemistry. The assistance of A. Rahman in the preparation of the manuscript is also gratefully acknowledged.

ISBN 0-12-222700-X

photochemical reactions which degrade dye and produce competitively absorbing products, thus influencing dye service life as well as other parameters. Among myriad possibilites, the number of organic compounds that meet most of these criteria is satisfactory, although not overwhelming. Drexhage (1973), in the first comprehensive review of laser-dye structure and properties, compiled a table of about three hundred dyes that had been found to lase measurably. In a later survey covering reports on laser dyes through 1980, Maeda (1984) identified 546 dyes from various classes of structure for which lasing parameters are reported.

With the advent of powerful laser or flashlamp pump sources and the needs for high output and repetition rate, the requirements for photochemical stability of lasing media are extreme. For pulsed lasers, projections as high as 100 W power, pulse frequency at 500 Hz and 200 mJ/pulse have been made. In a review of advances in laser-dye development, Schaefer (1983) notes that the barrier to wider industrial use of the dye laser is due in part to the unreliability associated with dye instability (photodegradation). The service life of a typical rhodamine dye in alcohol solvent on excitation with the unfiltered light of a bank of flashlamps is limited to a few flashes. More precisely measured quantum efficiencies for dye bleaching must not exceed 10^{-4} under routine conditions and be even more stringently limited in specialized applications. A further challenge to the photochemist is the array of pump wavelengths (e.g., 249 nm (KrF excimer), 337 nm (N_2 gas), 532 nm (Nd/YAG, $\lambda/2$), and Xe flashlamps (250–600 nm). Thus, a single dye with different applications may be subjected to excitation energies ranging from 40–120 kcal/Einstein (2–4 ev/photon).

A representative group of laser dyes with accompanying structures is listed in Table 7.1, including absorption and fluorescence emission wavelengths and peak lasing wavelengths under specified laser or flashlamp pumping conditions. These dyes are members of the major classes which include:

conjugated hydrocarbons: polyphenyls or stilbenes for the uv

oxazoles: diphenyloxazole(POP) and derivatives, generally for the 400-nm range

coumarins: the most common "blue-green" laser dyes

xanthenes: most notably, the rhodamines for the 600-nm region

cyanines: red and near-ir dyes

All of these classes of structure share the property of conjugation of π bonds to carbon not unlike that of the parent compound, benzene. Many

Table 7.1

Peak Wavelengths for Absorption, Spontaneous Fluorescence, and Lasing for Representatives of Major Classes of Laser Dyes

Dye	λ_a	λ_f	λ_{las} [a]
p-Terphenyl (I)	276	354	338 (KrF) 341 (Lamp)
POPOP (II)	358	415	419 (N_2) 419 (Lamp)
Coumarin 102 (III) (Coumarin 480)	390	468	480 (Nd/YAG) 480 (Lamp)
Rhodamine 6G (IV) (Rhodamine 590) (Chloride Salt)	528	547	563 (Nd/YAG) 585 (Lamp)
DOTC (V)	695	719	762 (Nd/YAG)

Note: Wavelengths in nanometers (nm). Hydrocarbon solvent for I, II; alcohol solvent for III–V. Alternate names ( ) provided for coumarin and rhodamine dyes.

[a] Mode of dye-laser pumping is shown in parentheses. Source: Drexhage (1973) and Duarte (1987).

dyes can be viewed in a simple way as having a benzene ring substituted with an electron withdrawing group (A) and an electron contributing group (D) as in VI. This familiar pattern tends to increase the transition moment for optical transitions and the rate of radiative decay for appropriately substituted dyes.

A singular advantage of dye lasers, documented in accompanying chapters in this volume, is the enormous selection of wavelengths for lasing due to the large variation in dye structure. Commercial dyes can be selected for output at wavelengths from ca. 300 nm now well into the infrared (>1.2 μ (Kopainsky *et al.*, 1982)). In addition, due to the breadth of emission bands that is typical of organic laser dyes (to be discussed shortly), each dye can be tuned over a 20–50-nm range.

It is the purpose of this review to summarize recent advances toward the understanding of photochemical mechanisms of dye degradation. Because photophysical properties are complementary to the photochemical pathways, important determinants of dye fluorescence and intersystem crossing (triplet formation) will also be reviewed. A feature to be highlighted is the role of medium and the opportunities to further tailor laser-dye properties with solvent or with sequestering agents. Where appropriate, recent advances in the development of new dye systems will be reported. These

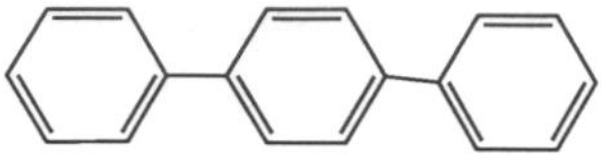

I

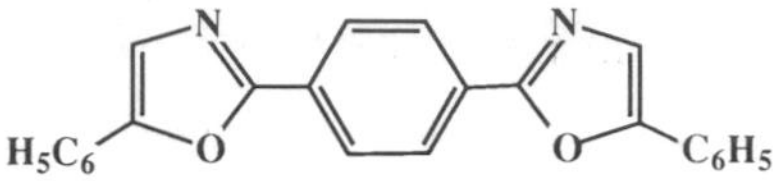

II

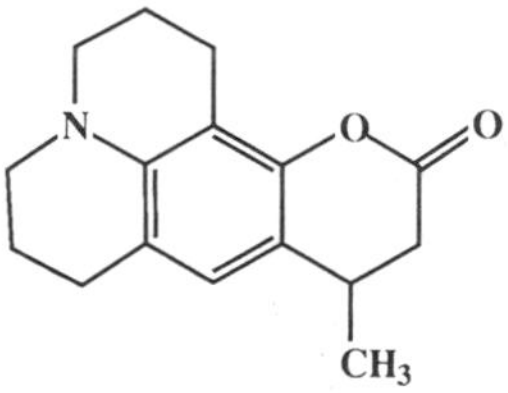

III

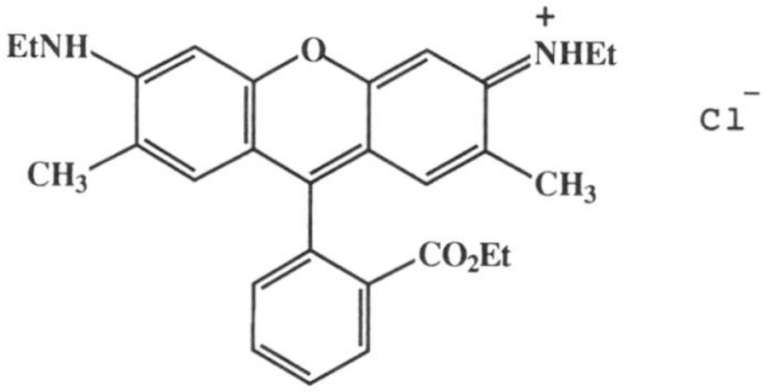

IV

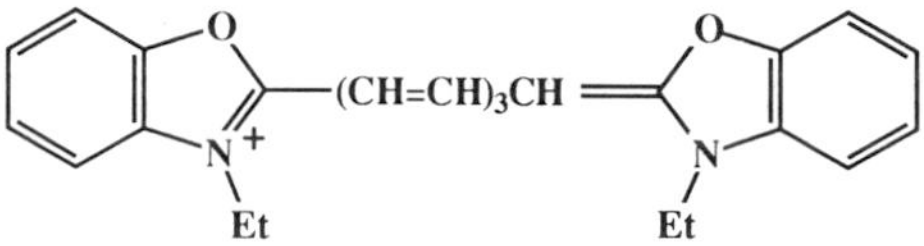

V

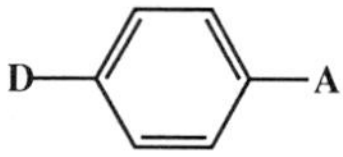

VI

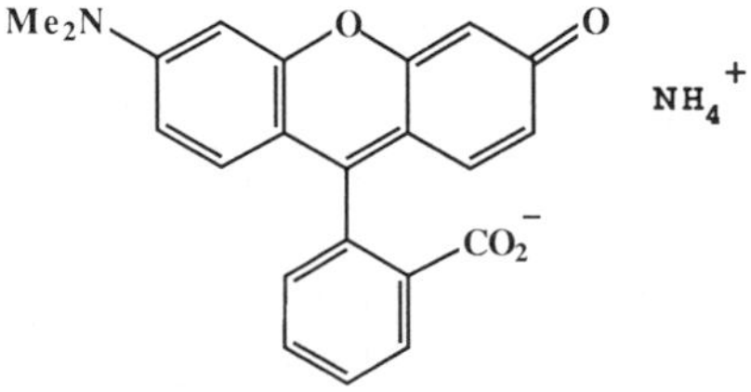

VII

include new structures for specified lasing frequencies, the appearance of water-soluble dyes, the linkage of dyes that provides novel means for energy transfer (down-shifting of output wavelengths), and dyes incorporated in the microstructures of detergent micelles, cyclodextrins, or other binding agents.

2. DYE PHOTOPHYSICS

2.1. Fundamental Photophysical Properties

Laser dyes typically show large absorptivities, as measured by their molar extinction coefficients, ϵ (units of liter/mole cm^{-1} or $M^{-1}\,cm^{-1}$). This feature is responsible for the high optical densities associated with moderate concentrations (10^{-4}–10^{-5} M) of dye and results from the high degree of delocalization and polarizability of π electrons in dye structures. The degree to which electron density in a dye molecule is "rearranged" or polarized as the result of absorption of photons is related to the transition dipole moment or oscillator strength (f) for the electronic transition. The parameter f is related to a first approximation, using classical theory (Turro, 1978), to the extinction coefficient (ϵ)

$$f = 4.3 \times 10^{-9} \int \epsilon\, d\nu = 4.3 \times 10^{-9}\, \epsilon_{max}\, \Delta\nu_{1/2} \tag{7.1}$$

where ϵ is the experimental extinction coefficient and ν is the energy (in wavenumbers) of the absorption in question. The integral of ϵ is more conveniently approximated by the product of extinction coefficient at the absorption maximum and the width of the absorption band at $1/2\epsilon_{max}$ (Eq. 7.1). The importance of oscillator strength is underscored on noting the relationship between f and the rate constant for radiative decay of an excited dye molecule (here k_f, the rate constant for fluorescence emission),

$$k_f = \nu(\text{max})^2 f \tag{7.2}$$

where ν(max) is the energy corresponding to the maximum wavelength of absorption. The probabilities of both spontaneous emission (as reflected in k_f) and stimulated emission (the latter responsible for "gain" or light amplification in a dye-laser cavity) are related to the size of the transition moment and the oscillator strength in the same way.

The most important electronic energy levels for organic dyes in fluid media are the ground state (S_0) and the first electronic excited state (S_1) as shown in Fig. 7.1. The absorption of light which raises molecules to the upper level is associated with promotion of an electron in a bonding orbital

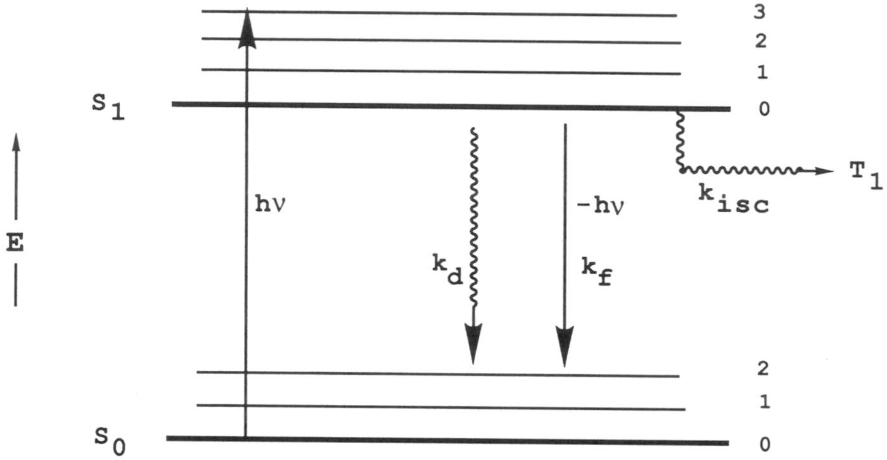

Fig. 7.1 Jablonski state energy diagram depicting molecular ground state (S_0), first excited singlet state (S_1), and lowest triplet state (T_1). Heavy arrows identify radiative (absorption and emission) processes, whereas curly arrows show radiationless transitions. Fluorescence, intersystem crossing, and nonradiative internal conversion occur with rate constants, k_f, k_{isc}, and k_d, respectively. Vibrational energy levels for ground and excited states are numbered.

to an antibonding (virtual) orbital of the molecule. This electronic rearrangement is accompanied by a displacement of the equilibrium nuclear coordinates of the molecule; the vibrational energy states associated with these nuclear coordinate changes are depicted by the subspacings of Fig. 7.1. Thus, the dye laser may be viewed as a "four-level" system in that the most probable excitations proceed to populate upper vibrational levels in S_1, a vibrational relaxation to the "zeroth" level in S_1 follows, and emission occurs to repopulate an upper vibrational level of the ground state (S_0).

Sauers *et al.* (1987) and Piechowski and Bird (1984) have noted that the most common laser dyes are not optimally designed for laser action with respect to the shaping of the absorption–emission bands of organic dyes which are broadened due to vibrational (rotational) relaxation. They reason that absorption and emission strength is associated with 0–0 transitions for which the probability of stimulated emission (to the overpopulated lowest ground state) is low. The absorption and emission spectra for the well-known xanthene dye, rhodamine B, are shown in Fig. 7.2 with reference to the position of the peak lasing wavelength. The Rutgers group further proposes that the "shaping" of spectral bands in favor of the "satellites" corresponding to 0–1 or 0–2 transitions can be predicted for synthetically attainable structures. The approach was illustrated with the

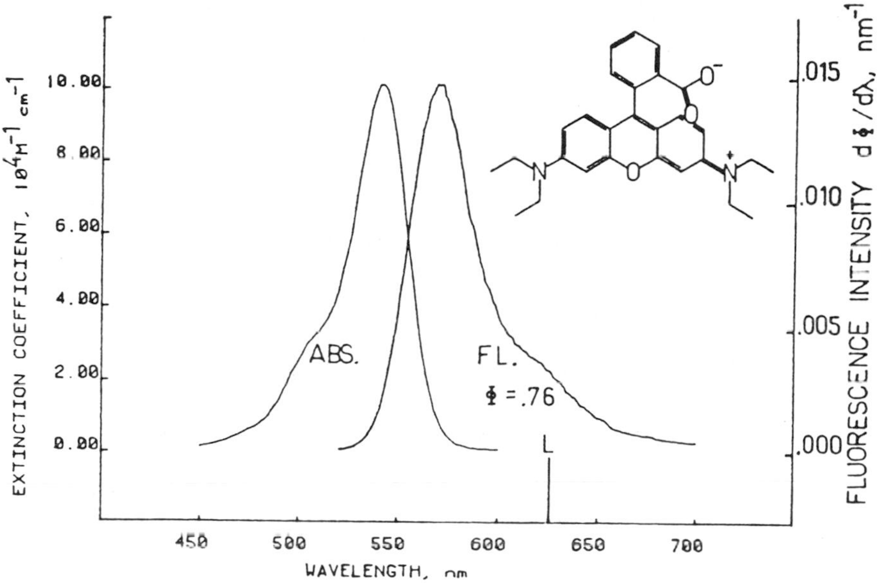

Fig. 7.2 Absorption and fluorescence spectra of rhodamine B in ethanol. Laser-dye spectra often show this mirror-image relationship. Note also that the peak lasing wavelength (*L*) is offset from the most intense band for spontaneous fluorescence [reprinted from Sauers *et al.*, 1987].

xanthenone ("rhodamone") dye VII, for which a slightly higher lasing output in the satellite region (559 nm) was observed (Sauers *et al.*, 1987).

The process of dye fluorescence (k_f) must compete with nonradiative decay of excited species by internal conversion to the ground state (k_d). For organic dye molecules, the interaction of electron spin and orbital motion is developed to such a degree that spin-prohibited processes are probable, involving transitions to other excited states (principally, the lowest triplet state). The formation of dye triplets through the process known as intersystem crossing is an interesting but often disturbing artifact of dye photochemistry. The problem has to do with the time evolution of excited species within a laser cavity. The S_1 state responsible for stimulated emission normally displays a lifetime of 1–5 nsec. Due to a spin prohibition for return to the ground state, the lifetime of the triplet (T_1) state can be much longer, typically 10–100 μsec in the absence of any quenching agent. The long lifetime of dye triplets not only ensures buildup of a species that may be chemically reactive (*vide infra*) but also provides a competitive absorber of light at lasing wavelengths.

The important competition among dye fluorescence (k_f), intersystem crossing (k_{isc}), and internal conversion (k_d) is reflected in quantum yields (Φ). For the emission process, the quantum efficiency is the fraction associated with the microscopic rate constants as follows

$$\Phi_f = k_f/(k_f + k_{isc} + k_d). \quad (7.3)$$

If a photochemical reaction proceeds from the same S_1 state for which preceding processes compete, a similar quantum yield expression involving a reaction rate constant, k_r, is derived:

$$\Phi_r = k_r/(k_f + k_{isc} + k_d + k_r). \quad (7.4)$$

Laser dyes almost by definition display low quantum yields for intersystem crossing (e.g., triplet yields, $\Phi_{isc} < 0.05$) due to the requirement for high efficiencies for spontaneous fluorescence. Typically, measured fluorescence yields, Φ_f, range from about 0.6 to near the optimum 1.0. However, most organic dyes which do have a nonnegligible triplet yield provide problems, since most triplet absorption (allowed transitions to higher triplet levels) is broad and overlaps the fluorescence (lasing) region (Section 2.3.2).

2.2. Solvent Effects on Dye-Emission Parameters: "Solvent Tuning" of Lasing Wavelengths and Emission Yields

For some time it has been known that the solvent may play a decisive role in laser-dye photophysics (Drexhage, 1973). An instructive example regarding solvent influences involves the series of coumarin dyes studied by Jones *et al.* (1980a,b,1985a). Substitution of coumarin with an amino group at the 7-position generates a merocyanine chromophore characterized by the conjugation of "push-pull" substituents (amine electron donor and carbonyl (C═O) electron acceptor groups). The pattern of substitution gives rise to an intramolecular charge-transfer (ICT) transition for which there is a large oscillator strength for absorption (S_0–S_1) and a high rate of fluorescence emission. Solvent influences for the coumarins can be understood in terms of the structures VIII (the common structure for coumarin 1) and IX which represent resonance contributors describing, in an approximate way, the electronic states of a typical molecule. A coumarin structure such as VIII is a faithful description of the ground state, whereas IX better represents the altered electron density pattern for the S_1 ICT state. Notably, a much higher dipole moment is predicted for S_1. Substitution of the dye chromophore with additional groups can either elevate or depress the excited-state dipole moment.

The role that solvent plays in stabilizing coumarin dipoles is illustrated in

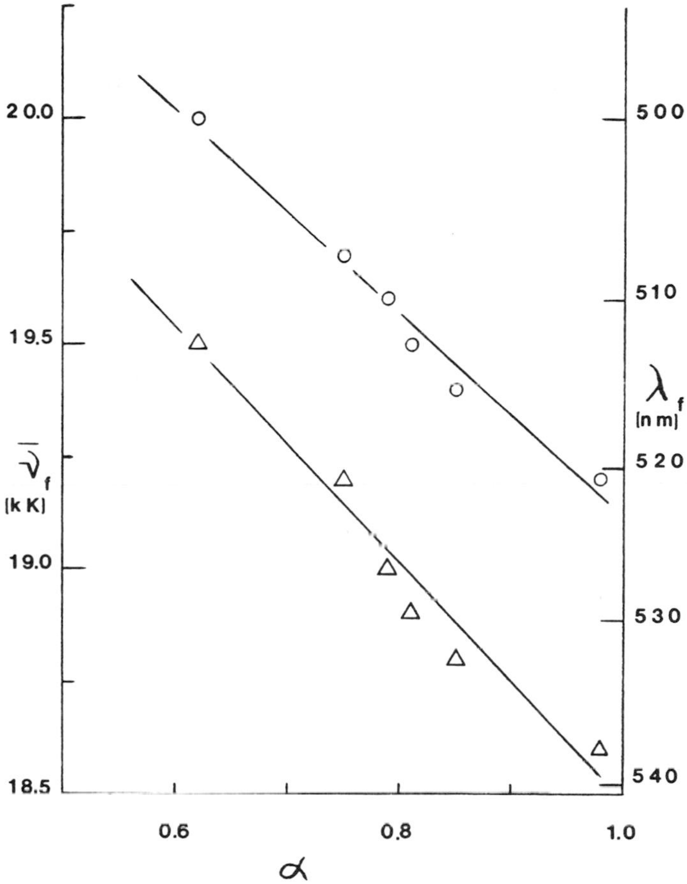

Fig. 7.3 Plots of peak fluorescence frequency (and wavelength) versus the solvent parameter, α. The latter is a measure of solvent hydrogen-bonding capacity for a series of alcohols ($\alpha = 0.62$ and 0.98 for t-butyl alcohol and methanol, respectively. 0, XIII; Δ, XII) [reprinted from Jones *et al.*, 1980a].

Fig. 7.3. The plot of frequencies employs the solvent parameter, α, which is a quantitave measure of the hydrogen bonding properties of a solvent developed by Taft and Kamlet (1976). Thus, for a series of alcohol solvents, the greater tendency to H-bond to dye correlates with a red shift of fluorescence. This smooth dependence which allows rather large adjustments in fluorescence (lasing) wavelengths for some dyes is consistent with the relative stabilization (lower relative energy) that must occur for coumarin ICT excited states. These findings (Jones *et al.*, 1980b) extend to a larger set of solvents including those not capable of H-bonding. A more

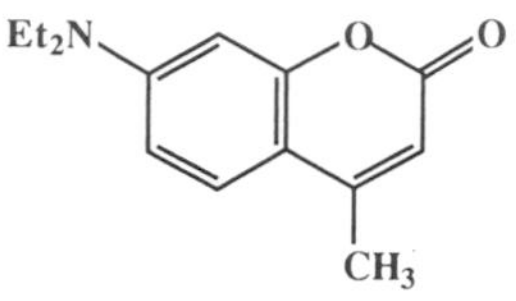

VIII

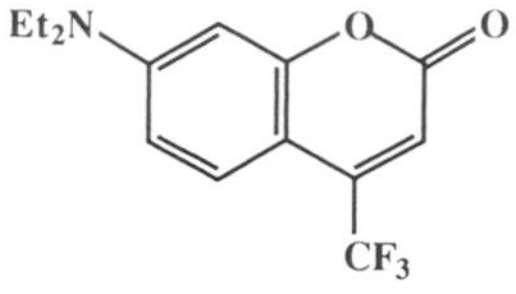

XI

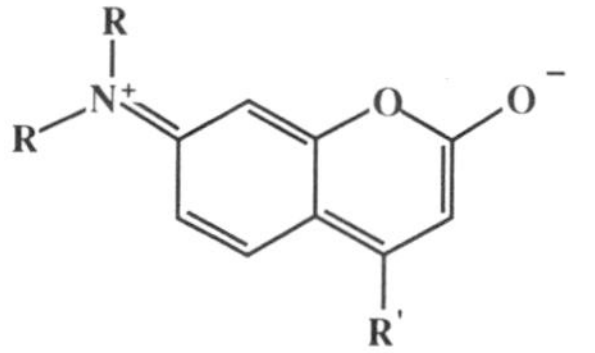

IX

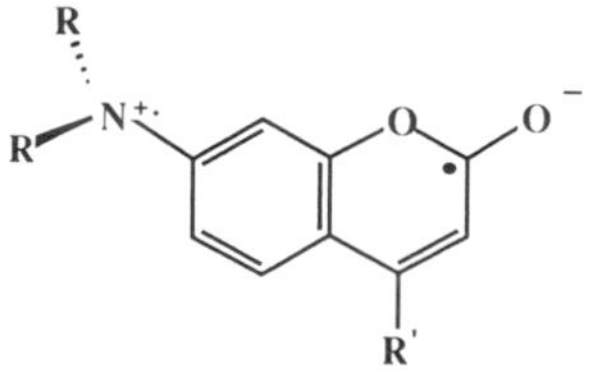

X

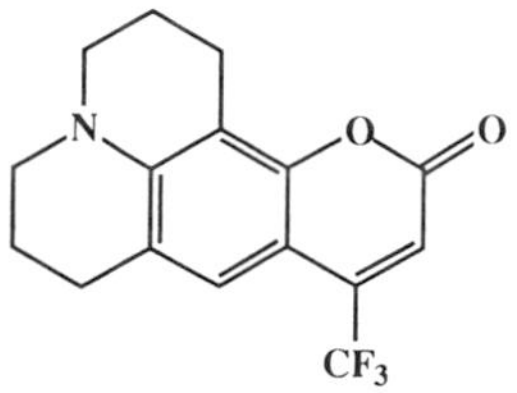

XII

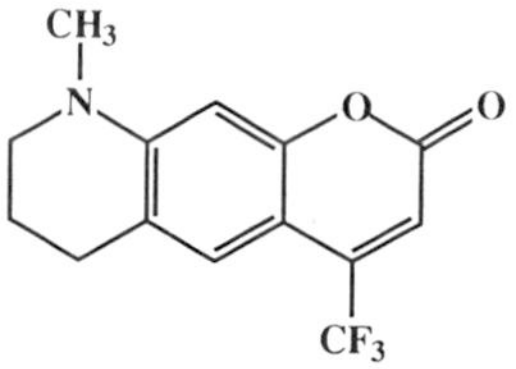

XIII

general solvent polarity–polarizability parameter, π^* (Kamlet *et al.*, 1977), was used along with α to correlate fluorescence data. Linear regression analysis showed dependences of the following type (e.g., for coumarin XI)

$$\nu_f = 21.9 - 3.28(\pi^* - 0.20\delta) - 1.52\alpha \qquad (7.5)$$

where ν_f is the frequency corresponding to the fluorescence maximum for a given solvent, π^* and α are the parameters for that solvent, and δ is employed in a correction term appropriate for chlorinated or aromatic solvents that show unusual degrees of polarizability (Kamlet *et al.*, 1983). This "tuning" of the emission band for coumarin dyes is superimposed on a distinct but smaller red shift of absorption. Thus, for those dyes showing push-pull substituent patterns, the Stokes shift of emission (which potentially influences lasing threshold and eficiency) as well as red shifts in lasing

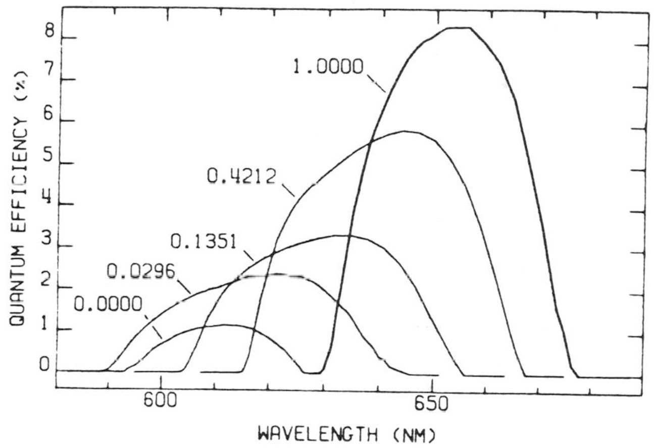

Fig. 7.4 Profile of stimulated emission (output of Nd-YAG laser-pumped dye laser) for 9-diethylaminobenzo [a] phenoxaz-5-one as a function of solvent polarity. Numbers refer to mole fraction of methanol in methanol-xylenes solvent mixtures [reprinted from Grenci *et al.*, 1986].

wavelengths are predictable and readily discerned from easily measured correlatable fluorescence frequencies. Grenci, *et al.* (1986) have demonstrated dramatically the effects of solvent tuning involving a xanthenone dye related to VII (9-diethylaminobenzo[a]phenoxaz-5-one, DBP). Shown in Fig. 7.4 is the red-shift and marked increase in dye-laser photon flux for Nd/YAG (532 nm) pumping of DBP solutions in toluene with increasing amounts of the more polar solvent, methanol (which could be added during laser operation!).

The selection of solvent can affect the performance of laser dyes in terms of alteration of fluorescence yield and lifetime as well as wavelength. This phenomenon is also related to the long-held observation (Drexhage, 1973; Reynolds and Drexhage, 1975) that dyes perform better when rigidly fixed in planar geometries. This restriction of allowed nuclear motion results from the structural constraints imposed by assembly of dye substituents in rings. These findings can now be understood in terms of a confluence of effects associated with the creation of dye dipoles (ICT states), the conformational mobility of substituent groups, and solvent stabilization of charged species. The effects of selected solvents on the fluorescence yield and lifetime parameters for coumarin XI are shown in Table 7.2 (Jones *et al.*, 1980b, 1985a). The data allow a dissection of the radiative and nonradiative decay parameters, k_f and k_d.

Revealed in this selection of data and confirmed in the survey of additional solvents (Jones *et al.*, 1985a) is the principal influence of increased

Table 7.2

Solvent-Dependence of Fluorescence Quantum Yield and Lifetime for Coumarin C1F (XI) and Associated Rate Constants for Radiative (k_f) and Nonradiative (k_d) Decay

Solvent	ϕ_f	τ_f	k_f	k_d
Cyclohexane	1.04	4.1	2.5	<0.1
Acetonitrile	0.091	0.60	1.5	15.0
Ethanol/H_2O, 50:50	0.032	0.45	0.71	22.0
Glycerol	0.18	2.7	0.67	3.0

Source: Jones *et al.* (1985a).
Note: Units of nsec for lifetime, τ_f and 10^8 s^{-1} for rate constants.

solvent polarity on k_d. Moreover, this undesirable reduction of emission yield and lifetime can be reversed (1) on adoption of the very viscous solvent glycerol, (2) with operation at lower temperatures (ideally, below room temperature), or (3) on "rigidization" of the 7-amino coumarin substituent. Thus, XII and XIII do not fall victim to reduction in fluorescence by polar solvent. These results are consistent with the postulate that nonradiative deactivation of the ICT fluorescent species (which is moderately polar) is dominated in more polar media by a twist (rotatory decay) mechanism involving the formation of twisted excited species of the type, X. The latter exhibits full charge separation (a twisted intramolecular charge transfer state, or TICT) and becomes important only for the most polar solvents, including some alcohols and water. The TICT species is also more important for the coumarins most adorned with charge-stabilizing donor-acceptor substituent groups (Jones *et al.*, 1985a). A part of this analysis is associated with some understanding of the alterations in electron density and bonding, which occur on excitation of the dye molecule. The wave functions (atomic orbital coefficients) associated with highest occupied (HOMO) and lowest unoccupied (LUMO) orbitals for a coumarin are shown, for example, in Scheme 1. The promotion of an electron from HOMO to LUMO results in an excited state for which the C-N π bond order has increased (the LUMO is less antibonding). Return to the ground state will be favored for a process that reverses this change (twisting about the C-N bond) (Zimmerman, 1982).

The ICT-TICT phenomenon and the solvent-induced molecular twisting in the excited state was first made prominent by Grabowski *et al.* (1979) and has since been pursued extensively by Lippert and Rettig and their coworkers (1987). An added feature in interpretation of fluorescence

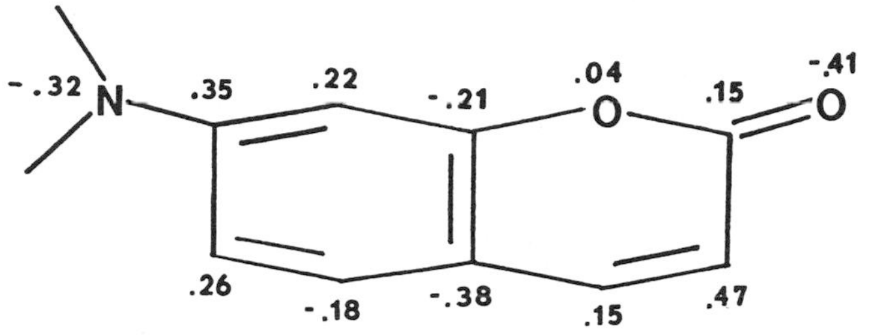

HOMO

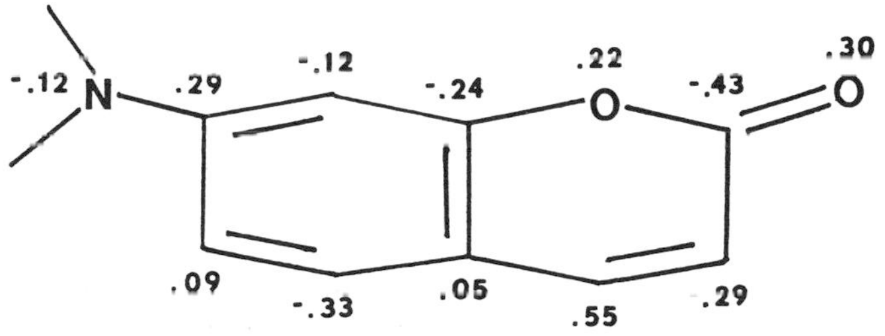

LUMO

SCHEME 1

changes involves the observation of fluorescence from TICT as well as ICT states, which is indicated for selected coumarins (Chu and Yangbo, 1987; Rettig and Klock, 1985). Reports by Vogel *et al.*, (1988), Arbeloa and Rohatgi-Mukherjee (1986), and Snare *et al.* (1982) extend the principles regarding solvent influences on fluorescence yields and lifetimes to the rhodamines (relatives of rhodamine B). Control of structural mobility has also been implicated in lasing studies that show a temperature-dependence of peak power for a coumarin dye, revealing an inhibition of nonradiative decay (twisting) in higher viscosity solvent and Mathews, (Peters 1980), and a reduction in maximum power for coumarin XI in more polar media (Halstead and Reeves, 1978).

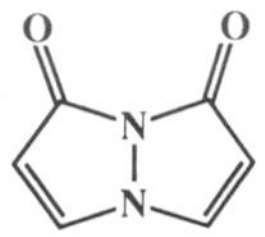

XIV

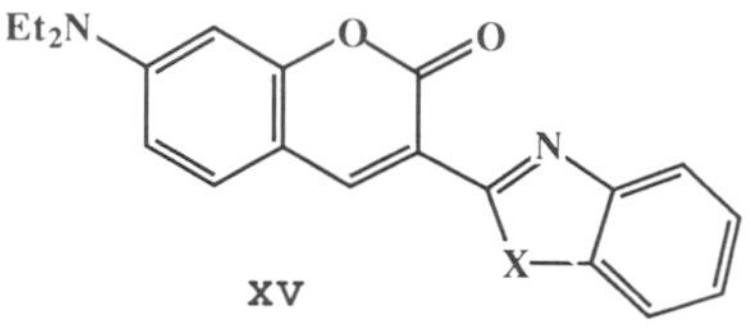

XV

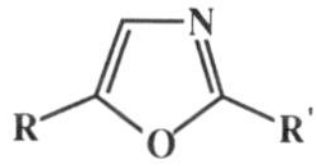

XVII

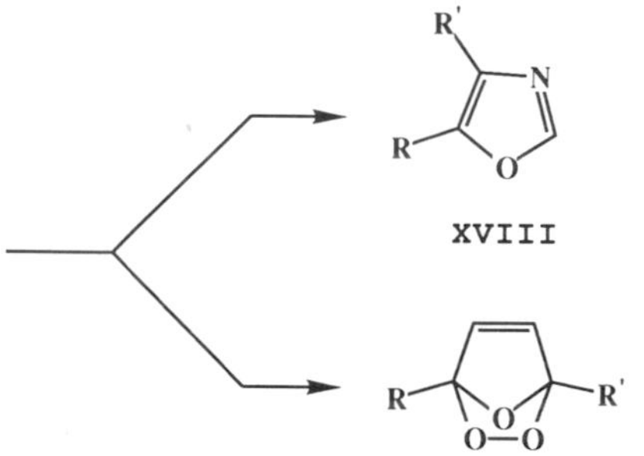

XVIII

XIX

It should be noted that the solvent influences on absorption wavelengths and emission yields need not be so regular or so intertwined. The popular dyes related to "bimane" (XIV) are also solvent-sensitive (fluorescence red shifts for more polar media at 390–430 nm) (Kosower *et al.*, 1986). However, their quantum yields of emission are not diminished in polar media although fluorescence lifetimes are lengthened (7–15 nsec). These data are consistent with a more subtle medium influence having to do with a solvent shift to lower fluorescence frequencies and a consequent reduction in radiative decay rate due to the interdependence of k_f and the transition frequency, ν (Eq.7.2). In the coumarin series, dyes such as XV for which the push-pull substituent pattern is compromised by additional π conjugation are less susceptible to solvent-induced shifts of absorption and emission bands or the enhanced nonradiative decay in polar (nonviscous) media (Jones *et al.*, 1985a).

A final note on the phenomenon of excited-state decay by molecular twisting should focus on the class of cyanine dyes. These structures, which include DOTC (V) and others (HITC, IR 140) form the important class of dyes suitable for lasing in the red and infrared. As polymethine structures, they are in principle susceptible to geometric isomerization about the carbon–carbon double bonds which typically link two heterocyclic rings. A parent structure in this series (pseudoisocyanine) having only one methine carbon linking two rings (as opposed to seven in V) in fact shows no fluorescence at room temperature in most solvents and an ultrashort excited singlet state lifetime (ca. 1 psec). Nonemissive geometric isomers that result from a twisting distortion for the cyanines have been identified by flash photochemical methods and show lifetimes as long as one msec and competitive absorption characteristics (Dietz and Rentsch, 1985; Fouassier *et al.*, 1979). The central point is that the elongated, more conjugated cyanine dyes appropriate for the red and infrared wavelengths represent exceptional cases of efficient lasing within a generally inappropriate nonfluorescent class of structures. The interesting lesson is that dyes that operate at especially long wavelengths, endowed with a much reduced energetic driving force for isomerization or other reactions, are suitable, even resilient, materials with respect to very efficient photochemical processes.

2.3. Intermolecular Interactions of Organic Dyes

2.3.1. *Dye Aggregation*

Aggregation that can occur for higher concentrations of dye in water or alcohol solutions is well known. Potential driving forces for this phenomenon include the freeing the solvent molecules from solvation (hydration) spheres surrounding individual dye structures. Weak intermolecular bonding involving dipole–dipole and other forces is also important. The results of aggregation of R6G in ethanol which is manifest at [dye] $> 5 \times 10^{-3}$ M have been summarized by Ojeda *et al.* (1988). The results of dimerization, which can be suppressed at higher temperatures, include subtle shifts in absorption due to exciton interaction (the coupling of the electronic states of individual molecules—"splitting" of energy levels—due to mutual overlap of orbitals). Fluorescence in dimers or higher aggregates is nearly completely suppressed. This unwelcome alteration in dye dynamics is connected to the reduced lifetime of aggregated forms of organic dyes. For example, using fast time resolved fluorescence and absorption methods, Smirl *et al.* (1982) have shown that rhodamine B monomer and dimer in water display excited-state lifetimes of 1.6 nsec and 100 psec, respectively, signaling a high rate of nonradiative decay for the dye dimer. Aggregation is not an important deterrent where very low concentrations can be accommodated ([dye] generally $<10^{-4}$ M) or where dye molecules are isolated in microdomains such as those provided by detergents (Section 4.1).

2.3.2. *Formation of Dye Triplets and Triplet Energy Transfer*

The intersystem crossing process by which dye molecules in their excited singlet states (S_1) evolve into triplets is shown as follows:

$$S_0(\uparrow\downarrow) \xrightarrow{h\nu} S_1(\uparrow\downarrow) \xrightarrow{k_{isc}} T_1(\uparrow\uparrow) \tag{7.6}$$

where arrows denote spins for electrons in high-lying orbitals, one of which is promoted on excitation. The $S_1 \rightarrow T_1$ conversion is a thermodynamically favorable process (T_1 below S_1; Fig. 7.1) due to reduced electron–electron repulsion in the spin-parallel triplet state. The efficiency of the spin-forbidden ISC step for laser dyes has been discussed at length (e.g., Drexhage, 1973) in terms of variation of triplet yields with dye structure and the presence of critical (heavy) atoms which provide enhancement of spin-orbit interaction. The quantum yield of fluorescence for rhodamine 6G (R6G) in ethanol is 0.88 (Olmsted, 1979). Other rhodamines show values in a similar range ($\Phi_f = 0.6$–1.0) with fluorescence lifetimes, $\tau_f =$ 3–4 nsec (Targowski, 1987), typical of many laser dyes. The yield of triplet dye is therefore generally modest, as reflected in parameters determined for R6G (EtOH): $\Phi_{isc} =$ ca. 10^{-3} (Dunne and Quinn, 1976) and $k_{isc} = 3.4 \times 10^6\ s^{-1}$ (Webb *et al.*, 1970). Photophysical parameters including triplet properties for a number of laser dyes can be found in the review of Chibisov and Korobov (1983).

Dye triplet states have the unfortunate characteristic of a long lifetime, usually 10–100 μsec in the absence of a quenching agent, due to spin prohibition on return to the ground state, S_o. This feature of dye excited-state population is coupled with a tendency toward broad absorption associated with transitions (T–T) within the manifold of triplet states (Pavlopoulos and Hammond, 1974). These properties are exemplified by R6G which in laser flash photolysis experiments displays a broad profile near 400 and 600 nm (Dempster *et al.*, 1974; Korobov and Chibisov, 1978) and an absorptivity, ϵ_{600}, of 13,000–18,000 $M^{-1}\ cm^{-1}$ near the optimum lasing wavelength. Numerous reports (e.g., Pappalardo *et al.*, 1970a,b, 1972) have alluded to the responsibility of dye triplets in reduced dye laser performance. In an early report, Goncharov *et al.* (1971) showed that lower fluorescence yields (higher triplet yields) resulted in reduced energy efficiency and increased threshold energies for lasing of five xanthene (rhodamine) dyes under laser pumping (Nd-glass laser at 530 nm).

Pavlopoulos and Hammond (1974) provided an early assessment of competitive triplet absorption from the point of view of designing dyes in which overlapping spectral regions could be minimized. They identified auxochromic groups (auxiliary substituents such as amine functions, -NR_2, which conjugate with the principal π chromophore) with which substitu-

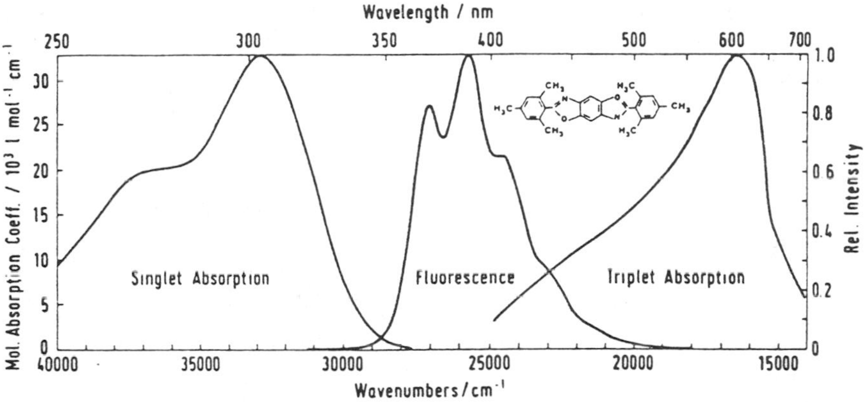

Fig. 7.5 Absorption, fluorescence, and triplet–triplet absorption spectra of a benzobisoxazole dye [reprinted from Guesten *et al.*, 1986].

tion in oligophenylene and oxazole dyes resulted in favorable red shifts of *T–T* absorption. They classified potential laser dyes in terms of the "constellations" of their *S* and *T* absorption and emission spectral profiles and identified substitution patterns on chromophores which should optimize laser performance (e.g., large shifts in triplet absorption due to -OMe). Other examples of molecular design that address the extent of competitive absorption by triplets and other photophysical parameters include the benzobisoxazoles studied by Guesten *et al.* (1986). These oxazole derivatives, which lase efficiently in the uv with an improved tuning range, display a very favorable displacement of the triplet spectrum as shown in Fig. 7.5, as well as a large Stokes shift of emission and other desirable photophysical properties.

The interception of dye triplet states by a "quenching" agent has been viewed both as a benefit in terms of depletion of unwanted triplets and as a source of potentially harmful photoproducts (*vide infra*). The best known of the triplet quenchers is molecular oxygen (Snavely and Schaefer, 1969) which is present at concentrations of 10^{-3}–10^{-4} M in dye solutions that are exposed to air. Marling *et al.* (1970) demonstrated that the lasing output for coumarin dye lasers is dramatically reduced on removal of oxygen from ethanol solutions. In another survey of environmental effects on the performance of coumarins, Fletcher *et al.* (1983) showed that removal of oxygen can result in the opposite effect (enhanced efficiency). This reversal was found for circumstances in which short-wavelength (flashlamp) pump light is filtered. The suggestion was that excitation at shorter wavelengths results in population of upper singlet states (e.g., an S_2 state not included in Fig. 7.1) opening up additional intersystem crossing pathways ($S_2 \rightarrow T_n$, population of an upper triplet) and an elevated role for oxygen quenching.

Notwithstanding the documented salutary effects that oxygen may have on dye performance, as a general proposition the aeration of samples is not desirable. This caution follows from the fact that, if oxygen is to play its role as a dye triplet quencher, the result is formation of the deleterious species, singlet oxygen (the $^1\Delta_g$ state of O_2) through an energy transfer process (Eq. 7.7).

$$\text{dye}\ (T_1) + {}^3O_2 \rightarrow \text{dye}\ (S_0) + {}^1O_2. \tag{7.7}$$

For this bimolecular step, electron spin is conserved and very high (diffusion-limited) rates of quenching are observed (k_q = ca. $10^{10}\ M^{-1}\ s^{-1}$). In the R6G example (Webb *et al.*, 1970), the dye triplet lifetime is reduced to 240 nsec for air-saturated ethanol solutions. The fate of dye and singlet oxygen will be discussed shortly in terms of photodegradation mechanisms (Section 3.3). An additional complicating feature for aerated dye samples is the process of quenching by oxygen of the dye *singlet state* for dyes having lifetimes greater than ca. 1.0 nsec (Kubin and Fletcher, 1983; Jones *et al.*, 1985a). The mechanism of singlet quenching by oxygen has not been established for laser dyes; an electron transfer process is likely (*vide infra*) and would result in formation of superoxide ($O_2^{-}\cdot$) which is itself a troublesome reactive species.

Another quencher, 1,3,5,7-cyclooctatetraene (COT), has achieved some popularity in dye-laser applications (Marling *et al.*, 1970; Pappalardo *et al.*, 1970a,b). The mechanism for COT–dye interaction, while thought to involve triplet energy transfer (Eq. 7.7) remains ambiguous, since the produce species, the COT triplet, has not been observed. In an examination of rhodamines and COT, Targowski *et al.* (1987) showed that the polyene acts as a quencher of dye *singlet states* (fluorescence). Quenching constants (k_q = ca. $10^8\ M^{-1}\ s^{-1}$) correlated for several dyes with singlet excitation energies for several dyes. This finding of a dependence on "driving force" (*vide infra*) can be interpreted again in terms of a mechanism of electron transfer for COT quenching. The effects of COT on R6G lasing (Pappalardo, 1970a,b) may have in part a more trivial origin in the absorption properties of the COT quencher which ensure that dye is shielded from harder excitation wavelengths (λ <280 nm). (Section 3.4).

A relative of COT, 1,3-cyclohexadiene, has been used in "triplet counting" experiments by Jones *et al.* (1984b). In this case the photoproducts of reaction of triplet cyclohexadiene were monitored in an unambiguous way. Appreciable triplet yields (5–30%) were measured for VIII, XI, and XII for the hydrocarbon solvent, cyclohexane, along with much reduced values (<1%) for ethanol solutions. A critical requirement for experiments of this type is that the energy transfer step (Eq. 7.7) be exergonic or nearly so. The cyclopolyenes such as COT, cyclohexadiene, and cycloheptatriene (CHT) have triplet energies that are not known precisely but that fall most likely in the range of 48–53 kcal/mol. Thus, they are suitable for triplet counting for the coumarins whose triplets lie at about 50–53 kcal/mol above the ground state (Specht *et al.*, 1982); they would not be acceptable for interception of rhodamine triplets whose triplet energies are lower (ca. 40–42 kcal/mol) (Chibisov and Korobov, 1983).

2.3.3. Singlet Energy Transfer between Dyes: Down-Shifting of Emission (Lasing) Frequency and Intramolecular Energy Transfer in Linked Dye Systems

Peterson and Snavely suggested in 1968 that dye-laser efficiency could be enhanced by the use of mixtures of dyes. An enhancement of energy output was demonstrated for the mixture of R6G and rhodamine B (RB). Benefits to accrue to such a system have to do with enhancement of what may be a poor absorptivity at pump wavelengths by the primary (active) dye by the secondary (booster) dye. In addition, an efficient means of excess energy disposal for dye mixtures may render photodegradation paths less important. Among other spectral features for dye mixtures (Pavlopoulos, 1978), the requirement that T–T absorption not provide interference applies only to the active dye fluorescence region (not both). An ideal dye mixture will provide fluorescence (lasing) only at the longer wavelength associated with the active dye. A very efficient process of energy transfer involving singlet states of booster and active dyes is invoked:

$$S_1 \text{ (booster)} + S_0 \text{ (active dye)} \rightarrow S_0 \text{ (booster)} + S_1 \text{ (active dye)}. \quad (7.8)$$

Requirements for such a system include strong absorption of the booster at pump wavelengths, an efficient mechanism for energy exchange (sufficient active dye for quenching), and a displacement of absorption and emission bands for the active dye (a shift to the red, relative to the spectrum of the booster). The critical bimolecular energy exchange step is not a dye-emission–reabsorption process, but employs a mechanism known as Foerster transfer (Turro, 1978). This well-documented process occurs at very high rates (bimolecular quenching constants, $k_q = >10^{10}$ M^{-1} s^{-1}) under favorable circumstances in which there is strong overlap between the emission band of the booster and the absorption band of the active dye (i.e., the transfer step, Eq. 2.8, is thermodynamically favorable) (Goodall and Roberts, 1985). In addition, the Foerster dipole–dipole transfer mechanism may operate over distances (>3.0 nm) much greater than encounter distances associated with the approach of two dye species

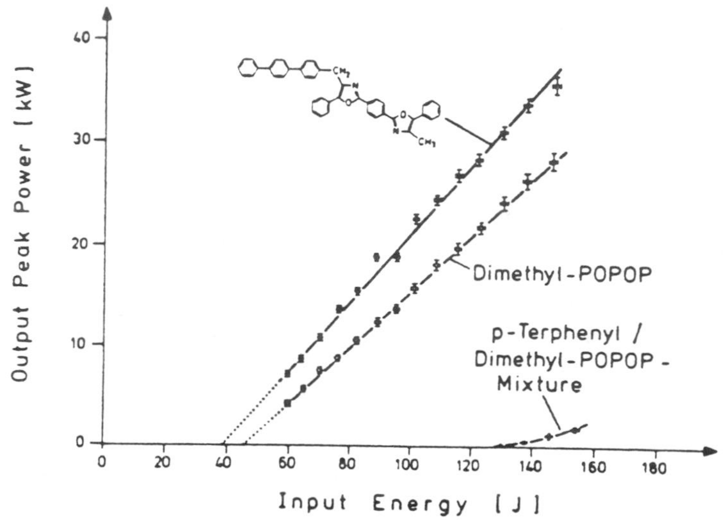

Fig. 7.6 Peak laser output power versus electrical input energy of dimethyl-POPOP, the *p*-terphenyl linked POPOP system (structure shown), and a mixture of *p*-terphenyl and dimethyl-POPOP [reprinted from Schaefer *et al.*, 1978].

in solution. However, the mixing of two dyes is not as simple as it appears, and a number of factors, including dye concentration and optical densities, and T–T absorption conspire so that only in rare cases does a dye mixture successfully lase (Pavlopoulos, 1978).

The prospects for energy transfer and frequency conversion are less dubious and in fact quite appealing for systems crafted so that dye interaction is an *intramolecular* process. Schaefer first proposed the covalent linkage of laser-dye molecules for the purpose of enhancing energy-transfer efficiency (for a review, see Schaefer (1983)). The concept was demonstrated through construction of a molecule in which dimethyl-POPOP and *p*-terphenyl dyes were joined at a single saturated carbon (Schaefer *et al.*, 1978). Lasing parameters for the linked dye system showed a 30% improvement in efficiency over use of the POPOP component alone; a simple mix of unlinked dyes was much less successful (Fig. 7.6). The improvement for the linked system was due to an improved absorption of (flashlamp) pump light and the efficient transfer expected for the selected chromophores. The energy transfer between linked chromophores occured in <1 psec according to a time-resolved fluorescence study (Kopainsky *et al.*, 1978).

Another novel system of linked coumarins has been studied by Mugnier and his coworkers (1985). These structures (XVI) deploy a higher-energy coumarinyl ether (donor) moiety fused with a lower-energy (acceptor)

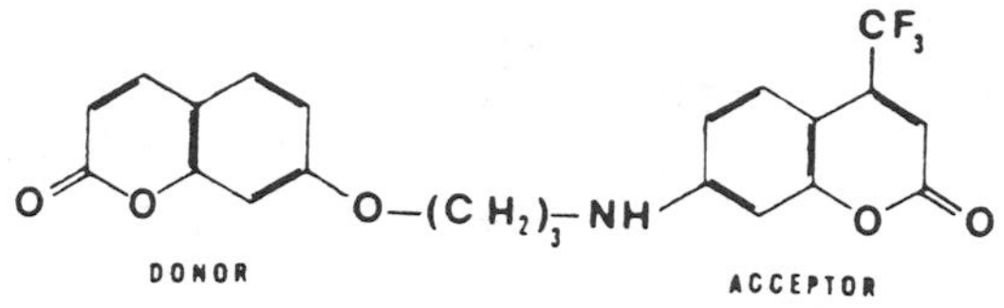

XVI

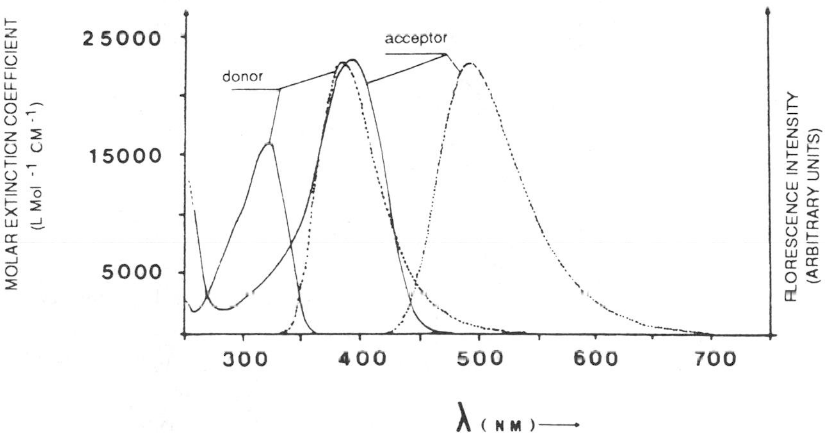

Fig. 7.7 Absorption (solid lines) and emission (dashed lines) spectra of *separated* donor and acceptor moieties associated with linked coumarin XVI. The linked system XVI displays only the fluorescence centered at 500 nm due to efficient energy transfer between donor and acceptor groups [reprinted from Bourson *et al.*, 1982].

fluorinated aminocoumarin component. Variations in the tether to which dye components are attached ($n = 3, 4, 8$, or 12) also provide a range of average distances for inspection (random chromophore orientations). Due to the very favorable overlap of donor emission and acceptor absorption (Fig. 7.7), high energy-transfer efficiencies (0.5–0.9) are observed, with small variations with temperature and a slight fall-off with the value of n (the length of the tether) in some solvents. High photochemical stability was also noted for the linked coumarin structures (Bourson *et al.*, 1982).

A second generation of linked dye structures has been designed by Liphardt *et al.* (1981). In this series an active dye is not attached to a booster but to a triplet quencher, for the purpose of disabling the competitive absorption or degradation due to triplets. A derivative of POPOP similar to that of Fig. 7.6 was modified with a stilbene chromophore. The strategy took into account the relative triplet excitation energies of the active POPOP component and the stilbene moiety (ca. 49 and 48 kcal/mol, respectively) which permitted energy transfer to the stilbene unit. In turn the stilbene triplet is shortlived and benign due to a facile isomerization mechanism of deactivation.

3. DYE PHOTODEGRADATION

3.1. Chemical Properties and Dynamics of Excited States

The stability of organic dyes under irradiation has received considerable attention due to the importance of technologies such as the dyeing of fabrics, color photography, and photoconductivity (e.g., Sinclair, 1980; Allen and McKellar, 1980). Photochemical reaction pathways for various classes of organic dyes have been reviewed in detail by Chibisov and Korobov (1983). Investigation of the behavior of laser dyes has taken two forms: (1) a reporting of the characteristics of dye solutions under actual or simulated lasing conditions and (2) studies devoted to identification of the mechanisms of photochemical degradation, most often *not* under conditions of actual dye lasing. The Berlin group (Weber, 1971, 1973; Beer and Weber, 1972) provided early demonstrations of dye photobleaching for rhodamine and other laser dyes. Their reports included intimations of the effects of oxygen on irradiated solutions and the addition of detergent additives (*vide infra*). A typical finding in the early studies as well as in numerous subsequent reports is the bleaching of the primary dye-absorption band. The loss of the lasing chromophore is usually accompanied by some increase in dye or photoproduct absorption at shorter wavelengths. In addition, diminution in dye fluorescence (the source of stimulated emission) is generally observed. Loss of fluorescence in fact can be a sensitive, although not totally unambiguous, measure of photochemical quantum yield (Gauglitz, 1978).

For photolysis occurring in dye lasers at high pump energies, one must also contend with multiple photon events which are not common to dye photochemistry under more normal irradiation conditions. High light fluxes lead to a buildup in population of a number of potentially reactive dye intermediates, including the triplet state and derived radicals. Even low–quantum-yield processes involving dye–dye interaction can be deleterious. As excited species compete for light a new array of high-energy forms of dye must be considered as potentially reactive intermediates. Upper state population through sequential two-photon excitation can be classified as follows:

1. triplet–triplet absorption (Section 2.3.2):

$$S_0 \xrightarrow{h\nu} S_1 \rightarrow T_1 \xrightarrow{h\nu} T_n;$$

2. excitation to higher excited singlet states (Magde, 1981):

$$S_0 \xrightarrow{h\nu} S_1 \xrightarrow{h\nu} S_n.$$

Speiser and Shakkour (1985) have characterized "photoquenching" (reversible loss of fluorescent (S_1) species) and the irreversible dye losses due to the intervention of upper singlet levels by the latter mechanism. The upper triplet mechanism is of low probability due to low intersystem crossing yields (Section 2.3.2), but triplets are long-lived (e.g., as long as one msec) and can rise to appreciable concentrations. The singlet mechanism, although burdened by the very short lifetime (1–5 nsec) for dye singlets, will play a role at high pump powers where S_1 species will have a nonnegligible absorption cross section. Along with dye-excited states, photochemical products such as radicals (lifetimes in the μsec–sec range) can build in concentration and be subject to photolysis. Knipe and Fletcher (1983) have provided a classification of multiple-photon phenomena in which absorption by transient or long-lived photoproducts leads to secondary chemical reactions.

3.2. Mechanisms of Photoinduced Electron Transfer

Although the tendency for the excited states of organic compounds to undergo electron transfer is increasingly well understood (Kavarnos and Turro, 1986), photoredox processes for laser dyes have been only partly realized. The notion that excited dye molecules are potent oxidizing or reducing agents is illustrated with the aid of Fig. 7.8. On promotion of an electron through excitation, a "hole" is created for repopulation by an external reducing agent. At the same time, the promoted electron is very readily oxidized. Both ionization potentials and electron affinities (solution oxidation and reduction potentials) for the excited dye molecule are enhanced by an equivalent in energy associated with the S_0–S_1 energy spacing (the HOMO-LUMO gap, Fig. 7.8)

An established protocol allows one to predict whether a dye will interact with a quencher (or even with itself) through an electron-transfer mechanism. The procedure involves a model in which an encounter between a dye molecule (M) in an excited state and an electron acceptor (A) results in transfer of an electron, producing a pair of radical-ions (Eq. 7.9). The alternative is that the dye will act as an electron acceptor and a donor (D) will be oxidized (Eq. 7.10):

$$[M^*, A] \rightarrow [M^+\cdot, A^-\cdot] \tag{7.9}$$

$$[M^*, D] \rightarrow [M^-\cdot, D^+\cdot] \tag{7.10}$$

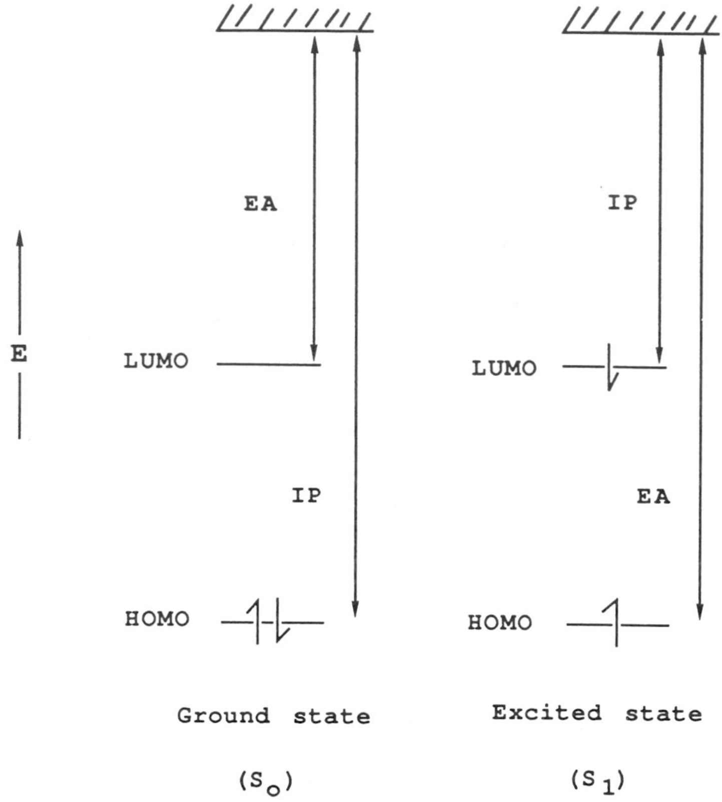

Fig. 7.8 Orbital energy level diagram for ground- and excited- state dye molecules, depicting the highest occupied (HOMO) and lowest unoccupied (LUMO) molecular orbitals in relation to the vacuum level. The energetics for loss or gain of an electron by each species are approximated by the spacings which correspond to ionization potentials (IP) or electron affinities (EA), respectively.

The analysis is illustrated with respect to the model involving dye oxidation (Eq. 7.9). The free energy change associated with this electron transfer step (ΔG_{et}, units of kcal/mol) is given by the Weller equation (Kavarnos and Turro, 1986) as follows:

$$\Delta G_{et} = 23.1\ (E_{ox} - E_{red}) - E_{oo} - e^2/\epsilon r. \qquad (7.11)$$

The parameters, E_{ox} and E_{red}, refer to electrochemical half-wave oxidation and reduction potentials (values in units of V versus a reference electrode) for grouund state dye (M) and the potential electron acceptor (A), respectively. The excitation energy, E_{oo}, corresponds to the separation between ground and the redox-active excited state of dye (S_1 or T_1,

Fig. 7.1) (or an upper state in rare circumstances). This "zero–zero" transition energy between lowest vibrational levels (see Fig. 7.1) is obtained from the frequency corresponding to the intersection of dye absorption and fluorescence bands.

The final term is a correction to these energies associated with the electrostatic attraction between nascent ionic species (e.g., Eq. 7.9) (alternatively, repulsion for products of like charge). If a distance of ion separation (r) of ca. 0.5 nm is assumed, the electrostatic term makes a small contribution for electron transfer in a solvent of moderately high dielectric constant (e.g., 1.7 kcal/mol or 0.08 eV for acetonitrile, $\epsilon = 38$). The physical meaning of the "Weller equation" (7.11) (Rehm and Weller, 1970) is that electron transfer will be thermodynamically feasible if the energy of the ion pair given by the electrochemical parameters (with electrostatic correction) lies below that of the dye excited-state energy.

A useful corollary to the Weller treatment is the prediction that for ΔG_{et} values which are more negative than about -5 kcal/mol, electron transfer is sufficiently thermodynamically driven that the bimolecular rate constant for quenching of excited dye by an external redox agent will approach a diffusion-limited value (10^9–10^{10} M^{-1} s^{-1} for moderate-viscosity organic solvents and water at room temperature). (The broader relationship between electron transfer rate and driving force, ΔG_{et}, is in fact much more complex (Kavarnos and Turro, 1986).)

The protocol is illustrated with reference to the quenching of fluorescence of coumarin dyes by either electron donors or acceptors (Jones *et al.*, 1984a). A quenching step involving an acceptor (A) and leading to a presumed ion-pair intermediate is shown in Eq. (7.12):

$$M(S_1) + A \xrightarrow{k_q} 1[M^+\cdot, A^-\cdot] \tag{7.12}$$

$$^1[M^+\cdot, A^-\cdot] \rightarrow M(S_0) + A. \tag{7.13}$$

Electrochemical data are shown in Table 7.3 for three selected coumarins (both oxidation and reduction potentials) and for prototype electron donor and electron acceptor quenchers (e.g., amines and nitriles, respectively). The finding of the study by Jones *et al.* (1984a) was that high rates of quenching (rate constants, k_q, Table 7.4) were indeed observed for those combinations of dye and redox agent for which large negative free energies were calculated.

Two distinctions were made concerning the primary product of electron transfer in the coumarin study. The photogenerated ions (Eq. 7.12) are not truly free agents until they have escaped the solvent cage in which they are born. A common finding is that for paired ions in which electron spins communicate and for which the electronic spin multiplicity remains singlet

Table 7.3

Redox potentials for coumarin dyes and electron donors and acceptors

Dye	Quencher	E_{ox}	E_{red}
Coumarin 102 (III)		0.72	<-2.3
Coumarin 1 (VIII)		1.09	−2.2
Coumarin 6 (XV)		1.02	−1.5
	Triethylamine	1.1	—
	Dimethylaniline	0.73	—
	Fumaronitrile	—	−1.3
	Methyl Viologen	—	−0.69
	Dicyanobenzene	—	−1.7

Source: Jones *et al.* (1989a).
Note: Values are half-wave potentials for electrochemical oxidation (E_{ox}) or reduction (E_{red}) in volts versus the saturated calomel electrode (SCE)

Table 7.4

Flourescence Quenching Data for Coumarins and Electron Acceptors

Dye	Quencher	k_q	ΔG_{et}
Coumarin 102 (III)	Fumaronitrile	24.	−23
Coumarin 1 (VIII)	Fumaronitrile	8.7	−17
	Methyl Viologen	29.	−31
Coumarin 6 (XV, X = S)	Fumaronitrile	9.7	−6.8
	Dicyanobenzene	<0.1	2.2

Source: Jones *et al.* (1984a).
Note: Rate constants in units of 10^9 M^{-1} s^{-1}; free energy changes calculated for electron transfer quenching (Eq. 7.11) in kcal/mol.

(note presuperscript), a back electron transfer step is dominant and free ions appear only in low yield. In a typical case, the energy-wasting nonradiative electron back transfer (Eq. 7.13), a step that is neither energy-nor spin-prohibited, will be 50–100 times more probable. In contrast, if it is the triplet state of the dye that is intercepted by an electron-transfer agent, a higher yield of free ion products is anticipated since cage escape of ions is faster than, or competitive with, a spin-prohibited back electron transfer to give ground-state singlet species. Triplet quenching by a dicationic electron acceptor, methyl viologen (MV^{2+}), and formation of a free radical ion, the product of viologen reduction (Eq. 7.14), were indeed observed in an investigation of coumarin dyes (Jones *et al.*, 1984a). Singlet (fluorescence) quenching was robust in terms of quenching constants (Table 7.3) but did not result in an appreciable yield of free ions. Similarly, Chibisov (1977) has found that for a carbocyanine dye related to DOTC (IV), substitution patterns can be deployed which lead to enhanced triplet formation; only triplets could be shown to engage in electron transfer with agents such as methyl viologen (MV):

$$\begin{aligned} M^*(T_1) + MV^{+}\cdot &\rightarrow {}^3[M^{+}\cdot, MV^{+}\cdot] \\ &\rightarrow M^{+\cdot} + MV^{+\cdot} \end{aligned} \tag{7.14}$$

Among the possibilities for redox reactivity for dyes is the process of electron transfer to solvent. Molecules such as pyrene and aromatic amines are well known to photoionize to yield radical cations of the organic chromophore and the solvated electron (Eq. 7.15) (Koehler *et al.*, 1985; Hirata *et al.*, 1982). The photoionization process depends on the ionization (oxidation) potential of the dye, the wavelength of excitation, and the medium. For molecules having oxidation potentials, E_{ox} = ca. 1.0 V (half-wave potentials in V versus SCE; compare coumarin data in Table 7.2), photoionization is measurable at excitation wavelengths of 300 nm or less for very polar media such as water. In the solvent-dependence study of coumarin VIII and its relatives (Jones *et al.*, 1985a), the coumarin triplet was observed as the principal phototransient (100-μsec time regime) for solvents such as acetonitrile; on photolysis of VIII in water, the triplet spectrum was replaced by the characteristic broad absorption due to the hydrated electron (ca. 700 nm) (Eq. 7.15):

$$M^* \rightarrow M^{+}\cdot + e^{-}_{solv}\cdot \tag{7.15}$$

The high energetic demands placed on the photoionization step (Eq. 7.15) dictate that either high-energy single photons (uv excitation >4 eV/photon) or multiple photon phenomena are required. On the basis of a study of the xanthene dye, fluorescein, in water, Lougnot and Goldschmidt (1980) proposed a pair of two-photon mechanisms for photoionization, one involving electron ejection to solvent after secondary excitation of the fluorescent excited singlet state ($S_1 \rightarrow S_n \rightarrow$ ionization). A second mechanism was proposed for photolysis of fluorescein with visible light, involving promotion to an upper triplet state ($T_1 \rightarrow T_n \rightarrow$ ionization).

3.3. Photochemistry of Oxazoles: The Singlet Oxygen Pathway

Oxazoles form an important class of dyes that operate at shorter wavelengths in the uv. Their photochemistry is illustrative of processes that may take place via excited states of relatively high energy. A novel rearrangement (XVII→XVIII, p.300) which is relatively common to five-membered heterocyclic compounds has been shown to be important for photolysis of simple diaryloxazoles in ethanol (Maeda and Kojima, 1977; Rendall *et al.*, 1986). This isomerization which alters the dye chromophore involves formation of strained-ring intermediates through the cross-addition of bonds within the rings. As of 1989, the reaction has not yet been observed for a laser dye such as POPOP (III) but remains a candidate for low-yield degradation. Further susceptibility of oxazole ring bonds to alteration is shown in the photodimerization of benzoxazoles on 300-nm photolysis in deaerated cyclohexane (Fery-Forgues and Paillous, 1986). In a related study, the fluorescence of benzoxazoles was shown to be quenched by amines via an electron-transfer mechanism (Fery-Forgues *et al.*, 1987).

A number of reports suggest that oxazoles may be particularly susceptible to degradation in the presence of oxygen. POPOP (III) and related compounds are known to undergo addition by singlet oxygen across bonds in the heterocyclic ring. The adducts (XIX), which are efficiently generated using a dye with triplet sensitizing capabilities for $^1O_2$ generation, are unstable and react further yielding carbonyl- (C=O) containing products (Wasserman, and Floyd, 1966). Dempster *et al.* (1974) have observed the triplet of 2,5-diphenyloxazole (POP) in ethanol on flash photolysis. A strongly absorbing 500-nm transient was observed at 200 μsec following flash excitation; the triplet yield was ca. 1%. Two reports on the POPOP (III) family of oxazole dyes showed the deleterious effects of oxygen on dye-laser performance (Fletcher *et al.*, 1987; Liphardt *et al.*; 1983). Liphardt *et al.* showed further that amine additives that serve as traps of singlet oxygen improve photostability significantly.

A peculiar variation on the theme of photodegradation via singlet oxygen has been reported by Gottschalk and Neckers (1986). They noted that for ionic cyanine dyes related to *V*, the counter-ion and solvent played a significant role in the formation of singlet oxygen (via energy transfer from dye). The proposal is that when dye is *ion-paired* with a heavy counter-ion (iodide as opposed to chloride) in a solvent of low polarity (dichloromethane), the yield of dye triplets and consequently singlet oxygen is enhanced. This finding is important in terms of the common appearance of laser dyes as salts and the likelihood that singlet oxygen mechanisms of degradation may be important in some solvents if heavy counter-ions induce dye intersystem crossing.

3.4. Photochemistry of Xanthene (Rhodamine) Dyes

The mechanism of photodegradation of the best known member of the family of laser dyes, rhodamine 6G (R6G) (IV) has been the focus of a number of photochemical studies which have deployed both steady-state and pulse photolysis techniques. Weber (1973) provided evidence that R6G decomposition on Xe flashlamp photolysis through glass (>320 nm) proceeded by way of dye triplets by showing that a photosensitizer (naphthalene or anthracene) which populates R6G triplets by an energy-transfer-mechanism (Section 2.3.2) enhanced degradation. These experiments included other tests in which triplet quenchers, oxygen and COT, slowed the rate of decomposition. Possible structures for rhodamine photoproducts in which the xanthene chromophore is lost are represented in XX, a generic rhodamine incorporating an addend, *R* (with variable aromatic substituent, Ar). Britt and Moniz (1973) reported that added acid or base has a moderate effect on quantum yield of R6G bleaching on 514-nm irradiation in ethanol (maximum $\Phi = 10^{-5}$ at pH = 9.3) and ascribed photolysis to formation of a hydrol (XX, R = OH) based on analogy with other dyes.

A more definitive study of mechanism appeared in 1976 after Yamashita and Kashiwagi investigated photolysis of R6G at 77K in glassy alcohol matrices. Using electron spin resonance (esr) techniques they were able to monitor directly the production of radicals on photolysis both at 514 nm (Ar laser) and on uv irradiation. The interesting finding is that the mechanisms of photolysis differ significantly according to wavelength of excitation with the more efficient process requiring uv light. The uv degradation, as reflected by the intensity of esr signals due to radical intermediates, was also dependent on the square of light intensity, demonstrating a requirement for two photons per R6G molecule. The mechanism proposed involves excitation to an upper triplet (T_n) (eq. 7.16) which is sufficiently energetic to abstract hydrogen from alcohol solvent to produce the radicals which could be identified by their esr characteristics (e.g., $CH_3CH{\cdot}OH$ from CH_3CH_2OH).

$$S_0 \xrightarrow{h\nu} S_1 \rightarrow T_1 \xrightarrow{h\nu} T_n \rightarrow \text{H abstraction from solvent.} \qquad (7.16)$$

Dye-related radical intermediates also result from H-atom abstraction from alcohol by R6G. Addition of a H-atom to R6G and loss of a proton yields the triaryl-type radical intermediate XXI. The bleaching mechanism is completed by abstraction of a second H-atom from alcohol by the meso (central) carbon. The leuco dye (XX, R = H) which results is the final stable photoproduct proposed by Yamashita and Kashiwagi (1976). While these steps constitute a possible decomposition mechanism, it should be noted that the key product XX(R = H) of R6G reduction in alcohol has not been unambiguously identified.

Laser flash photolysis of R6G in ethanol provides further evidence of reaction mechanism. Kestle *et al.* (1972) first identified the broad absorption at about 600 nm due to the R6G triplet and showed that the triplet

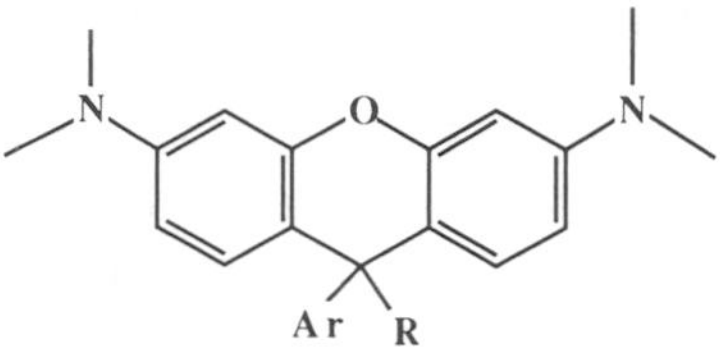

XX

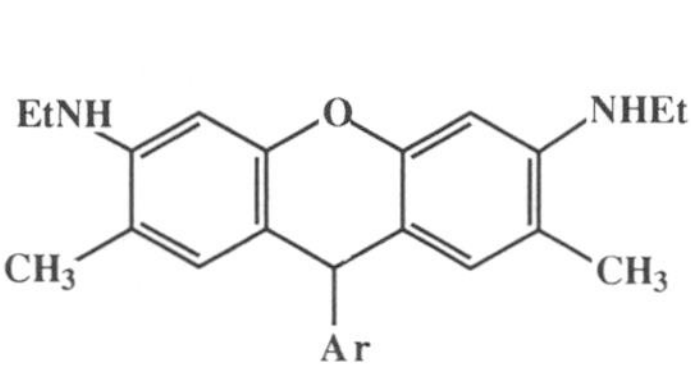

XXI

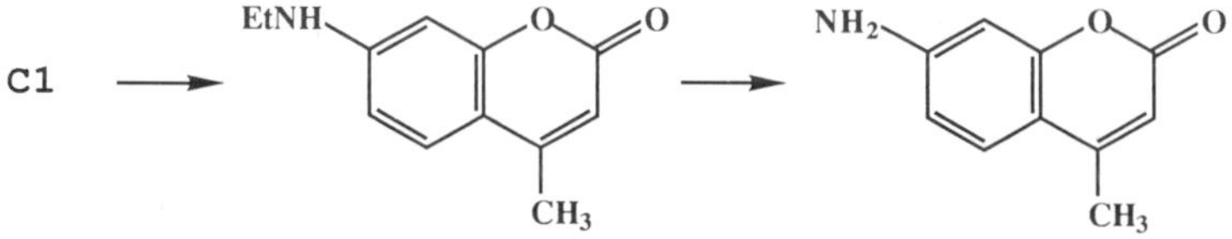

VIII XXII XXIII

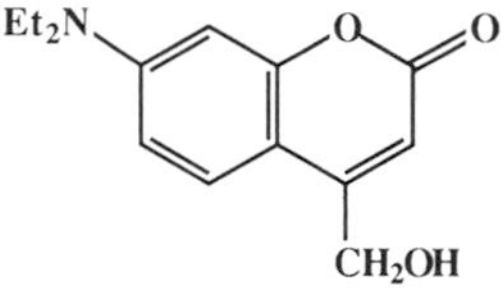

XXIV

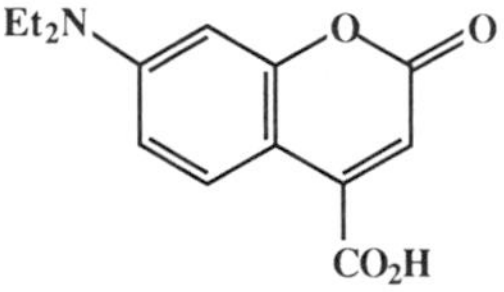

XXV

(T_1) energy for R6G resides at about 40 kcal/mol above the ground state, based on energy-transfer experiments. Dempster *et al.* (1974) observed quenching of a 200-μsec triplet transient by oxygen and COT and measured the R6G intersystem crossing efficiency (2×10^{-3}). A transient product, P, which lived as long as seconds and absorbed at 420 nm was observed and shown to have a quantum yield of ca. 10^{-3}.

Another study (Dunne and Quinn, 1976) suggests that the pathway from reactive excited state to product P for R6G photolysis in ethanol under conditions *of excitation with visible light* is not due to the mechanism of Eq. (7.16), proposed for uv photolysis. Dunne and Quinn did show that independent generation of radicals from ethanol and trapping by R6G leads to product P (Eq. 7.17). On the other hand, the origin of the state that produces solvent radicals was not clear as demonstrated in their double irradiation experiment. Thus, samples flashed initially with a 50-J Xe flashlamp pulse were subjected to another excitation from a dye laser at 460 nm at 500 μsec following the initial pulse. In this way, the triplet state is allowed to build in concentration and then undergo secondary photolysis (see Eq. 7.16). Under these conditions the R6G triplet was shown not to be depleted to product P, implicating some other more complicated route for generation of solvent radicals, at least for visible R6G photolysis of the dye and perhaps for the more destructive uv photolysis. It appears that the uv degradation mechanism employs multiple photon events to produce highly excited dye species or excited states of solvent which show a sensitivity to added agents not unlike the R6G triplet and which in turn are responsible for the heightened reactivity toward formation of solvent radicals.

$$CH_3CH \cdot OH + R6G \rightarrow P. \tag{7.17}$$

Korobov and Chibisov (1978) have further described the photoredox nature of R6G decomposition, especially in water (see also Chibisov and Slavnova (1978) and Korobov *et al.* (1977). Using a 400-J (15-μsec) Xe flash apparatus for examination of transients at ca. 200 μsec, they were able to observe intermediates at 475 and 410 nm, which were identified as one-electron–oxidation and one-electron–reduction products, respectively. The latter intermediate (XXI, very likely the Dempster intermediate, P) was characterized by its reaction with the electron acceptor, nitrobenzene. A mechanism for formation of reduced and oxidized species involving the annihilation of two triplets (two-photon mechanism) was proposed by Korobov and Chibisov (1978):

$$R6G\ (T_1) + R6G\ (T_1) \rightarrow R6G^{+\cdot} + R6G^{-\cdot} \tag{7.18}$$

Formation of semioxidized or reduced species on photolysis in water has implications for the mechanism of R6G degradation in alcohols. Korobov and Chibisov (1978) showed, for example, that $R6G^{+\cdot}$ (475-nm transient) reacts slowly with propanol solvent (Eq. 7.19). The remaining uncertainty has to do with the more general origin of the $R6G^{+\cdot}$ transient in solvent other than water. A likely explanation of many findings is that multiple photon events for uv photolysis of R6G lead to photoionization of the dye, yielding the critical semioxidized species and solvated electrons (Eq. 7.20) (Section 3.2). The semioxidized transient is responsible for solvent (alcohol) oxidation (Eq. 7.19), and subsequent solvent radicals, which are actually rather good reducing species, react further with R6G yielding *P*

($R6G^{-}\cdot$ or XXI) and other products (Eq. 7.17):

$$R6G^{+}\cdot + RCH_2OH \rightarrow R6G + RCH\cdot OH + H^{+} \quad (7.19)$$

$$R6G^{**} \rightarrow R6G^{+}\cdot + e_{solv}\cdot \quad (7.20)$$

It is possible that upper excited singlet states are important for uv photolysis of R6G and provide the origin of photoionization (Eq. 7.20). (Recall the example of secondary photolysis of fluorescein reported by Lougnot and Goldschmidt (1980), Section 3.2.) Aristov and Shevandin (1981) have reported a "two-color" laser-pump experiment in which R6G is subjected to 532-nm primary excitation along with 355-nm pumping which activates a (2-nsec) transient which absorbs in the uv (apparently S_1). This "double-barrel" photolysis, which results in enhanced irreversible loss of R6G, illustrates the interplay of multiple-wavelength (broadband) excitation, competitive absorption, and excited-state dynamics (upconversion).

Other routes leading to photooxidation of rhodamines have been reported. Rhodamine B (RB) has been the focus of several studies involving dye photooxidation. Ruiz *et al.* (1986) showed that oxidation sensitized by Fe(III) in water on uv excitation in water involves $\cdot$OH radicals derived from the metal complex. In another investigation of RB which was deposited on SnO_2 glass photoelectrodes, Austin *et al.* (1986) demonstrated that RB is deethylated to give rhodamine 110 through electron injection from dye to electrode. Dealkylation at amine functions is another common result of photooxidation, as shown for the coumarins in the next section.

3.5. Photochemistry of Coumarin Dyes

Winters *et al.* (1974) were the first to provide a detailed study of coumarin photodegradation products and mechanism. They reported that coumarin I (VIII) is decomposed in ethanol on Xe flashlamp irradiation to as many as five products which were isolated and identified through analysis of spectra. The dealkylation occurring at the 7-position amine function could be followed in steps, involving XXII and XXIII. These photoproducts showed emission spectra similar to VIII and even participated in lasing. A more problematic degradative path was discovered on flashing oxygenated ethanol solutions. Three additional products emerged including XXIV and XXV, the result of sequential oxidation at the 4-methyl substituent. The appearance of the latter product, a carboxylic acid, raised dramatically the threshold for C1 lasing due to competitive absorption at the lasing wavelength (ca. 460 nm). Several strategies were recommended in this early study: (1) replacement of the vulnerable site of oxidation, the

—CH_3 (methyl) substituent, (2) use of a triplet quencher other than oxygen, and (3) removal of the offending photoproduct, XXV, by filtration of a flowing dye solution. The latter option could be taken with use of an alumina filter which showed a selective attraction to XXV. The observation concerning the site of oxidation led to the synthesis of a series of trifluoromethyl ($-CF_3$) substituted dyes (e.g., XI–XIII) which showed improved photochemical stability under flashlamp excitation (Schimitschek *et al.*, 1973, 1976).

Other evidence of mechanism for coumarin dealkylation and other decomposition pathways was provided by Jones *et al.* (1984b). They identified another product of photodegradation (XXVI from VIII) in which a coumarin ring bond is reduced. The appearance of this product, which is itself unstable and is reliably followed only at very low conversions of the starting dye, depends on the concentration of starting VIII. This finding implicated a bimolecular step in which an excited state of dye is intercepted by a ground-state dye molecule. For interception of fluorescing dye singlet states (M^*), the important processes are "concentration quenching" (k_{cq}), a new mode of nonradiative deactivation of dye (Eq. 7.21) and bimolecular reaction leading to intermediates or products (k_r) (Eq. 7.22). These steps, which depend on the concentration of dye ([M]), compete with the familiar forms of radiative deactivation via fluorescence (k_f) and unimolecular nonradiative decay (k_d).

$$M^* + M \rightarrow M + M \tag{7.21}$$

$$M^* + M \rightarrow \text{products}. \tag{7.22}$$

The mechanism requires a linear relationship between the reciprocal of quantum yield for product (XXVI) and the reciprocal of dye concentration:

$$1/\Phi = (k_r + k_{cq})/k_r + (k_d + k_f)/k_r[M]. \tag{7.23}$$

An estimate could be obtained of the limiting (maximum) quantum yield for product formation, 3×10^{-4}, from the intercept of a linear plot of $1/\Phi$ and $1/[M]$. In addition, the analysis provided a source of the sum of rate constants, $k_{cq} + k_r$, representing the rate of dye deactivation due to bimolecular encounter (Eqs. 7.21 and 7.22). In a separate study (Jones and Bergmark, 1984) bimolecular deactivation of the coumarin fluorescent state was demonstrated for three dyes in three solvents with the sum of bimolecular rate constants ranging $1 - 17 \times 10^9$ M^{-1} s^{-1} (i.e., near the maximum rate dictated by molecular diffusion).

The process of bimolecular encounter of dye was understood in somewhat more detail to involve formation of an excited dimer species [an

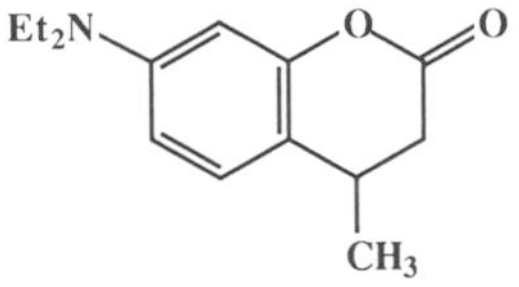

XXVI

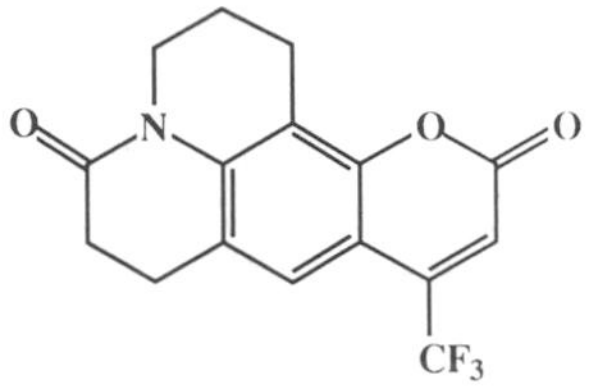

XXVII

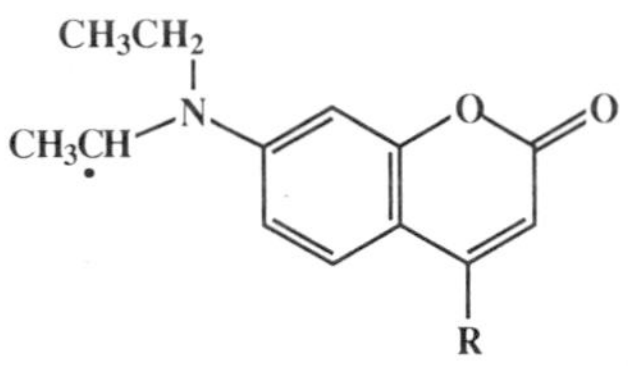

XXVIII

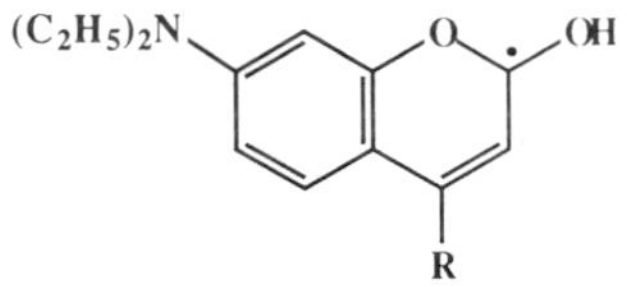

XXIX

"excimer," $(M—M)^*$] which is prone to transfer between dye partners a hydrogen atom (or equivalently, in steps, an electron and a proton). This reaction of paired dye results in formation of radicals, XXVIII and XXIX, which in turn are responsible, through a series of coupling, disproportionation, and other steps, for amine group dealkylation and formation of XXVI. Radical intermediates are also appropriate for that portion of dye degradation that results in uncharacterized polymeric material.

Radicals such as XXVIII and XXIX are further implicated for coumarins according to the labeling study conducted by von Trebra and Koch (1983). They observed that addition of radical chain transfer agents, a combination of a thiol (RSH) and a disulfide (RSSR), was successful in suppressing the photodegradation of coumarin 311 (XXX). They further observed through a nuclear magnetic resonance analysis that photolysis of XXX in the presence of the divalent sulfur compounds that were labeled with deuterium (RSD) resulted in D incorporation in various coumarin substituent or ring positions. The results of this mechanistic probe are consistent with formation of radicals by bimolecular encounter of unexcited dye and excited dye (Eq. 7.22) (the triplet state according to von Trebra and Koch

(1983)). The intermediate radicals are trapped by transfer agent and shielded from further degradation (returned to their native state).

An important detail concerning the reactivity of substituent groups on coumarins involves the wavelength and intensity of the excitation source. Von Trebra and Koch (1987) have determined that the decomposition of coumarin 311 (XXX) at high XeCl excimer laser flux is dependent on the intensity of the pump beam (10–40 mJ/10 ns pulse at 10 Hz). The high intensities were particularly destructive, leading to products indicative of decomposition of solvent. From ethanol a yellowish oligomeric material was obtained along with gaseous products (e.g., CH_3CHO and H_2) and an adduct in which coumarin dye and solvent are incorporated (XXXI).

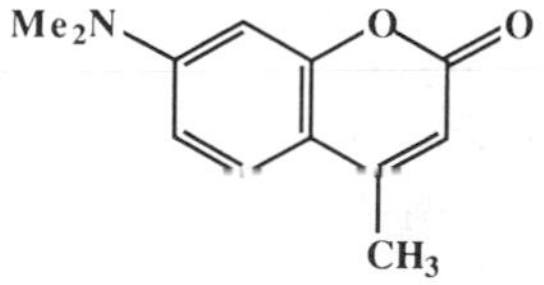

XXX

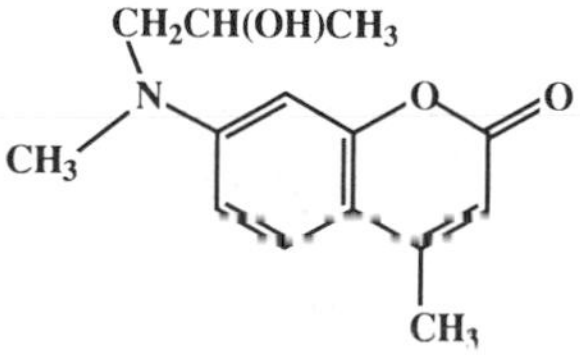

XXXI

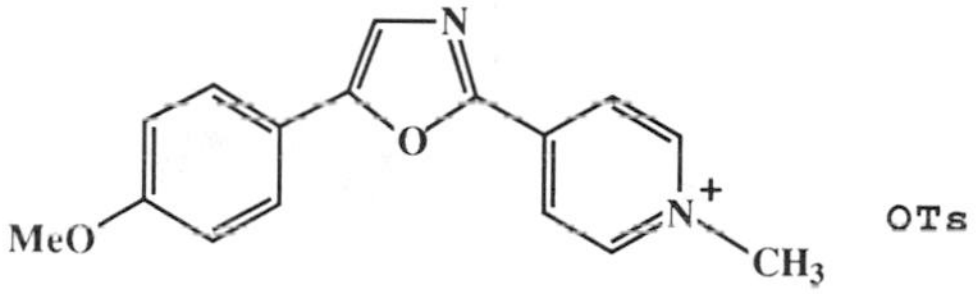

XXXII

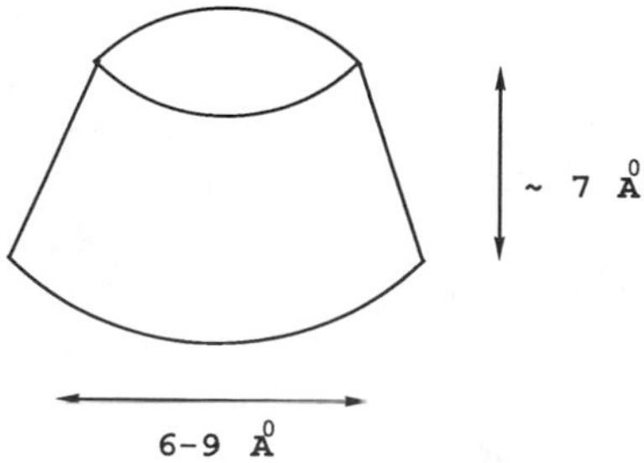

XXXVI

Several aspects of photodegradation of coumarins occurring on irradiation of *aerated* solutions were reported in the 1980s. For coumarin XII, a structure lacking the vulnerable $-CH_3$ group which is oxidized in the VIII → XXV sequence, oxygen is incorporated in a pair of isomeric amides (e.g., XXVII) (Jones *et al.*, 1984b). Coumarin 120 (XXIII) has been modified on photolysis in oxygen-saturated methanol yielding 7-amino-3-hydroxy-4-methylcoumarin (Kunjappu and Rao, 1987). An important determination has also been made as to whether it is the species, singlet oxygen ($^1O_2$) that is responsible for these transformations. Von Trebra and Koch (1986) examined the photolysis of coumarin 311 (XXX) showing that products analogous to the five derivatives of coumarin 1 (VIII) were obtained. They conducted an independent assay of the feasibility of formation of singlet oxygen through energy transfer from coumarin 311 (Section 2.3.2). The dye was successful in providing a low yield of $^1O_2$ through sensitization; the reactive form of oxygen is in fact responsible for coumarin degradation. However, the more deleterious product of the type, XXV, was not obtained via singlet oxygen, so that another route involving other species was invoked. A mechanism was suggested, which will be shown shortly in terms of a generic reactive position on dye (D—CH_3) (Eq. 7.24). The steps include oxygen quenching of dye singlet states through electron transfer plus subsequent proton transfer between the resultant superoxide and dye radical-cation species. The coupling of radicals leads to hydroperoxide intermediates which in turn are likely sources of products of the type, XXIV, XXV, and XXVII:

$$\begin{aligned} \mathrm{D{-}CH_3}^* + \mathrm{O_2} &\rightarrow \mathrm{D{-}CH_3}^{+}\cdot + \mathrm{O_2}^{-}\cdot \\ &\rightarrow \mathrm{D{-}CH_2}\cdot + \cdot\mathrm{OOH} \\ &\rightarrow \mathrm{D{-}CH_2OH} \rightarrow \text{products.} \end{aligned} \tag{7.24}$$

3.6 Photodegradation of Dyes under Lasing Conditions

Compared to the more sedate conditions of routine cw lamp photolysis, the process of dye lasing invites numerous complications of photochemical mechanism. In addition to the loss of emitting species, transient or stable photoproducts compete for the relatively intense pump light. A number of circumstances conspire to intercept the dye-excited (fluorescent) state, including the act of stimulated emission itself. Indeed, the hard-won dye parameters involving the quantum yields of emission and triplet formation (Φ_f, Φ_{isc}, k_f, and k_{isc}; Section 2.1), obtained through methods that employ low-intensity photophysical probes, are less informative with regard to the pertinent events, as multiple photon processes become important. Complication is also introduced due to the distribution of pump (excitation) energy which is not uniform through the cross section of a dye cell. There will be a "mode volume" associated with high-density regions of stimulated emission. Photodegradation is likely to show a nonuniformity since excited-state lifetime in the lasing (stimulated) region will be reduced vis-a-vis destructive competitive paths (i.e., higher quantum yields of degradation are expected for light absorption outside the mode volume).

Several of the variety of studies regarding the photodegradation of laser dyes under lasing conditions (or close stimulations) will be summarized here. Dorko *et al.* (1980, 1982) reported on the decomposition of sulforhodamine B (Kiton Red S or Kiton Red 620). For flashlamp-pumped solutions in a triaxial cavity configuration, they noted a beneficial effect of oxygen in terms of rate of degradation and a much less efficient photolysis when light of $\lambda < 240$ nm was excluded. This series of studies also showed that photodecomposition is also more pronounced at higher pump energies and further provided evidence for a different mechanism of dye decomposition under cw Hg lamp photolysis conditions (a route that is accelerated by oxygen). Ippen *et al.* (1971) noted subtle structural influences (R6G versus other dyes) on the rate of rhodamine photobleaching in alcohols.

In research on coumarin dyes, Fletcher *et al.* have contributed an extensive series of reports (e.g., 1977, 1978, 1983) documenting the dye degradation and lasing characteristics of solutions on flashlamp pumping. In these studies emphasis was placed on the spectral appearance of products that absorb at the lasing wavelength, the nature of solvent and cover gas, and the selection of pump wavelengths (use of optical filters). Some of the findings include a dependence of dye bleaching rate on coumarin concentration, the more deleterious effects of short pump wavelengths (<220 nm) on dye decomposition (especially the fluorinated coumarins, e.g., XI–XIII), and the undesirability of oxygen as a cover gas for most but not all coumarins. As part of these investigations, a dye degradation equation (7.25) was developed to compare different dyes and conditions. The relationship (Fletcher and Knipe, 1982) shows the dependence of laser output (ϕ) on the electrical input energy per flash (I) and the total input energy per unit volume of dye solution (T). The initial lasing slope efficiency is taken as equivalent to constant b for the system under study, the initial lasing threshold $= -a/b$, and the dye lifetime constant is $1/c$ (units of J dm^{-3}).

$$\phi = a + bI \,/\, (1 + cT). \tag{7.25}$$

Antonov and Hohla (1983a) have reported the most extensive survey of dye stability under conditions of excimer laser pumping. For 35 dyes representing different classes, they measured relative stabilities for XeCl excimer laser pumping (308 nm, 200 mJ/pulse at 10 Hz) of methanol or dioxane solutions in a flow sytem. Photoproduct absorption at both lasing and pump (308-nm) wavelengths was observed and, in general, laser performance (output energy) deteriorated more rapidly than did the absorption of dye itself. For 308-nm pumping, the uv dyes of the polyphenyl class proved to be most stable, due apparently to the fact that they are promoted directly to the S_1 (fluorescent or lasing) excited state. The rhodamines and cyanines (excited to 308 nm to upper S_2 states) were moderately stable with some exceptions: the coumarins were least resilient with strong photoproduct absorption at both pump and lasing wavelengths. In a second communication, Antonov and Hohla (1983b) reported the degradation of cyanine dyes with excimer laser pumping (again XeCl at 308 nm). They noted the importance of testing dyes under conditions in which the pump laser beam is split in such a way as to separately excite transversely the dye solution of a saturated amplifier cell. In this way, a true test of dye under conditions of high power (light flux) is achieved. The observation was also made that the relatively short excimer laser pulses (ca. 20 nsec) held a distinct advantage over flashlamp excitation (ca. 500 nsec) since the absorption of pump light by longer-lived dye triplets would be much less important.

The effect of "treatment" of dye solutions for the purpose of stabilization against degradation has been reported. For example, von Trebra and Koch (1982) have used the saturated amine DABCO (1,4-diazabicyclo [2.2.2] octane) as a stabilizer for coumarin dyes. Output power could be sustained up to three times longer for several dyes under nitrogen laser or flashlamp pumping when 0.01 M DABCO is present in ethanol solutions. The stabilization by DABCO was proposed to involve its participation as a quencher of dye triplets or (for aerated solutions) a quencher of singlet oxygen. The apparent degradative effects of solvent decomposition that appear to play a decisive role in rhodamine and coumarin photolysis (Sections 3.4 and 3.5) can be reversed to a degree. Mostovnikov *et al.* (1976) showed that the photoproducts of ethanol decay with flashlamp pumping of R6G solutions could be selectively removed by filtration with alumina and the service life of dye solutions extended.

4. PROPERTIES OF DYES IN UNUSUAL MEDIA: DETERGENTS, AMPHIPHILES, AND INCLUSION AGENTS

4.1. Detergent Micelles

Since the outset of the evaluation of organic dyes and their lasing properties, the role of the medium in controlling the rate of photodegradation has been appreciated. The need for solvent to provide a stable environment for laser dyes in an intense radiation field is superimposed on other solvent influences on photophysical parameters (Section 2.2). The solvent ethanol has been the medium of choice for most dyes due to the reduced tendency for dye aggregation and the good photochemical stability associated with this medium. In several studies, other organic solvents have proven to be less worthy. The lasing of dyes in solvents such as carbon tetrachloride or dimethylformamide (DMF) is particularly destructive. Solvents such as benzyl alcohol that absorb more uv light, even as mixtures with ethanol, are also less successful (data for R6G, Rosenthal (1978)). More viscous solvents such as glycerol show mixed results on comparison with ethanol, whereas mixtures of alcohol and water provide some improvement over pure alcohol in some cases (data for coumarins and stilbenes, Tuccio *et al.* (1973) and Mukherjee *et al.* (1986, 1987)).

On inspection of the optical properties of solutions, water emerges as an ideal solvent for laser dyes. During the process of dye pumping, heating of the solution is not uniform and thermally induced gradients of refractive index result. Unfortunately, for the wider array of organic solvents, the refractive index is strongly temperature-dependent. For water and its isotopically substituted analog, D_2O, the refractive index change with temperature is fourfold-reduced (for $\lambda = 589$ nm near room temperature). However, many laser dyes, particularly those that do not carry formal charges, have very poor water solubility. To address this limitation, investigators have constructed dyes with solubilizing groups (e.g., $-CO_2H$, $-SO_3^-$, or CH_2OH) that permit the use of pure water as the lasing medium (Drexhage *et al.*, 1975). Lee and Robb (1980) have provided an informative account of the selection of dyes for use in water and introduced the series of pyridinylphenyloxazoles and their pyridinium salts (e.g., structure XXXII, as its *p*-toluenesulfonate, OTs^-, salt).

An alternative approach that addresses both the strategy of solubilization of dyes in water and, in principle, provides "protective" media of high photostability involves microencapsulation of dye in small colloid-like structures. The important distinction is that pure solvents or simple mixtures provide a simple *homogeneous* environment for solubilized dye. *Microheterogeneous* environments result when *organized assemblies* of molecules or ions such as those associated with surfactants (detergents) are created in water. Detergents are referred to as *amphiphiles* due to their tendency to interact favorably with both organic solvents and water (both hydrophilic and lipophilic properties). The structures of detergents provide clues to their duplicitous nature. Sodium dodecyl sulfate (SDS, XXXIII) displays a polar, charged "head group" ($-OSO_3^-$) and a lipophilic (fatlike) hydrocarbon tail. Alternatively, the amphiphile may carry a positive charge, as with cetyltrimethylammonium bromide (CTAB, XXXIV). A nonionic structure, usually a polyether, is also possible (e.g., Triton X-100 or octoxynol, XXXV).

CH_3-$(CH_2)_{10}$-CH_2-OSO_3- Na^+ (SDS) XXXIII

CH_3-$(CH_2)_{14}$-CH_2-$N(CH_3)_{3+}$ Br− (CTAB) XXXIV

$CH_3C(CH_3)_2CH_2C(CH_3)_2 - C_6H_4 - O(CH_2CH_2O)_nH$ XXXV.

The duality of structure in surfactants provides a driving force for self-assembly in water. It is quite generally observed that these materials show unusual (discontinuous) behavior in solution in terms of light scattering, surface tension, or spectroscopic properties (Fendler and Fendler, 1975). These changes result from the formation of assemblies of ions or molecules called micelles, structures that appear above a certain surfactant concentration, the "critical micelle concentration," or cmc. Values of the molar concentration of surfactant at the cmc range from 10^{-2} to 10^{-4} M. Micelle structures will typically consist of 50–100 units of structure of the type SDS or CTAB and will have shapes that are (very roughly) spherical or rod-like, as shown schematically in Fig. 7.9. Shapes and dimensions are actually quite variable, but the size of a 10-nm sphere provides a rough guide. Forces that bind micelles together include the attractive (van der Waals) interactions among the hydrocarbon chains of surfactants in addition to a favorable entropy that is associated with the freeing of water molecules from surfactant solvation shells on micellar assembly (Evans and Ninham, 1986). Amphiphilic structures other than micelles that provide organized assemblies in water include the donut-shaped vesicles, microemulsions, and surfactant monolayers (Fig. 7.9) (Fendler, 1980).

The important effects of incorporation of chromophores in micelles on the photochemical and photophysical properties of molecules bearing

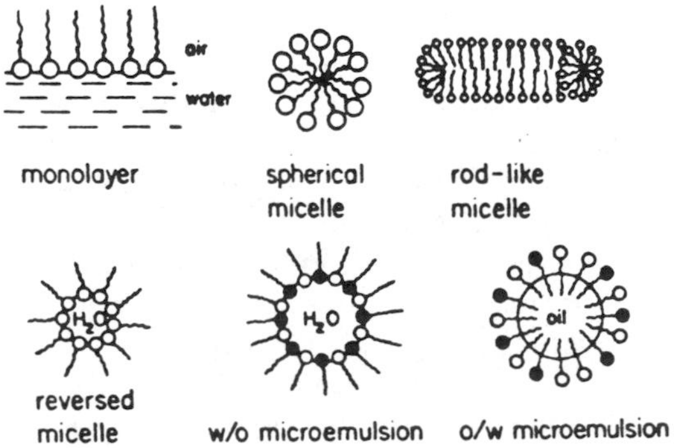

Fig. 7.9 Schematic representations of various organized assemblies of amphiphiles (reprinted from Fendler, 1980).

chromophores have been extensively investigated (Turro *et al.*, 1980). On considering the modification of photochemical properties for laser dyes, the dynamics of micelles must be taken into account. A number of probes have been used to determine that the time scale for the exit of organic molecules (in fact, the exit of surfactant monomers themselves) from a micellar assembly is 10^{-6} to 10^{-3} sec. The micellar aggregate as a whole is broken down in milliseconds to seconds. These time scales are long with respect to the lifetime of dye singlet states (nanoseconds), but are more comparable to the lifetime of dye triplets (Section 2.1). An organic molecule will be partitioned between an aqueous environment associated with the bulk water phase and some portion of the micelle interior or interface. The equilibrium of dye in water versus the more hydrophobic phase will favor the latter, commonly by a factor of $>10^2$ (Fendler and Fendler, 1975). Dye-excited states will reside during a typical lifetime almost entirely in the surfactant domain. Micelles are thus organized assemblies in which dye molecules or ions are solubilized (Mukerjee, 1980) and, in principle, sequestered from external agents that are more likely to reside in the water domain. A demonstration of this form of hydrophobic protection is the observation that triplet states of various organic species are longer lived and give rise to room-temperature phosphorescence due to the inhibition of reaction with adventitious oxygen or other quenching agents (Turro *et al.*, 1978; Love *et al.*, 1980).

Just as micelles provide hydrophobic environments for lipophilic organic dyes, the dyes themselves serve as "reporters" of their local surroundings. For example, Humphry-Baker *et al.* (1980) showed that in addition to inhibiting the formation of aggregates in water for a cyanine dye related to DOTC (V), the use of SDS and other salts as surfactants increased the fluorescence quantum yield tenfold. They also found that incorporation of dye in micelles also led to a longer lifetime for dye fluorescence, along with a higher value for the emission anisotropy (polarization), consistent with the more rigid environment for dye in the interior of a micelle. Ebeid *et al.* (1986) have shown that an enhanced fluorescence in micellar media permits a weak laser action in the blue for a weakly luminescent stilbene-type pyridinium dye.

In an extension of studies of the coumarins for which "solvent tuning" was reported (Section 2.2), the effects of dye incorporation in surfactants were investigated (Jones *et al.*, 1985b) Solubilities for the dyes in water were shown to be increased over fiftyfold (up to 1×10^{-3} M) with added surfactant. For SDS and CTAB micelles, regular blue shifts in absorption and emission were recorded for the coumarin reporters on alteration of the medium from water to surfactant media. The solvatochromic comparison of the coumarins in micelles versus other media revealed an interesting detail that is corroborated in other work concerned with the probing of micelle interiors. The coumarin C6F (XII) which shows a special sensitivity for the hydrogen bonding properties of a solvent (Section 2.2) displayed a spectroscopic signature (an α value equivalent to 1.0 to 1.1, Fig. 7.3) consistent with micelle solubilization sites that are very polar. That is, dye is not incorporated in the hydrophobic core of surfactant assemblies, but is associated more with the interfacial region in which surfactant chains are significantly hydrated and near the "head groups" that constitute roughly the micelle periphery (Fig. 7.9). Some water molecules in fact penetrate deeply into micelle interiors (Menger, 1979), so that the boundary between hydrophobic and hydrophilic regions is not as well defined as once believed.

The fluorescence yield of coumarin C1F (XI), which shows a striking sensitivity to viscosity (the rotatory decay mechanism, Section 2.2), was used to reveal also that micelle interiors have large microviscosities (Jones *et al.*, 1985b). Using as a similar probe the emission yield of a merocyanine dye, Law (1981) showed that microviscosities of SDS, CTAB, and Triton X-100 micellar domains fall in the range 23–31 cP (about that of ethylene glycol, $\eta = 26$ cP, 25°). Importantly, the *bulk* viscosities of dilute micellar solutions are not greatly altered from the value obtained for pure water.

The photochemistry of laser dyes in surfactant solutions has not been studied extensively. Several early findings show mixed results regarding photodegradation. Britt and Moniz (1973) reported that R6G on cw Ar-ion laser irradiation (514 nm) has a photostability in water with Triton X-100 only slightly reduced from that found for ethanol solutions, whereas another study in which R6G solutions were irradiated at the same wavelength but in capillary guides showed that dye was degraded faster (16X) in the presence of the same detergent. In retrospect, triton X-100 and other uv-absorbing surfactants are poor choices for solubilization–stabilization of laser dyes due to the high probability that aromatic groups will photoionize in water (Section 3.2).

The alteration of photochemical properties for chromophores other than laser dyes in micelle environments is well known (Turro *et al.*, 1980). For example, a neutral micelle-solubilized species will be less likely to undergo electron transfer with a quencher in the aqueous phase (Atik and Thomas, 1981). In their study of surfactant effects on coumarin dyes, Jones *et al.* (1985b) showed that coumarin 1 (VIII), when incorporated in SDS micelles, is less prone to eject electrons on flashlamp irradiation in water. Levin *et al.* (1986) also showed that several rhodamines display improved stability in aqueous micellar solution (SDS/H_2O), even compared to alcohol solutions, but under circumstances in which another additive such as ascorbic acid is present.

On close inspection, the electron-transfer properties of assemblies of charged micelle and dye appear to be largely governed by electrostatics. For example, the molecule pyrene reacts with the hydrated electron at a rate in water that is 2000 times that found for a medium in which pyrene is incorporated in the negatively charged SDS surfactant. On the other hand, if pyrene is associated with CTAB that is laden with positive charges, the rate of reaction with the solvated electron is 200-fold increased (Kalyanasundaram and Thomas, 1977). The monophotonic electron ejection occurring on uv photolysis of tetramethylbenzidine (TMB), solubilized in water, is dependent on micelle surface charge (higher yield of photoionization (Eq. 7.15) for cationic surfactant) (Narayana *et al.*, 1982). A more detailed examination of photoionization in detergent media of a hydrocarbon and TMB shows that photoionization can be suppressed by "swelling" micelles with a hydrophobic (hydrocarbon) agent or through alteration of the micelle–water interfacial electrical potential by the addition of an ancillary electrolyte (salt) (Bernas *et al.*, 1988).

A growing number of reports indicate that exciting possibilities exist for co-solubilization of two (or more) dyes for wavelength shifting by energy transfer. Marszalek *et al.* (1980) reported that aqueous solutions of coumarin 1 and R6G display dual fluorescence but that the emission from rhodamine is enhanced on addition of triton X-100 due to the co-binding of dyes and energy transfer. A dye co-binding mechanism is apparently important for several systems that employ cationic dyes and anionic surfactants. Kasatani *et al.* (1985) reported that excitation of R6G resulted in emission from a cationic cyanine, HITC; their model of the data suggested that both dyes are bound near micelle surfaces for efficient energy transfer. The same group examined energy transfer between R6G and another cyanine, pinacyanol, and discovered an unusual feature. Highly efficient transfer was observed for dye co-binding to "premicelles" of SDS, small surfactant aggregates that are in fact induced through interaction with hydrophobic dye counter-ions (Sato *et al.*, 1983). Experiments using fast time-resolution (subnanosecond) techniques have been used to investigate energy transfer between co-bound dye molecules (R6G, malachite green, and SDS); inferences could be made about micelle size and shape and dye solubilization sites (Choi *et al.*, 1988).

4.2. Cyclodextrins

Cyclodextrins are oligosaccharides consisting of glucose molecules linked end to end in a circular fashion. These interesting molecules assume the shape of a torus (structure XXXVI) and provide a central cavity of about 0.5–1.0 nm in which other molecules may reside. Like detergents the cyclodextrins provide a vestibule for housing organic structures; they are not strictly amphiphiles (compare XXXIII–XXXV) and are better classified as inclusion compounds (Davies *et al.*, 1988; Bender and Komiyama, 1978). They are highly soluble in water and are therefore candidates as solubilizing agents for laser dyes. The truncated cones of cyclodextrins feature a hydrophilic periphery of -OH functions and a more hydrophobic region for the torus interior in which primary sugar hydroxyl functions are located (XXXVI, p. 321).

The photophysical and photochemical properties of a variety of organic substrates in cyclodextrin (CD)-water media have been investigated thoroughly (Ramamurthy and Eaton, 1988). These studies show the effects of removal of organic moieties from a homogeneous aqueous phase (e.g., enhancement of fluorescence) and the potential for aggregation of species within cavities of proper dimension. Investigations of laser dyes in this context have been more limited. Degani *et al.* (1984) showed that the fluorescence yield for rhodamine B, which is much diminished in pure water, is increased eight fold on addition of ca. 10-mM concentrations of the β oligomer of cyclodextrin (β-CD) (seven linked glucose units). Other subtle spectral features pointed to an intimate contact of the rhodamine dye with the β-CD host. Lasing could be observed with nitrogen laser pumping but with an intensity about one third of that displayed by RB in ethanol. The yield of fluorescence for the dye in CD-water was shown to be 0.25 (versus 0.5 for RB in ethanol). The investigators observed that RB is sufficiently large to preclude a perfect "fit" into the β-CD cavity (having a 0.7-nm interior radius), so that exclusion of dye from the water medium is incomplete.

The deployment of γ-cyclodextrin and an oxazole chromophore led to quite different results (Agbaria and Gill, 1988). This oligomer of somewhat greater dimension (eight glucose units and a 0.85-nm interior) was effective in solubilizing 2,5-diphenyloxazole but produced large aggregates in aqueous solution that produced turbidity. In this case intriguing structures apparently resulted from the protrusion of aromatic rings from the periphery of the cyclodextrins such that a number of these units could be joined through a linkage of hydrophobic groups projecting from one cavity into the space of another. Scypinski and Drake (1985) reported the inclusion of the condensed-ring coumarin 540 (C6F, XII) in β- and γ-CD and noted an enhancement of fluorescence over that found for weakly soluble XII in water. They proposed that two dye-CD complexes are important, complexes that differ subtly in their fluorescence properties and in their mode of orientation of the coumarin hydrophobe. A later study by Bergmark *et al.* (1987) has suggested that the anomalous results regarding XII are more readily explained in terms of the intervention of organic co-solvent which tends to assist in the inclusion process. Thus, for coumarins VIII, XI, and XII, it could be shown that addition of very low concentrations (0.3% v/v) of solvents such as ethanol or acetonitrile to aqueous dye solutions, including β- or γ-CD, result in even greater effects on fluorescence yield (Φ_f up to 0.62 for XII in γ-CD with 0.3% ethanol). The results are consistent with the notion that some water inhabits CD interiors, even when dye is cavity-bound. Addition of another organic species (co-solvent) results in a displacement of residual water molecules and a more successful fit of dye into a hydrophobic pocket of proper size.

4.3. Polymers and Plastics

Simple organic polymer hosts or plastics would appear to hold a number of advantages as media for laser dyes. Dye-doped laser rods can be fashioned into a variety of shapes and produced at low cost. The plastics employed in fluorescence concentrators constitute a related application in which laser dyes have been used (Drake *et al.*, 1982). Decades ago, Peterson and Snavely (1968) reported stimulated emission from 0.5 × 8.5-cm rods of polymethyl methacrylate (PMMA) doped with rhodamine dyes under flashlamp pumping. In most of the reported doping procedures, one need only carry out a conventional free radical polymerization of a common organic monomer (e.g., methyl methacrylate) in the presence of dye. Dye structures are entrapped in a more or less random fashion among highly irregular coiled segments of growing polymer chains. A demerit is introduced in that organic glasses are plagued by poor optical properties associated with large coefficients of thermal expansion, thermal gradients, stress birefringence, and photochemical instability.

Improvements in photostability have been sought through alteration of the polymer host structure (e.g., substitution of the more familiar PMMA with an acrylic, polyacrylonitrile) (Reich and Neumann, 1974). The durability of rhodamine dye in PMMA is also improved at low temperatures (e.g., 200 K) (Fork and Kaplan, 1972). The stability of dye-in-plastic appears to be a complicated function of length of irradiation time, pump energy, and polymer structure. In one study the mechanism of degradation of three different dyes appears to involve one-photon processes; impurities in plastic matrices may also play an important role in the photochemistry (Kaminow *et al.*, 1972). Higuchi and Muto (1981) demonstrated that PMMA plastic is improved significantly on copolymerization of MMA with methacrylic acid (MA) (up to 10%) due to a special affinity of polymer segments containing the latter monomer for R6G. Their study of photodegradation of rhodamine in copolymer suggested that dye is not uniformly bound within the plastic so that some dye structures are found in specific sites more prone to photodegradation. Photoproduct absorption (ca. 400 nm, not unlike product P for solutions (Section 3.4)) was also identified (Higuchi and Muto, 1983). Gromov *et al.* (1984) have also reported the effects of adding ethanol (a co-solvent) to solid solutions of R6G in PMMA. They noted that alcohol additive inhibits photodegradation due to inhibition of the formation of dye aggregates or ion-pairs.

Solid materials other than organic plastics, which have received less attention but which may hold some promise, include the silicas, porous glasses, and clays. For example, Endo *et al.* (1986) have reported that rhodamines can be incorporated in a smectite material, montmorillonite clay. An X-ray structure could be determined for dye, intercalated (sandwiched) in the regular spacings of aluminosilicate layers. The regularity and rigidity of structure for dye in solid carriers of this type have been underscored by Avnir *et al.* (1984), who investigated R6G in silica gel glass. Several advantages were associated with this structure, including high fluorescence yield, large Stokes shift of dye emission, transparency in the uv, and a high degree of isolation of dye molecules which improved durability.

5. SUMMARY AND PROSPECTUS

Since the incisive review by Drexhage of early work on laser dyes in 1973, significant advances have been made in terms of the elaboration of dye structure and the understanding of how the solvent medium influences dye performance. In addition, reaction mechanisms have been elucidated for several dye prototypes, and new media with "protective" capabilities have been developed. A number of these advances were highlighted in a symposium devoted to dye lasers (Bentley, 1987).

Examples of *new or highly modified dye structures* include derivatives of bimane (XIV, Section 2.2) which show high photostability under flashlamp-pump conditions (Pavlopoulos *et al.*, 1986). Quarterphenyls have been developed for use in the uv (350–390 nm) with excimer laser pumping. To optimize efficiency and dye stability for the series, substitution patterns that involve ring bridging, steric hindrance to conjugation, and the placement of auxochromic groups have been investigated (Rinke *et al.*, 1986). The new substituted quarterphenyls include sulfonated derivatives which show improved solubility and exceptionally high energy conversion (18.5%) on excimer laser pumping (Hueffer *et al.*, 1980). The XeCl excimer laser (Telle *et al.*, 1981) has proved to be valuable new pump source (308 nm) for certain dyes (e.g., oxazoles, Sections 3.3 and 3.6) for which flashlamp pumping is much less successful.

Improved coumarin dyes for the blue-green have been reported (Chen *et al.*, 1988). These new structures (e.g., coumarin "314T") are related to the compounds showing the annulation of rings at the 7-nitrogen position which is important in reducing nonradiative decay (as in III and XII) (Section 2.2). They are distinctive in having further substitution by methyl groups (T = tetramethyl) in these saturated rings. This substitution removes sites for photooxidation (XII → XXVII, Section 3.5) (Jones *et al.*, 1984b), an effect that could play a role in the improved lasing efficiency and photostability for this family of dyes under coaxial flashlamp excitation.

A new family of cationic dyes displaying a merocyanine type of conjugation (Section 2.1) operates beyond 500 nm (e.g., the oxazole XXXII, referred to as 4PyMPO). These structures constitute the most photostable dyes yet investigated for flashlamp excitation (Fletcher *et al.*, 1987, Kaufman and Bentley, 1988). The oxazoles are also the prototype for elaboration of a strategy in which dye structures are modified by linkage to various types of proactive groups (Schaefer, 1983). Thus, synthesis of dye derivatives allows the direct attachment of a light harvesting (booster) dye as a pendant or, alternatively, a triplet quenching agent or a chemical (radical) scavenger (Section 2.3.3).

The "proton transfer laser" has been proposed in which a dye molecule (salicylamide or 3-hydroxyflavone) undergoes an intramolecular proton transfer after excitation and emission occurs from the rearranged product (Acuna *et al.*, 1986; Kasha, 1986). The concept is reminiscent of an earlier proton-transfer–based dye system, 4-methylumbelliferone (Drexhage, 1973) and has the advantage that remarkable Stokes shifts of emission can be observed.

Dye-medium interactions can extend the tuning range of dyes, particularly those of the merocyanine class (e.g., coumarins, oxazoles, or xanthenes displaying push-pull substitution) (Section 2.2). Influences on dye fluorescence yield and life-time as well as the probability of formation of dye triplets are very subtle functions of dye structure (rigidity) and medium polarity and viscosity. A somewhat overlooked medium phenomenon has to do with ion-pairing. Most of the superior dyes of the rhodamine, cyanine, and oxazole classes are cations which require negatively charged counter-ions for charge balance. These charged species may be tightly paired, especially in organic solvents. One effect of ion-pairing due to a "heavy-atom" counter-ion which results in enhanced triplet formation has been noted (Section 3.3). A study (Kauffman and Bentley, 1988) on the promising oxazole family related to 4PyMPO (XXXII) revealed effects on lasing wavelength and output energy associated with the choice of counter-ion or the substitution of the dye structure with a negatively charged group (formation of zwitterions).

Dyes chosen for discussion here illustrated by class the regimes of *photodegradation mechanism* to which most dyes are susceptible. Simple oxazoles show a heightened sensitivity to light associated with addition reactions with the excited singlet oxygen molecule (Section 3.3). Because of the likelihood for generation of this or other reactive species (superoxide, $O_2^{-}\cdot$), the deployment of oxygen remains a dilemma in terms of the quenching of unwanted dye triplets. It cannot be generally recommended as a cover gas for very efficient, sustained dye laser applications. *Reversible* electron transfer quenchers of triplets, which are also poor absorbers, offer new opportunities regarding the triplet problem (the amine, DABCO, is a candidate) (von Trebra and Koch, 1982).

The coumarins are exemplary dyes which engage in "self-quenching," a mechanism involving dye–dye interaction which results in formation of free radicals (Section 3.5). Radical trapping agents show some promise in improving dye performance in terms of this class of reactivity. The xanthene (rhodamine) class of dyes displays a reactivity that is most easily understood in terms of electron-transfer processes. The driving force for loss or gain of an electron by dye is obtained via the encounter of two excited species (two triplets, Eq. 7.8) or through the participation of upper excited states of a single dye molecule (S_n or T_n). Multiple excitation events or short-wavelength irradiation are required (Section 3.1 and Eq. 7.16). Use of high pump energies or unfiltered light promote these deleterious processes. A mechanism of particular importance due to its potential generality for dye lasing in polar media involves the ejection of an electron to solvent from a highly energized dye molecule, followed by oxidation of solvent by the electron-deficient dye intermediate (Eqs. 7.19 and 7.20). A number of papers now report in varied detail evidence suggesting that decomposition of solvent via radical intermediates is important. The photoionization process can display a complicated dependence on the intensity and width of excitation pulses (Hall, 1978).

A general note of warning concerns the radical ion species associated with reduction or oxidation of organic dyes. These very reactive intermediates will engage in expulsion of groups (e.g., transfer of protons), in radical coupling, or in reaction with species such as oxygen. Moreover, they absorb strongly in the visible and exhibit lifetimes in the μsec–msec range. It is further the rule, and not the exception, that highly conjugated organic dye structures display redox potentials that are favorable to photoinduced electron transfer if the medium can support charge separation. Any additive (indeed, the native dye itself) is a potential redox quenching agent. The important point is that among the remaining low–quantum-yield processes that will inevitably plague to some degree the best dye-laser systems, electron transfer will be among the most difficult to defeat.

The importance of electron transfer or photoionization involving solvent can be directly assessed using flash photolysis or electron-spin (magnetic) resonance techniques. A great many details regarding critical dependences of dye photodegradation on excitation wavelength and intensity await further studies of this type. It will be particularly challenging to thwart photodecomposition and sustain lasing in organic media when hard uv excitation is employed (e.g., <250 nm) from excimer or flashlamp sources at high pump energies and repetition rates.

Unusual media for laser dyes have been explored in some detail. Amphiphiles (detergents, surfactants) provide a means for solubilizing dye molecules (without dye aggregation) in water. For this medium, dye structures are incorporated in relatively large surfactant aggregates called micelles. These structures or the cyclic oligasaccharides (cyclodextrins) provide opportunities for dye "isolation," a reduction of photochemical paths that depend on dye diffusion, and the potential for control of the yield of processes such as photoionization. Investigations of these organizing structures, which may include organic polymers or plastics, encourage the design of dye assemblies in which the principal chromophore may interact with an energy-transfer agent for wavelength shifting or a trap for triplets or reaction intermediates.

These systems will depend for their function on a proximity of components enforced by hydrophobic or electrostatic interaction and, in advanced designs, will likely incorporate combinations of reagents, such as surfactants and alcohols or cyclodextrins and amphiphiles (Bernas *et al.*, 1988; Herkstroeter *et al.*, 1986). These new media hold some advantage in terms of practical use since they fall within the tradition of deployment of "additives" and for simple use do not require direct dye modification. However, much more is to be learned about the "unusual" media from the point of view of optical quality (light scattering) and the potential for introduction of new unwanted photochemical mechanisms.

This chapter represents necessarily a selective overview of a developing field which, in its larger dimension, addresses issues of photostability for dyestuffs that intersect many technologies. These include not only dye laser applications but also new systems for information and computer technology (imaging, photoconductivity, and molecular electronics). Advances in these important areas will depend significantly on the development of robust, reliable dye media in which operation can be substantially prolonged.

REFERENCES

Acuna, A.U., Costela, A., and Munoz, J.M. (1986). A Proton-transfer laser. *J. Phys. Chem.* **90,** 2807–2808.

Agbaria, R.A., and Gill, D. (1988). Extended 2,5-diphenyloxazole-γ-cyclodextrin aggregates emitting 2,5-diphenyloxazole excimer fluorescence. *J. Phys. Chem.* **92,** 1052–1055.

Allen, N.S., and McKellar, J.F. (1980). *Photochemistry of Dyed and Pigmented Polymers.* Applied Science, London.

Antonov, V.S., and Hohla, K.L. (1983a). Dye stability under excimer-laser pumping. II. Visible and UV dyes. *Appl. Phys. B* **32,** 9–14.

Antonov, V.S., and Hohla, K.L. (1983b). Dye stability under excimer-laser pumping. I. Methods and modelling for infrared dyes. *App. Phys. B* **30,** 109–116.

Arbeloa, I.L., and Rohatgi-Mukherjee, K.K. (1986). Solvent effects on the photophysics of the molecular forms of Rhodamine B. Internal conversion mechanism. *Chem. Phys. Lett.* **129,** 607–614.

Aristov, A.V., and Shevandin, V.S. (1981). Spectroscopic evidence of reversible and irreversible transformation of Rhodamine 6G molecules under two-stage optical excitation. *Opt. Spectrosc. (USSR)* **48,** 638.

Atik, S.S., and Thomas, J.K. (1981). Photoinduced electron transfer in organized assemblies. *J. Am. Chem. Soc.* **103,** 3550–3555.

Austin, J.M., Harrison, I.R., and Quickenden, T.I. (1986). Electrochemical and photoelectrochemical properties of Rhodamine B. *J. Phys. Chem.* **90,** 1839–1843.

Avnir, D., Levy, D., and Reisfeld, R. (1984). The nature of the silica cage as reflected by spectral changes and enhanced photostability of trapped rhodamine 6G. *J. Phys. Chem.* **88,** 5956–5959.

Beer, D., and Weber, J. (1972). Photobleaching of organic laser dyes. *Optics Commun.* **5,** 307–309.

Bender, M.L., and Komiyama, N. (1978). *Cyclodextrin Chemistry.* Springer-Verlag, New York.

Bentley, J.H. (ed.) (1987). *Proceedings: Dye Laser/Laser Dye Technical Exchange Meeting, 1987.* U.S. Army Missile Command, Redstone Arsenal, Ala.

Bergmark, W.R., Davis, A., York, C., and Jones II, G. (1987). Dramatic fluorescence effects associated with small amounts of additives and cyclodextrin-complexed coumarin laser dyes. *Proceedings of the National Meeting of the American Chemical Society, New Orleans, La., Sept. 1987.*

Bernas, A., Grand, D., and Hautecloque, S. (1988). On ion-pair photoseparation and recombination in micellar aggregates. *Radation Phys. Chem.* **32,** 309–314.

Bourson, J., Mugnier, J., and Valeur, B. (1982). Frequency conversion of light by intramolecular energy transfer in bifluorophoric molecules. *Chem. Phys. Lett.* **92,** 430–432.

Britt, A.D., and Moniz, W.B. (1972). The effect of pH on photobleaching of organic laser dyes. *IEEE J. Quant. Elect.* **QE-8,** 913–914.

Britt, A.D., and Moniz, W.B. (1973). Reactivity of first-singlet excited xanthene laser dyes in solution. *J. Org. Chem.* **38,** 1057–1059.

Chen, C.H., Fox, J.L., Duarte, F.J., and Ehrlich, J.J. (1988). Lasing characteristics of new coumarin-analog dyes: Broadband and narrow-linewidth performance. *Applied Optics* **27,** 443–445.

Chibisov, A.K. (1977). Triplet states of cyanine dyes and reactions of electron transfer with their participation. *J. Photochem.* **6,** 199–214.

Chibisov, A.K., and Korobov, V.E. (1983). Primary photoprocesses in colorant molecules. *Russ. Chem. Rev.* **52,** 27.

Chibisov, A.K., and Slavnova, T.D. (1978). The study of the photophysics and primary photochemistry of rhodamine 6G aggregates. *J. Photochem.* **8,** 285–297.

Choi, K.J., Turkevich, L.A., and Loza, R. (1988). Picosecond resonant energy transfer studies of aqueous anionic micellar solutions. *J. Phys. Chem.* **92,** 2248–2256.

Chu, G., and Yangbo, F. (1987). Solvent and substituent effects on intramolecular charge transfer of selected derivatives of 4-trifluoromethyl-7-aminocoumarin. *J. Chem. Soc. Faraday Trans. 1* **83,** 2533–2539.

Davies, J.E.D., Kemula, W., Powell, H.M., and Smith, N.O. (1988). Inclusion/ Compounds—past, present, and future. *J. Inclusion Phenomena* **1,** 3.

Degani, Y., Willner, I., and Haas, Y. (1984). Lasing of rhodamine B in aqueous solutions containing β-cyclodextrin. *Chem. Phys. Lett.* **104,** 496–499.

Dempster, D.N., Morrow, T., and Quinn, M.F. (1974). The photochemical characteristics of rhodamine 6G–ethanol solutions. *J. Photochem.* **2,** 343–359.

Dietz, F., and Rentsch, S.K. (1985). On the mechanism of photoisomerization and the structure of the photoisomers of cyanine dyes. *Chem. Phys.* **96,** 145–151.

Dorko, E.A., O'Brien, K., Rabins, J., and Johnson Jr., S (1980). Kinetic studies of Kiton Red S photodecomposition under continuous working and flash photolytic conditions. *J. Photochem.* **12,** 345–356.

Dorko, E.A., Briding, A.J., and Johnson Jr., S. (1982). Analysis of Kiton Red S dye degradation effects on the output energy of the pulsed organic dye laser. *J. Photochem.* **18,** 251–267.

Drake, J.M., Lesiecki, M.L., Sansregret, J., and Thomas, W.R.L. (1982). Organic dyes in PMMA in a planar luminescent solar collector: A performance evaluation. *Appl. Optics* **21,** 2945–2952.

Drexhage, K.H. (1973). Structure and properties of laser dyes. In *Topics in Applied Physics, Vol. 1: Dye Lasers.* Springer-Verlag, New York.

Drexhage, K.H., Erikson, G.R., Hawks, G.H., and Reynolds, G.A. (1975). Water-soluble coumarin dyes for flashlamp-pumped dye lasers. *Optics Commun.* **15,** 399–403.

Duarte, F.J. (1987). Kodak Laser Dyes. Eastman Kodak Company, Rochester, New York.

Dunne, A., and Quinn, M.F. (1976). Photochemical studies of rhodamine R 6G in ethanol solutions using laser flash photolysis. *J.C.S. Far. Trans. 1* **72,** 2289–2295.

Ebeid, E.M., Issa, R.M., Ghoneim, M.M., and El-Daly, S.A. (1986). Emission characteristics and micellization of cationic 1,4-bis(β-pyridyl-2-vinyl)benzene laser dye. *J. Chem. Soc. Faraday Trans. 1* **82,** 909–919.

Endo, T., Sato, M., and Shimada, M. (1986). Fluorescent properties of the dye-intercalated smectite. *J. Phys. Chem. Solids* **47,** 799.

Evans, D.F., and Ninham, B.W. (1986). Molecular forces in the self-organization of amphiphiles. *J. Phys. Chem.* **90,** 226–234.

Fendler, J.H. (1980). Microemulsions, Micelles, and Vesicles as Media for Membrane Mimetic Photochemistry. *J. Phys. Chem.* **84,** 1485–1491.

Fendler, J.H., and Fendler, E.J. (1975). *Catalysis in Micellar and Macromolecular Systems.* Academic Press, New York.

Fery-Forgues, S., and Paillous, N. (1986). Photodehalogenation and photodimerization of 2-(4-halophenyl)benzoxazoles. Dependence of the mechanism on the nature of the halogen atom. *J. Org. Chem.* **51,** 672–677.

Fery-Forgues, S., Lavabre, D., and Paillous, N. (1987). Electron transfer on the photodehalogenation of 2-(4-chlorophenyl)benzoxazole assisted by amines. *J. Org. Chem.* **52,** 3381–3386.

Fletcher, A.N. (1977). Laser dye stability. Part 3. Bicyclic dyes in ethanol. *Appl. Phys.* **14,** 295–302.

Fletcher, A.N. (1978). Laser dye stability. Part 4. Photodegradation relationships for bicyclic dyes in alcohol solutions. *Appl. Phys.* **16,** 93–97.

Fletcher, A.N. (1983). Laser dye stability. Part 9. Effects of a Pyrex UV filter and cover gases. *Appl. Phys. B* **31,** 19–26.

Fletcher, A.N., Bliss, D.E., and Kauffman, J.M. (1983). Lasing and fluorescent characteristics of nine, new, flashlamp-pumpable, coumarin dyes in ethanol and entanol: water. *Optics Commun.* **47,** 57–61.

Fletcher, A.N., Henry, R.A., Pietrak, M.E., and Bliss, D.E. (1987). Laser dye stability, part 12. The pyridinium salts. *Appl. Phys. B* **43,** 155–160.

Fork, R.L., and Kaplan, Z. (1972). Increased resistance to photodegradation of rhodamine 6G in cooled solid matrices. *Appl. Phys. Lett.* **20,** 472–474.

Fouassier, J.-P., Lougnot, D.-J., and Faure, J. (1979). Photoisomers and behavior of polymethine dyes in lasers. *Nouveau J. Chim.* **3,** 359–363.

Gauglitz, G. (1978). Determination of photochemical quantum yields using fluorescence data, demonstrated by laser dyes. *Z. Phys. Chem. Neue Folge* **113,** 217–229.

Goncharov, V.A., Zverev, G.M., and Martynov, A.D. (1971). Effect of triplet levels on the energy characteristics of some laser-pumped xanthene dye lasers. *Optics and Spectroscopy* **30,** 78–79.

Goodall, D.M., and Roberts, D.R. (1985). Energy transfer between dyes. *J. Chem. Ed.* **62,** 711–714.

Gottschalk, P., and Neckers, D.C. (1986). 3,3'-dialkylthiacarbocyanine iodide: A versatile alternative to non-polar rose bengals for singlet oxygen generation in organic solvents. *Photochem. Photobiol.* **43,** 379–383.

Grabowski, Z.R., Rotkiewica, K., Siemiarczuk, A., Cowley, D.J., and Baumann, W. (1979). Twisted intramolecular charge transfer states (TICT). A new class of excited states with a full charge separation. *Nouveau J. Chim.* **3,** 443–454.

Grenci, S.M., Bird, G.R., Keelan, B.W. and Zewail, A.H. (1986). Practical broad-band tuning of dye lasers by solvent shifting. *Laser Chem.* **6,** 361–371.

Gromov, D.A., Dyumaev, K.M., Manenkov, A.A., Maslyukov, A.P., Matyushin, G.A., Nechitailo, V.S., and Prokhorov, A.M. (1984). Efficient lasers based on dyes implanted in polymeric matrices. *Izvest. Akad. Nauk SSSR* **48,** 1364.

Guesten, H., Rinke, M., Kao, C., Zhou, Y., Wang, M., and Pan, J. (1986). New Efficient laser dyes for operation in the near UV range. *Optics Commun.* **59,** 379–384.

Hall, G.E. (1978). Comment on the communication "Photoionization by green light in micellar solution." *J. Am. Chem. Soc.* **100,** 8262–8263.

Halstead, J.A., and Reeves, R.R. (1978). Mixed solvent systems for optimizing output from a pulsed dye laser. *Optics Commun.* **27,** 273–276.

Herkstroeter, W.G., Martic, P.A., Evans, T.R., and Farid, S. (1986). Cyclodextrin inclusion complexes of 1-pyrenebutyrate: The role of coinclusion of amphiphiles. *J. Am. Chem. Soc.* **108,** 3275–3280.

Higuchi, F., and Muto, J. (1981). Photo and thermal bleaching of rhodamine 6g (Rh6G) in the copolymer of methyl methacrylate (MMA) and methacrylic acid (MA). *Phys. Lett.* **86A,** 51–53.

Higuchi, F., and Muto, J. (1983). On the photobleaching quantum yields of heat-treated rhodamine 6G (Rh6G) molecules in the copolymer of methyl methacrylate (MMA) with methacrylic acid (MA). *Phys. Lett.* **99A,** 121–124.

Hirata, Y., Mataga, N., Sakata, Y., and Misumi, S. (1982). Electron photoejection from the relaxed fluorescence state of bis(dimethylamino)tetrahydropyrene in acetonitrile solution. Importance of solute-solvent exciplex interactions. *J. Phys. Chem.* **86,** 1508–1511.

Hueffer, W., Schieder, R., Telle, H., Raue, R., and Brinkwerth, W. (1980). New efficient laser dye for pulsed and CW operation in the UV. *Optics Commun.* **33,** 85–88.

Humphry-Baker, R., Graetzel, M., and Steiger, R. (1980). Drastic fluorescence enhancement and photochemical stabilization of cyanine dyes through micellar systems. *J. Am. Chem. Soc.* **102,** 847–848.

Ippen, E.P., Shank, C.V., and Dienes, A. (1971). Rapid photobleaching of organic laser dyes in continuously operated devices. *IEEE J. Quant. Elect.* **QE-7,** 178–179.

Jones II, G., and Bergmark, W.R. (1984). Photodegradation of coumarin laser dyes. An unexpected singlet self-quenching mechanism. *J. Photochem.* **26,** 179–184.

Jones II, G., Jackson, W.R., Kanoktanaporn, S., and Halpern, A.M. (1980a). Solvent effects on photophysical parameters for coumarin laser dyes. *Optics Commun.* **33,** 315–320.

Jones II, G., Jackson, W.R., and Halpern, A.M. (1980b). Medium effects on fluorescence quantum yields and lifetimes for coumarin laser dyes. *Chem. Phys. Lett.* **72,** 391–395.

Jones II, G., Griffin, S.F., Choi, C., and Bergmark, W.R. (1984a). Electron donor-acceptor quenching and photoinduced electron transfer for coumarin dyes. *J. Org. Chem.* **49,** 2705–2708.

Jones II, G., Bergmark, W.R., and Jackson, W.R. (1984b). Products of photodegradation for coumarin laser dyes. *Optics Commun.* **50,** 320–324.

Jones II, G., Jackson, W.R., Choi, C., and Bergmark, W.R. (1985a). Solvent effects on emission yield and lifetime for coumarin laser dyes. Requirements for a rotatory decay mechanism. *J. Phys. Chem.* **89,** 294–300.

Jones II, G., Jackson, W.R., Kanoktanaporn, S., and Bergmark, W.R. (1985b). Photophysical and photochemical properties of coumarin dyes in amphiphilic media. *Photochem. and Photobiol.* **42,** 477–483.

Kalyanasundaram, K., and Thomas, J.K. (1977). Solvent-dependent fluorescence of pyrene-3-carboxaldehyde and its applications in the estimation of polarity at micelle-water interfaces. *J. Phys. Chem.* **81,** 2176–2180.

Kaminow, I.P., Stulz, L.W., Chandross, E.A., and Pryde, C.A. (1972). Photobleaching of organic laser dyes in solid matrices. *Appl. Optics* **11,** 1563–1567.

Kamlet, M.J., Abboud, J.L., and Taft, R.W. (1977). The solvatochromic comparison method. 6. The π^* scale of solvent polarities. *J. Am. Chem. Soc.* **99,** 6027–6038.

Kamlet, M.J., Abboud, J.M., Abraham, M.H., and Taft, R.W. (1983). Linear solvation energy relationships. 23. A comprehensive collection of the solvatochromic parameters, π^*, α, β, and some methods for simplifying the generalized solvatochromic equation. *J. Org. Chem.* **48,** 2877–2887.

Kasatani, K., Kawasaki, M., Sato, H., and Nakashima, N. (1985). Micellized sites of dyes in sodium dodecyl sulfate micelles as revealed by time-resolved energy transfer studies. *J. Phys. Chem.* **89,** 542–545.

Kasha, M. (1986). Proton-transfer spectroscopy. Perturbation of tautomerization potential. *J. Chem. Soc. Faraday Trans 2* **82,** 2379–2392.

Kauffman, J.M., and Bentley, J.H. (1988). Effect of various anions and zwitterions on the lasing properties of a photostable cationic laser dye. *Laser Chem.* **8,** 49–59.

Kavarnos, G.J., and Turro, N.J. (1986). Photosensitization by reversible electron transfer: Theories, experimental evidence, and examples. *Chem. Rev.* **86,** 401–449.

Ketsle, G.A., Levshin, L.V., Slavnova, T.D., and Chibisov, A.K. (1972). Triplet state of rhodamine 6G molecules. *Soviet Phys.—Doklady* **16,** 986–987.

Knipe, R.H., Fletcher, A.N. (1983). Effects of secondary chemical reactions upon the performance of dye lasers. *J. Photochem.* **23,** 117–130.

Kubin, R.F., and Fletcher, A.N. (1983). The effect of oxygen on the fluorescence quantum yields of some coumarin dyes in ethanol. *Chem. Phys. Lett.* **99,** 49–52.

Koehler, G., Getoff, N., Rotkiewica, K., and Grabowski, Z.R. (1985). Electron photoejection from donor-aryl-acceptor molecules in aqueous solution. *J. Photochem.* **28,** 537–546.

Kopainsky, B., Kaiser, W., and Schaefer, F.P. (1978). Ultrafast energy transfer within bifluorophoric molecules. *Chem. Phys. Lett.* **56,** 458–462.

Kopainsky, B., Qiu, P., Kaiser, W., Sens, B., and Drexage, K. (1982). Lifetime, photostability, and chemical structure of IR heptamethine cyanine dyes absorbing beyond 1 μm. *Appl. Phys. B* **29,** 15–18.

Korobov, V.E., and Chibisov, A.K. (1978). Primary processes in the photochemistry of rhodamine dyes. *J. Photochem.* **9,** 411–424.

Korobov, V.E., Shubin, V.V., and Chibisov, A.K. (1977). Triplet state of rhodamine dyes and its role in production of intermediates. *Chem. Phys. Lett.* **45,** 498–501.

Kosower, E.M., Giniger, R., Radkowsky, A., Hebel, D., and Shusterman, A. (1986). Bimanes. 22. Flexible fluorescent molecules. Solvent effects on the photophysical properties of syn-bimanes (1,5-diazabicyclo[3.3.0]octa-3,6-diene-2,8-diones). *J. Phys. Chem.* **90,** 5552–5557.

Kubin, R.F., and Fletcher, A.N. (1983). The effect of oxygen on the fluorescence quantum yields of some coumarin dyes in ethanol. *Chem Phys. Lett.* **99,** 49–52.

Kunjappu, J.T., and Rao, K.N. (1987). Separation and identification of the products of the photodegradation of 7-amino-4-methylcoumarin and the mechanism of its photodegradation. *J. Photochem.* **39,** 135–143.

Law, K.Y. (1981). Fluorescence probe for microenvironments: A new probe for micelle solvent parameters and premicellar aggregates. *Photochem. and Photobiol.* **33,** 799–806.

Lee, L.A., and Robb, R.A. (1980) Water soluble blue-green lasing dyes for flashlamp-pumped dye lasers. *IEEE J. Quant. Elect.* **QE-16,** 777–784.

Levin, M.B., Cherkasov, A.S., and Krasnov, I.V. (1986). Photostability of aqueous micellar solutions of rhodamines under flash lamp excitation. *Opt. Spektrosk.* **60,** 732–737.

Liphardt, B., Liphardt, B., and Luettke, W. (1981). Laser dyes with intramolecular triplet quenching. *Optics Commun.* **38,** 207–210.

Liphardt, B., Liphardt, B., and Luettke, W. (1983). Laser dyes III: Concepts to increase the photostability of laser dyes. *Optics Commun.* **48,** 129–133.

Lippert, E., Rettig, W., Bonacic-Koutecky, V., Heisel, F., and Miehe, J.A. (1987). Photophysics of internal twisting. *Adv. Chem. Phys.* **68,** 1–173.

Lougnot, D.-J., and Goldschmidt, C.R. (1980). Photoionization of fluorescein via excited triplet and singlet states. *J. Photochem.* **12,** 215–224.

Love, L.J.C., Skrilec, M., and Habarta, J.G. (1980). Analysis by micelle-stabilized room temperature phosphorescence in solution. *Anal. Chem.* **52,** 754–759.

Maeda, M. (1984). *Laser Dyes. Properties of Organic Compounds for Dye Lasers.* Academic Press, New York.

Maeda, M., and Kojima, M. (1977). Photorearrangement of phenyloxazoles. *J.C.S. Perkin* **1,** 239–247.

Magde, D., Gaffney, S.T., and Campbell, B.F. (1981). Excited state singlet absorption in blue laser dyes: Measurement by picosecond flash photolysis. *IEEE J. Quant. Elect.* **QE-17,** 489–495.

Marling, J.B., Gregg, D.W., and Wood, L. (1970). Chemical quenching of the triplet state in flashlamp-excited liquid organic lasers. *Appl. Phys. Lett.* **17,** 527–530.

Marszalek, T., Baczynski, A., Orzeszko, W., and Rozploch, A. (1980). Energy transfer in dye solutions of micellar structure. *Z. Naturforsch.* **35a,** 85–91.

Menger, F.M. (1979). On the structure of micelles. *Acc. Chem. Res.* **12,** 111–117.

Mostovnikov, V.A., Rubinov, A.N., Ginevich, G.R., Anufrik, S.S., and Abramov, A.F. (1976). Recovery of lasing properties of dye solutions after their photolysis. *Sov. J. Quantum Electron.* **6,** 1126–1128.

Mugnier, J., Pouget, J., Bourson, J., and Valeur, B. (1985). Efficiency of intramolecular electronic energy transfer in coumarin bichromophoric molecules. *J. Lumin.* **33,** 273–300.

Mukerjee, P. (1980). Solubilization in micellar systems. *Pure and Appl. Chem.* **52,** 1317–1321.

Mukherjee, T., Rao, K.N., and Mittal, J.P. (1986). Photodecomposition of coumarin laser dyes in different solvent systems. *Indian J. Chem.* **25A,** 993–1000.

Mukherjee, T., Rao, K.N., and Mittal, J.P. (1987). Photodecomposition of stilbene-3 laser dye in different solvent systems. *Indian J. Chem.* **26A,** 12–16.

Narayana, P.A., Li, A.S.W., and Kevan, L. (1982). Electron spin resonance and electron spin-echo studies of the photoionization of tetramethylbenzidine in frozen aqueous anionic, cationic, and nonionic micellar solutions. Effect of micelle type and anionic micelle size. *J. Am. Chem. Soc.* **104,** 6502–6505.

Ojeda, P.R., Amashta, I.A.K., Ochoa, J.R., and Arbeloa, I.L. (1988). Excitonic treatment and bonding of aggregates of rhodamine 6G in ethanol. *J.C.S. Faraday Trans. 2* **84,** 1–8.

Olmsted III, J. (1979). Calorimetric determinations of absolute fluorescence quantum yields. *J. Phys. Chem.* **83,** 2581–2584.

Pappalardo, R., Samelson, H., and Lempicki, A. (1970a). Long-pulse laser emission from rhodamine 6G using cyclooctatetraene. *Appl. Phys. Lett.* **16,** 267–269.

Pappalardo, R., Samelson, H., and Lempicki, A. (1970b). Long-pulse laser emission from rhodamine 6G. *IEEE J. Quantum Elect.* **QE-6,** 716–725.

Pappalardo, R., Samelson, H., and Lempicki, A. (1972). Calculated efficiency of dye lasers as a function of pump parameters and triplet lifetime. *J. Appl. Phys.* **43,** 3776–3787.

Pavlopoulos, T.G. (1978). Laser dye mixtures. *Optics Commun.* **24,** 170–174.

Pavlopoulos, T.G., and Hammond, P.R. (1974) Spectroscopic studies of some laser dyes. *J. Am. Chem. Soc.* **96,** 6568–6579.

Pavlopoulos, T.G., Boyer, J.H., Politzer, I.R., and Lau, C.M. (1986). Laser action from syn-(methyl,methyl)bimane. *J. Appl. Phys.* **60,** 4028–4030.

Peters, D.W., and Mathews, C.W. (1980). Temperature dependence of the peak power of a Haensch-type dye laser. *Appl. Opt.* **19,** 4131–4132.

Peterson, O.G., and Snavely, B.B. (1968). Stimulated emission from flashlamp-excited organic dyes in polymethyl methyacrylate. *Appl. Phys. Lett.* **12,** 238–240.

Piechowski, A.P., and Bird, G.R. (1984). A new family of lasing dyes from an old family of fluors. *Optics Commun.* **50,** 386–392.

Ramamurthy, V., and Eaton, D.F. (1988). Photochemistry and photophysics within cyclodextrin cavities. *Acc. Chem. Res.* **21,** 300–306.

Rehm, D., and Weller, A. (1970). Kinetics of fluorescence quenching by electron and H-atom transfer. *Isr. J. Chem.* **8,** 259.

Reich, S., and Neumann, G. (1974). Photobleaching of rhodamine 6G in polyacrylonitrile matrix. *Appl. Phys. Lett.* **25,** 119–121.

Rendall, W.A., Torres, M., Lown, E.M., and Strausz, O.P. (1986). Photoisomerization of furan, pyrrole, and thiophene: The intermediate formation of their dewar and ring contracted Forms. *Reviews Chem. Intermed.* **6,** 335–364.

Rettig, W., and Klock, A. (1985). Intramolecular fluorescence quenching in aminocoumarins. Identification of an excited state with full charge separation. *Can. J. Chem.* **63,** 1649-1653.

Reynolds, G.A., and Drexhage, K.H. (1975). New coumarin dyes with rigidized structures for flashlamp-pumped dye lasers. *Optics Commun.* **13,** 222–225.

Rinke, M., Guesten, H., and Ache, H.J. (1986a). Photophysical properties and laser performance of photostable UV laser dyes. 1. Substituted p-quarterphenyls. *J. Phys. Chem.* **90,** 2661–2665.

Rinke, M., Guesten, H., and Ache, H.J. (1986b). Photophysical properties of photostable UV laser dyes. 2. Ring-bridged p-quarterphenyls. *J. Phys. Chem.* **90,** 2666–2670.

Rosenthal, I. (1978). Photochemical stability of rhodamine 6G in solution. *Optics Commun.* **24,** 164–166.

Ruiz, T.P., Lozano, M., and Tomas, V. (1986). Study of photo-oxidation of the rhodamine dyes. Kinetic determination of iron (III). *Quim. Anal.* **5,** 180.

Sato, H., Kawasaki, M., and Kasatani, K. (1983). Energy transfer between Rhodamine 6G and pinacyanol enhanced with sodium dodecyl sulfate in the premicellar region. Formation of dye-rich induced micelles. *J. Phys. Chem.* **87,** 3759–3769.

Sauers, R.R., Husain, S.N., Piechowski, A.P., and Bird, G.R. (1987). Shaping the absorption and fluorescence bands of a class of efficient, photoactive chromophores: Synthesis and properties of some new 3H-xanthen-3-ones. *Dyes and Pigments* **8,** 35–53.

Schäfer, F.P. (1983). New developments in laser dyes. *Laser Chem.* **3,** 265–278.

Schäfer, F.P., Bor, Zs., Luettke, W., and Liphardt, B. (1978). Bifluorophoric laser dyes with intramolecular energy transfer. *Chem. Phys. Lett.* **56,** 455–457.

Schimitschek, E.J., Trias, J.A., Taylor, M., and Celto, J.E. (1973). New improved laser dye for the blue-green spectral region. *IEEE J. Quant. Elect.* **QE-9,** 781–782.

Schimitschek, E.J., Trias, J.A. Hammond, P.R., Henry, R.A., and Atkins, R.L. (1976). New laser dyes with blue-green emission. *Optics Commun.* **16,** 313–316.

Scypinski, S., and Drake, J.M. (1985). Photophysics of coumarin inclusion complexes with cyclodextrin. Evidence for normal and inverted complex formation. *J. Phys. Chem.* **89,** 2432–2435.

Sinclair, R.S. (1980). The light stability and photodegradation of dyes. *Photochem. Photobiol.* **31,** 627–629.

Smirl, A.L., Clark, J.B., Van Stryland, E.W., and Russell, B.R. (1982). Population and rotational kinetics of the rhodamine B monomer and dimer: Picosecond transient spectrometry. *J. Chem. Phys.* **77,** 631–640.

Snare, M.J., Treloar, F.E., Chiggino, KP., and Thistlethwaite, P.J. (1982). The photophysics of rhodamine B. *J. Photochem.* **18,** 335–346.

Snavely, B.B., and Schaefer, F.P. (1969). Feasibility of CW operation of dye lasers. *Phys. Lett.* **28A,** 728–729.

Specht, D.P., Martic, P.A., and Farid, S. (1982). Ketocoumarins. A new class of triplet sensitizers. *Tetrahedron* **38,** 1203.

Speiser, S., and Shakkour, N. (1985). Photoquenching parameters for commonly used laser dyes. *Appl. Phys. B* **38,** 191–197.

Taft, R.W., and Kamlet, M.J. (1976). The solvatochromic comparison method. 2. The α-scale of solvent hydrogen-bond donor (HBD) acidities. *J. Am. Chem. Soc.* **98,** 2886–2894.

Targowski, P., Zietek, B., and Baczynski, A. (1987). Luminescence quenching of rhodamines by cyclooctatetraene. *Z. Naturforsch.* **42a,** 1009–1013.

Telle, H., Hueffer, W., and Basting, D. (1981). The XeCl excimer laser: A powerful and efficient UV pumping source for tunable dye lasers. *Optics Commun.* **38,** 402–406.

Tuccio, S.A., Drexhage, K.H., and Reynolds, G.A. (1973). CW laser emission from coumarin dyes in the blue and green. *Optics Commun.* **7,** 248–252.

Turro, N.J. (1978) *Modern Molecular Photochemistry*. Benjamin/Cummings, Reading, Mass., chapter 5.

Turro, N.J., Liu, K., Chow, M., and Lee, P. (1978). Convenient and simple methods for the observation of phosphorescence in fluid solutions. Internal and external heavy atom and micellar effects. *Photochem. and Photobiol.* **27,** 523–529.

Turro, N.J., Graetzel, M., and Braun, A.M. (1980). Photophysical and photochemical processes in micellar systems. *Angew. Chem. Int. Ed. Engl.* **92,** 712–734.

Vogel, M., Rettig, W., Sens, R., and Drexhage, K.H. (1988). Structural relaxation of rhodamine dyes with different N-substitution pattern. *Chem. Phys. Lett.* **147,** 452–460.

Von Trebra, R., and Koch, T.H. (1982). DABCO stabilization of coumarin dye lasers. *Chem. Phys. Lett.* **93,** 315–317.

Von Trebra, R., and Koch, T.H. (1983). Chemical stabilization of the coumarin 1 dye laser. *Appl. Phys. Lett.* **42,** 129–131.

Von Trebra, R., and Koch, T.H. (1986). Photochemistry of coumarin laser dyes: The role of singlet oxygen in the photo-oxidation of coumarin 311. *J. Photochem.* **35,** 33–46.

Von Trebra, R., and Koch, T.H. (1987). Two-photon laser photochemistry of a coumarin laser dye. *J. Photochem. Photobiol, A: Chem.* **41,** 111–120.

Wasserman, H.H., and Floyd, M.B. (1966). The oxidation of heterocyclic systems by molecular oxygen, IV. The photosensitized autooxidation of oxazoles. *Tetrahedron, Suppl. No. 7,* 441–448.

Webb, J.P., McColgin, W.C., Peterson, O.G., Stockman, D.L., and Eberly, J.H. (1970). Intersystem crossing rate and triplet state lifetime for a lasing dye. *J. Chem. Phys.* **53,** 4227–4229.

Weber, J. (1971). Photobleaching of rhodamine 6G dye laser by flashlamp exciting. *Phys. Lett.* **37A,** 179–180.

Weber, J. (1973). Study of the influence of triplet quencher on the photobleaching of rhodamine-6G. *Optics Commun.* **7,** 420–422.

Winters, B.H., Mandelberg, H.I., and Mohr, W.B. (1974). Photochemical products in coumarin laser dyes. *Appl. Phys. Lett.* **25,** 723–725.

Yamashita, M., and Kashiwagi, H. (1976). Photodegradation mechanisms in laser dyes: A laser irradiated ESR study. *IEEE J. Quant. Elect.* **QE-12,** 90–95.

Zimmerman, H.E. (1982). Some theoretical aspects of organic photochemistry. *Acc. Chem. Res.* **15,** 312–317.

Chapter 8

INDUSTRIAL APPLICATIONS OF DYE LASERS

David Klick

Massachusetts Institute of Technology,
Lincoln Laboratory
Lexington, Massachusetts

1. INTRODUCTION

Because dye lasers usually involve a circulating liquid laser medium that degrades over time, there has been reluctance to use them in industrial applications, where reliability is paramount. However, in some applications only a dye laser can provide the desired light tuned to the wavelength required by a process. Other applications make use of the wide pulse-width range available with dye lasers, from millisecond (flashlamp-pumped dye

ISBN 0-12-222700-X

lasers) to femtosecond (mode-locked dye lasers). Initially, the techniques used to adapt dye lasers for industrial applications will be discussed. High power is usually the key laser characteristic required by industry, except in diagnostic applications. The largest, most powerful dye lasers have been built for industrial processes, such as laser isotope selection (LIS). (See, for example, Paisner (1988). LIS is beyond the scope of this chapter.) Early high-power dye lasers were flashlamp-pumped, but problems with such lasers have led to increased use of dye lasers pumped by other lasers. One of these problems is photochemical degradation of the dye. In an industrial setting, the drop in power associated with dye degradation makes continuous processing difficult and can lead to high costs from dye replacement. Because dye lasers may contain a circulating, toxic, flammable liquid, safety considerations are vital in industrial applications. In a scientific laboratory, the dye laser is usually surrounded by trained people familiar with its safety problems, but the opposite may be true in a factory. Methods for making dye lasers safer and more reliable will be discussed, examples being by improving the dye flow cell, installing dye solution clean-up systems, changing solvents, and eliminating liquid flow altogether with a solid dye medium.

Following the section on technology, there are sections that address dye-laser applications from process to product, i.e. from the use of dye lasers to power manufacturing processes, through their use in industrial diagnostics, to their role in monitoring unintended factory products (pollution) at the smokestack and in the environment. Two case studies are presented in depth. The first is a description of a dye-laser–based chemical manufacturing process (UV curing of pigmented polymer coatings) that proceeds from initial laboratory measurements, through development of a dye laser, to design of a pilot plant, and finally to real-world considerations of costs, markets, and competing processes. This illustrates the sequence one must follow to succeed with an industrial application of a laser. Science is only the beginning! The second case study is of probably the most common use of dye lasers for industrial measurements, combustion diagnostics. Examples are provided of techniques with acronyms such as LIF, TRISP, and CARS (not to mention the whimsically named BOXCARS and USED CARS)!

2. TECHNOLOGY FOR INDUSTRIAL DYE LASERS

To achieve high power from a dye laser, one is tempted to pump the medium directly with a flashlamp. The main problem is the poor conversion efficiency of pump light to output light (up to 3% for a flashlamp but up to 50% for a laser pump). The excess light absorbed from the flashlamp pump causes thermal instabilities and dye degradation. Furumoto (1986) reviews the attributes of flashlamp-pumped dye lasers. Depending on the beam characteristics desired, one can employ stable or unstable resonators, injection-locked oscillators, or master-oscillator–power-amplifier configurations. Electrical to light output efficiency is as high as 1.6%. Pulse energies of 100 J and average powers of hundreds of watts are at the state of the art for flashlamp-pumped dye lasers. Average powers of a kilowatt have been achieved in a burst mode (Morton and Dragoo, 1981) with a flashlamp-pumped dye laser operated at a high repetition rate. The typical pulse width of 1 μs can be stretched to as long as 1 ms for applications where peak power must be limited. Janes *et al.* (1987) achieved long pulse widths (100 μs) at high peak power (50 kW) with good beam quality by using a two-stage dye laser. The pump was a flashlamp-pumped long-pulse dye laser at 100 kW whose beam quality deteriorated with time due to thermal inhomogeneities. This pump beam was scanned through the dye medium of the second laser at a rate faster than the propagation of sound, so that thermal distortions were left behind. Such a configuration also avoids triplet-state accumulation and window damage.

Changing the temperature, viscosity, or heat capacity of the dye solution can have a marked effect on the output power of a flashlamp-pumped dye laser (Fletcher *et al.*, 1985). Beam propagation through the lasing medium is sensitive to refractive index fluctuations caused by uneven thermal input from the flashlamps. A zigzag geometry has been proposed to avoid this problem (Dearth *et al.*, 1987). The intense blackbody radiation to which a flashlamp-pumped dye laser is subjected also causes dye degradation and generation of absorbing photoproducts. These reactions are affected by dye, solvent, and oxygen concentration. Everett and Zollars (1987) solve this problem without a huge reservoir of dye solution by using automatic control loops that remove photoproducts (and dye) from the solution with a carbon filter according to a measurement of photoproduct absorption, and add more dye according to a measurement of dye absorption. Dye-laser stability has been treated by Antonov and Hohla (1983) and in a series of papers by Fletcher (bibliography given in Fletcher *et al.* (1985)).

Dye lasers pumped by other lasers can provide high peak and average power with fewer thermal and dye-degradation problems than flashlamp-pumped dye lasers. Nitrogen-laser-pumped dye lasers (Dunning and Stebbings, 1974) can be made miniature and inexpensive to be included in, for example, analytical equipment or as a microscope light source (Thomson, 1988). However, they are incapable of high power for use in manufacturing. More powerful pump lasers are harmonics of the Nd:YAG laser (Kortz, 1982; Bernhardt *et al.*, 1982), excimer lasers (Wheeler, 1982;

Hohla, 1982a,b), and copper-vapor lasers (Hargrove and Kan, 1977; Paisner, 1988). For near-IR output, a dye laser pumped with a ruby laser (Seibert and Johnson, 1981) or the fundamental of a Nd:YAG laser (Johnson and Johnson, 1986) is preferred. Limitations on dye-laser output due to high pump-power densities have been noted for Nd:YAG lasers (Moore and Decker, 1978) and excimer lasers (Tomin *et al.*, 1978). However, Rottler *et al.* (1987) were able to scale up an excimer-pumped dye laser while maintaining 20% efficiency (output power/pump). They produced 800-J pulses from coumarin 480 dye pumped by a 4 kJ long-pulse (650-ns) XeCl laser. A lively discussion of the relative merits of excimer and Nd:YAG-pumped dye lasers has appeared in the literature (Hohla, 1982a,b; Kortz, 1982; Bernhardt *et al.*, 1982; Bernhardt, 1983; Asher, 1984). In some regions of the spectrum, the cost of an industrial system (i.e., one that runs constantly) is dominated by dye-replacement costs rather than capital costs.

Circulating systems for dye lasers are critical to their optimum performance. Flow rate through the dye cell must be sufficient to replace the solution between pulses. Commercial systems are now available to match excimer pump laser pulse-repetition rates of 1 kHz (Lambda Physik, 1988d). Still faster dye flows are required for multikilohertz copper-vapor pump lasers (Lavi *et al.*, 1985; Paisner, 1988). For coverage of a wide range of wavelengths, dye lasers with multiple circulating solutions have been developed. Bos (1981b) covered the wavelength range from 335 to 910 nm by means of an excimer-pumped dye laser with a rotating turret containing eight independent dye cells and circulating loops. The design of the dye cell can strongly affect the beam quality of a dye laser. Bethune (1981) and Brink and van der Hoeven (1984) use cylindrical dye volumes pumped equally from all sides to avoid cell-wall diffraction effects and achieve near-TEM_{00} beam quality, as well as high output power without cell damage.

A dye solution can present hazards as it circulates through a flow system. Many dyes are toxic and sometimes carcinogenic, although the dye concentration is small. Dyes are commonly dissolved in organic solvents, because suitable water-soluble dyes are often unavailable. The solvents may also be toxic and are usually flammable. Sometimes experimenting with mixtures of solvents can eliminate the flammability or toxicity problems. For example, Everett and Zollars (1987) were able to live with acetamide in water, rather than a more flammable and toxic methanol mixture. Choice of solvent, however, affects both dye-laser output and dye stability (Cassard *et al.*, 1981). Some workers (Pacheco *et al.*, 1987; Erickson, 1987) have solved the liquid–health-hazard problem by using dyes imbedded in solid polymer rods as the lasing medium. This has not been very successful due to the poor optical and thermal properties of plastics as well as the rapid degradation of the dye. Thermally induced inhomogeneities in the plastic are reduced if the solid dye medium is in the form of a rotating disk (Bondar *et al.*, 1987). Use of a plastic optical fiber doped with dye as the gain medium of a laser has even been demonstrated (Muto *et al.*, 1986).

3. MANUFACTURING APPLICATIONS OF DYE LASERS

3.1. Materials Processing

Most industrial applications of dye lasers to materials processing come from the microelectronics field. Some examples follow. In an IBM patent (Burns *et al.*, 1987), an excimer laser was used to dry-chemical etch a chromium-clad copper substrate in the presence of chlorine gas. This is a technique used to expose electrodes before soldering. However, when the chromium is etched away, the copper etches several orders of magnitude more rapdily than the chromium. In order to stop etching as soon as the copper is exposed, a dye laser was used to excite laser-induced fluorescence (LIF) (see Section 6) by tuning to a gas-phase CuCl absorption line. The excimer laser is stopped immediately after a pulse reaches the copper. Another group at IBM (Feder *et al.*, 1978) built a device for connecting the two conductive layers in a metal-oxide-semiconductor (MOS) sandwich. A hole is drilled and the metal is melted, all by the same laser, a nitrogen-pumped pulsed dye laser. Both pulse energy and pulsewidth are critical in this application, and the dye laser is useful because of its variable pulsewidth. Pulsewidth is adjusted by spectrally tuning on the wings of the dye-power–versus–frequency curve and changing the UV focusing into the dye cell. This process could allow integrated circuits to be "personalized" or repaired (Schuster and Cook, 1976). A method for repairing defective photomasks with a pulsed dye laser tuned to the correct wavelength for the photographic chemicals is given in Miyauchi *et al.* (1984). Another group (Contolini and Osinski, 1986) used the tunable high power of a pulsed dye laser to fabricate gratings on InP substrates for distributed feedback lasers. Tuning the dye laser affects the spacing of the holographic grating. Finally, the long 1-μs pulse from a flashlamp-pumped dye laser was employed (Tuckerman and Schmitt, 1985) for planarization of MOS films, a required step in device fabrication. The 1-μs pulse was long enough to uniformly penetrate and melt the metal layer, but not long enough to penetrate to the oxide layer. The metal solidified to a planar layer with <0.1-μm surface roughness.

3.2. Chemical Production

The use of lasers for manufacture of chemicals has been reviewed by others (e.g., Hall, 1982; Woodin and Kaldor, 1984; Wolfrum, 1988; Schäfer, 1988). An early study of dye-laser-induced photochemistry was that of Reddy *et al.* (1978). Dye lasers for photochemistry were discussed by Schäfer (1976). The ability of dye lasers to selectively populate a particular molecular state is not very useful in a typical industrial reaction for two reasons: 1) industrial reactions are usually at high pressures or in the liquid phase, and the excitation is quickly thermalized before reactions occur, and (2) if one laser photon is required for each molecular bond, the process is uneconomic. Thus for dye lasers to be used in industry, an atypical reaction must be found, and it must have high quantum yield or high value-added. In the early days of laser chemistry research, Hall (1982) speculated that the following reactions might be suitable for laser production: isotope selection (which is beyond the scope of this chapter), novel materials, purification, surface chemistry, catalyst production, and chain reactions. Researchers focused especially on photochemical reactions carried out in industry with lamps (see Bauer (1984) for a list of these). One such reaction (a polymeric chain reaction similar to those now initiated by Hg lamps in the coatings industry) is the subject of the case study in Section 4. Schäfer (1988) makes a detailed comparison of the most common lamps and lasers (including an archetypal 10-W dye laser) and finds that lasers are cost-competitive. In general, though, successful industrial applications of dye lasers for chemical production will most likely not come from processes powered by lamps, but rather from novel processes that can only be powered by dye lasers. The biotechnology industry provides fertile ground for such innovative processes. Wolfrum (1988) gives examples of the use of a UV dye laser coupled to a microscope for genetic engineering by cell fusion, genetic transformation of cells, and microdissection of chromosomes. If a desired product can be achieved in the biotechnology industry, the replicability of biological systems gives it an infinite quantum yield!

4. MANUFACTURING CASE STUDY: DYE-LASER CURING OF PIGMENTED COATINGS

This work was performed by the author and his colleagues M.A. Akerman, G.L. Paul, D. Supurovic, and H. Tsuda while affiliated with the Industrial Laser Photochemistry Laboratory at the University of New South Wales, Sydney, Australia in 1984–1985. The use of the dye laser as a source of energy for photoassisted curing of industrial polymeric coatings was investigated. UV Hg vapor lamps are often used to excite a photoinitiating molecule mixed with the starting monomers and oligomers of a coating. The resulting polymeric chain reaction multiplies the effect of the initial photons, making economical use of the light source. The high cost of laser photons may thus be justifiable if lasers provide advantages over lamps. Pigmented coatings (20-μm TiO_2 mixtures typical of appliance or automotive finishes) are not easily cured with UV lamps because it is difficult for light to penetrate the absorbing and scattering pigmented layer. However, economically viable cure rates were achieved with certain photoinitiators using a tunable excimer-pumped UV dye laser. A prototype of such a laser suitable for factory use was built and used to cure these coatings. Results are scaled to a factory situation, and costs are calculated to show the advantages of the laser method over processes being used as of 1989.

The potential for UV lasers in the photoassistance of polymerization was recognized early (e.g., Wilson, 1975). Previous work on laser curing of polymers usually stressed the high-power, single-pulse, focused-beam curing that can be accomplished with a laser (Decker, 1983, 1984a; Fouassier *et al.*, 1984). The emphasis was often on curing fine lines for microcircuit photoresists by either scanning a focused beam or imaging a mask (Jain *et al.*, 1982). In our work, the opposite regime of laser curing is investigated for the first time. We diverge the beam greatly and use many pulses to cure. The coherent beam nature of the laser is no longer utilized, except as it facilitates routing the light to the coating. We also describe to our knowledge the first laser curing of pigmented coatings (a previous work had speculated on the possibility without demonstrating it (McGinniss, 1974)). The advantage of the dye laser in this work derives from its tunability and monochromaticity.

Polymer coatings are most heavily used in the furniture and printing industries (clear coatings and inks) and for autos and appliances (pigmented paints). The standard method is to apply a solvent-based coating and dry the film. Problems with this are the time consumed, the space required for drying racks, and atmospheric pollution. A common solution to these problems is the drying oven with a burner to eliminate solvent emission. But this creates new problems: energy usage is profligate, and ovens cost ~\$5M and are ~100 m long. An alternative is electron-beam curing (~\$1M and 5 m/unit). UV curing by lamps emerged as another alternative with the following advantages: small capital costs and space requirements (~\$50K and 2 m/lamp unit), no solvent waste (all of the input materials remain in the cured coating with no emissions), nearly instantaneous cure, low energy usage, and low process temperature for sensitive substrates. In this section, we show that further advantages accrue from using UV lasers in place of lamps for radiation curing. The laser shares the same advantages as the Hg lamp (including capital costs and space requirements of the same order), but the laser can cure clear coatings at an order of magnitude higher throughput than a single lamp, and it can cure pigmented coatings where the lamp has great difficulty.

4.1. Laboratory Results

Chemical formulations consisted of a monomer to be polymerized, a reactive oligomer (a medium–molecular-weight polymer) to increase the formulation viscosity, a photoinitiator (PI) to absorb light (and to begin and sustain the chain reaction), and a tertiary amine to donate a proton to the excited PI, thus creating a free radical. Although various formulations were evaluated, the formulation used for the results in this section was a mixture of three parts oligomer (epoxy acrylate), two parts diacrylate monomer (HDDA or hexanedioldiacrylate), and equal percentages of PI and amine. Percentages were determined by weight. The amine was *N*-methyldiethanolamine from Union Carbide. Photoinitiators that were used for the results in this chapter are DETX (2,4 Diethylthioxanthone from UCB), T12 (Trigonal 12 or *p*-phenyl benzophenone from Akzo Chemie), and APO (Acylphosphine Oxide from BASF). When pigmented coatings were tested, rutile TiO_2 was added to the preceding formulation. The formulations were stable unless exposed to UV light. The formulations had the consistency of varnish or paint. They were spread into coatings of a given thickness by one of two methods: 1) For thin coatings (<40 μm) wire-wound rods were used to apply the coating to the substrate. The coating thickness was determined by the wire thickness. 2) For thick coatings (>40 μm), tape was applied to the substrate such that a region of the correct thickness was enclosed by tape. The interior region was filled with the formulation. The substrate was typically cardboard, precoated to avoid absorption into the paper. Other substrates were used, such as metal and wood. Curing was performed under a nitrogen atmosphere, as this greatly improved curing speed and coating quality. The effects of oxygen inhibition are well known (e.g., Decker and Jenkins, 1985) and we determined that it would be more economical for a coating process to include an oxygen-free curing chamber than to suffer the reduced curing rates and low photon utilization of atmospheric curing.

Lasers available to us included Lambda Physik 150 ETS and EMG 203 excimer lasers operating at 193 nm (ArF), 248 nm (KrF), 308 nm (XeCl), or 351 nm (XeF), which could pump a Lambda Physik Fl 2002 dye laser or a Raman cell to reach other wavelengths. All of these lasers had pulse-widths of the order of 30 ns. The maximum repetition rate available was

250 Hz for the EMG 203, giving it a 100-watt average power with KrF or XeCl gas fills. Laser beams were diverged and attenuated to reach the low fluences required for best curing efficiency. The excimer laser wavelengths were used to cure clear coatings (Klick *et al.*, 1988), while the excimer-pumped dye laser or Raman shifter were used to reach the 410–430-nm region best suited to curing pigmented coatings. A dye laser was built (see Section 4.2) as a prototype for the pigmented curing process. The result of interest was generally the amount of light required to achieve complete curing (defined as a commercially acceptable coating). Degree of polymerization was generally monitored with a scratch test. More sophisticated diagnostic equipment was also used, including a rapid-scan FTIR spectrometer, with which we could monitor both the dynamics and degree of polymerization (Klick *et al.*, 1989).

Coatings manufacturers minimize raw-materials costs by applying the thinnest pigmented coating that hides the substrate. The pigment of choice for maximum hiding power is rutile TiO_2, although other powders are sometimes used as extenders. In this work we considered only white coatings pigmented with rutile TiO_2. A single crystal of TiO_2 is visibly transparent with an absorption edge at about 420 nm. Powdered TiO_2 provides the scattering in the visible that gives a pigmented coating its white appearance and hides the substrate. Figure 8.1 shows the transmission spectrum of a 35-μm–thick inert (no PI) pigmented coating at two TiO_2 concentrations. Note that very little light penetrates the coating in the scattering regime ($\lambda > 420$ nm), and even less in the absorbing regime ($\lambda < 400$ nm). UV curing of a pigmented coating is accomplished by exciting a PI at a wavelength longer than the absorption edge. However, if the PI absorbs visible light at wavelengths sufficiently longer than 420 nm, the coating will appear yellow, not white. For best results, the illumination wavelength is near the absorption edge, and here the tunability of laser light provides an advantage. We used either Raman-shifted excimer light (first Stokes of H_2 shifts 351 nm to 411 nm, while second Stokes shifts 308 nm to 414 nm) or an excimer-laser-pumped dye laser (dyes POPOP, PBBO, Bis-MSB, and DPS work well in the 420-nm region). Figure 8.2 displays the drop in transmission with increased TiO_2 concentration at a single wavelength at the absorption edge. Interestingly, the coating displays near-perfect Beer's-law behavior in this wavelength regime.

The 24% TiO_2 transmission curve of Fig. 8.1 is repeated on a log scale in Figs. 8.3 and 8.4. The absorption edge is evident as transmission falls by $\times 100$ between 420 and 400 nm. Figures 8.3 and 8.4 also plot the molar absorption coefficient ϵ for two PIs: DETX in Fig. 8.3 and APO in Fig. 8.4. In both cases, the absorption rapidly drops as wavelength increases because these are nonyellowing PIs. By combining the two curves (TiO_2

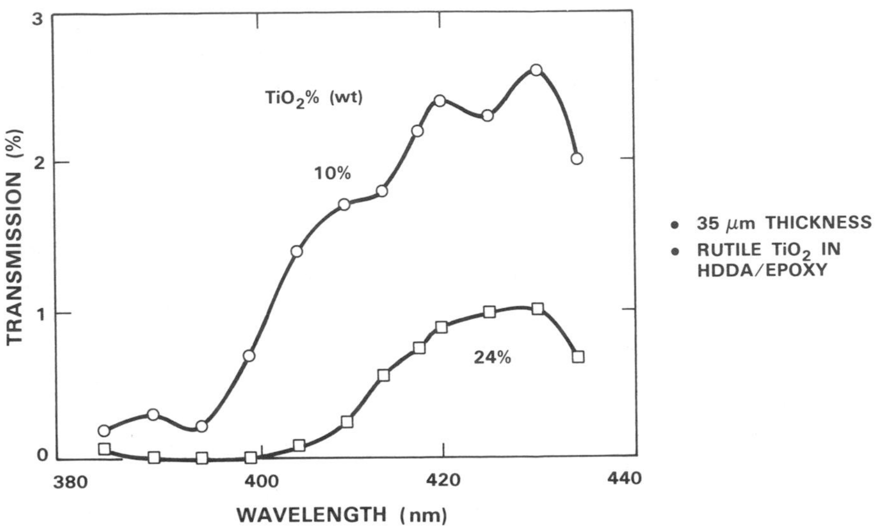

Fig. 8.1 Transmission spectrum near the absorption edge of an inert coating with 10 or 24% (by weight) of TiO_2.

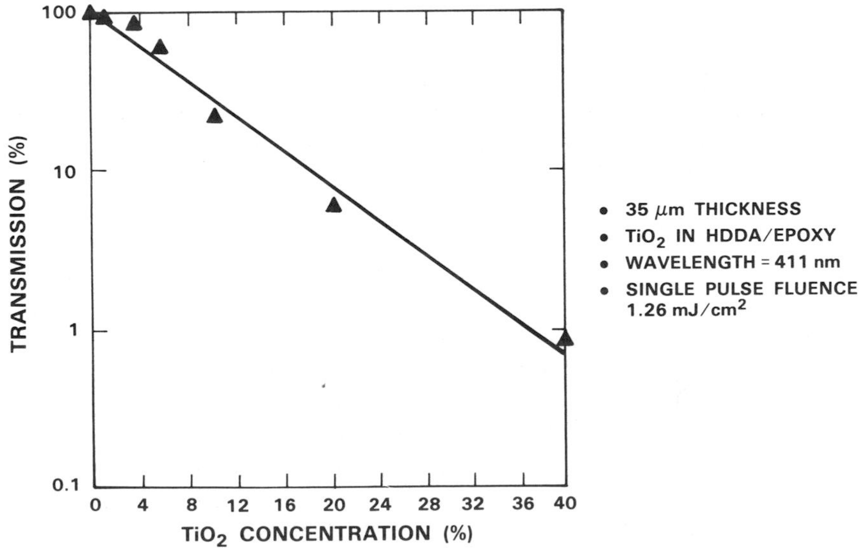

Fig. 8.2 Log transmission versus TiO_2 concentration at the absorption edge for an inert coating. The line is a best fit to Beer's law.

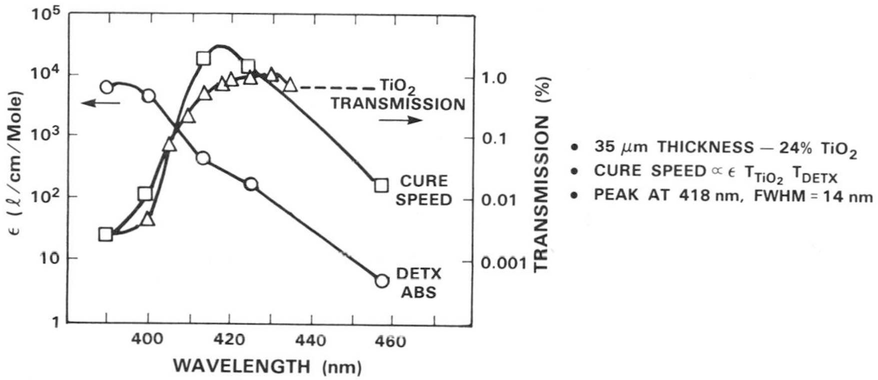

Fig. 8.3 Cure speed (in arbitrary units) for a pigmented coating containing the photoinitiator DETX is predicted to peak at $\lambda = 418$ nm with FWHM = 14 nm. Cure speed is calculated from the DETX molar absorption coefficient ϵ and the transmission of TiO_2, which are also plotted.

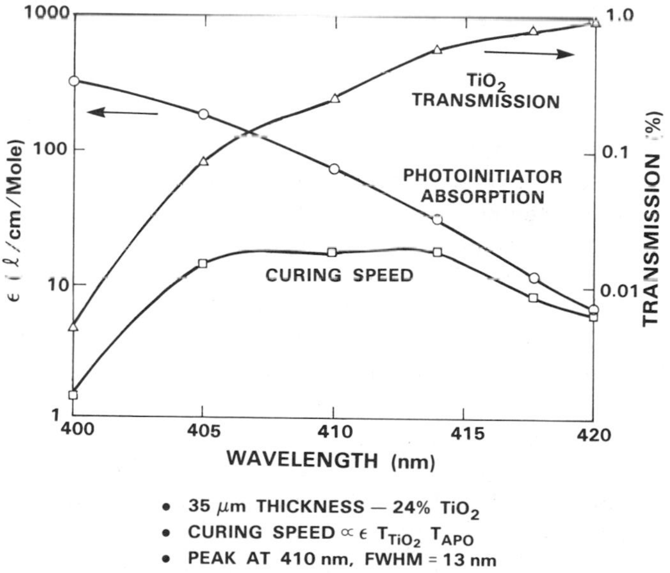

Fig. 8.4 Cure speed (in arbitrary units) for a pigmented coating containing the photoinitiator APO is predicted to peak at $\lambda = 410$ nm with FWHM = 13 nm. Cure speed is calculated from the APO molar absorption coefficient ϵ and the transmission of TiO_2, which are also plotted.

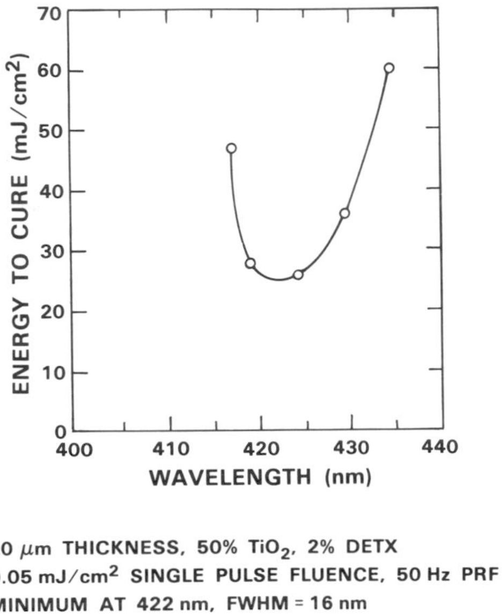

Fig. 8.5 Energy-to-cure (the inverse of cure speed) measured for a pigmented coating containing DETX. Optimum $\lambda = 422$ nm with FWHM = 16 nm (compare to Fig. 8.3).

transmission rising and Pl absorption falling with increasing wavelength), one can arrive at an overall curing efficiency curve that shows an optimum at a certain wavelength. To predict the curing efficiency, it was assumed that complete cure is limited by ability to cure the coating near the substrate, since pigmented coatings are nearly opaque. The light absorbed by the thin layer at the bottom is proportional to epsilon, and the light reaching the bottom is given by the product of transmissions of TiO_2 and the Pl. We multiplied ϵ by these two transmissions for a 35-μm coating to arrive at a quantity (marked "cure speed" in Figs. 8.3 and 8.4) presumed indicative of the ability to cure the bottom layer. By these arguments, we predict a peak curing speed for DETX at 418 nm (FWHM = 14 nm) and for APO at 410 nm (FWHM = 13 nm). Actual measurements of energy-to-cure (the inverse of curing speed) are plotted in Fig. 8.5 for DETX and Fig. 8.6 for APO. The measured peaks are at 422 nm (FWHM = 16 nm) for DETX and 415 nm (FWHM ~ 12 nm) for APO. Prediction and experiment agree, except that for both Pls the predicted peak is 4–5 nm to shorter wavelength, due probably to the heavier TiO_2 loading used in the experiments. APO (Jacobi and Henne, 1983) gave both faster curing speeds and whiter coatings than DETX, so we concentrated on APO as the Pl for pigmented coatings. For maximum curing efficiency of coatings with

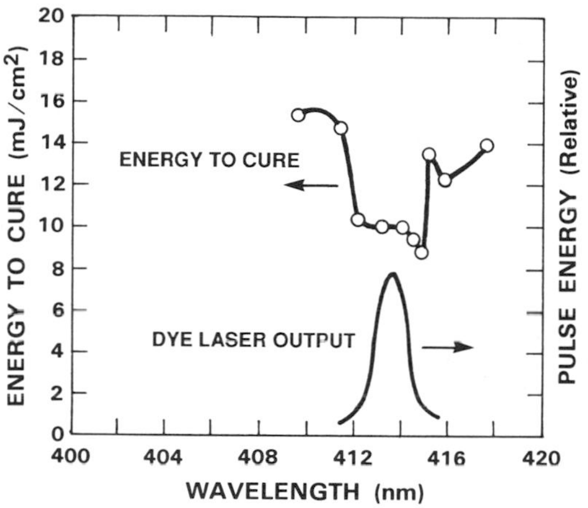

- 35 μm THICKNESS, 32% TiO_2, 3% APO
- 0.05 mJ/cm^2 SINGLE PULSE FLUENCE, 50 Hz PRF
- MINIMUM AT 415 nm, FWHM ~ 12 nm

Fig. 8.6 Energy-to-cure (the inverse of cure speed) measured for a pigmented coating containing APO. Optimum $\lambda = 415$ nm with FWHM ~ 12 nm (compare to Fig. 8.4). Also, the output spectrum of a typical broadband dye laser pulse.

APO, light was required at 415 nm. Figure 8.6 shows that the broadband dye laser described in Section 4.2 fulfilled this requirement.

By the method of tuning the illumination near the absorption edge, even very thick or heavily loaded pigmented coatings can be cured. Figure 8.7 shows that a 35-μm coating with 50% by weight TiO_2 loading (20% pigment volume concentration or PVC) and a 60-μm coating with 32% by weight loading can be cured with moderate quantities of UV light. An interesting aspect of the data in Fig. 8.7 is that more energy per unit thickness is required to cure the thinnest coating than the thick ones. The counterintuitive implication is that the surface cure is actually slower than the bulk cure for these pigmented coatings. A possible explanation is that the pigment serves to trap the light that reaches the bulk until it is absorbed by a PI molecule. Some corroboration of this idea is provided by the points at 7-μm thickness in Fig. 8.7, where increased pigment loading actually speeds curing. To test the idea that the surface limits the cure speed in pigmented coatings, we illuminated with 308-nm excimer laser light (which is absorbed by TiO_2 near the surface) as well as the 414.5 nm light that can penetrate the bulk. We also added 3% T12 (a PI that absorbs at 308 nm) to the 3% APO pigmented formulation. Energy-to-cure improved twofold. In another test, this same dual-PI formulation with and without TiO_2 was

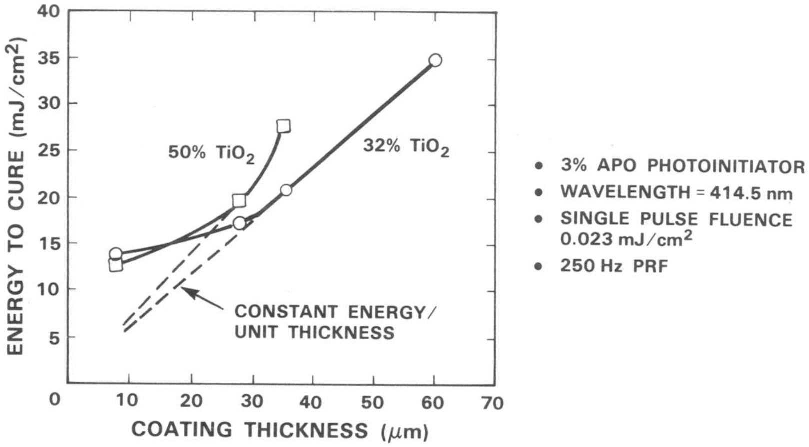

Fig. 8.7 Energy-to-cure versus coating thickness for pigmented coatings with APO. Note that thinner coatings require more energy-to-cure per unit thickness, implying that the surface limits the cure speed.

illuminated by the same dual-wavelength light at the same low fluence. The pigmented coating required only 50% more energy-to-cure. While the addition of a scattering pigment causes greatly reduced transmission through the coating, it appears that the pigment may serve to trap the radiation until it is usefully absorbed. Dual-wavelength, dual-PI curing was used for our subsequent pigmented coating results and in the pilot plant design.

Figure 8.8 shows that energy-to-cure falls with single pulse fluence. The quantities plotted are for light at 414.5 nm. The amount of light at 308 nm varied, but was always less than half of that at 414.5 nm. The improvement in energy-to-cure as the fluence drops is predicted by the theory of bimolecular radical recombination (Phillips, 1978; Decker, 1984b). In the regime where this theory holds, chain reactions are terminated by the meeting and mutual annihilation of two radicals. Since recombination rises as the square of the radical concentration, one would expect polymerization rate to rise only as its square root and, therefore, polymerization efficiency to fall as the square root of the radical concentration. In our case, radical concentration is proportional to the single pulse fluence, and the energy-to-cure (the inverse of polymerization efficiency) is expected to rise as the square root of the single pulse fluence. The measured slope of Fig. 8.8 is 0.58, close to the expected 0.5.

Consideration of Fig. 8.8 leads to the conclusion that for maximum utilization of expensive laser photons, one should work in the regime of high polymerization efficiency. This implies a low density of radicals, each

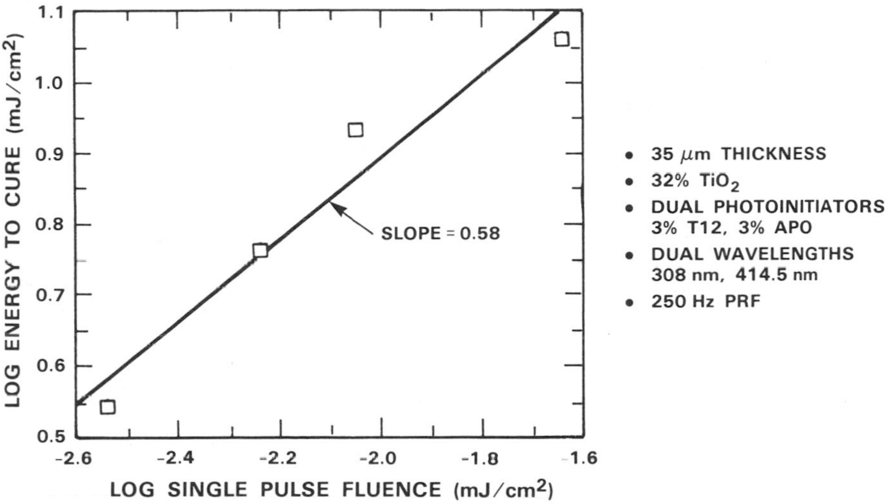

Fig. 8.8 Log energy-to-cure versus log single pulse fluence for a dual-photoinitiator pigmented coating under dual-wavelength illumination. The best fit line to the data points has a slope close to the predicted 0.5.

initiating a chain that becomes very long before recombination or another process terminates it. Figure 8.9 illustrates this principle of high photon economy as it applies to free-radical polymerization. A photoinitiator absorbs a UV photon, becoming a radical. It continues to join monomer molecules until termination. The initial photon can thus cause many reactions, a situation called high quantum yield or high photon economy. Note that the radical acts as a catalyst and the chain reaction would continue until depletion of the reactants in the absence of separate terminating reactions. If we assume that each photon creates a product molecule of equal size, we can estimate the final molecular weight (g/mole) with the following formula:

$$MW_f = \rho t \mathrm{N}/\Phi, \tag{8.1}$$

where ρ is the mass density (~2g/cm^3 for pigmented coatings), t is the coating thickness, N is Avogadro's number, and Φ is the photon fluence to cure. For Fig. 8.8, the point of highest polymerization efficiency gives a final molecular weight of about 10^6. Since the initial molecular weights are about 300 for the monomer and 2000 for the oligomer, the photon economy (number of reactions per photon) is on the order of 1000. Illumination must be very weak (.003 mJ/cm^2) to reach this regime. One must diverge a dye laser beam (typically of initial fluence one J/cm^2) by about three orders of magnitude in a linear dimension (i.e., several mm goes to

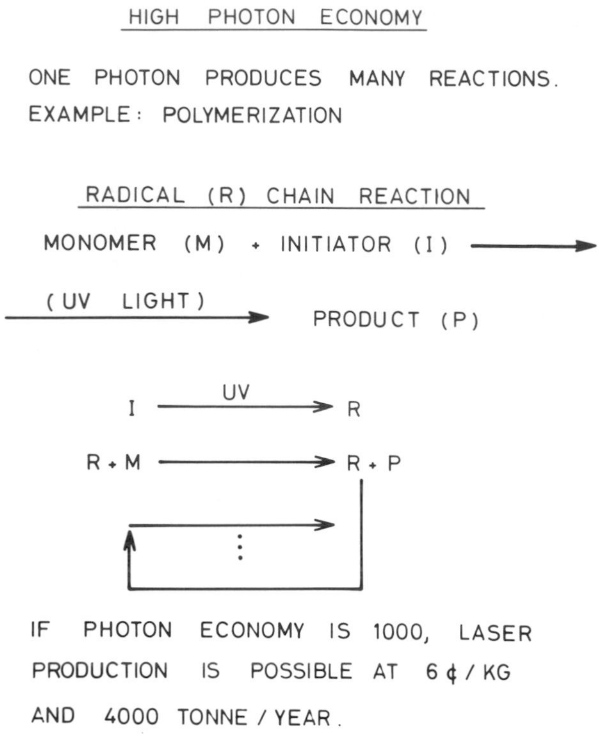
HIGH PHOTON ECONOMY
ONE PHOTON PRODUCES MANY REACTIONS.
EXAMPLE: POLYMERIZATION
RADICAL (R) CHAIN REACTION
MONOMER (M) + INITIATOR (I)
(UV LIGHT)
PRODUCT (P)
I
UV
R
R + M
R + P
IF PHOTON ECONOMY IS 1000, LASER
PRODUCTION IS POSSIBLE AT 6 ¢ / KG
AND 4000 TONNE / YEAR.

Fig. 8.9 Free-radical polymerization as a process with high photon economy. Production figures are derived later (see Fig. 8.26).

several meters) to reach the high-efficiency, low-fluence regime. One is limited in practice by the largest illumination area that is tolerable in the process equipment.

The laser pulse repetition frequency (prf) for most of the preceding graphs was 250 Hz, the highest available for our laser, producing the highest average power. Fig. 8.10 is a plot of energy-to-cure versus prf for formulations with APO at both high and low fluence. As Fig. 8.10 shows, there is a gradual drop in polymerization efficiency with increased prf. This repetition rate effect dovetails well with the results (Hoyle *et al.*, 1987) of gel permeation chromatography (GPC) measurements for polymer mixtures cured at different laser-repetition rates. Those measurements show a decrease in product molecule size (as well as the reduced polymer production per unit energy input that we see) as prf increases. Other GPC results (Olaj *et al.*, 1987) find a peak in product molecular weight at the size expected if many chains initiated by one laser pulse are terminated at the

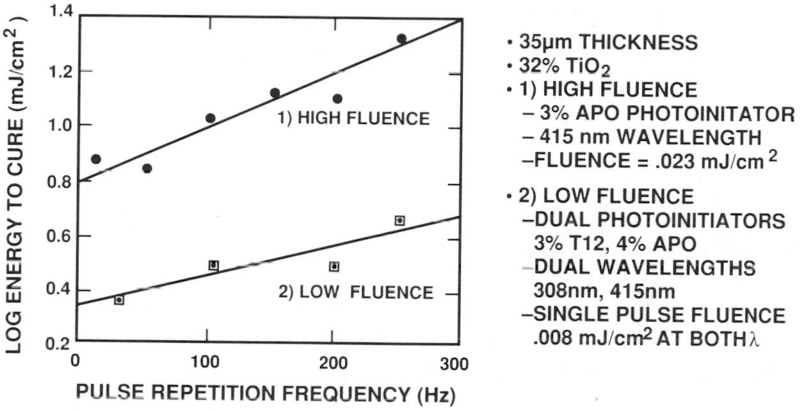

Fig. 8.10 Log energy-to-cure versus pulse-repetition frequency for pigmented coatings.

next pulse. Clearly, the time delay between laser pulses is a determinant of the final chain size, at higher prf, polymerization efficiency is lowered because the chain reaction initiated by a laser photon has less time to proceed before termination.

The implication of the prf effect for the industrial process, however, may not be significant. We employ excimer lasers because they are by far the highest–average-power UV laser source. This average power increased rapidly over the 1980s, not from large increases in instantaneous power, but from increased prf. Excimer-laser manufacturers upgraded their charging circuitry and flow systems to support prf's up to one kHz without reduced pulse energy. We must use high prf's to achieve high throughput in the industrial process, and high throughput pays for the capital investment in lasers. To avoid the decreased polymerization efficiency at high prf, one would have to use a laser with much higher pulse energy at a low prf. If employed in our process, the beam of such a laser would have to be expanded to a prohibitively large area or suffer decreased polymerization efficiency due to high fluence. Thus, we opt for using an excimer laser of the highest prf available.

Our conclusions from laboratory curing of pigmented coatings were:

(1) Illumination is required at an optimum wavelength (415 nm for one PI, 422 nm for another).
(2) This light must have a relatively narrow bandwidth (<10 nm).
(3) Theory predicted the optimum wavelength and bandwidth with good accuracy.
(4) Using light at the optimum wavelengths, curing a pigmented coating is nearly as efficient as curing a clear coating.

(5) Dual-wavelength illumination of a dual-PI coating enhances curing speed by providing surface and bulk cure.
(6) Lower illuminating fluence leads to higher polymerization efficiency at a rate predicted by theory.

It is shown in the next section how the dye laser provides a light source that satisfies these process requirements.

4.2. Dye-Laser Design

For curing pigmented coatings then, light is required at either 415 or 422 nm (depending on the PI), with little margin for error in wavelength. Higher-energy photons are insufficiently transmitted, and lower-energy photons insufficiently absorbed. We require no special properties of the light source, such as coherence or low divergence. High power (for maximum process output) and reliability (for factory use) are important. As of 1989, mercury lamps are the most widely used light source for industrial photochemistry. Most photochemical reactions will proceed under illumination by one or more of the mercury lines distributed through the UV spectrum. In our case, however, the two closest mercury lines are 405 and 436 nm, outside the desired wavelength range. The reactions proceed very slowly under illumination by typical industrial mercury lamps. The preponderance of deep-UV radiation tends to overcure the surface, leaving it wrinkled and yellow, while the bulk remains uncured. Hg lamps can be coated with fluorescing material or supplied with additives to the gas fill that shift some of their UV output into the desired wavelength range (Aldridge *et al.*, 1984). Such lamps have reportedly been used for curing pigmented coatings (Nakabayashi, 1975; Mibu *et al.*, 1980). Unfortunately, power and lamp life are reduced by these methods, and the output has too broad a wavelength range (>30 nm) for optimal curing. For these reasons the process has not been economically viable as of the 1980s.

We considered lasers for the light source. Nitrogen lasers can be made to lase at 428 nm, but not at high power. The strontium-vapor laser (Piper and Little, 1989) at 430 nm has the potential for high-power operation, but it is in the developmental stage. We experimented with stimulated Raman scattering (SRS) to shift excimer laser beams to longer wavelengths (Stokes shift). This is potentially a very efficient process, with conversion efficiencies reported at 45% in hydrogen (Fulgham *et al.*, 1984) and higher in liquid nitrogen (Brueck and Kildal, 1982). The XeCl line at 308 nm shifts to 414 nm (second Stokes in hydrogen). The XeF line at 351 nm shifts to 411 nm (first Stokes in hydrogen) or 420 nm (second Stokes in liquid nitrogen). Efficient SRS generation requires excellent beam quality with low divergence (Loree *et al.*, 1979). Our beam quality was evidently not good enough, despite the low divergence in our unstable resonator configuration, as we were unable to achieve more than a few percent conversion in 40 atm of hydrogen at low repetition rate. Because of thermal lensing, conversion efficiency declined at higher repetition rates. Thus, for high-power operation, a Raman shifter must apparently contain a circulating high-pressure gas or cryogenic liquid, a difficult apparatus to build. For this reason, and the fact that only discrete wavelengths are available, lessening its general utility, we decided against the Raman shifter, despite its potentially high-power output near desired wavelengths.

We turned to dyes, which are continuously tunable in the range of interest. Flashlamp-pumped dye lasers can have high power, but due to intense illumination at all wavelengths, dye photodegradation occurs too rapidly for extended use (Furumoto, 1986). We considered using a nonlasing dye cell pumped by mercury lamps to generate tuned CW light, but as can be seen at the top of Fig. 8.11, fluorescent bandwidths for typical dyes are on the order of 70 nm, 10 times the desired bandwidth. The output spectrum of a narrowband dye laser would appear as a vertical line which could be tuned over the range shown in the middle of Fig. 8.11. These are commercially available and reasonably efficient (our Lambda Physik FL 2002 is 10% efficient at 414 nm), but they are too complicated and sensitive to misalignment for factory use. We decided to develop a broadband dye laser for maximum simplicity and efficiency. It can be tuned to either desired wavelength, and its output (bottom of Fig. 8.11) is sufficiently narrow to meet the bandwidth specifications.

In some ways, the dye laser we built was similar to the very first dye lasers built in the 1960s (Sorokin and Lankard, 1966; Schäfer *et al.*, 1966; Sorokin *et al.*, 1967). Like them, our laser is simply a broadband oscillator pumped transversely by another laser; the oscillator cavity is formed by a rear total reflector and the front wall of the quartz dye cell (no separate output coupler). Our laser is tuned to the desired wavelengths by changing the solvent for a single dye (Schäfer *et al.*, 1966). The first excimer-pumped dye laser (Sutton and Capelle, 1976) was demonstrated soon after the discharge-pumped excimer laser was invented (Burnham *et al.*, 1976). It was later found (Uchino *et al.*, 1979) that the XeCl excimer laser operating at 308 nm gave the best results among excimer lasers as a dye pump. Since then, a number of studies (Telle *et al.*, 1981; Bos, 1981b; Cassard *et al.*, 1981; Bethune, 1981; Antonov and Hohla, 1983; Brink and van der Hoeven, 1984) have focused on increasing the power, wavelength coverage, and dye longevity of XeCl-pumped dye lasers operating in the violet. In general, the novelty of the work described in this section lies in the

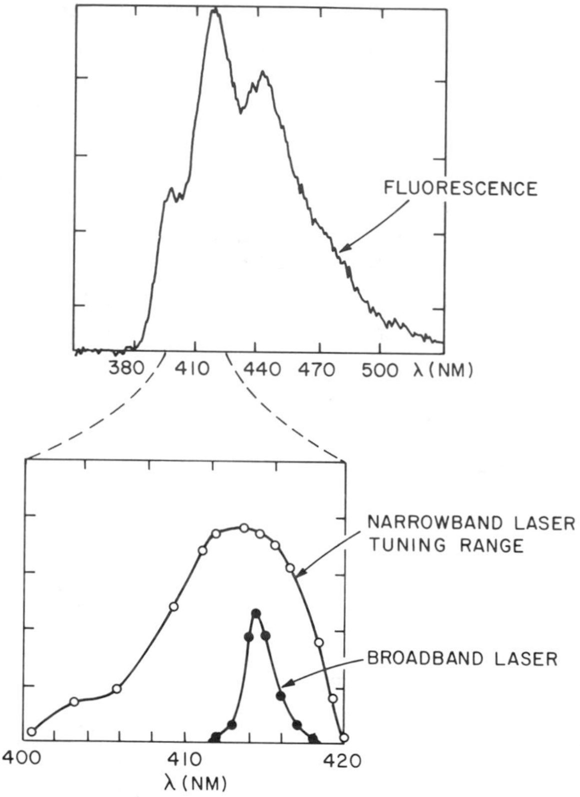

Fig. 8.11 Emission spectra for the dye POPOP: wideband fluorescence, broadband dye-laser output, and output over the tuning range for a narrowband dye laser.

application of the scientific research cited previously to make possible an industrial photochemical process of commercial significance.

Several dyes that lase around 420 nm (DPS, PBBO, Bis-MSB, POPOP) were tried with various solvents in a broadband dye laser. On the basis of output power, POPOP was superior at both wavelengths of interest. In addition, it is relatively inexpensive and stable (as will be quantified later). POPOP is an abbreviation for phenyl-oxazolyl-phenyl-oxazolyl-phenyl and is also known as 1,4-bis[2-(5-phenyloxazolyl)] benzene or as 2,2′-*p*-phenylene-bis(5-phenyloxazole). The chemical structure is given by Naboikin *et al.* (1970). A number of studies have investigated the performance of POPOP in a broadband laser with various solvents (see Table 8.1). Our own measurements given in Fig. 8.12 agree well with the literature values if the curves are extrapolated to the 0 and 100% endpoints. In

Table 8.1

Literature Values for POPOP Lasing Wavelength in Various Solvents

Solvent[a]	Output λ (nm)[b] Peak ± HWHM	Input Wavelength[c]	Laser Type[d] (Harmonic)	Reference
Ethanol	421 ± 5	347	Ruby (2)	Deutsch and Bass, 1969
Toluene	419 ± 2.5	353	YAG (3)	Naboikin *et al.*, 1970
Toluene	417 ± 1.5	353	YAG (3)	Kotzubanov *et al.*, 1968
Cyclohexane	410.5 ± 2.5	347	Ruby (2)	Deutsch and Bass, 1969
Dioxane	419 ± 1.5	308	XeCl	Uchino *et al.*, 1979

[a] We found solvent acidity to have no effect on tuning. Thus, changing solvent is the preferred tuning method for this broadband dye laser. Solvent tuning comes about when the ground and excited states of the dye have different dipole moments (Schäfer, 1977).

[b] Due to the large Stokes shift in POPOP, there is no self-absorption, so the output wavelength is unaffected by dye concentration or cavity configuration (Deutsch and Bass, 1969).

[c] Input wavelength is not expected to affect dye-laser emission, since the exited state reaches "equilibrium" in a few psec (Schäfer, 1977).

[d] Input power has no effect on POPOP output wavelength (Naboikin *et al.*, 1970).

addition, Fig. 8.12 shows measured lasing wavelengths for POPOP in mixed solvents, which we used to tune precisely to the desired wavelengths, 414 ± 4 nm and 423 ± 7 nm. It was found that POPOP in 2:3 dioxane: cyclohexane lased at 414 ± 1 nm (peak ± HWHM) and POPOP in 1:1 toluene: ethanol lased at 423 ± 1.5 nm, so that all the laser light was emitted at the most efficient photochemical wavelengths. Lasing efficiency remained high for POPOP in the solvents mixtures. The only reported efficiency value (output: input energy) for POPOP in a broadband dye laser was 20% with XeCl pumping (Uchino *et al.*, 1979). We were able to reach 25% efficiency. Such a value is typical for UV dyes, though it would be low for visible dyes. There is increasing pump absorption by the excited singlet state of dyes as their emission approaches the UV (Schäfer, 1977). Fig. 8.13 demonstrates the effect qualitatively for our dye laser. The 100-watt pump beam at 308 nm can pump many dyes, filling a range from 330 to 950 nm. Despite the fact that more energy is lost to the Stokes shift, output power increases going from the UV toward the blue (due to increasing dye efficiency) until power begins to drop at wavelengths longer than 480 nm.

Figure 8.14 shows the dye laser that was designed and tested. In an effort to realize a homogeneously pumped volume of dye, we employed two-sided pumping (Hargrove and Kan, 1977). The excimer laser beam was 10 mm high by 25 mm wide; a mirror reflected half of the beam to the back

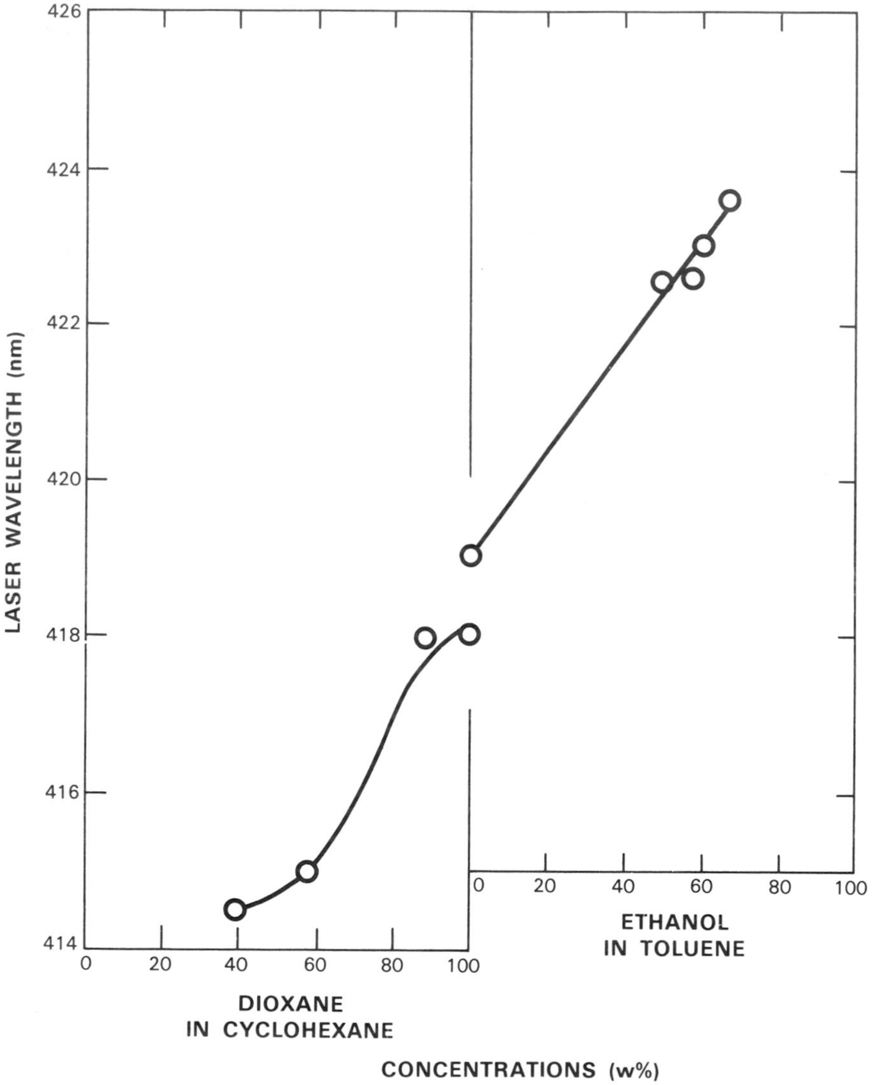

Fig. 8.12 Peak output wavelength for a broadband dye laser using POPOP in the noted solvent mixtures. Such a laser can be solvent-tuned over a 10-nm range.

side of the dye cell. Thus, the dye cell width was set at 13 mm. Other dimensions were established by a desire to avoid cell damage by keeping incident intensity on quartz surfaces below 1 J/cm^2. (End or longitudinal pumping was not employed due to the high pump intensity that would be incident on the end face.) For a 13-mm–wide cell, the 0.4 J pump pulse

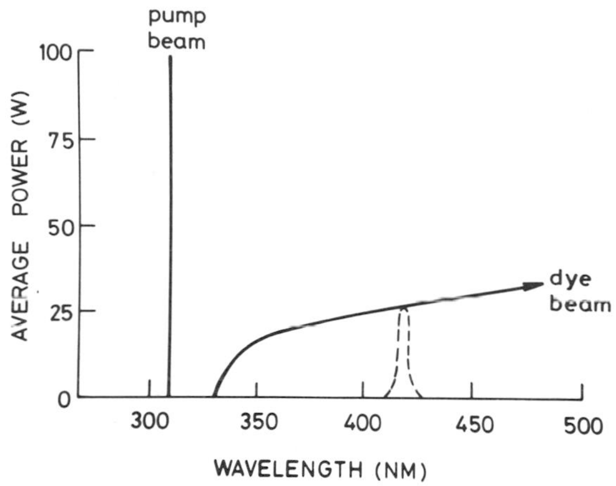

Fig. 8.13 Schematic of power output by a broadband dye laser at wavelengths longer than the 100-watt pump laser. This is an expected composite curve using various available dyes and solvents.

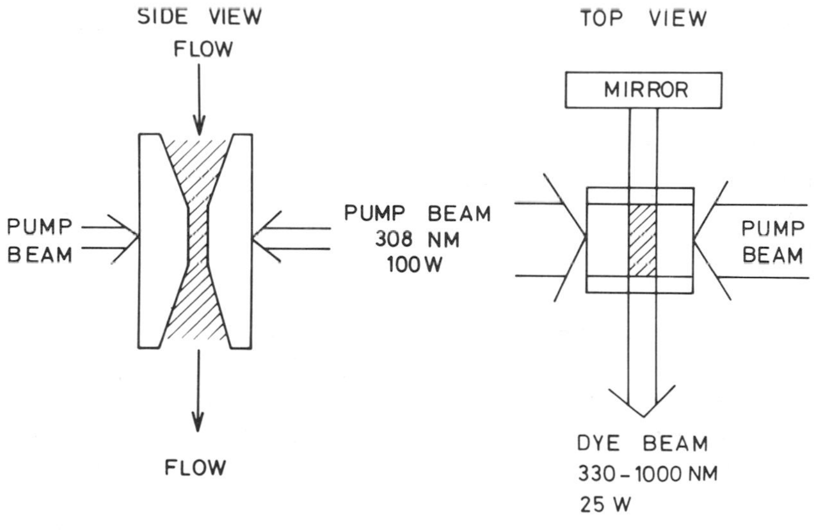

Fig. 8.14 Design of the broadband dye laser of Section 4.2, employing two-sided pumping.

could then be focused to no less than 1.5 mm high. It was found experimentally (Fig. 8.15) that peak power was achieved at a 5-mm pump height. The cell depth then had to exceed 2 mm to avoid damage to the end faces by a 25%-efficient dye beam. The depth was set to 3 mm. Ideally

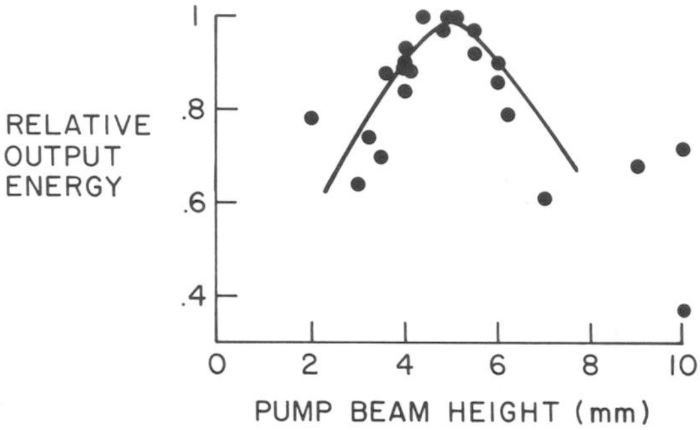

Fig. 8.15 Dye-laser output energy versus height of pumping beam. Optimum height of pump beam on dye cell (see left side of Fig. 8.14 for geometry) was found to be 5 mm at any of several dye concentrations (each set of data normalized).

then, the pump beam would have a uniform 5 × 13-mm rectangular cross section on each side face, and the dye beam would have a close to uniform 5 × 3-mm cross section at the output.

The molar absorption coefficient for POPOP at 308 nm was measured at 13,430 liter/(cm mole). For 90% absorption of each pump beam halfway into the cell, a dye concentration of 5e-4 M or 0.18 g/liter (MW = 364) was required. It was found, however, that optimum efficiency was achieved at a higher concentration, such that each side of the dye cell was lasing independently. The top curve of Fig. 8.16 shows that at low repetition rate the efficiency continues to improve (albeit weakly) at nearly three times the concentration specified in the design. The bottom curve in Fig. 8.16 demonstrates the need for a higher flow rate of dye through the active region. At low dye concentration, much of the decrease in output energy in going to the highest repetition rate is due to a drop in pump energy from 0.4 to 0.35 J above about 150 Hz. At high concentration, however, the active regions are quite near the walls where flow velocity falls off, so that the dye solution is not being completely replaced between pulses. Thermal lensing results and output energy falls. Volume flow rate for the mechanical pump was chosen by specifying that the active volume be replaced five times between laser pulses. The purchased pump fell far short of its specifications, but we still replace the active dye twice between pulses, assuming plug flow. However, if we assume a parabolic velocity distribution across the channel and calculate the average replacement of dye in the active region near the wall, we find that dye in the active region is not being fully replaced at high dye concentrations. The flow system consisted of a one-liter dye solution reservoir that was cooled by circulating water. The free flow rate of the mechanical pump was specified at 22 liter/min but was measured at 8 liter/min. Flow tubes were teflon, as was the pump interior.

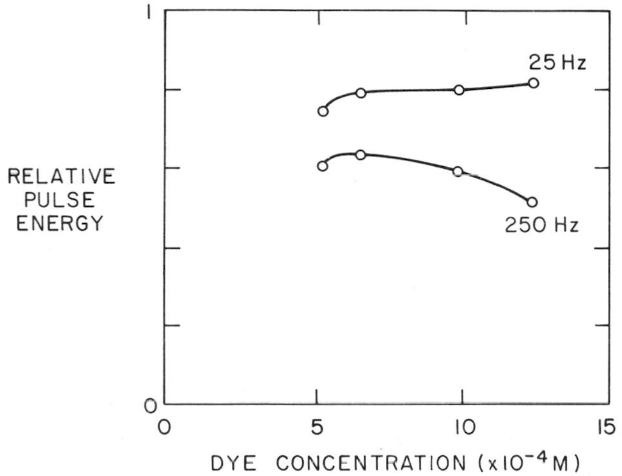

Fig. 8.16 Output pulse energy versus dye concentration at low and high repetition rates.

A stainless-steel wire ran through the tubes to prevent static charge buildup due to circulating nonpolar solvents.

Figure 8.17 shows that an efficiency of 25% is attained with this laser at high concentration and low repetition rate. (Quoted dye-laser efficiency and power values assume all surfaces are dielectrically coated for maximum reflection or transmission.) Use of a higher–flow-rate mechanical

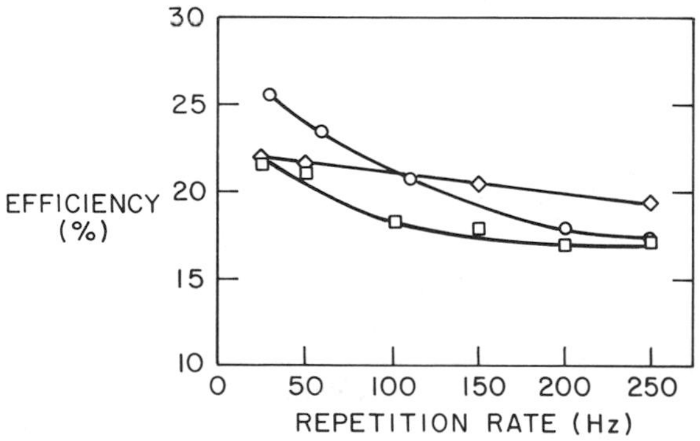

Fig. 8.17 Dye-laser output efficiency (output/input pulse energy) versus pulse repetition rate under various operating conditions:

Symbol	*Dye concentration (e-4 M)*	*Output coupler*
circle	10	none
square	5	none
diamond	5	30% reflector

pump should allow the same efficiency to be attained at the highest repetition rate. The highest measured power was recorded at lower concentration, where the efficiency at 250 Hz was 20%. Since the best pump-laser power we could reach (with fresh gases and clean windows) was 88 watts, the maximum measured dye-laser power was 18 watts. Elimination of thermal lensing should allow a power of 22 watts to be attained. We tried several design changes in search of increased efficiency. Installation of an output coupler offered a small improvement (a 30% reflector increased output <10%), but the increased sensitivity to misalignment was deemed inappropriate for a factory dye laser. Polarizing the normally unpolarized pump laser (three intracavity quartz plates at Brewster's angle gave a 90% vertically polarized beam without loss of ouput power) gave only an 8% rise in dye-laser power. The small gain from polarization was not sufficient to justify the added complexity.

The present design results in a comparatively large divergence. Large divergence angles are not a problem in our application, but could be in others (e.g., when the beam must be transported some distance). The divergence would increase if the dye concentration were raised to achieve better efficiency, because the active region would then be closer to the diffracting cell wall. With a smaller active region, the dye beam also becomes more intense, and the possibility for end window damage increases. We experienced several damage incidents, as have others (e.g., Cassard *et al.*, 1981). To avoid such incidents, it would be best to utilize a low dye concentration at the expense of a small loss in efficiency, especially in a factory setting. It may also help to introduce some detergent to the dye solution to prevent the formation of damage initiation sites due to dye deposition on the walls (Bos, 1981b). An alternative design that might be tried (Bethune, 1981; Brink and van der Hoeven, 1984) has the dye flowing in a cylindrical tube that is transversely pumped from all directions. Such a design could be freer from the problems of optical damage and high divergence, while retaining the features of simplicity and high power that we require.

Summarizing our laboratory work, we developed a tunable dye laser that should make the pigmented curing process commercially feasible, according to our projections. Using a single dye, the dye laser can be solvent-tuned to either of the desired wavelengths, and its output is entirely within the high-curing-efficiency limits. Indeed, these "narrow" spectral bands that we cannot match with lamps are sufficiently wide that a broadband dye laser (one with no narrowing element) matches them easily. The dye laser can then be of the simplest design (dye cuvette plus rear total reflector), since narrow linewidth and low divergence are not required. High output power, the main goal, is not sacrificed in this simple design.

In the violet region, the highest dye-laser power is achieved using an excimer laser as the pump (Wheeler, 1982; Kortz, 1982; Hohla, 1982a, b; Bernhardt *et al.*, 1982; Bernhardt, 1983, Asher, 1984), and we used the one that produced the highest commercially available output power at that time, 100 watts (0.4 J pulses at 250 Hz) at 308 nm, the preferred excimer pumping wavelength. Another design goal, reliability for factory use, is enhanced by the simple design, which features tolerance to misalignment and a small number of parts. It is recognized that the leap from lamps to lasers is a huge one for the chemical industry, and the reliability of both the dye laser and pump laser must be excellent before they will be introduced in chemical processing, whatever the projected economies for the process.

4.3. Pilot Plant Design, Costs, and Market

As part of a pilot plant plan, a second-generation prototype dye laser was designed, keeping in mind that for factory use it should be simple, reliable, and foolproof. It consists of a box that bolts to the front of the excimer laser. The box has two openings, one to drain the reservoir and the other to fill it. There are no external adjustments except for a swinging beam splitter and a meter to read input and output power. A top view of the dye laser is found in Fig. 8.18. This design readily allows for the dual-wavelength output that is desired for curing both the surface and bulk

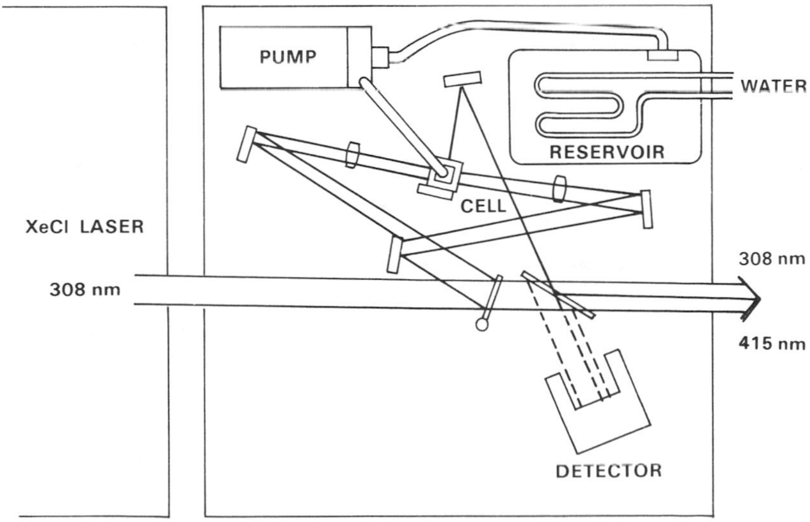

Fig. 8.18 Top view of prototype dye laser designed for a pilot pigmented coatings plant.

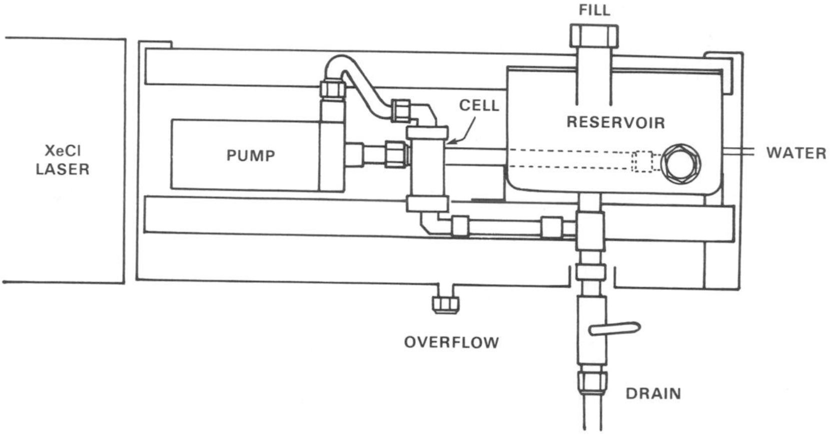

Fig. 8.19 Side view of the prototype dye laser showing plumbing.

of pigmented coatings. For curing clear coatings, the swinging beam splitter is out of the beam, allowing all the excimer light through to the curing chamber, except for a small amount sampled by the detector. To cure pigmented coatings, the beam splitter sends 88% of the excimer beam to the dye cell, which generates a dye-laser beam. The dye beam is reflected from the dichroic to be collinear with 12% of the original excimer beam, and both beams are sampled at the detector. A side view of the dye-laser plumbing is shown in Fig. 8.19.

The rest of the pilot plant design is shown in Fig. 8.20, Laser light is routed to a set of scatter plates spaced 0.6 m above a conveyor belt. There are 16 scatter plates at 0.5-m intervals. Above them are beam splitters with splitting ratios such that each scatter plate receives equal power. The laser light is confined within a curing chamber that also confines the inert atmosphere. Nitrogen gas is forced in at the inlet end of the chamber. Short lengths of purged chamber at inlet and outlet provide for oxygen removal from the coating before curing and for postillumination curing. At the chamber inlet, a roll coater applies the liquid coating to the substrate and places the material on the belt. At the outlet, the material is removed from the belt and stacked. An optical bench provides a station for small-scale experiments.

Figures 8.21–8.23 plot the expected illumination pattern at the belt within the chamber. Figure 8.21 shows results for one and two scatter plates (or beams) only. The curve marked "one beam" is the spatial distribution of intensity measured for our scatter plate, which was a piece of quartz rough on one side and polished on the other, and antireflection coated on both sides. The curve marked "two beams" is the combination

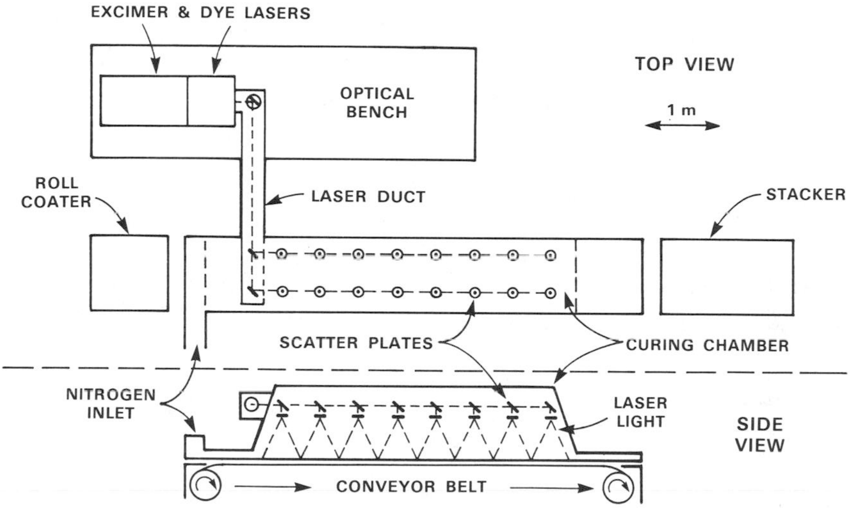

Fig. 8.20 Top and side views of a pilot plant designed for laser curing of coatings.

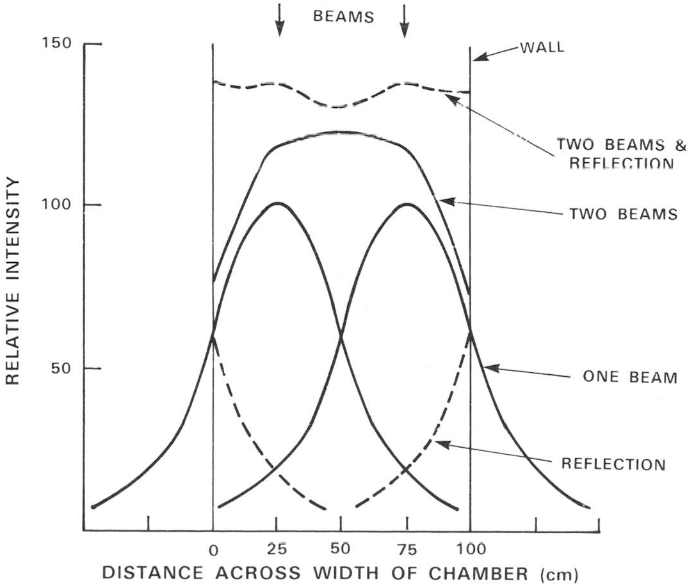

Fig. 8.21 Expected intensity at the belt across the width of the curing chamber under two scatter plates.

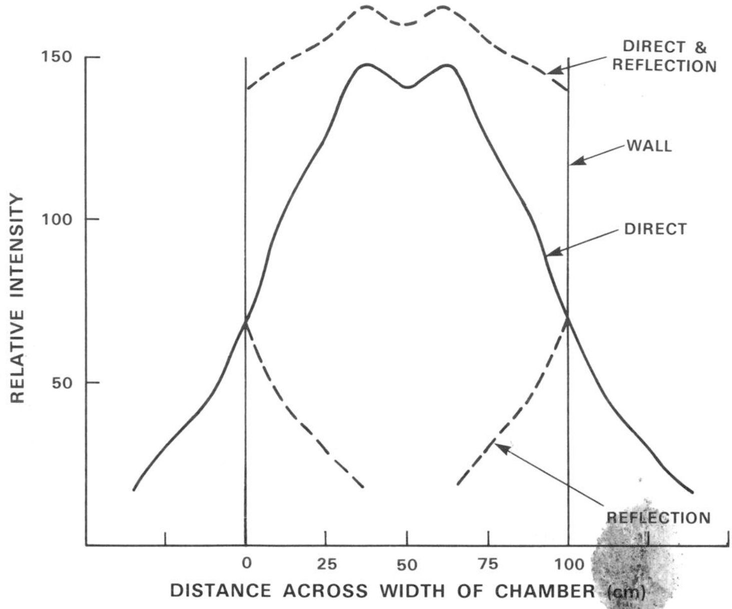

Fig. 8.22 Expected intensity at the belt across the width of the curing chamber between scatter plates.

of the direct light from two plates. If the chamber wall provides specular reflections, these will add as shown to give the nearly uniform curve marked "two beams & reflection." This is the expected distribution of intensity across the width of the chamber under two scatter plates. Halfway between two pairs of plates, the widthwise distribution of intensity is expected to be as in Fig. 8.22, which again is very uniform. The distribution of intensity along the belt center for the ensemble of beams is shown in Fig. 8.23. The intensity is constant within ±10%.

The laboratory results of Section 4.1 can be readily scaled to a factory situation. If one chooses a laser to be operated at a given power P (watts) and prf R(Hz), one can determine the chamber size required to reach the desired single pulse fluence from the following equation,

$$W \times L = P/(10FR), \tag{8.2}$$

where W and L are the width and length in meters and F is the single pulse fluence in mJ/cm^2. The speed of the belt S in m/s that moves the pieces just slowly enough to reach the required energy-to-cure E(mJ/cm^2) is given by

$$S = P/(10EW) = FRL/E. \tag{8.3}$$

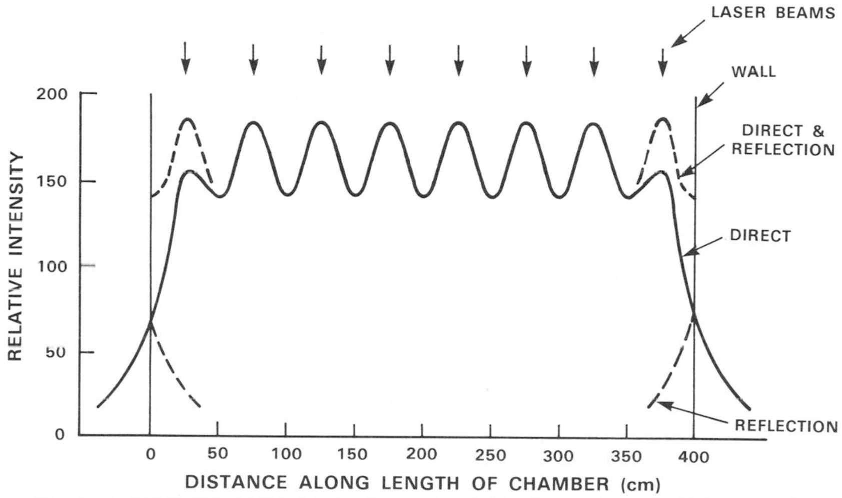

Fig. 8.23 Expected intensity at the belt along the length of the curing chamber between scatter plates.

For pigmented coatings, the best result we achieved was an energy-to-cure of 2.6 mJ/cm^2 at 415 nm in a 35-μm coating with 32% TiO_2.(From Fig. 8.7, we expect the same or better energy-to-cure for the more industrially useful coating of 20-μm thickness and 50% TiO_2 loading.) The best pigmented result was achieved with a formulation of 4% APO and 3% T12 and single-pulse fluences of .0015 mJ/cm^2 at 308 nm and .003 mJ/cm^2 at 415 nm at a prf of 250 Hz. Dual-wavelength cure at the "best" fluence requires that 12% of the 308-nm pump beam be split off and directed to the coating, while the remainder pumps a 25%-efficient dye laser. Using the belt-length equation, we find that the required curing chamber for operation at these fluences is 1m $\times$ 3m. Both wavelengths are spread over the same area and the same number of pulses (867) is used. The material is in the chamber for 3.5 s, so the belt is running at 0.85 m/s.

This facility has another use, to cure clear coatings (Klick *et al.*, 1988). It was found that for 7-μm clear coatings, "best" fluence for a 100-watt XeCl laser was .008 mJ/cm^2, requiring 100 pulses at 250 Hz to cure. For a standard 1-m width, the belt length is then 5 m. The belt speed is 12.5 m/s, which is on the order of 20 times faster than for a typical Hg lamp. The belt speed is 15 times slower for pigmented than for clear coatings more because of the increased thickness of the former and the power losses incurred in shifting wavelength, than because the curing efficiency is reduced due to the pigment. In the pilot plant design shown in Fig. 8.20, we

Fig. 8.24

Comparison of Typical Throughout, Space Requirements, and Capital Costs for Curing Processes Employed in the Late 1980s and for the Proposed Laser Curing Method

	Oven	Electron beam	Single UV Hg Lamp	Single laser
Cure speed (m/s)				
Clear	4	7	0.6	12.5
Pigmented	4	1.6	—	0.85
Chamber length (m)	100	5	2	3–5
Capital cost ($M)	5	1.5	0.05	0.1

compromised between the 3-m chamber desired for pigmented curing and the 5-m chamber desired for clear curing.

UV photoassisted polymer curing has advantages over traditional industrial methods using ovens in terms of capital costs, space required, and polluting emissions. Line speeds reach 4 m/s curing clear coatings with either ovens or banks of Hg lamps. Lamps cannot cure pigmented coatings satisfactorily, but ovens reach the same speeds curing pigmented coatings. A single excimer laser can cure 7-μm clear coatings at speeds of 12.5 m/s, with capital costs of $100K. A set of five excimer-pumped dye lasers could cure 20-μm pigmented coatings with 20% PVC TiO_2 loading at the 4-m/s belt speed achieved by an oven, with capital costs on the order of 10 times lower. Figure 8.24 is a comparison of various curing methods. Efficient laser curing in our method is accomplished at low fluence by diverging the beam to illuminate large areas of coating. Low fluence leads to lower total energy-to-cure requirements, at least ten times less than previously reported for laser curing of clear coatings (Decker, 1983, 1984; Fouassier *et al.*, 1984), and 1000 times less than previously reported for curing of pigmented coatings by various lamps (Aldridge *et al.*, 1984; Mibu *et al.*, 1980).

Our estimate (see Figs. 8.25 and 8.26) of capital, maintenance, and operating costs for a factory equipped with five excimer lasers processing a generic reaction with a photon economy of one, was $60/kg, producing 4,000 kg/year (assuming 50% downtime, yielding one kg/hr). This is only practical for precious chemicals such as isotopes or pharmaceuticals. The chemicals considered here sell at about $10/kg. We showed in Section 4.1 that the photon economy achieved in our process is on the order of 1000, so that laser processing would cost only $0.06/kg (negligible compared to

Fig. 8.25

Assumptions for Determining the Costs of Laser Chemistry
(Photon Economy of One)

Product molecular wt = 100, photon economy = 1
2 factory shifts, 3/4 year = 50% downtime
Electricity costs, 12¢/kWh, 1% efficient lasers
5 excimer lasers, each 1J × 200 Hz = 200 W
10^7 shots/gas fill
Laser cost, $100,000 5-year amortization
Operation and maintenance, $16,000/laser/year

Fig. 8.26

Laser-Chemistry Cost Determination
Using the Assumptions of Fig. 8.25
(Photon Economy of One)

Electricity	$13/kg
Fill gas	4
Laser purchase	23
Operation and maintenance	20
Total	60

Notes: $60/kg product or $60/kWh laser power.
• 4,000 kg produced per year.

chemical costs), and production of 4×10^6 kg/year is possible. Other cost estimates (Klauminzer, 1988; Holmes, 1988; Schäfer, 1988) are in reasonable agreement with ours. Over the five-year amortization period, each laser would fire about 10^{10} times, which has been demonstrated (Lambda Physik, 1988a).

Figures 8.25 and 8.26 are concerned with costs for an excimer-laser process. If the excimer lasers must pump dye lasers, as when curing pigmented coatings, additional costs are incurred from building and maintaining the dye lasers. A major issue for industrial use of the dye laser is operating cost, since capital cost is relatively low (10–20% of the excimer-laser cost). The main operating cost is replacement of degraded dye solution. Cost and degradation rate of POPOP are sufficiently low that dye consumption is a negligible problem for this process. The value quoted by Lambda Physik for POPOP stability in dioxane is 250 watt-hr. Thus, a dye laser with one liter of dye solution pumped by a 100-watt excimer laser will lose half its power in 2.5 hrs. We used the least expensive POPOP (from

Sigma Chemicals Co.) and solvents available, with no adverse effects on performance. In that case, the laser-dye solution consumed by a five-laser factory producing pigmented coating would cost \$6/hr, which is only 10% of the excimer laser cost. Total costs for the dye-laser process were estimated at \$75/hr.

Energy-to-cure can be reduced (i.e., photon economy can be increased) to a level where the laser cost becomes small compared to the cost of chemicals input to the process. Using laboratory results from Section 4.1 along with the preceding estimates of capital, operating, and maintenance costs for a five-laser factory curing 7-μm clear coatings at a 12.5 m/s belt speed, 1600 kg/hr (\$16,000 worth) of material would be processed using \$60 worth of laser photons (\$.04/kg processing). For 20-μm pigmented coatings, an excimer-pumped dye laser producing 22 watts illuminates a $1\text{m} \times 3\text{m}$ chamber at a belt speed of 50 m/min. In a five-laser factory, 300 kg/hr (\$3,000 worth) of chemicals are processed, using \$75 worth of laser photons (\$0.25/kg processing). Figure 8.27 summarizes these results. The conclusion is that laser-processing costs are a negligible fraction of raw-materials costs (0.4% for clear coatings and 2.5% for pigmented). This situation is almost a prerequisite for a successful chemical process. As Table 8.2 shows, the cost to manufacture paint is also a small fraction of the raw-materials cost.

A determination of the true savings that a coatings manufacturer could derive from the laser process would require an analysis of time and space

Fig. 8.27

Capabilities of a Proposed Factory Employing Five 100-Watt Lasers to Cure Clear and Pigmented Coatings

	Coating type	
	Clear	Pigmented
Thickness (μm)	7	20
Materials input (kg/hr)	1,600	300
(\$/hr)	16,000	3,000
5 coating chambers (m × m)	1 × 5	1 × 3
5 belts at speed (m/min)	750	50
Laser costs (\$/hr)	60	75
(\$/kg)	\$0.04	\$0.25

Table 8.2

Cost Breakdown for a Typical Can of Paint

Research and development	3%
Administration and finance	7%
Warehousing and distribution	8%
Marketing and sales	13%
Manufacturing	13%
Raw materials and containers	56%

use that is beyond the scope of this chapter. For example, in coil or can coating, a full-production paint line could cost \$30M, with most of that cost in the 50–100-m–long air dry ovens. The laser process would not only reduce the cost of the line, but also free up substantial production space. Similarly, most of the plant operating space in the paper and wood-coating industries is taken up with storage of air-drying stacks of work-in-process. Use of an instantaneous cure process saves not only space in this case, but also time. Finally, the elimination of polluting carrier solvents from coatings by use of a solventless, 100% reactive formulation (as is the case in UV curing) may allow a manufacturer to comply with government regulations.

The laser method shares these advantages with UV curing by lamps. As of the 1980s, the total world market for UV-curable resins is 23×10^6 kg/year, a small fraction of the total world OEM coatings market of 175×10^6 kg/year (Layman, 1985). It is our contention that use of the laser process could boost the market share for UV curing by allowing it to penetrate previously untapped markets, chiefly for pigmented coatings. The advantages of laser curing derive from the laser's ability to provide all of its light at a desired wavelength (where nearly all the light is efficiently absorbed by the photoinitiator and little by the bulk or by a pigment). Illuminating at only useful wavelengths not only is more efficient, but also avoids difficulties presented by the multiline Hg lamp, namely, faster surface cure than bulk cure (leading to wrinkling) and generation of unwanted photoproducts. For clear coatings, it is not difficult to find an excimer laser line and a corresponding nonyellowing photoinitiator. However, for coatings pigmented with TiO_2, the laser light must be relatively narrowband (<10-nm bandwidth) and precisely tuned for optimum curing efficiency. Only dye lasers exhibit the required tunability and bandwidth. The result is a higher throughput of clear coatings in the laser process compared to the lamp, plus the ability of the laser to cure pigmented coatings where the lamp cannot.

Table 8.3

Breakdown by Market of World OEM Coatings Sales

Automotive	20%
Metal	15%
Wood	10%
Coil	9%
Industrial lacquers	8%
Miscellaneous machinery and equipment	6%
Marine	4%
Appliances	4%
Other	24%

Note: original equipment manufacturers (OEM) sales exclude architectural as well as special-purpose and aftermarket sales.

The worldwide OEM coatings market we aim for is valued at $4 billion annually (Layman, 1985). Table 8.3 supplies a breakdown of the industry by market. Perhaps the best initial application of the process is to pigmented coatings on simple, flat metal plates, such as in the appliance or coil-coating industry. The process (Akerman *et al.*, 1986) awaits an entrepreneur!

5. DIAGNOSTICS APPLICATIONS OF DYE LASERS

5.1. Materials Diagnostics

Dye lasers have been used in materials diagnostics chiefly in the area of microelectronics. These applications fall into two general classes-measurements on devices and measurements of process chemicals during fabrication. In the former class, Johnson and Johnson (1986) used a near-IR (1.13-μm) pulsed dye laser to probe carrier lifetimes deep within a silicon ingot and in a solar cell. Cooper *et al.* (1987) employed a mode-locked dye laser tuned above the band edge in CdTe. Band-edge emission spectra were used to select samples before growth of HgCdTe epilayers for IR detectors. Researchers at IBM (Yu, 1978; Ketchen *et al.*, 1986) have built photoconductive switches into microelectronic devices that use the short pulses available from a mode-locked dye laser to excite a circuit and measure its response with subpicosecond time resolution. Weiner *et al.* (1987) can measure signals on any semiconductor device with picosecond time resolution by probing a metal strip on the device with an 80-fs dye-laser pulse. The electrical-signal waveform is deduced by analysis of the energy characteristics of the photoelectron emission. Use of a dye laser tuned to some spectral feature (say, the absorption edge) of a material in order to cause heating by absorption is the basis of two techniques for measuring properties of materials. Frosch *et al.* (1981) detect the thermal lens so formed by deflection of a CW probe laser to measure molecular relaxation and thermal diffusivity in the material. Tam and Sullivan (1983) detect transient blackbody emission from the heated spot to measure the absorption coefficient and sense physical and thermal properties of the sample.

In the area of using lasers to monitor the gas-phase reactants during processing of microelectronics, there have been a number of review articles (Karlicek *et al.*, 1983; Wormhoudt *et al.*, 1983). Specifically examining the use of dye lasers, the technique of laser-induced fluorescence (see Section 6) is prominent. Karlicek *et al.* (1983) give a list of 30 species of interest in either chemical-vapor deposition (CVD), plasma deposition (PD), or plasma etching (PE) that have been detected by LIF or a related technique, laser-induced photofragment emission. Species distributions and temperature during semiconductor processing have been measured with these techniques, made possible by tuning the dye laser selectively to an absorption line of the species under study. Wormhoudt *et al.* (1983) show a table listing 25 radical species important for CVD, PD, and PE, and give minimum detectable densities (ranging from 10^{11} cm^{-3} for CH_2 to 10^5 cm^{-3} for Si) using LIF.

Other less-established techniques have been used for microelectronic diagnostics. Selwyn (1987) has used two-photon LIF to measure species important in PE (F, Cl, and O) that require VUV excitation. CVD has been enhanced to laser-assisted CVD, and the influx of C_2 radicals onto the substrate after excimer-laser photodissociation of C_2H_2 was measured by LIF using a dye laser (Okada *et al.*, 1987). Zhang *et al.* (1987) used tunable dye laser time-of-flight mass spectroscopy to detect photofragments from laser-assisted CVD. Schäfer (1988) shows that using a picosecond dye-laser pulse to cause multiphoton ionization in a mass spectrometer provides not only spectral, but also temporal, selectivity. Only the selected molecule is ionized during the ps pulse, as there is no energy available afterwards to dissociate it into the fragments seen with other methods. Perry *et al.* (1987) employed CARS (see Section 6) to measure silane depletion in an rf plasma during deposition of amorphous silicon. O'Brien *et al.* (1987) used intracavity laser spectroscopy to measure SiH_2 and C_2 concentrations during CVD. In this sensitive technique, the sample gas is within the cavity of a multimode dye laser, and its absorption lines become impressed on the dye-laser output spectrum.

A number of other industrial uses of dye lasers, unrelated to microelectronics, fall under the category of materials diagnostics. In one study (Hunt and Pappalardo, 1985) of the red phosphor $Y_2O_3:Eu^{3+}$ found in some fluorescent lamps, a pulsed dye laser was tuned to 254 nm to simulate the mercury line. The short pulse available from the dye laser allowed measurements of decay times, and tuning the dye laser produced the excitation spectrum. It was hoped that better understanding of the mechanisms and kinetics of lamp fluorescence would lead to reduced europium usage. In another study (Podmaniczky, 1983), more accurate readings (to better than 0.1 μm) of optical fiber diameter were achieved using a dye laser, compared to previous methods. The backscatter from the dye-laser–illuminated optical fiber showed Fabry-Perot resonances as the wavelength was tuned, from which the diameter was extracted.

5.2. Chemical Diagnostics

The subject of diagnostics of chemical reactions can be divided in two parts, the first dealing with in-situ diagnostics of chemical processes (El-Sayed, 1986, 1987; Wolfrum, 1988), and the second with off-line analytical chemistry (Duley, 1983; Omenetto, 1988). Most of the dye-laser–based diagnostic techniques applicable to chemical processes (e.g., CARS and LIF) are discussed in Section 6 in the context of combustion diagnostics. One interesting use of dye-laser–induced fluorescence (Horvath and Semerjian, 1987) is in the pulp-and-paper industry. It was found that LIF, with selectivity in both the UV absorption and emission wavelengths, has advantages compared to absorption as an in-line process monitor of dissolution of lignin during a sulfate cook. An application in the oil-and-coal industry of time-resolved LIF using a short-pulse dye laser (Pleil *et al.*, 1987) found that fluorescence lifetime could be correlated with sample maturity for both crude oil and coal. In the uranium mining and processing industry, a kinetic phophorimeter based on a nitrogen-laser-pumped dye laser is being marketed for monitoring uranium contamination. Time-resolved emission of the uranyl ion in environmental samples provides detectability of uranium at the 10-ppt level (Thomson, 1988). A dye-laser spectrometer with wide tunability is at the heart of an instrument sold to the food-and-agriculture industry. Controlled by an expert system, this device is capable of changing the excitation wavelength to monitor blending of constituents in a process while searching for contaminating chemicals (Thomson, 1988).

In the area of highly accurate analytical chemistry measurements, Omenetto (1988) finds that dye-laser-based analytical techniques such as laser-induced fluorescence (see Section 6) and optogalvanic spectroscopy (see Goldsmith (1987)) can be made significantly more sensitive than conventional tools. The limit of detection using laser methods, combined with conventional atmospheric-pressure atomizers (e.g., flames, plasmas, or graphite furnaces), can be in the range of femtograms. Such accuracy will be most likely required in the following industries: high-purity materials, semiconductors, biomedicine, and the nuclear industry. Excitation by double-resonance provides signal and selectivity advantages over single-resonance excitation. Combining the dye-laser–induced techniques of fluorescence and ionization results in an even more powerful tool. Photoacoustic (see Tam (1986)) and photodeflection (see Gupta (1986)) spectroscopy, which also employ dye lasers, are cited as performing measurements that are impossible with conventional methods, because the chemicals (e.g., actinide ions) are strongly absorbing and nonfluorescing. Another ultrasensitive analytical technique employing dye lasers is the combination of a capillary column gas chromatograph with a laser ionization mass spectrometer (Reilly, 1987). Polyaromatic hydrocarbons were detected with parts per trillion sensitivity (multifemtogram level). In liquids, fluorimetry is commonly used in industry to measure the concentration of amino acids, vitamins, aromatic hydrocarbons, and fluorescently labeled molecules. While the conventional technique is sensitive, the use of a dye-laser source was found to provide a hundredfold increase in sensitivity (Duley, 1983). Dye-laser–induced fluorimetry has also been applied to the study of pharmaceuticals (Srojny and de Silva, 1980). Further improvements are realized if the dye laser is pulsed (especially mode-locked) and time-resolved fluorescence is measured, or if the dye laser has high spectral resolution and the sample is suspended in a cryogenic matrix.

6. DIAGNOSTICS CASE STUDY: DYE LASERS IN COMBUSTION DIAGNOSTICS

This work was performed by the author and his colleagues (E.W. Kaiser, K.A. Marko, L. Rimai, and T. Kushida) while affiliated with the Ford Motor Co. Scientific Research Lab, Dearborn, Michigan, in 1980–1986. We investigated the use of dye-laser–based diagnostics to measure quantities of interest in combustion, particularly as related to automotive applications. Combustion in industry is typically sooty, turbulent, time-variant, and in a complex geometry without optical access. Its details are usually poorly understood, and any in-situ measurements that can be made are helpful in gaining insight into and perhaps improving the quality of combustion. Over the years, there have been changes in what combustion elements are considered in need of urgent improvement, as society's priorities shifted from pollution control to fuel economy to performance.

However, the ultimate measurement desired is usually a time-and-space–resolved mapping of the thermodynamic and chemical distributions around the flame. Quantities of interest include the concentration of stable species (hydrocarbons, O_2, CO_2, CO, H_2, NO) and of free radicals (OH, O, etc.). Information about temperature can be gained simultaneously by examining the spectroscopic information obtained for a particular molecule.

The optical techniques employed to gain this information are spectroscopically selective and therefore can advantageously use dye lasers to tune a probe beam to particular molecular lines. The advantages of laser techniques over others that have been employed to probe combustion include the high space and time resolution that can be achieved and the nonintrusive nature of the laser probe. Spatial resolution is often given by the size of a focal spot (typically 100 μm), since the secondary radiation generated at the focus usually carries the required information. Time resolutions of 1 μs to 10 ns are routine and ps resolutions are achievable, assuming that the instrument is capable of measurements during single laser pulses. At power levels up to the point of laser-induced breakdown, laser probes can generally be made nonperturbing to the system under study.

The techniques to be discussed here are the ones we used that employed dye lasers. Many other dye-laser–based combustion diagnostic probes have been described (Crosley, 1980; Eckbreth, 1988), and some of these will be noted at the end of the section. We have explored the following techniques: laser-induced fluorescence, time-resolved infrared spectral photography (TRISP), and, especially, coherent anti-Stokes Raman scattering (CARS). All of these are spectroscopic techniques that are affected by the populations of molecular levels, and so can be used to measure species concentration, temperature, and density. Fluorescence is very sensitive to selected species (e.g., OH and other radicals), and LIF has been used to map out 2-D species distributions with space and time resolution. Its main drawback lies in interpreting the data due to competing relaxation processes. Infrared absorption is a standard technique for probing molecules (except for homopolar diatomics such as O_2 and N_2) with high sensitivity. The time-resolved version described here, TRISP, still lacks space resolution, as the measurement is integrated over the path. The Raman effect is active for all molecules of interest. Spontaneous Raman has limited utility in industrial measurements because its extremely weak signal is often overwhelmed by interfering radiation. However, the nonlinear Raman variant, CARS, shows great promise as a practical diagnostic, since it provides strong signals and can be made spatially and temporally precise. The main difficulty with CARS is its limited sensitivity to species at low concentration (below 1000 ppm).

In LIF, a dye laser is tuned to a line in the excitation spectrum of the

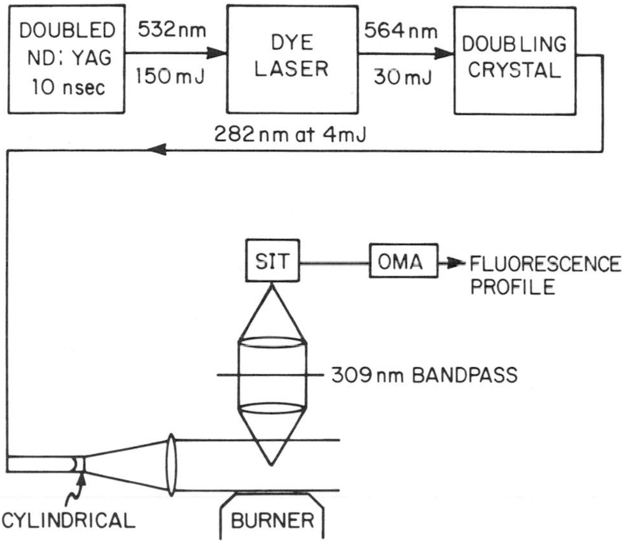

Fig. 8.28 Block diagram of apparatus for measuring laser-induced fluorescence (LIF) profiles of OH above a flat flame burner.

molecule of interest, and the resulting fluorescence is detected. For time resolution and instantaneous power, a pulsed dye laser is usually used. The detector typically views at right angles to the laser beam, so spatial resolution is achieved by focusing the laser (to either a point, a line, or a sheet) along the detector line of sight. The fluorescence is focused onto a small-element detector for spatial resolution (or a 1-D or 2-D array of such elements). Figure 8.28 is a diagram of the apparatus we used, employing a sheet of light and effectively a 1-D array. The dye laser contained rhodamine 590 and was pumped by the second harmonic of a Nd:YAG laser with a 10-ns pulse. Doubling the dye-laser output with a KD*P crystal gave us 4 mJ of pulse energy at 282 nm (0.3 cm^{-1} linewidth). Fluorescence from the OH radical was passed with a 309-nm filter (bandpass ~1 nm) into an imaging detector, so that the vertical variation of fluorescence above a flat flame burner could be determined.

A typical plot of LIF signal versus height above the burner is shown in Fig. 8.29 for an equivalence ratio of 1.17 (i.e., a fuel-rich flame). The spike marked "burner" is a specular reflection from the burner surface and serves as a fiducial. The curve marked "flame" is fluorescence not induced by the laser (which was off), but rather from chemically excited OH produced by the flame. This indicates that OH production takes place in a

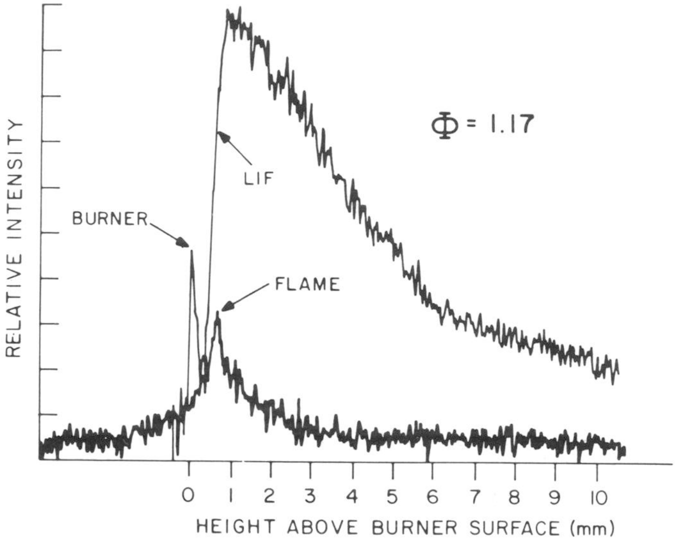

Fig. 8.29 OH fluorescence intensity versus height above a flat flame burner with the laser on ("LIF") and off ("flame").

narrow zone about 1 mm above the surface. The ground-state OH population (measured by the curve marked "LIF") decays slowly (1-mm height is equivalent to 1.4-ms residence time at this flow rate) from the superequilibrium concentration induced by production in the flame toward an equilibrium concentration governed by the burned gas temperature (Klick and Kaiser, 1984). Absorption measurements in the UV using a cw dye laser were performed on the same flame. Both fluorescence and absorption can be related to OH concentration as in Fig. 8.30. Both sets of data fit the theoretically predicted curve for a second-order kinetic dependence of OH radical decay from superequilibrium values in the postflame gases of rich hydrogen-air flames (Kaiser *et al.*, 1986).

Besides OH, other flame species are amenable to LIF measurements (e.g., C_2, CH, CN, and NO). In addition to species concentration, LIF measurements can be used to measure temperature and density (e.g., McKenzie and Gross, 1984). Flames of industrial interest are generally more complex than the flat flame we studied, and in this case, 2-D (or better, 3-D) imaging LIF methods are invaluable. Some of the early work in this area was performed by Dyer and Crosley (1982, 1984) and Kychakoff *et al.* (1982, 1984). More recently, 2-D LIF measurements of OH produced in the turbulent flame of an idealized internal combustion engine were obtained (Suntz *et al.*, 1988). Wolfrum (1988) used LIF of acetaldehyde as a tracer for the unburned gas in the same engine. In an engine

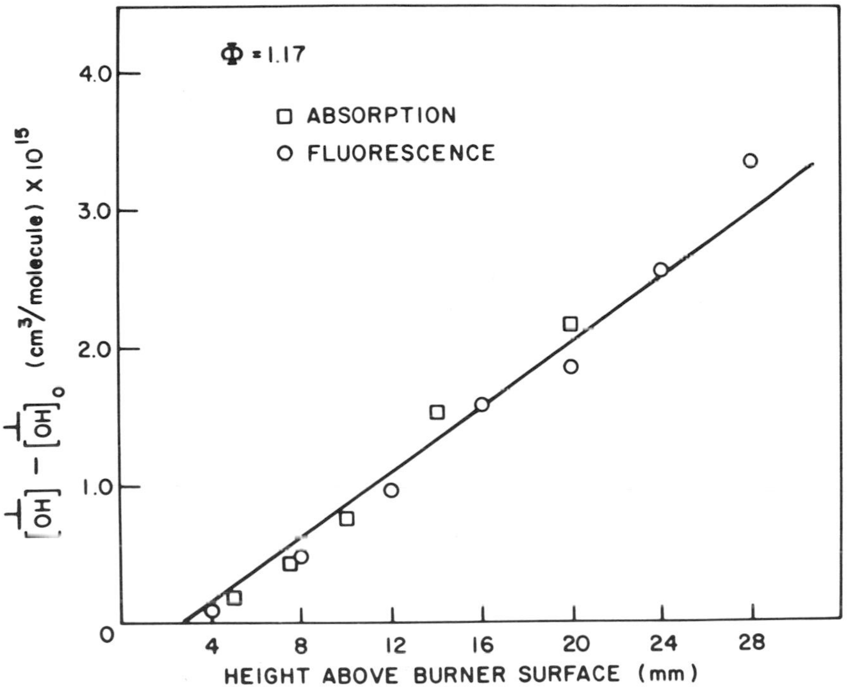

Fig. 8.30 Dye-laser–induced fluorescence and dye-laser absorption of OH above a flat flame. Both data sets obey second-order kinetics.

with a more realistic geometry, images of species distribution were obtained using LIF of NO and iso-octane as well as OH (Lambda Physik, 1988b). Application of the 2-D LIF technique to practical combustion systems provides measurements of the complex flames of interest to industry, which can be compared to finite element modeling results for the reacting flow field. The main obstacle to achieving quantitative LIF imaging measurements (correcting for competing processes such as collisional deexcitation) may be soluble even in highly absorbing or sooty flames (Hertz and Alden, 1987).

Time-resolved infrared spectral photography is a technique that provides a complete infrared absorption spectrum during a short pulse. A broadband (~15 nm) UV dye laser is downshifted into the infrared using stimulated electronic Raman scattering in an alkali metal vapor. Maintaining its original energy bandwidth, the pulse now has an IR bandwidth (~1000 cm^{-1}) that spans a significant portion of the IR spectrum. Such a powerful, tunable IR dye-laser–based source is novel in itself (Hodgson *et*

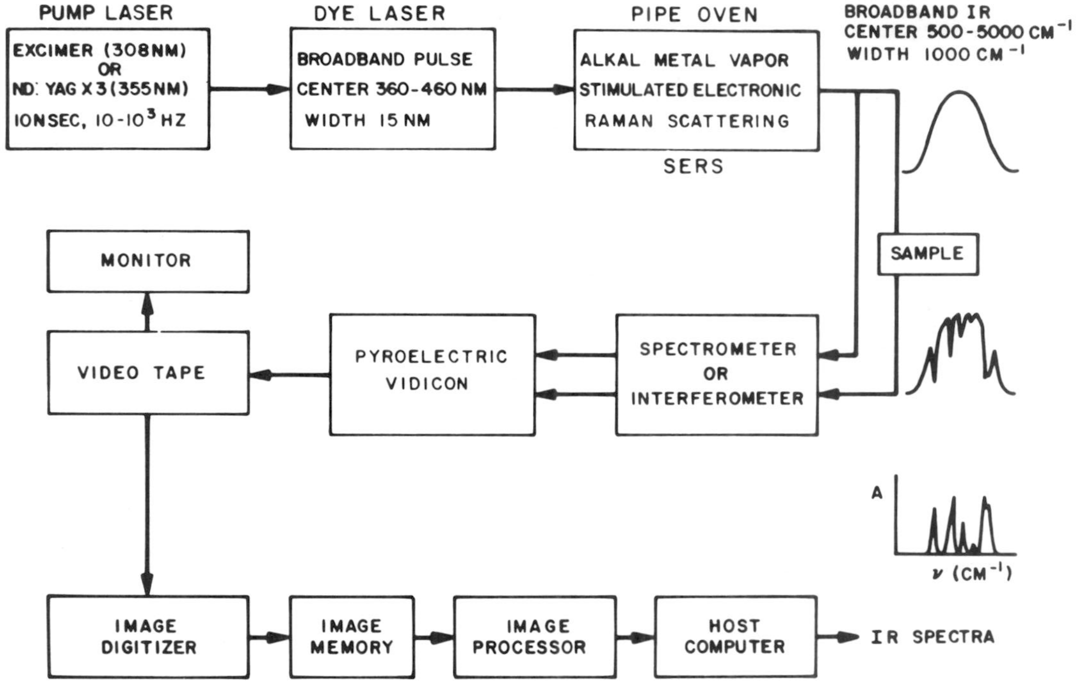

Fig. 8.31 Apparatus for time-resolved infrared spectral photography, a dye-laser–based instrument for transient IR absorption measurements.

al., 1975). Its application to infrared absorption (Bethune *et al.*, 1979) allowed single-pulse IR spectra to be acquired with 10-ns time resolution. More recently, the time resolution has been improved to the sub-picosecond regime (Glownia *et al.*, 1987). This technique has been applied to transient species concentration measurements of a number of reactions of industrial importance, particularly explosions (see Trott and Renlund (1987)). Our apparatus designed for combustion diagnostics (featuring a pyroelectric vidicon as the IR detector) is shown in Fig. 8.31. By choice of dye and alkali metal, the center wavelength of the broadband IR pulse can be tuned across the "fingerprint" region of the IR spectrum (2–20 μm or 500–5000 cm^{-1}). Two channels, reference and sample, are recorded on the pyroelectric vidicon after passing through the spectrometer. The ratio is taken and the result is the transient absorption spectrum.

Another nonlinear optical technique that is farther along the path to industrial usefulness is coherent anti-Stokes Raman scattering. Hall and Eckbreth (1984) review CARS, and a brief description of it will be given here, with an emphasis on industrial applications. As seen in Fig. 8.32a, in spontaneous Raman scattering the incident pump photon (energy ω_1, wave vector $\mathbf{k}_1$) is scattered from a medium such that it either loses (Stokes shift, ω_s, $\mathbf{k}_s$) or gains (anti-Stokes, ω_{as}, $\mathbf{k}_{as}$) vibrational or rotational energy to or from the medium. Spontaneous Raman scattering is weak, isotropic, and often masked by interfering light. It is often not suitable for measurements in industrial systems, whereas CARS data have been acquired on aircraft and automotive engines. CARS is strong, unidirectional,

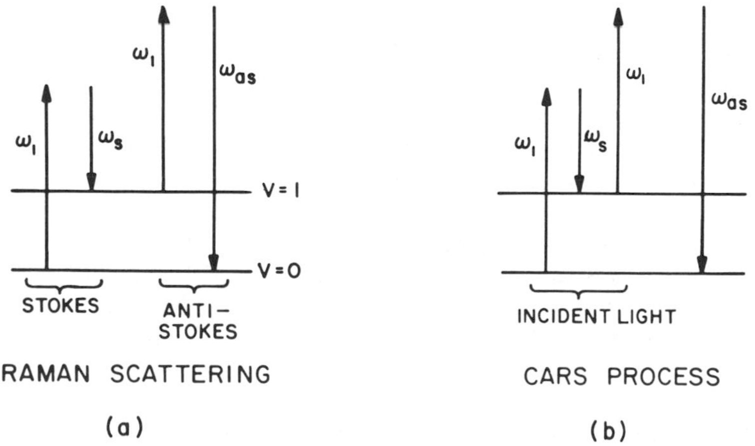

Fig. 8.32 Energy-level diagrams for (a) spontaneous Raman scattering and (b) coherent anti-Stokes Raman scattering (CARS).

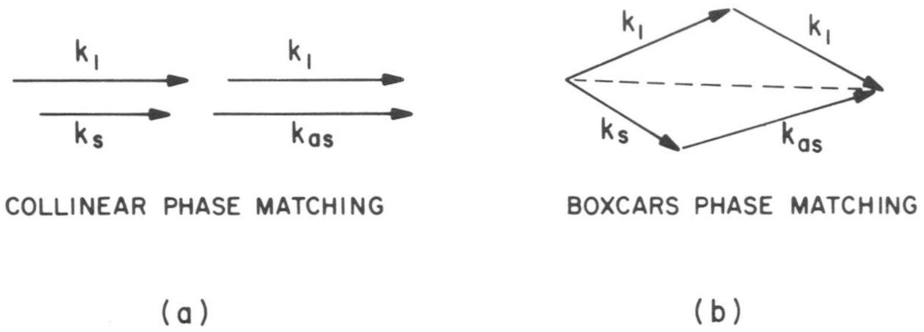

Fig. 8.33 Wave vector phase-matching diagrams for CARS: (a) collinear and (b) BOXCARS.

and not masked by interfering light (e.g., fluorescence, soot incandescence, and Rayleigh or Mie scattering). The acronym CARS indicates the manner in which these interferences are rejected. The coherent nature of CARS dictates that it emerges from the sample as a beam, whereas the interfering light is more isotropic and can be spatially filtered. The anti-Stokes nature of CARS means that the signal emerges at higher energy than the pump laser, whereas any potentially interfering LIF is generated at lower energy. As a Raman effect, CARS is active for all the species of interest in combustion, and the spectral features are generally not overlapping.

Figure 8.32b is an energy-level diagram for CARS, showing the similarity with spontaneous Raman. The same four photon energies are involved, but three photons (two pump and one probe at ω_s) are incident, and the signal appears at ω_{as}. The incident probe laser beam is generally a dye laser, since it must be tuned to a resonant condition (that is, the level spacing $\Delta V = \omega_1 - \omega_s$). Not only energy has to be conserved, but also wave vector, a condition called phase-matching. In Fig. 8.33a, the simplest phase-matching is depicted, with all the wave vectors collinear. The spatial resolution here is poor (even if the beams are focused) because CARS signal can be generated anywhere along the path. In Fig. 8.33b, the wave vectors form a phase-matching box, a technique providing high spatial resolution that was called BOXCARS (Eckbreth, 1978). When the three incident beams cross at the correct angles, the signal beam emerges in the prescribed direction. Maintaining beam alignment in the turbulent combustion environment found in most industrial systems is difficult with classical BOXCARS, due to thermal beam deflection.

We invented a BOXCARS variant (depicted in Fig. 8.34) and used it for in-cylinder temperature measurements in an internal combustion (IC) engine (Klick *et al.*, 1981; Davis *et al.*, 1981). Our phase-matching scheme

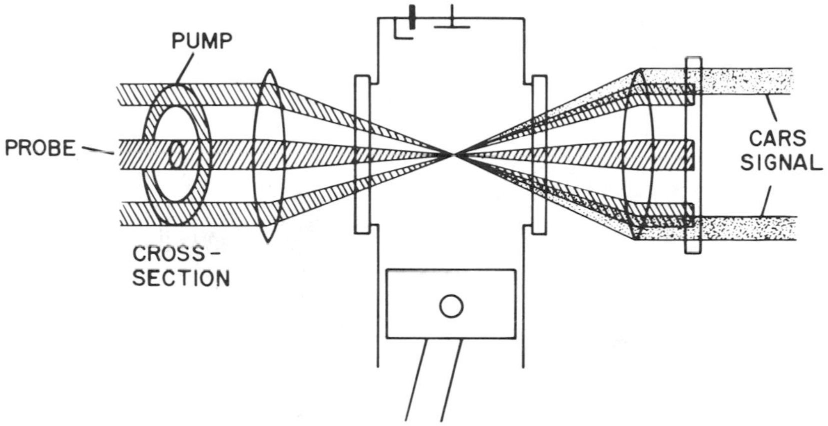

Fig. 8.34 USED CARS phase-matching in an internal combustion engine. The Gaussian probe beam is coaxial with an annular pump beam.

has the probe beam centered within an annular pump beam, so that whatever beam-steering occurs will tend to affect pump and probe similarly. This arrangement is also convenient, since the most common pump laser is a doubled Nd:YAG annular beam produced by an unstable resonator cavity, and the most common probe is a transversely pumped dye laser which has a narrow Gaussian beam. The annular CARS signal beam is produced at the focus, and is passed by an edge filter. Our phase-matching technique of Fig. 8.34 is now a standard geometry for industrial CARS measurements (see Snelling *et al.* (1985), Eckbreth and Anderson (1986, 1987), Wolfrum (1988), and Eckbreth *et al.* (1988)), and was dubbed USED CARS (unstable-resonator spatially enhanced detection CARS) by others (Eckbreth *et al.*, 1984).

Besides high spatial resolution, we require high time resolution, implying single-pulse measurements. Since the system under study would change while we tuned the probe dye laser to scan a spectrum, we instead acquire an entire spectrum with a single pulse of a broadband dye laser. The technique is outlined in Fig. 8.35. The broadband probe is tuned to overlap the Raman lines of the molecule of interest. For temperature measurements of fuel-air flames, it is common to analyze nitrogen spectra since N_2 is an abundant molecule that does not take part in combustion. Temperature is deduced from the spectral broadening that accompanies higher temperatures as the upper ro-vibrational levels are populated. A CARS signal is produced which reflects the Raman spectrum modulated by the shape of the broadband dye spectrum. To correct for the modulation, the beams are recrossed in a reference cell containing a high-pressure gas

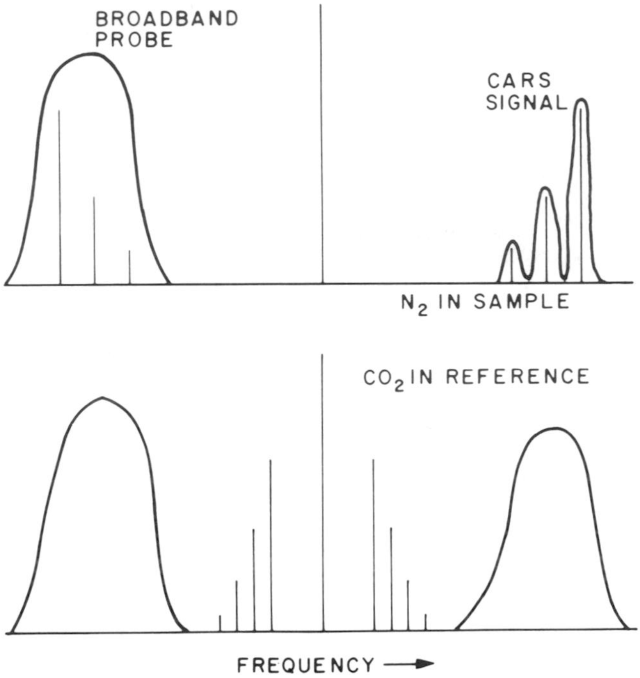

Fig. 8.35 Spectra of probe and pump input beams and CARS output beam for N_2 in the sample channel (top) and CO_2 in the reference channel (bottom). Center line is monochromatic pump spectrum.

with no lines in this spectral region. Characteristic of CARS, a nonresonant background signal is produced at the anti-Stokes wavelength that has the dye-laser spectral shape. The corrected spectrum is the ratio of signal to reference.

Our apparatus to acquire these spectra in an IC engine is depicted in Fig. 8.36. Both CARS signals, reference and engine, are dispersed in a spectrometer and viewed with an intensified TV camera. A typical video frame is shown in Fig. 8.37. Frames are recorded on videotape, to be digitized and processed off-line. Figure 8.38 depicts a view up from the piston of the industry-standard single-cylinder engine we used, fueled with either propane or gasoline. A 10-Hz laser repetition rate provided a pulse each time the four-stroke engine cycle repeated as it ran at 1200 rpm. The standard 30-Hz framing rate of the videotape was sufficient to record independent data from each pulse. As such, the apparatus was capable of recording a great deal of data, which is important in order to achieve the statistics required for accurate temperature determinations. We measured

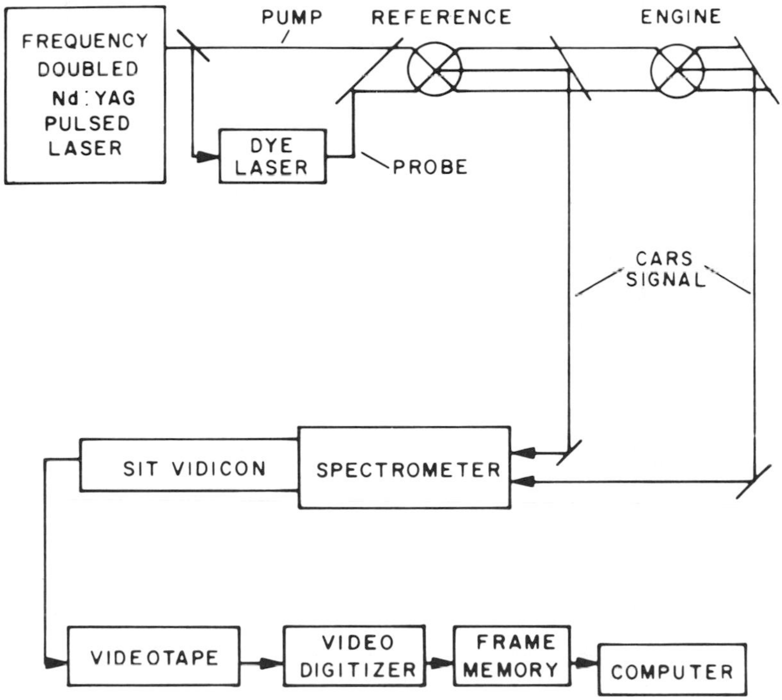

Fig. 8.36 Apparatus for CARS measurements in an IC engine using video techniques for acquiring many cycles of single-pulse spectra.

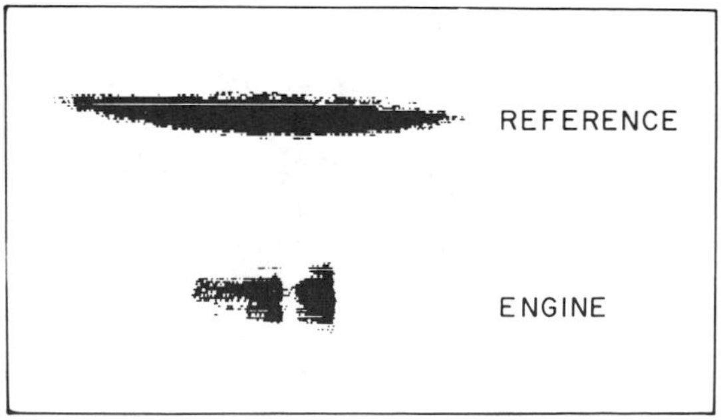

Fig. 8.37 Typical frame of digitized video data showing engine N_2 and reference spectra acquired by the apparatus of Fig. 8.36.

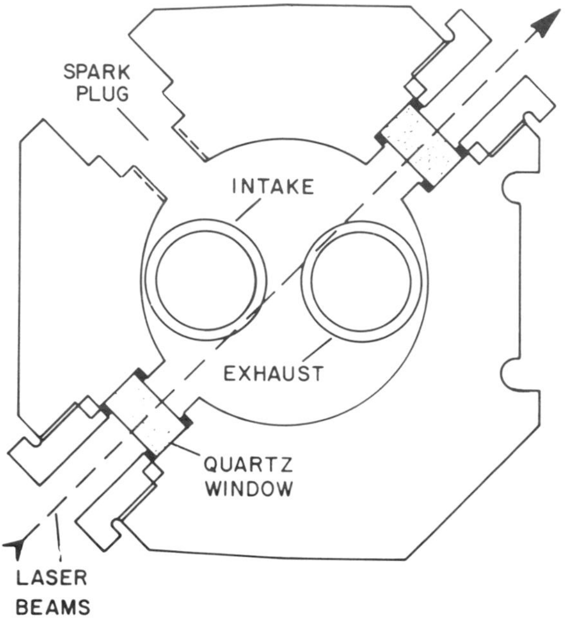

Fig. 8.38 Single-cylinder IC engine for laser diagnostics.

(Klick *et al.*, 1984) a wide distribution of temperatures at any given time in the cycle, which is known as cyclic variation.

Figure 8.39 displays the marked variation of single-pulse CARS nitrogen spectra with time in the engine cycle (expressed in degrees of crankshaft rotation after piston top dead center or °ATDC). At −30° ATDC, the intake gases are being compressed, but the temperature is low enough that the N_2 molecules are in the ground state. Burn occurs near TDC, and at 30° ATDC two higher vibrational bands and all the rotational lines within them are populated. At the later times shown, the gases have cooled somewhat during the exhaust stroke, so most of the rotational lines of the lowest vibrational band are populated. In order to derive temperatures from the spectra, they must be fit to calculated spectra. This turns out to be a major difficulty with CARS, as the spectra are very complex. There is a vast number of rotational lines which interact with each other and the density-dependent nonresonant background. For the high pressures and temperatures of an IC engine environment, one must include linewidth changes with p and T, as well as collisional narrowing.

However, the effects are sufficiently well understood that, with effort, the calculated spectra can match the data. Figure 8.40 shows data and a

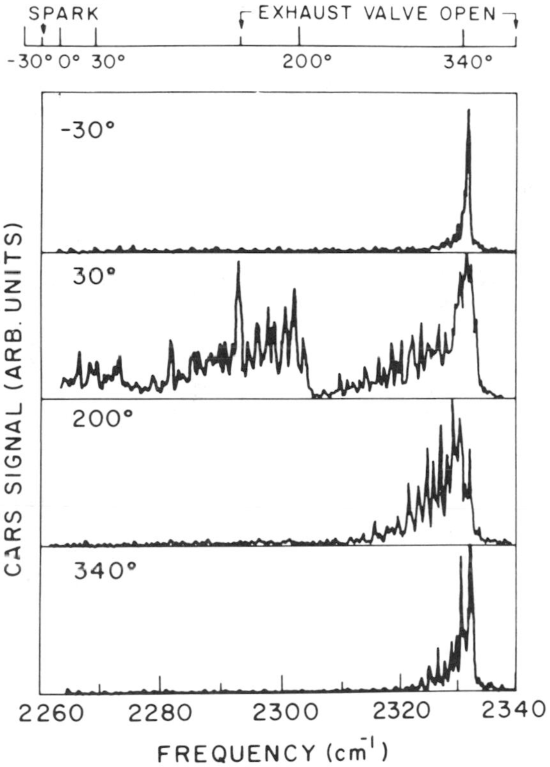

Fig. 8.39 At top is a timeline of events versus IC engine cycle time (expressed in degrees of crankshaft rotation). In-cylinder N_2 spectra are plotted at four cycle times.

corresponding calculation from a motored engine (no combustion) at $-100°$ ATDC. These are cycle-averaged data in orthogonal polarization to bring out the weak O bands of N_2. A transducer measures the pressure for input to the calculation, which has no free parameters other than temperature. Agreement with the data is excellent, and the usefulness of having high-resolution (0.5-cm^{-1}) spectra for comparing narrow lines is obvious. The dispersion-like character to the lines in Fig. 8.40 is due to interaction with the nonresonant background, which is significant for these weak lines. The effect is explained in Fig. 8.41. The third-order susceptibility of the medium χ, which controls the spectral shape for CARS, can be decomposed into a nearly constant nonresonant real part χ_{NR} (due to resonances at other energies), and a complex part $\chi_1 + i\chi_2$ (having the shapes indicated in Fig. 8.41) due to the resonance of interest. The CARS spectral

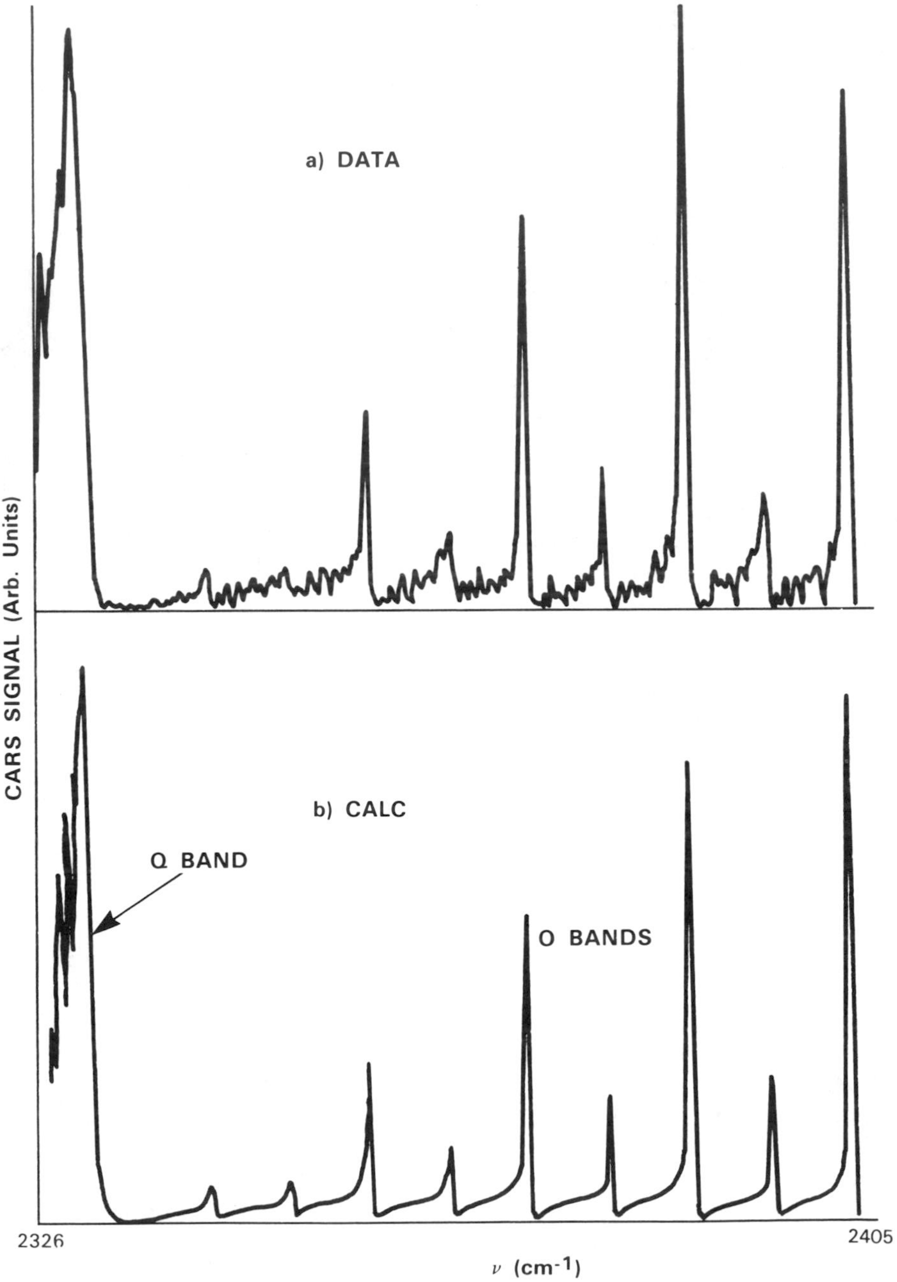

Fig. 8.40 Cycle-averaged N_2 O-band spectra (a) measured in a motored IC engine and (b) calculated at the measured pressure.

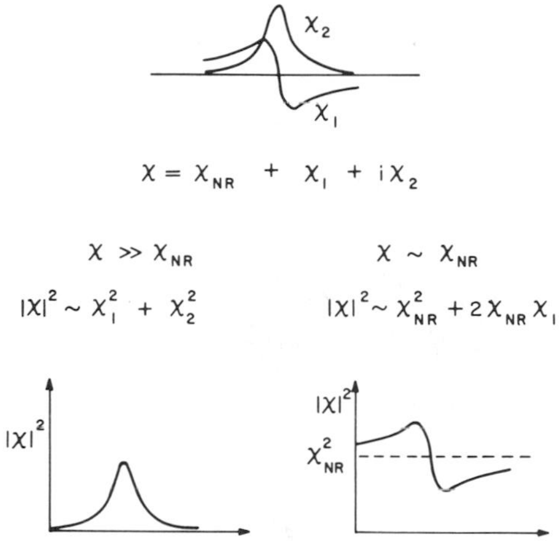

Fig. 8.41 Motivation for the CARS spectral shapes seen in Figs. 8.40 and 8.42. Shape can be explained by the relative size of resonant and nonresonant terms in the susceptibility χ.

shape factor $|\chi|^2$ is either peaked or dispersive, depending on whether χ_{NR} dominates χ.

A dispersive shape is also seen in the Q bands of N_2 for the fired engine data of Fig. 8.42. Two single-pulse spectra are shown, both taken at 30° ATDC at a pressure of 35 atmospheres. The spectra have relatively low resolution so that the nonresonant background can be displayed at high and low energy. Rhodamine 640 in the broadband dye laser is tuned with pH and solvent to the two energy extremes for the two measured spectra. The background is seen to be constant on one side of the resonance and slowly rising from zero on the other side. Our calculation at $T = 3000$K shows similar features, as well as very good agreement with the Q-band shape. Shape differences in the two measured Q-band spectra demonstrate the magnitude of cyclic variation in temperature.

The small-scale structure of the measured spectra in Fig. 8.42 is largely determined by fluctuations (mode-hopping) in the broadband dye laser. Since the interaction of beams in the reference cell and sample cannot be made identical, taking the ratio does not perfectly normalize the sample spectra. A large part of the temperature imprecision (±80K rms) for single-pulse spectra measured by us in a stable burner (Klick *et al.*, 1982) can be ascribed to dye-laser noise (Klick *et al.*, 1984). Others have also investigated the precision of single-pulse CARS temperature measurements. Barton and Garneau (1987) find a standard deviation of 73K for results in a stable burner. Greenhalgh *et al.* (1984) find that single-pulse

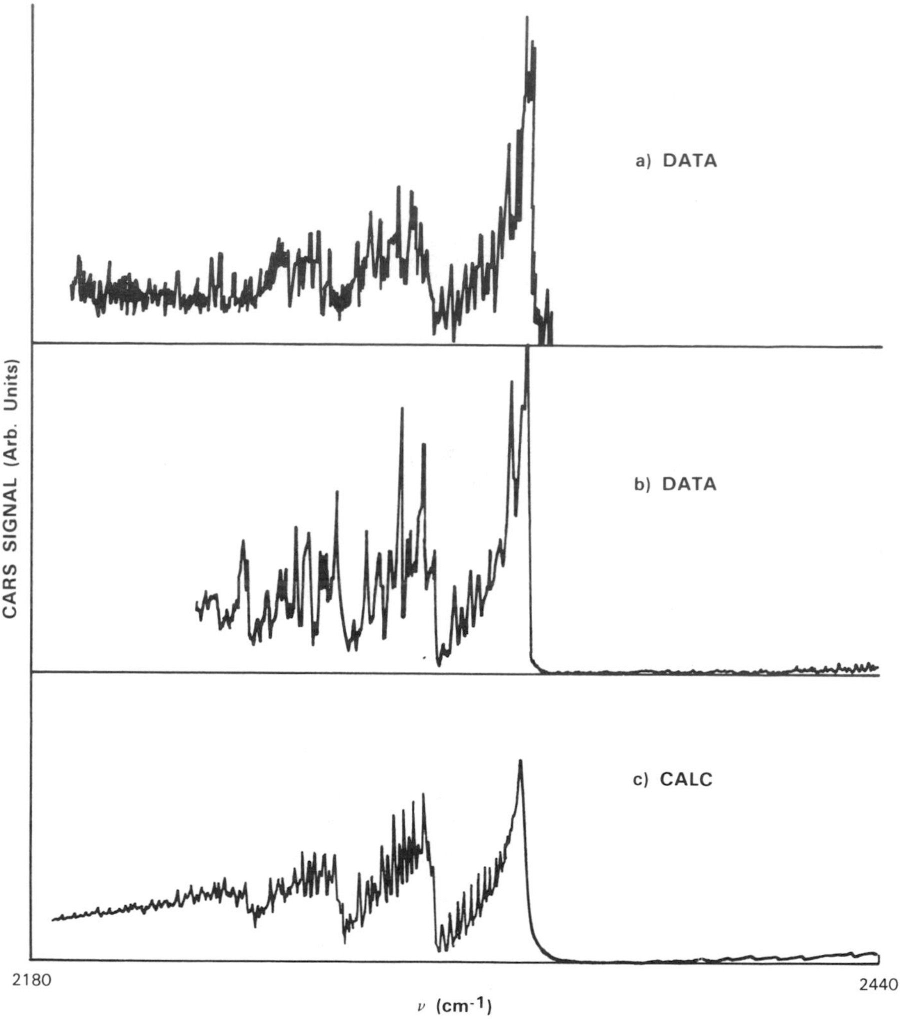

Fig. 8.42 Single-pulse N_2 Q-band spectra measured at the same cycle time in a firing IC engine at (a) low and (b) high wavenumber, and (c) calculated at the measured pressure (35 atm.) and a temperature of 3000K.

temperature measurements are accurate to 5 to 6%. Pealat *et al.* (1985) find an accuracy of 3 to 5%. Snelling *et al.* (1987) have performed an exhaustive study of this question and improved the standard deviation to 42K. Broadband dye-laser noise (spectral, spatial, and temporal) has received much attention as a source of imprecision in CARS measurements (see Greenhalgh and Whittley (1985) and Kroll *et al.* (1987)). The former suggest the novel technique of using a corner cube resonator in the dye laser to reduce mode noise. Li *et al.* (1987) have used a saturated dye amplifier with a nonsaturated absorber to temporally smooth broadband dye-laser pulses. It has only recently become clear that the characteristics of the seemingly simple broadband dye laser (first demonstrated by Sorokin and Lankard (1966)) have profound effects on CARS measurements.

There have been other CARS measurements carried out in IC engines, besides the work discussed here. Stenhouse *et al.* (1979) employed a scanned, collinear apparatus, which is not very appropriate for turbulent combustion. Rahn *et al.* (1982) also scanned the spectrum, but in an engine with little cyclic variation. Alessandretti and Violino (1983) and Marie and Cottereau (1987) have also made in-cylinder CARS measurements. Lucht (1987) employed broadband CARS to measure unburned gas temperatures versus distance from the cylinder wall for knock studies. He also measured local fuel-air ratios using three-laser CARS. Researchers at Harwell (Williams *et al.* 1986) have acquired CARS data using limited optical access in production engines. And Kajiyama *et al.* (1982) even carried out CARS measurements in the sooty environment of a diesel engine. CARS studies of aircraft engines are even more numerous (see Switzer and Goss (1982), Taran and Pealat (1982), and Goss *et al.* (1983)). The group at United Technologies is in the forefront of this field (see Eckbreth (1980) and Eckbreth *et al.* (1984, 1988)). Industrial furnaces have been probed by the CARS technique (see Alden and Wallin (1985) and Beiting (1986)). Fleming *et al.* (1984) were able to make CARS measurements during rocket propellant explosions. CARS experiments have also been performed on incandescent lamps (see Trebino and Farrow (1986)) and discharges (see Taran (1986)).

A number of clever techniques have been incorporated to facilitate the practical application of CARS. Ways to traverse the measurement point across the combustion system under study (Eckbreth and Stufflebeam, 1985) and to collect the CARS signal with fiber optics in harsh environments (Eckbreth, 1979) have been demonstrated. Transportable CARS devices have been built (Anderson *et al.*, 1986). The use of two dye lasers has allowed stimultaneous monitoring of the two species at the different dye frequencies, and other species in the overlap region (see Alden *et al.*, (1984), Eckbreth and Anderson (1985, 1986, 1987), Lucht (1987), and Eckbreth *et al.* (1988)). This last example demonstrates that innovative approaches to the use of dye lasers are vital to improvements in CARS, which will lead to its widening application. In summary, CARS is widely used as an industrial diagnostic because its strong signal beam provides immunity from interfering effects and because it can be employed remotely with relatively simple optical access. High space and time resolutions are achievable, which are important in the inhomogeneous and rapidly evolving systems under study. Among the chief difficulties are the complexity of equipment and data processing required, and extracting the signal from the nonresonant background for dilute molecules (<1000 ppm). Enhancing the stability of broadband dye lasers would increase the precision of CARS measurements.

There are many dye-laser–based schemes for combustion diagnostics other than the ones discussed here. In photodeflection spectroscopy (see Gupta (1986)), absorption by the medium of the dye-laser pump beam causes heating, which deflects a low power probe beam. Photoacoustic spectroscopy monitors the same heating with a microphone (see Tam (1986)). In optogalvanic spectroscopy, absorption of the dye-laser photons leads to selective ionization, which can be sensed with electrodes (see Goldsmith (1987)). More exotic dye-laser–based combustion diagnostics (e.g., asynchronous optical sampling, Kneisler *et al.* (1989)) are discovered with regularity. It remains to be seen which of these will be applied to industrial measurements.

7. REMOTE SENSING APPLICATIONS OF DYE LASERS TO MONITOR INDUSTRIAL POLLUTION

Laser remote sensing of the environment has been reviewed by Melngailis (1972), Hall (1974), Duley (1983), Killinger and Mooradian (1983), and Measures (1984, 1988). A general term for single-ended methods for remote molecular detection (as opposed to double-ended path absorption) is lidar (light detection and ranging), where a pulsed laser provides the desired range information. The discussion here is confined to the use of dye lasers to remotely monitor industrial pollution. Such measurements have been made in air on smokestack emissions and in water on effluent plumes and oil slicks. Preferred techniques are differential absorption lidar (DIAL) and fluorescent lidar, both of which use a tunable laser (such as a dye laser) to precisely tune onto (or off) molecular absorption lines. In DIAL, scattering from distant particles is measured on and off the absorption line. The range-resolved difference between the signals allows one to map the concentration of the molecule versus distance. Thus the sensitivity of path absorption is combined with the convenience of single-ended lidar. The theory of DIAL was presented by Byer and Garbuny (1973), including calculations of sensitivity using reflection from particles and from solid surfaces or retroreflectors, which may be available for local measurements of industrial air pollution.

An early application of the DIAL technique (Rothe *et al.* 1974) used a dye laser and measured pollutant emissions in a smokestack plume and from a chemical factory. They were able to create a 2-D map with 100-m resolution of the NO_2 concentration over the factory area. Concentrations as low as 0.2 ppm were measured at 4 km. Konefal *et al.* (1981) made similar measurements of NO_2 above the chimney of a chemical factory. Fredriksson and Hertz (1984) used a mobile system based on a YAG-pumped dye laser to map out the NO_2 in plumes from plants producing nitric acid and saltpeter. SO_2 concentrations were measured in the smoke plumes of electric power plants burning fossil fuels (Gustafsson *et al.*, 1977; Adrian *et al.*, 1979). A flashlamp-pumped dye laser was doubled to the vicinity of 300 nm to provide an "eye-safe" UV pulse. Plumes were tracked to a distance of 2–3 km, and concentrations of 70 ppb were measured at a kilometer range. Edner *et al.* (1987, 1988) measured SO_2 and NO in a plume and NO_2 over an urban area during a temperature inversion using a mobile dye laser system. The same group (Egeback *et al.*, 1984) mapped SO_2 concentrations above or downwind from a paper mill, a metallurgical plant, an oil refinery, a heating plant, and chemical plants using a mobile YAG-pumped doubled–dye-laser system. Flashlamp-pumped dye lasers were used (Lahmann *et al.*, 1984) to measure SO_2 in plumes above a factory and urban NO_2 concentrations (at the 10-ppb level out to 6-km range). Other species with absorption lines in the UV-visible region accessible to dye lasers are O_3, Hg, NH_3, benzene, methyl ketone, phosgene, ethylene, propylene, and vinyl chloride (Dubinsky, 1988). Lines in the deeper UV (with wavelengths shorter than 250 nm) are not usable from remote locations due to O_2 absorption in the intervening atmosphere (Rothe and Walther, 1976). Korb and Kalshoven (1985), however, used dye-laser DIAL at the O_2 resonance to remotely measure temperature and pressure, which could be useful in industrial applications.

Dye lasers with high pulse energies are desired for lidar measurements. Comparing two types of lasers doubled to 300 nm, a flashlamp-pumped dye laser achieved a 140-mJ pulse output (Schotland, 1980), while a Nd:YAG-pumped dye laser reached 55 mJ (Bos, 1981a). The state of the art in mobile systems for measuring atmospheric industrial pollution with DIAL is represented by the Nd:YAG-pumped doubled–dye-laser system built into a truck by Edner *et al.* (1987), and the excimer-pumped dual–dye-laser system in a van at the Swiss Ecole Polytechnique Lausanne (Lambda Physik, 1988c). Dye lasers for DIAL ideally produce two pulses, one on and the other off of the absorption line, within a one-ms interval. Atmospheric turbulence is sufficiently stationary within one ms such that the two measurements to be compared are then made over the same path. By separating the pulses in time, the detector does not have to discriminate in wavelength between two spectrally nearby returns. Rather than using two synchronized lasers, the pulses can be derived from a double-pulse or high-prf pump laser and a single dye laser that switches between two tuning elements (Browell, 1976; Sage and Aubry, 1982; Hung and Brechignac, 1988). Saito *et al.* (1986) developed a dye laser emitting more than two wavelengths for simultaneous DIAL measurement of more than one species. Molecular absorption peaks are broadened at atmospheric pressure and may have significant structure. Dye-laser bandwidth and wavelength stability can affect the concentration measurement, as can interfering species. It may be advantageous to measure the spectrum near the peak using a broadband or scanned narrowband dye laser, rather than take on- and off-line readings (Woods and Jolliffe, 1978; Fujii and Masamura, 1978). DIAL is often performed with multiline IR gas lasers rather than with dye lasers because a wider variety of molecules can be accessed and because the beam is then eye-safe (Byer, 1975).

Fluorescent lidar, or laser-induced fluorescence, has been discussed earlier in Section 6. Andersson *et al.* (1987) review the application of this technique to remote measurements of pollutants in the air and in water. Davis *et al.* (1979) calculated that the molecules OH, NO, SO_2, and CH_2O are readily detectable in the atmosphere by LIF. More importantly, they identified a larger group of molecules (such as NO_3, HNO_2, NO_2, HNO_3, H_2O_2, and CS_2) that could be decomposed by an initial photolysis laser pulse, followed by a pulse that induces fluorescence from a photofragment. Bradshaw *et al.* (1982) built a system to measure NO and SO_2 concentrations at the ppt level for acid-rain studies. They mixed doubled–dye-laser radiation with the fundamental from the Nd:YAG pump laser to produce excitation at 222 and 226 nm and then measured single-photon LIF. Fluorescent lidar has been employed to detect, classify, and measure the thickness of oil slicks. It is particularly useful for detecting intentional oil spills at night (Dubinsky, 1988). An excimer-pumped dye laser allowed Cecchi *et al.* (1987) to tune to a wavelength (420 nm) where the light could penetrate thick oil layers. Then the magnitude of the fluorescence signal was indicative of the film thickness. Measuring the fluorescence spectra at two excitation wavelengths allowed them to identify the type of oil in the slick and, thus, potentially, the source. Monitoring other organic contamination of water and the mapping of effluent plumes may also be accomplished using fluorescent lidar (Dubinsky, 1988).

Dye-laser–based lidar is nearing routine industrial application. In Sweden, the severity of the acid-rain problem has led to a decision by one of the major polluting industries to install a DIAL system for the control of SO_2 emissions (Egeback *et al.*, 1984). Dial sensing of NO_2 has now achieved the sensitivity and accuracy of conventional point-monitoring instruments (Fredriksson and Hertz, 1984). The ease with which a mobile dye-laser–based system can be driven to a factory for remote measurements of pollution has led Edner *et al.* (1987) to promote a similar technique for in-plant measurements. The laser system can be protected in a properly designed environment, and the beam routed into the factory via fiberoptics for combustion studies, surface monitoring, chemical process diagnostics, or indeed any of the measurements discussed in this chapter.

REFERENCES

Adrian, R.S., Brassington, D.J., Sutton, S., and Varey, R.H. (1979). The measurement of SO_2 in power station plumes with differential lidar. *Opt. Quant. Electr.* **11,** 253–264.

Akerman, M.A., Klick, D.I., Mayrick, R., Paul, G.L., Reid, G., Supurovic, D., and Tsuda, H. (1986). Laser curing of coatings and inks. European Patent Application no. 86303326.2, date of filing 1 May 1986.

Alden, M., and Wallin, S. (1985). CARS experiments in a full-scale (10 × 10 m) industrial coal furnace. *Appl. Opt.* **24,** 3434–3439.

Alden, M., Fredriksson, K., and Wallin, S. (1984). Application of a two-color dye laser in CARS experiments for fast determinations of temperatures. *Appl. Opt.* **23,** 2053–2057.

Aldridge, A., Francis, P., and Hutchison, J. (1984). UV curing of TiO_2 pigmented coatings—Evaluation of lamps with differing spectral characteristics. *J. Oil Col. Chem Assoc.* **2,** 33–39.

Alessandretti, G.C., and Violino, P. (1983). Thermometry by CARS in an automobile engine. *J. Phys. D* **16,** 1583–1594.

Anderson, T.J., Dobbs, G.M, and Eckbreth, A.C. (1986). Mobile CARS instrument for combustion and plasma diagnostics. *Appl. Opt.* **25,** 4076–4085.

Andersson, P.S., Montan, S., and Svanberg, S. (1987). Remote sample characterization based on fluorescence monitoring. *Appl. Phys. B.* **44,** 19–28.

Antonov, V.S., and Hohla, K.L. (1983). Dye stability under excimer-laser pumping. II. Visible and UV dyes. *Appl. Phys. B* **32,** 9–14.

Asher, S. (1984). Some important considerations in the selection of a tunable UV laser excitation source. *Appl. Spectros.* **38,** 276–278.

Barton, S.A., and Garneau, J.M. (1987). Effect of pump-laser linewidth on noise in single-pulse coherent anti-Stokes Raman spectroscopy temperature measurements. *Opt. Lett.* **12,** 486–488.

Bauer, S.H. (1984). Overview—Laser applications to industrial chemistry. *Proc. SPIE* **458,** 2–10.

Beiting, E.J. (1986). Multiplex CARS temperature measurements in a coal-fired MHD environment. *Appl. Opt.* **25,** 1684–1689.

Bernhardt, A.F. (1983). Erratum: Excimer vs. YAG. *Laser Focus* (Jan.) 94–95.

Berhardt, A.F., Herbst, R.L., and Kronick, M.N. (1982). A comparison of excimer-pumped and YAG-pumped dye laser systems. *Laser Focus* (Oct.) 59–63.

Bethune, D.S. (1981). Dye cell design for high-power low-divergence excimer-pumped dye lasers. *Appl. Opt.* **20,** 1897–1899.

Bethune, D.S., Lankard, J.R., and Sorokin, P.P. (1979). Time-resolved infrared spectral photography. *Opt. Lett.* **4,** 103–105.

Bondar, M.V., Przhonskaya, O.V., Romanov, A.G., Tikhonov, E.A., and Khomenko, A.F. (1987). Polymer dye lasers with a 10 kHz repetition rate. *Sov. Phys. Tech. Phys.* **31,** 1439–1440.

Bos, F. (1981a) Versatile high-power single-longitudinal-mode pulsed dye laser. *Appl. Opt.* **20,** 1886–1890.

Bos, F. (1981b). Optimization of spectral coverage in an eight-cell oscillator-amplifier dye laser pumped at 308 nm. *Appl. Opt.* **20,** 3553–3556.

Bradshaw, J.D., Rodgers, M.O., and Davis, D.D. (1982). Single photon laser-induced fluorescence detection of NO and SO_2 for atmospheric conditions of composition and pressure. *Appl. Opt.* **21,** 2493–2500.

Brink, D.J., and van der Hoeven, C.J. (1984). Excimer-pumped dye laser with high beam quality. *Rev. Sci. Instrum.* **55,** 1948–1951.

Browell, E.V. (1976). Two-wavelength double-pulse tunable dye laser for DIAL applications. In *Proceedings of the Conference on Lasers and Electro-Optic Systems.* Optical Society of America, Washington, D.C., pp. 52–53.

Brueck, S.R.J., and Kildal, H. (1982) Efficient Raman frequency conversion in liquid nitrogen. *IEEE JQE* **18,** 310–312.

Burnham, R., Harris, N.W., and Djeu, N. (1976). Xenon fluoride laser excitation by transverse electric discharge. *Appl. Phys. Lett.* **28,** 86–87.

Burns, F.C., Dreyfus, R.W., and Susko, J.R. (1987) End point detection and control of laser induced dry chemical etching. U.S. Patent no. 4687539.

Byer, R.L. (1975). Review: Remote air pollution measurement. *Opt. and Quant. Electron.* **7,** 147–177.

Byer, R.L., and Garbuny, M. (1973). Pollutant detection by absorption using Mie scattering and topographic targets as retroreflectors. *Appl. Opt.* **12,** 1496–1505.

Cassard, P., Corkum, P.B., and Alcock, A.J. (1981). Solvent dependent characteristics of XeCl pumped UV dye lasers. *Appl. Phys.* **25,** 17–22.

Cecchi, G., Pantani, L., Radicati, B., Barbaro, A., and Mazzinghi, P. (1987). Oil film detection and characterization by lidar fluorosensors. In *Topical Meeting on Laser and Optical Remote Sensing: Instrumentation and Techniques Technical Digest Series, vol. 18.* Optical Society of America, Washington, D.C., pp. 106–108.

Contolini, R.J., and Osinski, J.S. (1986). Novel tunable pulsed dye laser holography and subsequent processing for fabrication of optoelectronic device gratings. *Elec. Lett.* **22,** 971–973.

Cooper, D.E., Bataj, J., Newman, P.R., Rentzepis, P.M., and Hutchinson, J.A. (1987). Laser diagnostics of cadmium telluride: Crystal quality and exciton dynamics. In *Technical Digest of the Topical Meeting on Lasers in Materials Diagnostics, vol. 7.* Optical Society of America, Washington, D.C., pp. 118–125.

Crosley, D.R., ed. (1980). *Laser Probes for Combustion Chemistry.* ACS Symposium Series. American Chemical Society, Washington, D.C.

Davis, D.D., Heaps, W.S., Philen, D., Rodgers, M., McGee, T., Nelson, A., and Moriarty, A.J. (1979). Airborne laser induced fluorescence system for measuring OH and other trace gases in the parts-per-quadrillion to parts-per-trillion range. *Rev. Sci. Instrum.* **50,** 1505–1516.

Davis, L.C., Marko, K.A., and Rimai, L. (1981). Angular distribution of coherent Raman emission in degenerate four-wave mixing with pumping by a single diffraction coupled laser beam: Configurations for high spatial resolution. *Appl. Opt.* **20,** 1685–1690.

Dearth, J.J., Vaughn, V.V., McGowan, R.B., Erlich, J., and Conrad, R.W. (1987). A flashlamp pumped zig-zag slab dye laser. In *Proceedings of the International Conference on Lasers '87* (Duarte, F.J., ed.). STS Press, McLean, Va., pp. 320–329.

Decker, C. (1983). Ultra-fast polymerization of epoxy-acrylate resins by pulsed laser irradiation. *J. Polymer Sci.* **21,** 2451–2461.

Decker, C. (1984a). UV curing of acrylate coatings by laser beams. *J. Coating Technol.* **56,** 29–34.

Decker, C. (1984b). Laser-induced polymerization. In *Materials for Microlithography*. ACS Symp. Series **266,** 207–223.
Decker, C., and Jenkins, A.D. (1985). Kinetic approach of O_2 inhibition in ultraviolet- and laser-induced polymerizations. *Macromolecules* **18,** 1241–244.
Deutsch, T.F., and Brass, M. (1969). Laser-pumped dye lasers near 4000 Å. *IEEE JQE* **5,** 260–261.
Dubinsky, R.N. (1988). Lidar moves toward the 21st century. *Lasers & Optronics* (April), 93–106.
Duley, W.W. (1983). *Laser Processing and Analysis of Materials.* Plenum, New York.
Dunning, F.B., and Stebbings, R.F. (1974). The efficient generation of tunable near UV radiation using a N_2 pumped dye laser. *Opt. Commun.* **11,** 112–114.
Dyer, M.J., and Crosley, D.R. (1982). Two-dimensional imaging of OH laser-induced fluorescence flame. *Opt. Lett.* **7,** 382–384.
Dyer, M.J., and Crosley, D.R. (1984). Rapidly sequenced pair of two-dimensional images of OH laser-induced fluorescence in a flame. *Opt. Lett.* **9,** 217–219.
Eckbreth, A.C. (1978). BOXCARS: Crossed-beam phase-matched CARS generation in gases. *Appl. Phys. Lett.* **32,** 421–423.
Eckbreth, A.C. (1979). Remote detection of CARS employing fiber optic guides. *Appl. Opt.* **18,** 3215–3216.
Eckbreth, A.C. (1980). CARS thermometry in practical combustors. *Combust. Flame* **39,** 133–148.
Eckbreth, A.C. (1988). *Laser Diagnostics for Combustion Temperature and Species.* Abacus Press, Tunbridge Wells, Great Britain.
Eckbreth, A.C., and Anderson, T.J. (1985). Dual broadband CARS for simultaneous multiple species measurements. *Appl. Opt.* **24,** 2731–2736.
Eckbreth, A.C., and Anderson, T.J. (1986). Dual broadband USED CARS. *Appl. Opt.* **25,** 1534–1539.
Eckbreth, A.C., and Anderson, T.J. (1987). Multi-color CARS for simultaneous measurements of multiple combustion species. *Proc. SPIE* **742,** 34–41.
Eckbreth, A.C., and Stufflebeam, J.H. (1985). Simple CARS measurement point traversing scheme. *Appl. Opt.* **24,** 1405–1410.
Eckbreth, A.C., Dobbs, G.M., Stufflebeam, J.H., and Tellex, P.A. (1984). CARS temperature and species measurements in augmented jet engine exhausts. *Appl. Opt.* **23,** 1328–1332.
Eckbreth, A.C., Anderson, T.J., and Dobbs, G.M. (1988). Multi-color CARS for hydrogen-fueled scramjet applications. *Appl. Phys. B* **45,** 215–223.
Edner, H., Fredriksson, K., Sunesson, A., Svanberg, S., Uneus, L., and Wendt, W. (1987). Mobile remote sensing system for atmospheric monitoring. *Appl. Opt.* **26,** 4330–4338.
Edner, H., Sunesson, A., and Svanberg, S. (1988). NO plume mapping by laser-radar techniques. *Opt. Lett.* **13,** 704–706.
Egeback, A.L., Fredriksson, K.A., and Hertz, H.M. (1984). DIAL techniques for the control of sulfur dioxide emissions. *Appl. Opt.* **23,** 722–729.
El-Sayed, M.A., ed. (1986). *Laser Applications in Chemistry and Biophysics,* **vol. 620.** SPIE, Bellingham, Wash.
El-Sayed, M.A., ed. (1987). *Laser Applications to Chemical Dynamics,* **vol. 742.** SPIE, Bellingham, Wash.
Erickson, G.F. (1987). Solid hosts for dye laser rods. In *Proceedings of the International Conference on Lasers '87* (Duarte, F.J., ed.). STS Press, McLean, Va., pp. 338–348.
Everett, P.N., and Zollars, B.G. (1987). Engineering of a multi-beam 300-watt flashlamp-pumped dye-laser system. In *Proceedings of the International Conference on Lasers '87* (Duarte, F.J., ed.). STS Press, McLean, Va., pp. 291–303.

Feder, M.P., Smith, J.F., and Liberman, H.E. (1978). Nitrogen-pumped dye laser tool for fabricating LSI connections. In *Proceedings of the Conference on Lasers and Electro-optic Systems*. Optical Society of America, Washington, D.C., pp. 92–93.

Fleming, J.W., Barber, W.H., and Wilmont, G.B. (1984). Optical diagnostics of high pressure solid propellant flames. *Proc. SPIE* **482,** 74–78.

Fletcher, A.N., Bliss, D.E., Pietrak, M.E., McManis, G.E. (1985). Improving the output and lifetime of flashlamp-pumped dye lasers. In *Proceedings of the International Conference on Lasers '85* (Wang, C.P., ed.). STS Press, McLean, Va., pp. 797–804.

Fouassier, J., Jacques, P., Lougnot, D., and Pilot, T. (1984). Lasers, photoinitiators, and monomers: A fashionable formulation. *Polymer Photochem.* **5,** 57–76.

Fredriksson, K.A., and Hertz, H.M. (1984). Evaluation of the DIAL technique for studies on NO_2 using a mobile lidar system. *Appl. Opt.* **23,** 1403–1411.

Frosch, R.A., Moacanin, J., Gupta, A., and Hong, S. (1981). Double-beam optical method and apparatus for measuring thermal diffusivity and other molecular dynamic processes in utilizing the transient thermal lens effect. U.S. Patent no. 4243327.

Fujii, Y., and Masamura, T. (1978). Detection of atmospheric pollutants by quantitative analytical spectroscopy using a continuously scanned tunable dye laser. *Opt. Eng.* **17,** 147–152.

Fulgham, S.F., Trainor, D.W., Duzy, C., and Hyman, H.A. (1984). Stimulated Raman scattering of XeF* laser radiation in H_2—Part II. *IEEE JQE* **20,** 218–221.

Furumoto, H. (1986). State of the art high energy and high average power flashlamp excited dye lasers. *Proc. SPIE* **609,** 111–128.

Glownia, J.H., Misewich, J., and Sorokin, P.P. (1987). Subpicosecond time-resolved infrared spectral photography. *Opt. Lett.* **12,** 19–21.

Goldsmith, J.E.M. (1987). Optogalvanic spectroscopy. In *Laser Spectroscopy VIII*, Springer Ser. Opt. Sci. **vol. 55** (Persson, W., and Svanberg, S., eds.). Springer, Berlin.

Goss, L.P., Trump, D.D., MacDonald, B.G., and Switzer, G.L. (1983). 10 Hz coherent anti-Stokes Raman spectroscopy apparatus for turbulent combustion studies. *Rev. Sci. Intrum.* **54,** 563–571.

Greenhalgh, D.A., and Whittley, S.T. (1985). Mode noise is broadband CARS spectroscopy. *Appl. Opt.* **24,** 907–913. Erratum. *Appl. Opt.* **26,** 768.

Greenhalgh, D.A., Porter, F.M., England, W.A., Williams, D.R., Whittley, S.T., Glass, D.H., Barton, S.A., Devonshire, R., and Dring, I. (1984). Quantitative CARS spectroscopy. In *Proceedings of the IXth International Conference on Raman Spectroscopy*, Tokyo, pp. 342–343.

Gupta, R. (1986). Pulsed photothermal deflection spectroscopy in fluid media: A review. In *Proceedings of the International Conference on Lasers '86* (McMillan, R.W., ed.). STS Press, McLean, Va., pp. 379–387.

Gustafsson, G., Hartmann, B., Spangstedt, G., and Steinvall, O. (1977). "Stenungsund-77": Smoke plume measurements with a pulsed dye laser. FOA Report no. C 30124-E1.

Hall, F.F. (1974). Laser systems for monitoring the environment. In *Laser Applications* (Ross, M., ed.). Academic Press, New York, pp. 161–225.

Hall, R.B. (1982). Lasers in industrial chemical synthesis. *Laser Focus* (Sept), 57–62.

Hall, R.J., and Eckbreth, A.C. (1984). Coherent anti-Stokes Raman spectroscopy (CARS): Application to combustion diagnostics. In *Laser Applications,* **Vol. 5** (Ready, J.F., and Erf, R.K., eds.). Academic Press, New York, pp. 213–309.

Hargrove, R.S., and Kan, T. (1977). Efficient, high average power dye amplifiers pumped by copper vapor lasers. *IEEE JQE* **13,** (Dec.), 28D.

Hertz, H.M., and Alden, M. (1987). Calibration of imaging laser-induced fluorescence measurements in highly absorbing flames. *Appl. Phys. B* **42,** 97–102.

Hodgson, R.T., Lankard, J.R., Sorokin, P.P., and Wynne, J.J. (1975). Tunable infrared/ultraviolet laser. U.S. Patent no. 3892979.

Hohla, K.L. (1982a). Excimer-pumped dye lasers—The new generation. *Laser Focus* (June), 67–74.

Hohla, K.L. (1982b). YAG vs. excimer-pumped dyes. *Laser Focus* (Dec.), 8–68.

Holmes, L.M. (1988). Industrial excimer lasers. *Laser Focus* (May), 80–85.

Horvath, J.J., and Semerjian, H.G. (1987). Laser-excited fluorescence studies of paper pulping liquors. In *Conference on Lasers and Electro-Optics Technical Digest Series,* **vol. 14.** Optical Society of America, Washington, D.C., p. 152.

Hoyle, C.E., Trapp, M., and Chang, C.H. (1987). Laser-initiated polymerization: The effect of pulse repetition rate. *Polym. Mater. Sci. Eng.* **57,** 579–582.

Hung, N.D., and Brechignac, P. (1988). Tunable alternate double-wavelength single grating dye laser for DIAL systems. *Appl. Opt.* **27,** 1906–1908.

Hunt, R.B., and Pappalardo, R.G. (1985). Fast excited-state relaxation of Eu-Eu pairs in commercial $Y_2O_3{:}Eu^{3+}$ phosphors. *J. Luminescence* **34,** 133–146.

Jacobi, M., and Henne, A. (1983). Acylphosphine oxides: A new class of UV initiators. *J. Rad. Curing* (Oct), 16–24.

Jain, K., Willson, C., and Lin, B. (1982). Ultrafast deep UV lithography with excimer lasers. *IEEE Elec. Dev. Lett.* **3,** 53–58.

Janes, G.S., Aldag, H.R., Pacheco, D.P., and Schlier, R.E. (1987). A long pulse duration, good beam quality, supersonically "scanned-beam" dye laser. In *Proceedings of the International Conference on Lasers '87* (Duarte, F.J., ed). STS Press, McLean, Va., pp. 356–363.

Johnson, S.M., and Johnson, L.G. (1986). Contactless measurement of bulk free-carrier lifetime in cast polycrystalline silicon ingots. *J. Appl. Phys.* **60,** 2008–2015.

Kaiser, E.W., Marko, K., Klick, D., Rimai, L., Wang, C.C., Shirinzadeh, B., and Zhou, D. (1986). Measurement of OH density profiles in atmospheric-pressure propane-air flames. *Combust. Sci. and Tech.* **50,** 163–183.

Kajiyama, K., Sajiki, K., Kataoka, H., Maeda, S., and Hirose, C. (1982). N_2 CARS thermometry in diesel engine. SAE Paper no. 821036.

Karlicek, R.F., Donnelly, V.M., and Johnston, W.D. (1983). Laser spectroscopic investigation of gas-phase processes relevant to semiconductor device fabrication. In *Laser Diagnostics and Photochemical Processing for Semiconductor Devices, Materials Research Society Symposium Proceedings,* **vol. 17** (Osgood, R.M., Brueck, S.R.J., and Schlossberg, H.R., eds.). North-Holland, New York, pp. 151–160.

Ketchen, M.B., Grischkowsky, D.R., Halbout, J.-M., Chi, C.C., Duling, I.N., Gallagher, W.J., Li, G.P., May, P.G., and Scheuermann, M. (1986). Sub-picosecond electrical signals for probing VLSI environments. In *Extended Abstracts of the 18th International Conference on Solid State Devices and Materials, Tokyo*, 733–734.

Killinger, D.K., and Mooradian, A. (1983). *Optical and Laser Remote Sensing.* Springer-Verlag, Berlin.

Klauminzer, G.K. (1988). Cost considerations for industrial excimer lasers. *Laser Focus* (Dec.), 108–111.

Klick, D.I., and Kaiser, E.W. (1984). Hydroxyl fluorescence profiles above a flat flame using an optical multichannel analyzer. *Appl. Opt.* **23,** 4184–4186.

Klick, D.I., Marko, K.A., and Rimai, L. (1981). Broadband single-pulse CARS Spectra in a fired internal combustion engine. *Appl. Opt.* **20,** 1178–1181.

Klick, D.I., Marko, K.A., and Rimai, L. (1982). Temperature measurements for combustion diagnostics from high-resolution single-pulse CARS N_2 spectra. In *Temperature: Its Measurement and Control in Science and Industry,* **vol. 5** (Schooley, J.F., ed). American Institute of Physics, New York, pp. 615–620.

Klick, D.I., Marko, K.A., and Rimai, L. (1984). Optical multichannel analysis with rapid mass storage of spectra: Application to CARS measurements of temperature fluctuations. *Appl. Opt.* **23,** 1347–1352.

Klick, D.I., Akerman, M.A., Paul, G.L., Supurovic, D., and Tsuda, H. (1988). Excimer laser curing of polymer coatings. *Proc. SPIE* **998,** 95–104.

Klick, D.I., Akerman, M.A., Paul, G.L., Supurovic, D., and Tsuda, H. (1989). Fourier transform infrared (FTIR) kinetics diagnostics of thin film polymerization photoinitiated by excimer laser pulses. *Proc. SPIE* **1056,** 56–63.

Kneisler, R.J., Lytle, F.E., Fiechtner, G.J., Jiang, Y., King, G.B., and Laurendeau, N.M. (1989). Asynchronous optical sampling: A new combustion diagnostic for potential use in turbulent, high-pressure flames. *Opt. Lett.* **14,** 260–262.

Konefal, Z., Szczepanski, J., and Heldt, J. (1981). NO_2 detection in the atmosphere using differential absorption lidar. *Acta Physica Polonica* **A60,** 273–278.

Korb, C.L., and Kalshoven, J.E. (1985). Method of and apparatus for measuring temperature and pressure. U.S. Patent no. 4493553.

Kortz, P.H. (1982). YAG-pumped dye lasers: The proven tunable source. *Laser Focus* (July), 57–63.

Kotzubanov, V.D., Naboikin, Y.V., Ogurtsova, L.A., Podgornyi, A.P., and Pokrovskaya, F.S. (1968). Laser action in solutions of organic luminophors in the 400–650 nm range. *Opt. Spectros.* **25,** 406–410.

Kroll, S., Alden, M., Berglind, T., and Hall, R.J. (1987). Noise characteristics of single shot broadband Raman-resonant CARS with single- and multimode lasers. *Appl. Opt.* **26,** 1068–1073.

Kychakoff, G., Howe, R.D., Hanson, R.K., and McDaniel, J.C. (1982). Quantitative visualization of combustion species in a plane. *Appl. Opt.* **21,** 3225–3229.

Kychakoff, G., Howe, R.D., Hanson, R.K., and McDaniel, J.C. (1984). Visualization of turbulent flame fronts with planar laser-induced fluorescence. *Science* **224,** 382–383.

Lahmann, W., Staehr, W., Weitkamp, C., and Michaelis, W. (1984). State-of-the-art DAS lidar for SO_2 and NO_2. In *Proceedings of IGARSS '84—Remote Sensing—From Research Towards Operational Use.* ESA SP-215, pp. 685–688.

Lambda Physik (1988a). Testing a XeCl laser: 10 billion laser pulses *Lambda Industrial* no. 3 (Sept.), 1–7.

Lambda Physik (1988b). Combustion optimization pushed forward by excimer LIF methods. *Lambda Highlights* no. 14 (Dec.), 1–5.

Lambda Physik (1988c). Excimer and dye lasers sensing the atmosphere. *Lambda Highlights* no. 12 (Aug.), 1–4.

Lambda Physik (1988d). Laser surface analysis. *Lambda Highlights* no. 9 (Feb.), 1–5.

Lavi, S., Amit, M., Bialolanker, G., Miron, E., and Levin, L.A. (1985). High-repetition-rate high-power variable-bandwidth dye laser. *Appl. Opt.* **24,** 1905–1909.

Layman, P.L. (1985). Paints and coatings: The global challenge. *Chem. Eng. News* (Sept. 30), 27-68.

Li, Z.W., Radzewicz, C., and Raymer, M.G. (1987). Temporal smoothing of multimode dye-laser pulses. *Opt. Lett.* **12,** 416–418.

Loree, T.R., Sze, R.C., Barker, D.L., and Scott, P.B. (1979). New lines in the UV: SRS of excimer laser wavelengths. *IEEE JQE* **15,** 337–342.

Lucht, R.P. (1987). CARS diagnostics of internal combustion engine processes. In *Conference on Lasers and Electro-Optics Technical Digest Series*, **vol. 14.** Optical Society of America, Washington, D.C., pp. 232–233.

Marie, J.J., and Cottereau, M.J. (1987). CARS thermometry in an internal combustion engine. *Entropie* **23,** 64–74.

McGinniss, V. (1974). UV and laser curing of pigmented polymerizable binders. U.S. Patent no. 3,847,771.

McKenzie, R.L., and Gross, K.P. (1984). Laser-induced fluorescence measures temperatures in turbulent flows. *Laser Focus (Oct.), 31–32.*

Measures, R.M. (1984). *Laser Remote Sensing.* Wiley, New York.

Measures, R.M. (1988). *Laser Remote Chemical Analysis.* Wiley, New York.

Melngailis, I. (1972). The use of lasers in pollution monitoring. *IEEE Trans. Geosci. Elec.* **GE-10,** 7–17.

Mibu, H., Ishihara, T., Tanaka, C., and Esaki, S. (1980). Two-step irradiation of pigmented coatings. U.S. Patent no. 4,182,665.

Miyauchi, T., Mizukoshi, K., Hongo, M., Mitani, M., Okunaka, M., Kawanabe, T., and Tanabe, I. (1984). Method and apparatus for redressing defective photomask. U.S. Patent no. 4,463,073.

Moore, C.A., and Decker, C.D. (1978). Power-scaling effects in dye lasers under high-power laser excitation. *J. Appl. Phys.* **49,** 47–52.

Morton, R.G., and Dragoo, V. (1981). Reliable high average power high pulse energy dye laser. *IEEE JQE* **17,** 2222–2227.

Muto, S., Ichikawa, A., Ando, A., Ito, C., and Inaba, H. (1986). Trial to plastic fiber dye laser. *Trans. IECE Japan* **E69,** 374–375.

Naboikin, Y.V., Ogurtsova, L.A., Podgornyi, A.P., Pokrovskaya, F.S., Grigoryeva, V.I., Krasovitskii, B.M., Kutsyna, L.M., and Tishchenko, V.G. (1970). Spectral and energy characteristics of organic molecular lasers in polymers and toluene. *Opt. Spectros.* **28,** 528–532.

Nakabayashi, S. (1975). UV curing of opaque film. SME Tech. Paper FC75-322.

O'Brien, J.J., Miller, D.C., and Atkinson, G.H. (1987). Intracavity laser spectroscopy: Analysis of chemical vapor deposition processes. In *Conference on Lasers and Electro-Optics Technical Digest Series,* **vol. 14.** Optical Society of America, Washington, D.C., pp. 216–217.

Okada, T., Nishigoori, T., Kajiyama, Y., and Maeda, M. (1987). LIF diagnostics of C_2 radical behaviour in a laser CVD environment. *Appl. Phys. B* **44,** 175–179.

Olaj, O., Bitai, I., and Hinkelmann, F. (1987). The laser-flash-initiated polymerization as a tool of evaluating (individual) kinetic constants of free-radical polymerization, 2) The direct determination of the rate constant of chain propagation. *Makromol. Chem.* **188,** 1689–1702.

Omenetto, N. (1988). The impact of several atomic and molecular laser spectroscopic techniques for chemical analysis. *Appl. Phys. B* **46,** 209–220.

Pacheco, D.P., Aldag, H.R., Itzkan, I., and Rostler, P.S. (1987). A solid-state flash-lamp-pumped dye laser employing polymer hosts. In *Proceedings of the International Conference on Lasers '87* (Duarte, F.J., ed.). STS Press, McLean, Va., pp. 330–337.

Paisner, J.A. (1988). Atomic vapor laser isotope separation. *Appl. Phys. B* **46,** 253–260.

Pealat, M., Bouchardy, P., Lefevre, M., and Taran, J.P. (1985). Precision of multiplex CARS temperature measurements. *Appl. Opt.* **24,** 1012–1017.

Perry, J.W., Shing, Y.H., Coulter, D.R., and Radhakrishnan, G. (1987). Coherent anti-Stokes Raman spectroscopy diagnostics of a silane radio frequency plasma. In *Conference on Lasers and Electro-Optics Technical Digest Series,* **vol. 14.** Optical Society of America, Washington, D.C., pp. 214–215.

Phillips, R. (1978). UV curing of acrylate materials with high intensity flash. *J. Oil Col. Chem. Assoc.* **61,** 233–240.

Piper, J.A., and Little, C.E. (1989). High-power violet Sr^+ recombination lasers. *Proc. SPIE* **1041,** 167–174.

Pleil, M.W., Gangopadhyay, S. Landis, C., and Borst, W. (1987). Fluorescence decay times of multicomponent microscopic materials by pulsed laser excitation. *Proc. SPIE* **743,** 86–93.

Podmaniczky, A. (1983). Diameter measurement of optical fibers and wires by laser. *Proc. SPIE* **398,** 274–280.

Rahn, L.A., Johnston, S.C., Farrow, R.L., and Mattern, P.L. (1982). CARS thermometry in an internal combustion engine. In *Temperature: Its Measurement and Control in Science and Industry,* **vol. 5** (Schooley, J.F., ed.). American Institute of Physics, New York, pp. 609–613.

Reddy, K.V., Bray, R.G., and Berry, M.J. (1978). Dye laser-induced photochemistry. In *Advances in Laser Chemistry* (Zewail, A.H., ed.). Springer-Verlag, Berlin, pp. 48–61.

Reilly, J.P. (1987). Applications of laser ionization to mass spectrometry and photoelectron spectroscopy. In *Conference on Lasers and Electro-Optics Technical Digest Series,* **vol. 14.** Optical Society of America, Washington, D.C., p. 308.

Rothe, K. W., and Walther, H. (1976). Remote sensing using tunable lasers. In *Tunable Lasers and Applications, Springer Series in Optical Sciences,* **vol. 3** (Mooradian, A., Jaeger, T., and Stokseth, P., eds.). Springer-Verlag, New York, pp. 279–293.

Rothe, K.W., Brinkmann, U., and Walther, H. (1974). Remote sensing of NO_2 emission from a chemical factory by the differential absorption technique. *Appl. Phys.* **4,** 181–182.

Rottler, C.L., Tang, K.Y., Treacy, E.B., White, C.J. (1987). An 800-J excimer-pumped dye laser. In *Conference on Lasers and Electro-Optics Technical Digest Series,* **vol. 14.** Optical Society of America, Washington, D.C.)., pp. 298–299.

Sage, J.P., and Aubry, Y. (1982). High power tunable dual frequency laser system. *Opt. Commun.* **42,** 428–431.

Saito, Y., Nomura, A., and Kano, T. (1986). Development of a simultaneous multi-wavelength dye laser for differential absorption technique. In *Thirteenth International Laser Radar Conference.* NASA Conference Publication 2431, pp. 306–309.

Schäfer, F.P. (1976). Dye laser technology. In *Tunable Lasers and Applications, Springer Series in Optical Sciences,* **vol. 3** (Mooradian, A., Jaeger, T., and Stokseth, P., eds.). Springer-Verlag, New York, pp. 50–59.

Schäfer, F.P., ed. (1977). *Dye lasers, Topics in Applied Physics,* **vol. 1.** Springer-Verlag, Berlin.

Schäfer, F.P. (1988). Lasers for chemical applications. *Appl. Phys. B* **46,** 199–208.

Schäfer, F.P., Schmidt, W., and Volze, J. (1966). Organic dye solution laser. *Appl. Phys. Lett.* **9,** 306–309.

Schotland, R.M. (1980). Efficient high-energy SHG using a triaxial flashlamp-pumped dye laser. *Appl. Opt.* **19,** 124–126.

Schuster, S.E., and Cook, P.W. (1976). Laser personalization of integrated circuits. *Proc. SPIE* **86,** 102–104.

Seibert, E.J., and Johnson, L.G. (1981). Operation of a high power, broadly tunable infrared dye laser. *Opt. Commun.* **39,** 186–189.

Selwyn, G.S. (1987). Spatially resolved detection of plasma etchant species by two-photon laser-induced fluorescence. In *Conference on Lasers and Electro-Optics Technical Digest Series,* **vol. 14.** Optical Society of America, Washington, D.C., pp. 214–215.

Snelling, D.R., Sawchuk, R.A., and Mueller, R.E. (1985). Single pulse CARS noise: A comparison between single-mode and multimode pump lasers. *Appl. Opt.* **24,** 2771–2778.

Snelling, D.R., Smallwood, G.J., Sawchuk, R.A., and Parameswaran, T. (1987). Precision of multiplex CARS temperatures using both single-mode and multimode pump lasers. *Appl. Opt.* **26,** 99–110.

Sorokin, P.P., and Lankard, J.R. (1966). Simulated emission observed from an organic dye, chloro-aluminum phthalocyanine. *IBM J. Res. Devel.* **10,** 162–163.

Sorokin, P.P., Lankard, J.R., Hammond, E.C., and Moruzzi, V.L. (1967). Laser-pumped stimulated emission from organic dyes: Experimental studies and analytical comparisons. *IBM J. Res. Devel.* **11,** 130–138.

Srojny, N., and de Silva, J.A.F. (1980). Laser induced fluorometric analysis of drugs. *Anal. Chem.* **52,** 1554–1559.

Stenhouse, I.A., Williams, D.R., Cole, J.B., and Swords, M.D. (1979). CARS measurements in an internal combustion engine. *Appl. Opt.* **18,** 3819–3825.

Suntz, R., Becker, H., Monkhouse, P., and Wolfrum, J. (1988). Two dimensional visualization of the flame front in an internal combustion engine by laser-induced fluorescence of OH radicals. *Appl. Phys. B* **47,** 287–293.

Sutton, D.G., and Capelle, G.A. (1976). KrF-laser-pumped tunable dye laser in the ultraviolet. *Appl. Phys. Lett.* **29,** 563–564.

Switzer, G.L., and Goss, L.P. (1982). A hardened CARS system for temperature and species-concentration measurements in practical combustion environments. In *Temperature: Its Measurement and Control in Science and Industry,* **vol. 5** (Schooley, J.F., ed.). American Institute of Physics, New York.

Tam, A.C. (1986). Applications of photoacoustic sensing techniques. *Rev. Mod. Phys.* **58,** 381–412.

Tam, A.C., and Sullivan, B. (1983). Remote sensing applications of pulsed photothermal radiometry. *Appl. Phys. Lett.* **43,** 333–335.

Taran, J.P. (1986). CARS diagnostics of discharges. In *Proceedings of the 10th International Conference on Raman Spectroscopy.* University of Oregon, Eugene, pp. 151–152.

Taran, J.P., and Pealat, M. (1982). Practical CARS temperature measurements. In *Temperature: Its Measurement and Control in Science and Industry,* **vol. 5** (Schooley, J.F., ed.). American Institute of Physics, New York.

Telle, H., Huffer, W., and Basting, D. (1981). The XeCl excimer laser: A powerful and efficient UV pumping source for tunable dye lasers. *Opt. Commun.* **38,** 402–406.

Thomson, B.N. (1988). The dye laser's broadening reach. *Photonics Spectra* (June), 163–166.

Tomin, V.I., Alcock, A.J., Sarjeant, W.J., and Leopold, K.E. (1978). Some characteristics of efficient dye laser emission obtained by pumping at 248 nm with a high power KrF* discharge laser. *Opt. Commun.* **26,** 396–401.

Trebino, R., and Farrow, R.L. (1986). CARS measurements of temperature profiles and pressure in a tungsten lamp. In *Proceedings of the 10th International Conference on Raman Spectroscopy.* University of Oregon, Eugene, pp. 1512–1513.

Trott, W.M., and Renlund, A.M. (1987). Optical methods for the study of explosive chemistry. In *Technical Digest of the Topical Meeting on Lasers in Materials Diagnostics,* **vol. 7.** Optical Society of America, Washington, D.C., pp. 32–35.

Tuckerman, D.B., and Schmitt, R.L. (1985). Pulsed laser planarization of metal films for multilevel interconnects. In *Proceedings of the 2nd International IEEE VLSI Multilevel Integrated Circuits Conference*, pp. 24–31.

Uchino, O., Mizunami, T., Maeda, M., and Miyazoe, Y. (1979). Efficient dye lasers pumped by a XeCl excimer laser. *Appl. Phys.* **19,** 35–37.

Weiner, A.M., Lin, P.S.D., Marcus, R.B., and Orloff, J. (1987). Photoemissive probing of high-speed signals. *Proc. SPIE* **774,** 78–80.

Wheeler, J.P. (1982). Excimer pumps—A new trend in dye lasers. *Electro-Optic Sys. Digest* (March), 37–41.

Williams, D.R., Baker, C.A., and Greenhalgh, D.A. (1986). CARS nitrogen thermometry at high pressure for I.C. engine applications. In *Proceedings of the 10th International Conference on Raman Spectroscopy*. University of Oregon, Eugene, pp. 1529–1530.

Wilson, R. (1975). Lasers and their utilization in the future of coatings. *J. Paint Technol.* **47,** 43–51.

Wolfrum, J. (1988). Laser spectroscopy for studying chemical processes. *Appl. Phys. B* **46,** 221–236.

Woodin, R.L., and Kaldor, A., eds. (1984). *Applications of Lasers to Industrial Chemistry,* **vol. 458.** SPIE, Bellingham, Wash.

Woods, P.T., and Jolliffe, B.W. (1978). Experimental and theoretical studies related to a dye laser differential lidar system for the determination of atmospheric SO_2 and NO_2 concentrations. *Opt. Laser Technol.* **10,** 25–28.

Wormhoudt, J., Stanton, A.C., and Silver, J.A. (1983). Spectroscopic techniques for characterization of gas phase species in plasma etching and vapor deposition processes. *Proc. SPIE* **452,** 88–99.

Yu, P.Y. (1978). Subpicosecond optoelectronic switch. *IBM Technical Disclosure Bulletin* **21,** 1270.

Zhang, Y., Stuke, M., Larciprete, R., and Borsella, E. (1987). Excimer laser photochemistry of Al-alkyls monitored by dye laser mass spectroscopy. In *Technical Digest of the Topical Meeting on Lasers in Materials Diagnostics,* **vol. 7.** Optical Society of America, Washington, D.C., pp. 110–111.

Chapter 9

DYE-LASER ISOTOPE SEPARATION

M.A. Akerman

Oak Ridge National Laboratory
Oak Ridge, Tennessee

1. INTRODUCTION

While this is primarily a grouping of references from 1978–1989 research in the area of dye-laser isotope separation, a few paragraphs will be devoted to a simple description of techniques and methods employed. The references include separation techniques applied not only to the lighter elements, but also the lanthanides and actinides. An entire section is devoted to a single actinide, uranium, since this is probably the most heavily researched single element.

The research includes the use of most of the major properties of lasers such as intensity, monochromaticity, collimation, pulselength, polarization, and tunability.

Basic methods of dye-laser isotope separation allow one to apply lasers in a number of different ways. First is the atomic vapor technique, which utilizes several narrowband lasers tuned to promote atoms to a specific intermediate excited state and ultimately to ionization. Second is a chemical technique in which a laser is used to excite an atom or molecule of a single isotope followed by a chemical reaction that occurs only with the laser-excited species. Third are hybrid techniques, in which a dye laser is used in conjunction with some other type of laser, or is used as a diagnostic in an isotope-separation process.

ISBN 0-12-222700-X

2. BIBLIOGRAPHY

The following bibliography pertains to isotope separation in which a dye laser is used. The dye laser may be used alone to cause separation, or it may be used with other types of lasers to cause separation, or it may be used as a diagnostic in an experiment pertaining to the general topic of isotope separation. There are many fine papers written in the 1960s and especially in 1970s, but the present bibliography concentrates on the advances between 1978 and 1989. While not complete, the listing represents the personal collection of several authors in this book as well as three major data bases. A still more complete listing could be generated by collecting the papers referred to in the papers themselves.

References are divided into several categories to simplify the search for specific information about isotope separation using dye lasers.

Isotope Separation as a Chemical Process includes articles dealing with the overall problem of separating isotopes. From these articles one could get an idea of the different isotopes that have come close to economic proof of principle in the 1980s. Cost of finished product compared to feedstock is considered. Capital costs and operating costs are derived based on individual isotope-separation schemes. The basics of lasers are discussed from the view point that one or more of the properties of laser light is used in each of a number of separation scenarios. Techniques rely heavily on the narrow linewidth generally associated with lasers, but also on the power level, directionality, coherence, polarization, and the short pulse characteristics that dye lasers exhibit.

Spectroscopic Separation or Observation of Various Isotopes is a collection of articles covering a very wide range of scientific investigations. A number deal with use of a dye laser to cause physical separation of an isotopic species from a beam of atoms. Resonance ionization mass spectroscopy couples the ability of a narrowband laser to selectively ionize a given isotope with a mass spectrometer for ultimately separating the different masses. The laser may be used to ionize the entire sample, or it may ionize only one or a few isotopic species. Still other applications concentrate on using a laser to examine products of nuclear reactions and to discover basic properties of nuclei.

Quantity Isotope Separation refers to papers where it appeared that a measurable quantity of some isotope could actually be separated or had been separated. Examples include lithium, iodine, chlorine, carbon, rare earths, tellurium, deuterium, rubidium, and beryllium. Mixed in with these are a number of papers where the laser was used to excite a molecule, thus stimulating chemical reactions to take place. Fewer than a half dozen examples of this type of separation turned up.

Uranium Isotope Separation includes papers that describe separation techniques applied specifically to uranium. Both two-step and three-step ionization are included, as well as information about dye lasers applied to uranium separation.

Laser Development for Isotope Separation includes papers about tunable dye lasers and excitation mechanisms for them. Copper-vapor laser-pumped dye lasers are heavily represented, but also flashlamp-pumped dye lasers, and combinations of dye lasers and nonlinear processes for wavelength conversion, such as into the midinfrared are included. One paper compares free-electron lasers to dye lasers for use in photochemistry, and a couple concentrate on the electronics and pulse power required for pump lasers.

3. REFERENCES

Isotope Separation as a Chemical Process

Chen, H., and Davis, J.I. (1982). Lasers in chemistry—nearing the breakthrough? *Photonics Spectra* **16**(10), 59–66.

Davis, J.I., Holtz, J.Z., and Spaeth, M.L. (1982). Status and prospects for lasers in isotope separation. *Laser Focus* **18**(9), 49–54.

Duley, W.W. (1983). *Laser Processing and Analysis of Materials.* Plenum, New York, pp. 195–232.

Hall, R.B. (1982). Lasers in industrial chemical synthesis. *Laser Focus* **18**(9), 57–62.

Laser Chemistry—Where and How the Laser Will be Used to Process Chemicals and Materials (undated). Technical Insights, New Jersey.

Letokhov, V.S. (1988). Laser-induced chemistry—basic nonlinear processes and applications. *Appl. Phys. B* **46**, 237–251.

Schäfer, F.P. (1988). Lasers for chemical applications. *Appl. Phys. B* **46**, 199–208.

Wanner, J. (1979). Laser chemical analysis of chemical reactions in the gas phase. *Acta Phys. Austriaca, Suppl. (Austria)* **20**, 133–155.

Spectroscopic Separation or Observation of Various Isotopes

Armstrong, D.P., McCulla, W.H., and Schweitzer, G.K. (1985). Separation of rare earth isotopes using resonance ionization time-of-flight mass spectrometry. In *Advances in Analytical Mass Spectrometry*. Lewis, Chelsea, Mich., pp. 119–125.

Bingham, C.R., Gaillard, M.L., Pegg, D.J., Carter, J.K., Mlekodaz, R.L., Cole, J.D., and Griffin, P.M. *International Conference on Fast Ion Beam Spectroscopy,* Quebec, Canada. NTIS CONF-810868-4.

Boesl, U., Neusser, H.J., and Schlag, E.W. (1979). Isotope selective molecular spectroscopy and production of isotopically pure molecules with a dye laser. In "Laser-Induced Processes in Molecules" *Springer Ser. Chem. Phys.*, **6**, Springer-Verlag, Berlin pp. 290–293.

Boesl, U., Neusser, H.J., and Schlag, E.W. (1979). Spectra of individual molecular isotopes in an unseparated natural mixture. *Chem. Phys. Lett.* **61**, 57–61.

Broglia, M., Zampetti, P., and Benetti, P. (1986). Single longitudinal mode interaction between a uranium atomic beam and a modified Hänsch-type pulsed dye laser. *Appl. Phys. B* **39,** 73–76.

Broyer, M., Chevaleyre, J., Delacretaz, G., and Woste, L. (1984). CVL-pumped dye laser for spectroscopic application. *Appl. Phys. B* **35,** 31–36.

Donohue, D.L., Christie, W.H., and Goeringer, D.E. (1985). Ion microprobe mass spectrometry using sputtering atomization and resonance ionization. In *International Conference on Ion Mass Spectrometry*, Washington, D.C., NTIS CONF-8509197.

Duong, H.T., Meyer, R.A., and Brenner, D.S. (1985). Laser spectroscopy of short-lived isotopes. In *American Chemical Society National Meeting*, Chicago, Ill., American Chemical Society, NTIS CONF-850942, pp. 380–385.

Gagne, J.M., Demers, Y., Dreze, C., and Pianarosa, P. (1983). ^{235}U isotope enrichment in the metastable levels of UI. *J. Opt. Soc. Am.* **73,** 498–499.

Gill, R.L., Meyer, R.A., and Brenner, D.S. (1985). Recent developments at TRISTAN: nuclear structure studies of neutron-rich nuclei. In *American Chemical Society National Meeting*, Chicago, Ill. American Chemical Society, NTIS-CONF-850942, pp. 420–425.

Kluge, H.J. (1984). The future of laser studies at isotope separators. In *Proceedings of the TRIUMF-ISOL Workshop*, NTIS-CONF-8406149, pp. 196–208.

MacGregor, I.J.D., Grant, I.S., and Walker, P.M. (1984). Current status of the Daresbury isotope separator. *Vacuum* (United Kingdom) **34,** 3–6.

Murnick, D.E., and Feld, M.S. (1979). Applications of lasers to nuclear physics. *Annu. Rev. Nucl. Sci.* **29,** 411–454.

Shaw, R.W., Young, J.P., and Smith, D.H. (1988). Removal of the samarium isobaric interference from promethium mass analysis. *Anal. Chem.* **60,** 282–283.

Skrotskii, G.V., and Solomakho, G.I. (1979). Isotope and isomer separation in an inhomogeneous magnetic field. *Sov. Tech. Phys. Lett.* **5,** 22–23.

Thibault, C., Guimbal, P., Klapisch, R., de Saint Simon, M., Serre, J.M., and Touchard, F. (1981). On-line spectroscopy with thermal atomic beams. *Nucl. Instrum. Methods Phys. Res. (Netherlands)* **186,** 193–199.

Touchard, F., Ejiri, H., and Fukuda, T. (1984). On-line laser spectroscopy of radioactive alkali isotopes. In *Conf. on the Changing Structure of the World Oil Industry.* World Scientific, Teaneck, N.J., pp. 115–132.

Whitaker, T.J., Bushaw, B.A., and Cannon, B.D. (1988). Laser-based techniques improve isotope analysis. *Laser Focus* **24**(2), 88–101.

Whitaker, T.J., Cannon, B.D., Bushaw, B.A., and Laing, W.R. (1985). High-resolution CW resonance ionization mass spectrometry. In *Oak Ridge National Laboratory Conference on Analytical Chemistry*, Knoxville, Tenn. Lewis, Chelsea, Mich., pp. 107–112.

Quantity Isotope Separation

Arisàwa, T., Maruyama, Y., Suzuki, Y., and Shiba, K. (1982). Lithium isotope separation by laser. *Appl. Phys. B* **28,** 73–76.

Boesl, U., Neusser, H.J., and Schlag, E.W., (1979). Production of isotopically pure molecules by dye laser excitation. *Chem. Phys. Lett.* **61,** 62–66.

Datta, S., Brauman, J.I., and Zare, R.N. (1979). Isotope enrichment and stereochemistry of the products from the reaction of electronically excited iodine monochloride with cis- and trans-1.2-dibromoethylene. *J. Am. Chem. Soc.* **101,** pp. 7173–7176.

Hedges, R.E.M., and Moore, C.B. (1978). Enrichment of ^{14}C and radiocarbon dating. *Nature* **276,** 255–257.

Hedges, R.E.M., Ho, P., and Moore, C.B. (1980). Enrichment of carbon-14 by selective laser photolysis of formaldehyde. *Z. Phys., A* (Fed. Rep. of Germany) **296,** 25–32.

Karlov, N.V., Krynetskii, B.B., Mishin, V.A., and Prokhorov, A.M. (1978). Laser isotope separation of rare earth elements. *Appl. Opt.* **17,** 856–862.

Kushawaha, V.S. (1980). Laser isotope separation of $^{129}I_2$. *Opt. Quant. Electron* **12,** 269–272.

Kushawaha, V.S. (1980). Laser-stimulated chemical reaction of $I_2 + C_2H_2$. *J. Am. Chem. Soc.* **102,** 256–258.

Larciprete R., and Stuke, M. (1986). Isotope selective ionization of tellurium dimers Te_2 formed by excimer laser photodissociation of an organometallic compound: $CH_3TeTecH_3$. *Appl. Phys. B* **41,** 213–215.

Latimer, C.J. (1979). Recent experiments involving highly excited atoms. *Contemp. Phys.* (United Kingdom) **20,** 631–653.

Lee, J.K.P., Crawford, J.E., Raut, V., Savard, G., Thekkadath, G., Duong, T.H., Pinard, J., and Talbert, W.L. (1986). Resonant ionization using synchronized laser pulses. In *Int. Conf. on Electromagnetic Isotope Separators and Techniques Related to Their Applications.* Elsevier Science, New York, pp. 444–447.

Li, L., Wang, Y., and Li, M. (1983). Separation of Li isotopes by laser deflection of atomic beams. *Chin. Phys.* **3,** 155–158.

Maillet, H. (1986). *Laser Principles and Application Techniques.* Technique et Documentation, Paris, 572 pages.

Mannik, L., Brown, S.K., and Keyser, G.M. (1981). Deuterium enrichment by ultraviolet laser dissociation of formaldehyde. *Chem. Phys. Lett.* **81,** 587–590.

Miller, C.M., Nogar, N.S., Gancarz, A.J., and Shields, W.R. (1982). Selective laser photoionization for mass spectrometry. *Anal. Chem.* **54,** 2377–2378.

Streater, A.D., Mooibroek, J., and Woerdman, J.P. (1987). Light-induced drift in rubidium: spectral dependence and isotope separation. *Opt. Commun.* **64,** 137–143.

Stuke, M., and Marinero, E.E. (1978). On-line computer controlled CW dye laser spectrometer for laser isotope separation. *Appl. Phys.* **16,** 303–308.

Travis, J.C., Lucatorto, T.B., Wen, J., Fassett, J.D., and Clark, C.W. (1987). Doppler-free resonance ionization mass spectrometry of beryllium. In *Topical Meeting on Laser Applications to Chemical Analysis,* Lake Tahoe, Nev., pp. TuB2.1–TuB2.4.

Yamashita, M., Kasamatsu, M., Kashiwagi, H., and Machida, K. (1978). The selective excitation of lithium isotopes by intracavity nonlinear absorption in a CW dye laser. *Opt. Commun.* **26,** 343–347.

Yappert, M.C., and Yeung, E.S. (1986). Selectivity in the laser-induced photochemistry of $I_2 + C_2H_2$ in the gas phase. *J. Am. Chem. Soc.* **108,** 7529–7533.

Uranium Isotope Separation

Clerc, M., and Rigny, P. (1985). Physics of uranium isotope separation by laser. *Rev. Gen. Nucl.* **6,** 513–523.

Clerc, M., Rigny, P., deWitte, O., and Maillet, J. (1984). Laser isotope separation. In *Laser: Principle and Application Techniques.* Technique et Documentation, Lavoisier, Paris, pp. 197–244.

de Rulter, W. (1985). Laser isotope separation. *La Recherche* **16,** 32–41.

De Witte, O. (1987). Lasers for isotopic separation. *Entropie* **23,** 19–32.

Gilles, L. (1987). Laser and uranium isotope separation. In *Int. Conf. on Materials*, Villeurbanne, France. NTIS CONF-87087140.

Hulsey, W.J. (1984). Atomic vapor laser isotope separation program. In *WATTEC Conference Proceedings.* LTM Consultants, Knoxville, Tenn., 11 pages.

Paisner, J.A. (1988). Atomic vapor laser isotope separation. *Appl. Phys. B* **46,** 253–260.
Paisner, J.A. (1989). High power dye laser technology (panel discussion). In *Proceedings of the International Conference on Lasers '88 (Sze, R.C., and Duarte, F.J., eds.). STS Press, McLean, Va., pp. 773–790.*
Moses, E.I. (1986). Lawrence Livermore National Laboratory's atomic vapor laser isotope separation program: laser technology and demonstration facilities. In SPIE Vol. 668, Laser Processing: Fundamentals, Applications, and Systems Engineering (Duley, W.W., and Weeks, R., eds.). SPIE, Bellingham, Wash., p. 347.
Wort, D.J.H. (1988). Scientific aspects of the UKAEA laser isotope separation programme. In *Proceedings of the International Conference on Lasers '87* (Duarte, F.J., ed.). STS Press, McLean, Va., pp. 875–881.

Laser Development for Isotope Separation

Arai, Y., Niki, H., Adachi, S., Takeda, T., Yamanaka, T., and Yamanaka, C. (1986). Development of a single-mode dye laser pumped by a copper vapor laser. *Technol. Rep. Osaka Univ.* **36,** 361–367.
Bernhardt, A.F., and Rasmussen, P. (1981). Design criteria and operating characteristics of a single-mode pulsed dye laser. *Appl. Phys. B* **26,** 141–146.
Bianchi, A., and Ferrario, A. (1980). Properties and applications of coherent tunable sources in the near and middle infra-red range. *Opt. Acta* **27,** 1077–1085.
Cahen, J., Clerc, M., and Rigny, P. (1978). A coherent light source, widely tunable down to 16 microns by stimulated raman scattering. INIS Rept. CEA-CONF-4009, 11 pages.
Duarte, F.J. (1988). Narrow-linewidth, multiple-prism, pulsed dye lasers. *Lasers and Optronics* **7**(2), 41–45.
Duarte, F.J., and Conrad, R.W. (1987). Diffraction-limited single–longitudinal-mode multiple-prism flashlamp-pumped dye laser oscillator: linewidth analysis and injection of amplifier system. *Appl. Opt.* **26,** 2567–2571.
Duarte, F.J., and Piper, J.A. (1984). Narrow linewidth, high prf copper laser-pumped dye-laser oscillators. *Appl. Opt.* **23,** 1391–1394.
Emmett, J.L., Krupke, W.F., and Davis, J.I. (1984). Laser R and D at the Lawrence Livermore National Laboratory for fusion and isotope separation applications. *IEEE J. Quant. Electron.* **QE-20,** 591–602.
Hargrove, R.S., and Kan, T. (1980). High power efficient dye amplifier pumped by copper vapor lasers. *IEEE J. Quantum Electron* **QE-16,** 1108–1113.
Hirth, A., Vollrath, K., and Gulitz, G. (1979). Investigation of a high power tunable dye laser. In *Congress on High-Speed Photography and Photonics.* Japan Society of Precision Engineering, Tokyo, Japan, NTIS CONF-780804, pp. 226–229.
Jethwa, J., Schäfer, F.P., and Jasny, J. (1978). A reliable high average power dye laser. *IEEE J. Quant. Electron.* **QE-14,** 119–121.
Kompa, K.L., Martellucci, S., and Chester, A.N. (1980). Some potential applications of free electron lasers to photochemistry. In *Proceedings of the Conf. on Physics and Technology of Free Electron Lasers,* Erice, Trapani, France. Plenum, New York, NTIS CONF-8008142, pp. 669–684.
Ross, R.I., Rose, M.F., and Martin, T.H. (1983). Small and light thyratron grid driver. In *IEEE Power Electronics Specialists Conference*, Albuquerque, N.M., NTIS CONF-830621, pp. 148–149.
Warner, B.E. (1987). An overview of copper-laser development for isotope separation. In *LASE-13: Pulse Power for Lasers Conference*, Los Angeles, Calif. NTIS CONF-870132-14, p. 13.

Chapter 10

DYE LASERS IN MEDICINE

Leon Goldman

University of Cincinnati
Cincinnati, Ohio

and

U.S. Naval Hospital
Balboa Park
San Diego, California

1. INTRODUCTION

Dye lasers in laser medicine and in laser surgery have been of interest and importance for some years; they will be even more so in the future. This colorful and fascinating phase started with the Ar^+ laser-pumped dye laser for cancer diagnosis and treatment; then followed the flashlamp-pumped dye lasers for such diverse bits as the incurable blood-vessel birthmarks of infants and children, the unsightly "sun bursts," small veins in thighs and legs, and then lithotripsy, breaking stones in the ureters and in the gallbladder. In addition, dye lasers join the vast armamentarium of those ambitious lasers that in conjunction with thermal delivery systems and fiber optics are utilized in angioplasty, hopefully even in the important coronary vessels of the heart. Finally, pulsed dye lasers are also being

ISBN 0-12-222700-X

considered for corneal ablation (see, for example, Moretti (1989)). Research has just been initiated on flashlamp-pumped dye lasers and copper-vapor-laser (CVL) pumped-dye lasers for vascular and cancer studies.

2. THE LASER SYSTEMS

Here we list some of the laser systems of use and potential use in exogenous chromophore technology: Ar^+ lasers, Ar^+ laser-pumped dye lasers (using rhodamine dyes), gold-vapor lasers (GVL) ($\lambda = 627.8$ nm), Kr^+ lasers (for diagnosis in the blue-green), Nd:YAG lasers (fundamental and 532-nm), pulsed dye lasers, HeNe lasers, free-electron lasers (FEL), excimer lasers, alexandrite lasers, and titanium sapphire lasers. The energy density needed is in the 100–400-J/cm^2 range.

For PDT (photodynamic therapy), as of 1989, the Ar^+ laser-pumped dye laser and the GVL are utilized in research and clinical programs. As emphasized by many researchers, the major advantage of the dye laser is its tunability (see, for example, McMahon and Fossati-Bellani (1988)). In addition, dye lasers offer a wide choice of output characteristics such as long pulse operation and high pulse-repetition frequencies (for further details, see Chapter 1).

The next system we consider for applications in medicine and surgery is the flashlamp-pumped dye laser. A detailed review on the subject has been provided by the armed forces. Ehrlich and Patterson (1988) list the potential advantages of the flashlamp-pumped dye lasers over other lasers in the visible range:

"i. UV to IR spectral coverage.
ii. Moderate tunability of individual dyes.
iii. Wide range of waveforms: pulses from picoseconds and microseconds to cw operation.
iv. Optical simplicity.
v. Potential reliability."

Areas of development not only for military use but also for applications in laser medicine and surgery are also listed by Ehrlich and Patterson (1988):

"i. Efficiency, weight, volume, and packageability.
ii. Dye lifetime and maintainability.
iii. Beam quality, spatial coherence, temporal coherence, and brightness.
iv. Scaling to higher energy/power in linear and coaxial flashlamp systems.

v. System complexity: electrical, fluid, maintainability, and reliability.
vi. Safety: flammable fluids, high voltages, and carcinogenic dyes."

There have been improvements in flashlamp-pumped dye lasers, linear and coaxial. Duarte and Conrad (1987) have shown that multiple-prism grating assemblies in flashlamp-pumped dye-laser systems can yield efficient narrow-linewidth emission and good beam quality. The multiple-prism techniques can be used also for other types of lasers. This is important in laser medicine and surgery, especially for copper-laser–pumped dye lasers. In cancer phototherapy, such narrow linewidth (<1 GHz) is usually not necessary, but for early localized small lesions, it may be necessary. Duarte *et al.* (1988) have commented on the spectral instabilities in flashlamp-pumped dye lasers when operating under conditions favorable for double-mode ($\Delta\nu \sim 250$ MHz) oscillation. Optical phase conjugation via SBS in a flashlamp-pumped dye laser was demonstrated by Russell *et al.* (1988). Other laser systems for excitation of pulsed dye lasers are excimer lasers, nitrogen lasers, copper lasers, and Nd:YAG lasers.

3. THE DYES FOR DYE LASERS

Now we discuss the dyes providing the active medium in dye lasers. The exogenous chromophores for PDT wil be reviewed later. Sorokin and Lankard (1966) pioneered the use of liquid organic dye solutions as a laser medium. For those involved with early flashlamp-pumped dye lasers, especially in laser medicine and surgery, it did appear often that laser dye chemistry was a very complex subject. In flashlamp-pumped dye lasers, not only must the temperature be regulated carefully, but also important vehicles and curious additives (caffeine, for example) must be used. Caffeine is used as a UV filter in flashlamp-pumped dye lasers.

The major dye suppliers are Eastman Kodak, Exciton, and Lambda Physik. Detailed information is easily available from these suppliers. Duarte (1988) reviews some aspects of laser dyes which should be of interest to laser photobiology as well as for the clinical applications of flashlamp-pumped dye lasers and for PDT.

> From the perspective of those who use dyes as the laser medium in an optically pumped device, the basic feature of interest is that dyes are large molecules (with molecular weights ranging from about 175 to 1000) that can absorb light from a wide spectral region and fluoresce in a spectral region involving longer wavelengths. For instance, rhodamine 590 (molecular weight = 479) in methanol absorbs from the ultraviolet up to about 550 nm and, depending on concentration and pump source, can lase at a variety of peak wavelengths from roughly 560 to 590 nm with tuning ranges in the 30-nm to 60-nm region.

Duarte (1988) goes on to explain that higher efficiencies and lower photochemical degradation can be achieved by using excitation sources whose emission wavelengths coincide with the $S_0 \rightarrow S_1$ absorption transition of the dye. In other words, direct excitation of higher singlet states (such as S_2 and S_3) is, on a relative basis, less efficient. This explains, for instance, the high efficiencies (in the 29–55% range) observed by Morey (1980) and Hargrove and Kan (1980) for copper-laser–pumped dye lasers using rhodamine dyes. Thus, according to Duarte (1988) one should ideally use pump sources "providing excitation wavelengths compatible with the $S_0 \rightarrow S_1$ electronic absorption transition." He concludes by stating:

> Finally, remember that selection of the particular pump source for the specific application depends on several factors. These factors include requirements on spectral region, pulse length, peak power, prf, environment and cost.

These bits are presented in much greater detail in the text (also, see Chapter 7 for details on dye chemistry). Neister (1988) reviewed the difficulties of laser manufacturers with regard to the dyes for dye lasers. This concerned classification of dye lifetimes, standards for reporting new dyes, and methods of testing. Duarte (1988) lists pump sources for dye lasers from UV and visible to 1064 nm (see Chapter 6, Table 6.1). One interesting fact Neister (1988) mentions is that "any dye will not lase at the same wavelength if it is excited or pumped by two different methods." Other factors include the "method of pumping, length of the cell or model of laser." Neister's method for rapid dye testing is to measure the amount of degradation after 200 shots at 50-J input energy per shot for a 200-ml solution. It is evident then that the flashlamp-pumped dye-laser system, although seemingly simple in laser construction, demands a lot of understanding and care of detail.

For the future, it is of interest to review in detail the proposals and hopes of Ehrlich and Patterson (1988) which apply also to our hopes for dye lasers in laser medicine and surgery. The U.S. Army dye-laser program includes the following items as future goals:

> DYE SYNTHESIS AND CHARACTERIZATION: Develop more efficient, longer lived dyes spanning the near UV to near IR spectrum.
>
> NON-LIQUID HOSTS: Develop potentially low-cost, simple sources of dye laser radiation exploiting proven solid-state technology.
>
> VAPOR DYE: Develop vapor dye laser radiation sources of high peak power.

4. THE SAFETY PROGRAM

For laser safety of the various laser systems, the known safety programs should be followed. For the vascular phases of the 577- and 585-nm flashlamp-pumped dye lasers, special protective glasses are needed for precision work. The pattern of small veins must be seen to use the yellow beam accurately along the course of the vessel. These glasses must provide accurate viewing even for small veins.

The dyes, their vehicles, and additives may add other hazards. So, dye spills should be avoided and, if this occurs, this should be cleaned up immediately. The laser-instrument room should be well ventilated. Smoking, of course, should be forbidden. If the dye vehicle includes flammable materials, a fire extinguisher should be mounted in the room for added safety. Fiber optics can transmit 577 and 585 nm from the laser-equipment room to the operation room and the operating table through a wall. Communication systems, varying from a protected window area to microphone communications, can be used for monitoring by the laser operator. Dye-laser systems are dynamic and laser-safety programs will change and continue to develop according to laser systems and wavelengths.

5. THE DYES FOR PDT

The start, for cancer diagnosis and treatment, was the development of laser systems that would energize chemicals associated with cancers, make them photosensitive, and then affect tissues with which they were associated. One such material was hematoporphyrin, $C_{34}H_{38}O_6N_4$. This is a reddish crystalline material from hemin, $C_{34}H_{32}N_4O_4FeCl$. Hematoporphyrin was known to be in cancer tissue in 1924. In 1961, Lipson found that hematoporphyrin added to cancer tissue caused destruction of this cancer tissue exposed to light. In the 1970s, Dougherty *et al.* (1978) of Roswell Park Memorial Institute of Buffalo, New York, studied the effects of a derivative of hematoporphyrin, HpD, in various forms of cancer exposed to both 400–500 nm for diagnosis and 625–635 nm for cytotoxicity, for destruction. It is to his credit that the current international program of such photodynamic therapy, called PDT, was developed. The complex derivative, dihematoporphyrin ether, or ester (DHE), is an additional photosensitizer. There was and continues to be considerable difficulty to attempt to develop a standard uniform product for animal and human experimentation. Dougherty (1982) was aware of this variability. Al-Watban and Harrison (1988) suggested biological testing of each batch of HpD using tissue cultures of murine fibrosarcoma cells (RIF-1) with identical PDT conditions.

The basic needs for PDT are:

1. A high-output light source, a laser.
2. A beam delivery by a fiber optical system.
   A. Uniform delivery.
   B. Long exposure time.
   C. Avoidance of fiber optic tissue contact.

The wavelength range of 700–900 nm for treatment is preferred, for in this range there is less disturbance from endogenous chromophores in the body such as melanin and hemoglobin. So, PDT introduces the fascinating technology of laser exogenous chromophores technology as proposed by Osserof. In this concept, chromophores are related to specific wavelengths of laser systems. When a specific chromophore, then, is coupled to a specific tissue such as cancer for PDT, the impact of a specific laser system for this specific complex completes this photosensitivity reaction with the consequent destruction of the mixture. So, with coupling of not only cancer, but toxins, antigens, antibodies, enzymes, etc., the range of laser exogenous chromophore reactions becomes extensive and very important.

It is assumed that this laser-photosensitivity reaction produces reactive forms of O_2 which lead to both a loss of cellular structure and, consequently, cellular function. The reaction method of the singlet oxygen on cells is thought to be heat or infrared light or "attacks on cells A to form AO_2" (Kimel, 1988).

Dye efficiencies as they relate to copper-vapor-laser-pumped dye lasers are covered in Morey (1980) and Hargrove and Kan (1980). Morey (1980) achieved more than 50% efficiency at 630 nm and the half-life of the dye, rhodamine 640, was reported to be more than 560 hours. For Hargrove and Kan (1980) an oscillator-amplifier system was used to provide about 18% efficiency using kiton red dye. Different oscillator designs were reviewed and evaluated by Duarte and Piper (1984). Duarte (1988) notes that the high efficiency and long lifetime of copper-laser-pumped dye lasers when using red dyes is due to coincidence of the green line of the copper laser and the strong $S_0 \rightarrow S_1$ absorption transition of most red dyes. A difficulty with dye efficiencies in user-friendly medical lasers has been the inability to maintain high persistent dye-laser efficiencies. In this regard, the importance of the CVL-pumped dye laser for work with exogenous chromophores is self-evident.

Now, we turn to the ever increasing list of those compounds to be photosensitized. Their properties are to be taken up by the specific area which is to be the target; in addition, they are to be retained longer than in any other tissues or the area around. Finally, they are to take up, as it were, the laser light intensity more than in the surrounding areas.

These photosensitizers include (some as listed previously):

1. HpD: hematoporphyrin derivative and DHE. Concern has been expressed as to the standardization of quality; may be retained in cancer tissue three to four weeks.
2. Phthalocyanines: especially the metallic compounds. These resemble the porphyrins.
3. Porphyrin C.
4. Rose bengal: a so-called nontoxic vital dye (not injurious to tissue). We have used this with both argon and GVL (Goldman *et al.*, 1985).
5. Purpurin: 600–800 nm.
6. Hypercin: a polycyclic plant pigment which we have used for the photosensitizing treatment for psoriasis.
7. Rhodamine dyes.
8. Arcridine orange: toxic photosensitizer which our laboratory has used in animal-skin studies. Japanese investigators have used this, topically, in gastric cancer with the argon laser.
9. Kiton red.
10. Chlorophyll derivatives.

For cancer treatment, the energy density with the laser and the photosensitizer is between 100 and 400 J/cm^2 for the local destruction; this depends on site, size, and type of cancer. When much tissue blocks the penetration of the beams, this barrier is removed by the CO_2 laser before the PDT laser is used. Thus, the PDT treatment is made much more effective. It is obvious, then, that there are many factors besides just mass size determining the amount of tissue ablation by the laser beam and dye deposition in tissue. Although it is controversial, most clinical evaluations are made by inspection tempered with experience, and by follow-up observations and repeated treatments if necessary. Often there are multiple fiberoptic transmissions controlled by thermal sensors in tissue. If the beam follows the chromophore into adjacent important vascular tissues, severe hemorrhage may result from the adjacent involved tissue. It is obvious that the real problem is the attempt to use PDT on cancers whose surface is not easily available. A cancer focus may be identified by tissue transillumination, CAT scan, or magnetic resonance imagery (MRI) deep in tissue; an exogenous chromophore such as DHE may be deposited here; other chromophores may have to be injected there. Then, how do you get beam transmission or direct fiberoptic transmission there? Should some types of PDT for certain viscera be part of open cancer surgery? This has been discussed before. With the help of biomedical engineers and laser-dye expertise, we have to go now from the laboratory directly to the cancer unit before the pronouncement of a gloomy verdict of an inoperable cancer. Hopefully the important studies of Preuss (1988) at Henry Ford Hospital in laser tissue fluorescence transmission will help us with the diagnosis of these metastases. The DHE for cancer treatment with PDT can be given internally 2.2–3.0 mg/kg or this may be applied locally as we have done for some time with Azone (Goldman *et al.*, 1985) as a vehicle. This mixture is used on areas where it may be absorbed easily into superficial underlying cancer. For deeper areas of cancer, this mixture of the photosensitizers and the vehicle may be injected locally. The vehicle used for injection also contains Azone, a superior agent for absorption through the skin. Azone is 1-dodecylazacycloheptan-2-one (Goldman *et al.*, 1985). Specific details of such therapy are found in the medical literature.

Now, the next factor is beam delivery of the laser so that the beam gets to the photosensitizer. Doiron (1987) has done detailed work in the development of laser light delivery systems to provide for uniform delivery and dispersion in tissue and, as indicated, even for the long exposures necessary in cancer treatment. There are various types of fiber systems for localized impact and dispersion and some even incorporate micro lenses of 1.9 mm OD. In cavities, avoidance of contact—especially with mucus and blood—is emphasized. Multiple fibers even may be used for uniform tissue beam distribution in an area. Fibers are provided also for instruments called endoscopes to go into cavities such as the lungs, stomach, colon, uterus, and gall bladder. In the bladder, with an instrument called the cystoscope, the fiberoptics delivers the laser beam through fluids directly to contact with foci of cancer in the bladder wall.

As of 1989, the problem for laser-beam delivery is how do you deliver laser beams not only on the surface of the skin directly but in cavities deep in tissues where cancer is spreading. These cancer foci are called metastases. How do we detect them? How do we reach them in such heterogeneous tissue masses? These problems have been reviewed with regard to implanting the diagnostic and treatment chromophores in the primary focus for conjugation with the metastases. Laser research in MRI now is related to the development of holographic imagery and laser spectroscopy for studies of function of an organ. Also for diagnoses, needle absorption impacts and scalpels are used to obtain biopsies. We have also been interested in optical phase conjugation to attempt to detect early cancer masses just near the surface of body cavities. How do we reach those foci, deep in tissue, to detect the fluorescence of the photosensitizer? Then how do we transmit fiber optics to these cancer foci (Goldman, 1988; Preuss, 1988)? Here, we certainly cannot avoid fiber optics tissue contact and contamination of the tips of the fiber optic. In the realm of laser cardiac catheters, MIT has developed a computer-programmed multiple, 19-bundle fiber optics system for the removal of the blocking atheroma in the vessel. Here each fiber is programmed separately, and the end result is the complete destruction of the blocking tissue.

Since HpD absorbs strongly around 413 nm (see, for example, Profio (1984)), there have been suggestions that the Kr^+ laser be used for diagnosis of the cancer focus. This is done clinically now by Doiron (1987) with the Kr^+ laser in an instrument called the bronchoscope. With the use of this complex instrument, which also tests the patient's sputum, it is possible to diagnose early lung cancer with image intensifiers and then to treat that early focus with DHE of PDT and clear it. With the very gloomy statistics of lung cancer and the recent increases of this cancer, especially in women, this is important news. Most PDT treatment of lung cancer as of 1989 is only palliative, such as temporary relief from obstruction to breathing and temporary relief to attempt to prevent continued progressive spread.

Complicating the problem of detection of induced fluorescence for the diagnosis of cancer is the autofluorescence of normal adjacent tissues and the subtraction of this fluorescence from the measurement of the total fluorescence found. Potter (1987) has developed detection instruments initially with an audio signal. He emphasizes background subtraction for the localization of tumors. IR thermal imaging diagnostics is also used. The SLO (scanning laser ophthalmoscope) can be adapted to scanning for induced fluorescence after surgery for cancer, especially after laser surgery for solid brain tumors. Scanning of fluorescence after surgery can show residual induced fluorescence of the base of the excised tumor and also any local spread.

6. LOCALIZED HYPERTHERMIA INSTRUMENTATION

As indicated previously, the problem of the mechanism of PDT at the cellular level is often mentioned as reactive singlet O_2 but there still is concern that beyond photochemical reactions, there indeed may be thermal reactions. So, this has led to local hyperthermia as an adjunct to PDT. Dougherty (1987) has shown this definitely in cancer in dogs and cats and our veterinary oncologists have confirmed this. Generalized hyperthermia treatment of cancer is already an established program in centers of oncology. Effective localized cancer hyperthermia research is a recent development. Localized hyperthermia treatment modalities include:

1. Hot water baths.
2. IR boxes as developed years ago by Kettering at the University of Cincinnati Hospital.

3. Localized IR rods as developed by Nath (1984), which we have used with PDT on basal cell carcinoma.
4. Ultrasonics.
5. RF veterinary tissue probe (Levine, 1988).
6. Microwaves.
7. Nd:YAG (used now after PDT). A Nd:YAG system for laser thermia is considered by Daikuzono *et al.* (1987).

One of the most detailed investigative programs in PDT and hyperthermia has been done in Lund, Sweden (Andersson-Engels *et al.*, 1988). These authors employed a dye laser for the PDT and a near-infrared incoherent source for the hyperthermia. For the destruction of cancer tissue, 42–43°C for 30 minutes is needed. Interesting laser Doppler studies were done by these investigators to determine blood perfusion since blood circulation is so important in cancer-tissue research. Detailed controlled studies are necessary in this important field of hyperthermia, as an adjunct to PDT, which has so many practical applications in the cancer program.

To evaluate the effectiveness of the PDT reaction in tumors, temperature sensors have been inserted in various portions of the tumor. We have used the Bailey apparatus. These sensors at least will sense any thermal activity in tissue, especially at a distance from the direct-impact laser-beam area.

7. THE PDT PROGRAM FOR CANCER

For the purpose of this review for physicists and engineers, the PDT program is listed only briefly. The dream continues to be the detection and treatment of early cancer and the palliative treatment of mostly inoperable cancers. Lung cancer, early cancer of the stomach, skin and soft-tissue cancers, with prejudice belong in the early-cancer programs. To avoid the severe photosensitivity reactions of the hematoporphyrin derivatives (Wooten *et al.*, 1988), we have used DHE locally as Spinella has done (Goldman *et al.*, 1985; Spinella, 1986). This has been done also by intratumoral injections by Moseng *et al.* (1985). For certain solid brain tumors, laser surgery is obligatory and carries along with it, controlled PDT. Superficial transitional cell cancer of the bladder is an excellent model for PDT. With the cystoscope instrument mentioned, the whole bladder inside is exposed to 628–630 nm in PDT. Only the superficial cancer sites are destroyed. Although exposed, none of the normal lining of the bladder is affected. The DHE is taken up only by cancer masses. The problem here, as elsewhere, is: has the local focus been destroyed completely and deeply? Continued controlled observation is necessary.

So for PDT, as for any applications in laser medicine and in laser surgery, the eternal questions are:

1. Do you need the laser?
2. What are your controls?

In addition, there are many questions about every single phase. Now, the dye-laser technology extends beyond the PDT cancer program. The photosensitizers of PDT, DHE, and also the antibiotic tetracycline fluorescence under the argon beam used in laser cardiovascular research. DHE and tetracycline are absorbed by the substances in the atheromamasses blocking peripheral vessels and in the coronary vessels of the heart. So, the argon laser can not only recognize early areas but can also remove them. This is called laser angioplasty. The concerns are damage to the blood vessels, incomplete removal of the fatty-fibrous-calcium tissues called atheromas, and, finally, perforation of the blood vessel. Lasers of interest for this application include ArF, pulsed dye, and Er: YAG (Moretti, 1989). The 193-nm excimer with its beam precision is a strong rival of the argon laser. So, the great challenge to the laser industry is to develop flexible safe, effective instruments that can do this. The great potential of the laser market impact is also there to stimulate this project.

The difficulty of supplying properly controlled amounts of DHE will be only temporary and the PDT program will be active again with the continued search for new chromophores. Recent developments include the observation of some immune suppression, reversible, induced by intraperitoneal PDT injection in mice by Jolles *et al.* (1988). This is not strange for medications for the treatment of cancer. However, this is the first investigative report on this phase. As of 1989, this has not been found clinically. If true, this would be of concern in the PDT programs for cancers in immunosuppressed patients including those with kidney and heart transplants.

8. 577-NM FLASHLAMP-PUMPED DYE LASER

Some instrumentation details of the 577-nm flashlamp-pumped dye laser have been listed already. This laser, in spite of its relatively short introduction into laser surgery, has established, with modifications, the treatment of blood-vessel types, the treatment of stones in a portion of the kidney apparatus called the ureter (which is not in the kidney itself) and finally its potential use in the removal of obstruction in blood vessels called atheromata.

The flashlamp-pumped dye laser at 577 nm is excellent for the treatment of vascular lesions if they are not too deep (see, for example, Nakagawa *et al.* (1985), Tan *et al.* (1986)). The port wine mark, that wretched incurable birthmark, shows that although all are created equal, some, who are not born with the disfiguring port wine mark, are created more equal. The pain of treatment is minimal with the 577-nm flash pumped dye laser. Infants can be treated, still with laser eye protection, when the mark is early in its development. Wellman Research Laboratory of the Massachusetts General Hospital introduced this laser system and has continued to study and develop it. If the mark is deep, argon or preferably CVL are used also. Modifications in the 577- and 585-nm lasers will be used in the future for deeper spots and also for blue-colored superficial veins. We have used 577-nm flashlamp-pumped dye lasers for several years for superficial disturbing veins of the thigh and legs. Controls included saline sclerotherapy; intravenous and paravenous #30 gauge electric needle; intravenous and paravenous fiberoptic argon; small-spot short-pulse argon; and 532-nm lasers. Preoperative and postoperative skin chilling is necessary to prevent heat radiation of the surface and scars. Postoperative special pressure Velcro bandages are necessary for heating, for prevention of collateral circulation, and for prevention of recanalization of the blood vessels.

The 577-nm flashlamp-pumped dye laser is also used for stones in the structure called the ureter (Dretler, 1987; Dretler *et al.*, 1987; Watson and Wickham, 1986; Watson *et al.*, 1987; Bua and Webster, 1987), not the kidney, and also in the gallbladder for gallstones. Colin Whitehurst of the Physics Department of the University of Manchester, Manchester, Great Britain found that "The peak wavelength absorption of individual stones varies according to their composition (Whitehurst, 1987). By determining the optimum wavelength, and tuning the dye laser accordingly, he hopes to break stones with a minimum amount of energy and therefore less risk to the patient. Whitehurst (1987) predicts that it may be possible to develop a compact instrument for the surgeon to spectrally analyze kidney stones before treating patients with the laser.

This concept of laser lithotripsy is an important application of flashlamp-pumped dye lasers. Duarte and Conrad (1987) describe flashlamp-pumped dye-resonator designs that can yield excellent beam quality for transmission in a near-diffraction limited beam. The Nd:YAG laser is also considered for ureteral calculi and, especially, for gallstones.

The popular 577- and 585-nm flashlamp-pumped dye laser can be used also in the treatment of atheroma in blood vessels. Beta carotene is given to help make the fatty portion of the atheroma more colorful and thus more effective for the subsequent laser treatment. Again, with the development of the laser instruments going into blood vessels, laser angioscopy will fulfill the dreams of those who make the laser cardiovascular programs of the future. CVL lasers are used also for chromophore atherolysis. CVL pumped dye lasers have great potential for biomedical applications.

9. CONCLUSION

Dye lasers offer wavelength agility and a wide variety of output characteristics. The versatile nature of these tunable lasers make them a valuable source of coherent light for medical applications. So, dye lasers of all types are important parts of laser biology, laser medicine, and laser surgery. The contributions of the dye laser to laser medicine in the future should be expected to become even more significant.

REFERENCES

Al-Watban, F.A.H., and Harrison, W.R. (1988). Biological activities of preparations of HPD. In *L. I. A. Vol. 60 ICALEO (1987)*. Laser Institute of America, Toledo, Ohio, pp. 75–81.

Andersson-Engels, S., Johansson, J., Killander, D., Kjellen, E., Svaasand, L.O., Svanberg, K., and Svanberg, S. (1988). Photodynamic therapy and simultaneous near-infrared light-induced hyperthermia in human malignant tumors. A methodological case study. In *L. I. A. Vol. 60 ICALEO (1987)*. Laser Institute of America, Toledo, Ohio, 67–74.

Bua, D., and Webster, K. (1987). The pulsed dye laser in medicine. *Lasers Appl.* **6**(4), 69–70.

Daikuzono, N., Joffe, S.N., Tajiri, H., Suzuki, S., Tsunekawa, H., and Ohyama, M. (1987). Laserthermia: a computer-controlled contact Nd:YAG system for interstitial local hyperthermia. *Med. Instrum.* **21,** 275–277.

Doiron, D. (1987). Laser light delivery systems. Presented at ICALEO, San Diego, Calif.

Dougherty, T.J. (1982). Variability in hematoporphyrin derivative preparations. *Cancer Res.* **42,** 1188.

Dougherty, T.J. (1987). Private communication.

Dougherty, T.J., Kaufman, J.E., Goldfarb, A., Weishaupt, K.R., Boyle, D., and Mittelman, A. (1978). Photoradiation therapy for the treatment of malignant tumors. *Cancer Res.* **38,** 2628–2635.

Dretler, S.P. (1987). Laser photofragmentation of ureteral calculi: analysis of 75 cases. *J. Endourology* **1,** 9–14.

Dretler, S.P., Watson, G., Parrish, J.A., and Murray, S. (1987). Pulsed dye laser fragmentation of ureteral calculi: initial clinical experience. *J. Urol.* **137,** 386–389.

Duarte, F.J. (1988). Narrow-linewidth, multiple-prism, pulsed dye lasers. *Lasers and Optronics* **7** (2), 41–45.

Duarte, F.J., and Conrad, R.W. (1987). Diffraction-limited single–longitudinal-mode multiple-prism flashlamp-pumped dye laser oscillator: linewidth analysis and injection of amplifier system. *Appl. Opt.* **26,** 2567–2571.

Duarte, F.J., Ehrlich, J.J., Patterson, S.P., Russell, S.D., and Adams, J.E. (1988). Linewidth instabilities in narrow-linewidth flashlamp-pumped dye laser oscillators. *Appl. Opt.* **27,** 843–846.

Duarte, F.J., and Piper, J.A. (1984). Narrow linewidth, high prf copper laser-pumped dye-laser oscillators. *Appl. Opt.* **23,** 1391–1394.

Ehrlich, J.J., and Patterson, S.P. (1988). Dye laser/laser dye research at the U.S. Army Missile Command. In *Proceedings of the International Conference on Lasers '87* (Duarte, F.J., ed.). STS Press, McLean, Va., pp. 304–307.

Goldman, L. (1988a). Chromophores in tissue for laser medicine and surgery. Presented at Optical Society of America Optcon, Santa Clara, Calif.

Goldman, L., Gregory, R.O., and LaPlant, M. (1985). Preliminary investigative studies with PDT in dermatologic and plastic surgery. *Lasers Surg. Med.* **5,** 453–456.

Hargrove, R.S., and Kan, T. (1980). High power efficient dye amplifier pumped by copper vapor lasers. *IEEE J. Quantum Electron.* **QE-16,** 1108–1113.

Jolles, C.J., Ott, M.J., Straight, R.C., and Lynch, D.H. (1988). Systematic immunosuppression induced by peritoneal photodynamic therapy. *Am. J. Obstet. Gynecol.* **158,** 1446–1453.

Kimel, S.O. (1988). Beckman Laser Institute. Unpublished report.

Levine, N. (1988). Private communication.

McMahon, M., and Fossati-Bellani, V. (1988). Tunable laser systems for photodynamic therapy. In *L. I. A. Vol. 60 ICALEO (1987).* Laser Institute of America, Toledo, Ohio, pp. 58–60.

Moretti, M. (1989). Medical-laser technology responds to user needs. *Laser Focus World* **25**(3), 89–102.

Morey, W.W. (1980). Copper vapor laser pumped dye laser. In *Proceedings of the International Conference on Lasers '79* (Corcoran, V.J., ed.). STS Press, McLean, Va., pp. 365–373.

Moseng, D., Volden, G., Midelfart, K., Kavli, G., Christensen, T., and Moan, J. (1985). Treatment of mouse carcinomas with intratumoral injections of hematoporphyrin derivative and red light. *Photodermatol.* **2,** 107–110.

Nakagawa, H., Tan, O.T., and Parrish, J.A. (1985). Ultrastructural changes in human skin after exposure to a pulsed laser. *J. Invest. Dermatol.* **84,** 396–400.

Nath, G. (1984). Private communication.

Neister, S.E. (1988). Dyes for dye lasers: a manufacturer's viewpoint. In *Proceedings of the International Conference on Lasers '87* (Duarte, F.J., ed.). STS Press, McLean, Va., pp. 971–984.

Potter, W. (1987). Fluorescence detection with background subtraction for localization of tumors. Presented at ICALEO, San Diego, Calif.

Preuss, D. (1988). Private communication.

Profio, A.E. (1984). Laser excited fluorescence of hematoporphyrin derivative for diagnosis of cancer. *IEEE J. Quantum Electron.* **QE-20,** 1502–1507.

Russell, S.D., Ehrlich, J.J., Patterson, S.P., Conrad, R.W., Klimek, D.E., and Fulghum, S.F. (1988). Stimulated Brillouim scattering experiments with flashlamp-pumped dye lasers. In *Proceedings of the International Conference on Lasers '87* (Duarte, F.J., ed.). STS Press, McLean, Va., pp. 680–684.

Sorokin, P.P., and Lankard, J.R. (1966). Stimulated mission observed from an organic dye, chloro-aluminum phthalocyanine. *IBM J. Res. Develop.* **10,** 162–163.

Spinella, P. (1986). Private communication.

Tan, O.T., Carney, J.M., Margolis, R., Seki, Y., Boll, J., Anderson, R.R., and Parrish, J.A. (1986). Histologic responses of port-wine stains treated by argon, carbon dioxide and tunable dye lasers. A preliminary report. *Arch. Dermatol.* **122,** 1016–1022.

Watson, G., Murray, S., Dretler, S.P., and Parrish, J.A. (1987). The pulsed dye laser for fragmenting urinary calculi. *J. Urol.* **138,** 195–198.

Watson, G.M., and Wickham, J.E.A. (1986). Initial experience with a pulsed dye laser for ureteric calculi. *Lancet* **1,** 1357–1358.

Whitehurst, C. (1987). Private communication.

Wooten, R.S., Smith, K.C., Ahlquist, D.A., Muller, S.A., and Balm, R.K. (1988). Prospective study of cutaneous phototoxicity after systemic hematoporphyrin derivative. *Lasers Surg. Med.* **8,** 294–300.

Appendix

LASER DYES

Appendix of Laser Dyes

Name	Molecular Weight (au)	Maximum Absorption λ(nm)	Maximum Fluorescence λ(nm)	Maximum Lansing λ (nm) (Pump Laser)	Solvents	Molecular Structure
p-Terphenyl (PTP)	230.31	276	354	–	cyclohexane	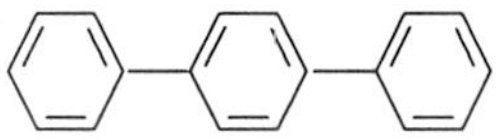
p-Quaterphenyl (PQP)	306.41	298	363	–	cyclohexane	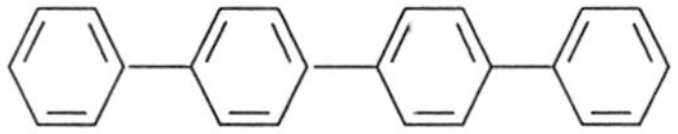
Carbostyril 124 (Carbostyril 7; 7-Amino-4-methylcarbostyril)	174.20	349	405	417 (N_2)	methanol	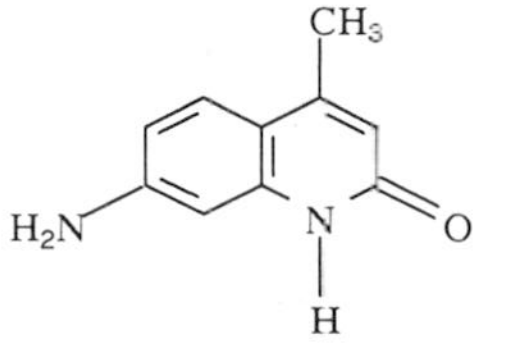
POPOP	364.40	358	415	419 (N_2)	cyclohexane	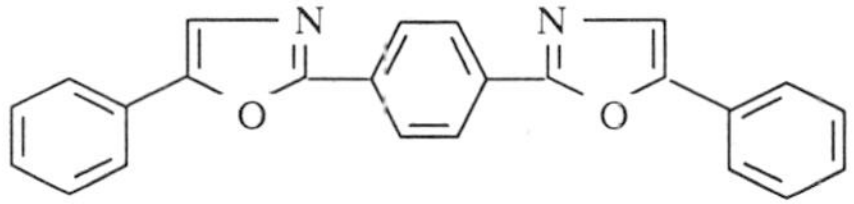
Coumarin 152 (Coumarin 485; 7-Dimethylamino-4-trifluoromethylcoumarin)	257.21	394	496	430 (N_2)	methanol, ethanol	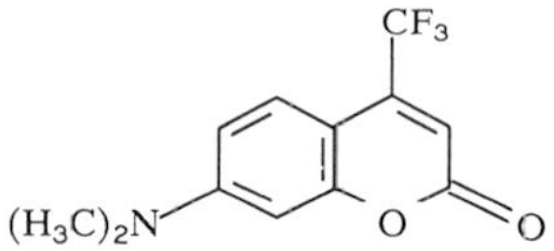

(continues)

Appendix of Laser Dyes

Name	Molecular Weight (au)	Maximum Absorption λ(nm)	Maximum Fluorescence λ(nm)	Maximum Lansing λ (nm) (Pump Laser)	Solvents	Molecular Structure
Coumarin 120 (Coumarin 440; 7-Amino-4-methylcoumarin)	175.19	352	428	444 (N_2)	methanol, ethanol	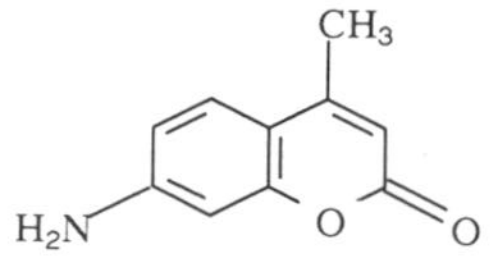
Coumarin 2 (Coumarin 450; 4,6-Dimethyl-7-ethylaminocoumarin)	217.27	365	435	450 (N_2)	methanol, ethanol	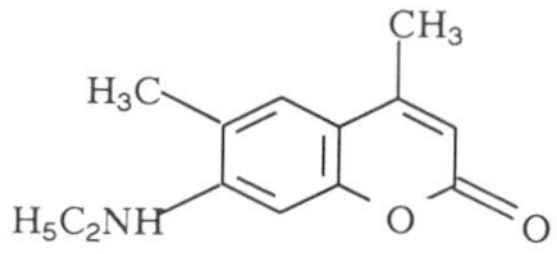
Coumarin 339	215.25	377	447	460 (N_2)	methanol	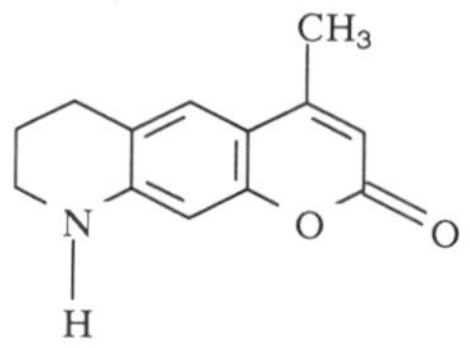
Coumarin 1 (Coumarin 47; Coumarin 460; 7-Diethylamino-4-methylcoumarin)	231.30	374	450	464 (N_2)	methanol, ethanol	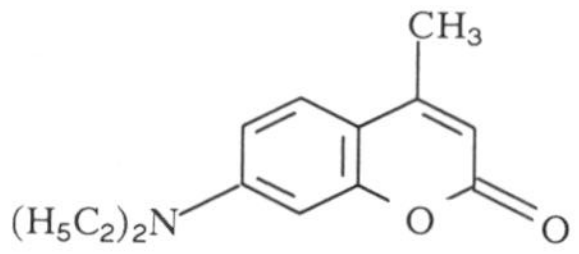

Coumarin 138 (7-Dimethylamino cyclopenta[c]-coumarin)	229.28	365	447	464 (N_2)	methanol	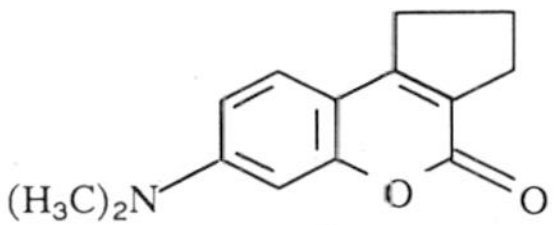
Coumarin 102 (Coumarin 480)	255.32	390	458	481 (N_2)	methanol, ethanol	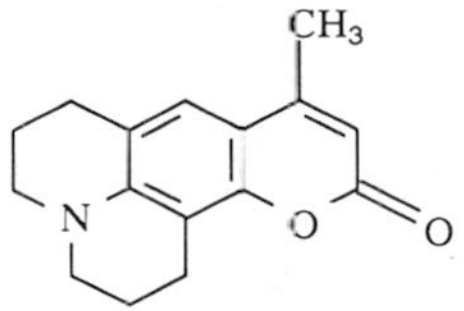
Coumarin 151 (Coumarin 490; 7-Amino-4-trifluoro-methylcoumarin)	229.16	377	479	493 (N_2)	methanol, ethanol	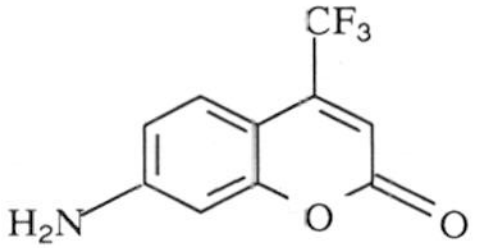
Coumarin 4 (Umbelliferon 47; 7-Hydroxy-4-methylcoumarin)	176.17	322	386	494 (N_2)	methanol	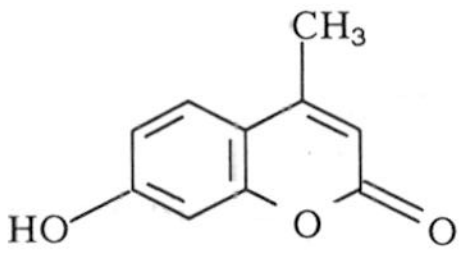

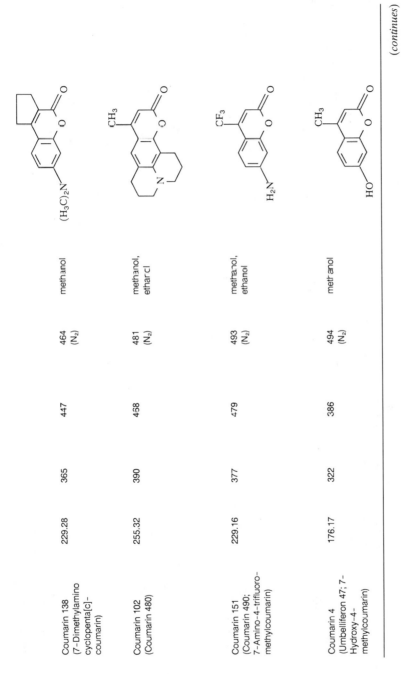

(*continues*)

Appendix of Laser Dyes

Name	Molecular Weight (au)	Maximum Absorption λ(nm)	Maximum Fluorescence λ(nm)	Maximum Lansing λ (nm) (Pump Laser)	Solvents	Molecular Structure
Coumarin 314 (Coumarin 504)	313.35	437	478	495 (N_2)	methanol, ethanol	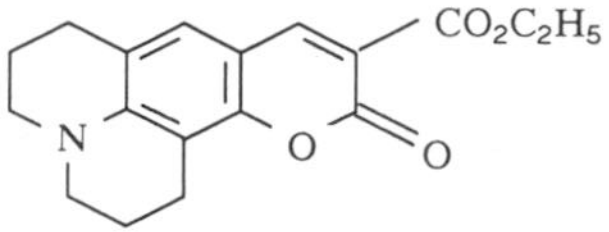
Coumarin 30 (Coumarin 515)	347.42	413	478	497 (N_2)	methanol, ethanol	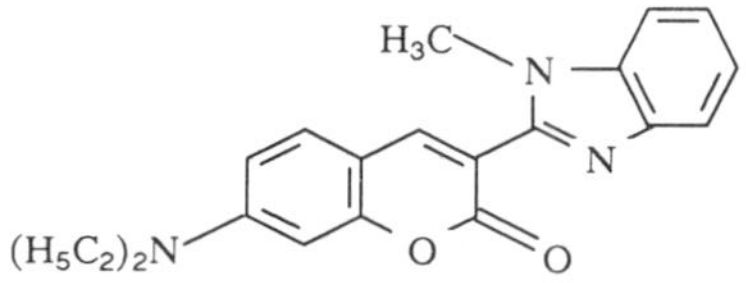
Coumarin 314T (Coumarin 504T)	369.35	435	–	506 (XeCl)	ethanol	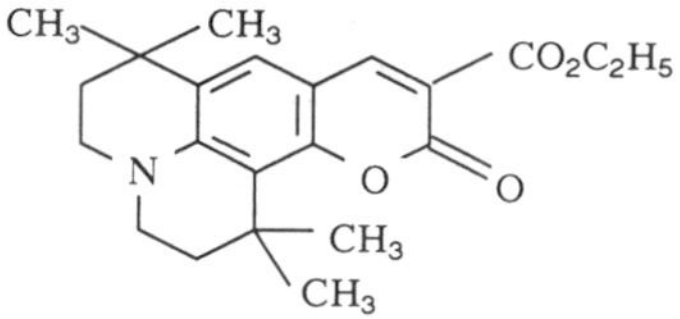
Coumarin 307 (Coumarin 503; 7-Ethylamino-6-methyl-4-trifluoro methylcoumarin)	271.24	395	488	510 (N_2)	methanol, ethanol	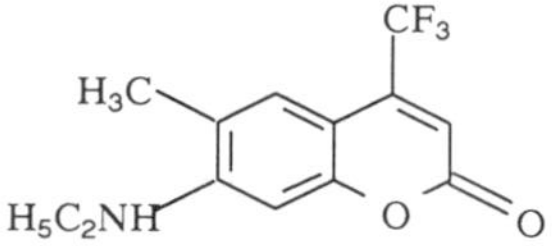

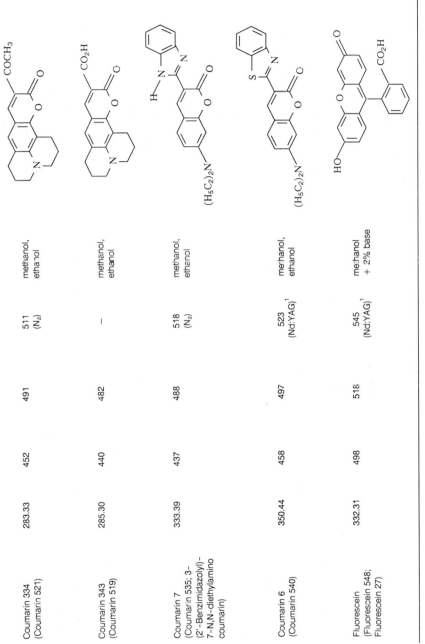

Coumarin 334 (Coumarin 521)	283.33	452	491	511 (N_2)	methanol, ethanol	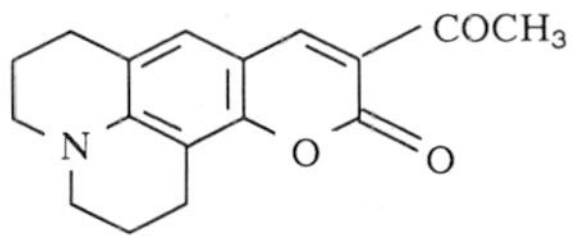
Coumarin 343 (Coumarin 519)	285.30	440	482	–	methanol, ethanol	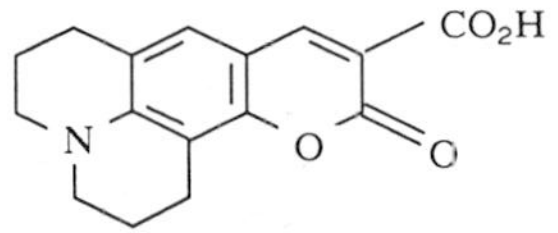
Coumarin 7 (Coumarin 535; 3-(2'-Benzimidazolyl)-7-N,N-diethylamino coumarin)	333.39	437	488	518 (N_2)	methanol, ethanol	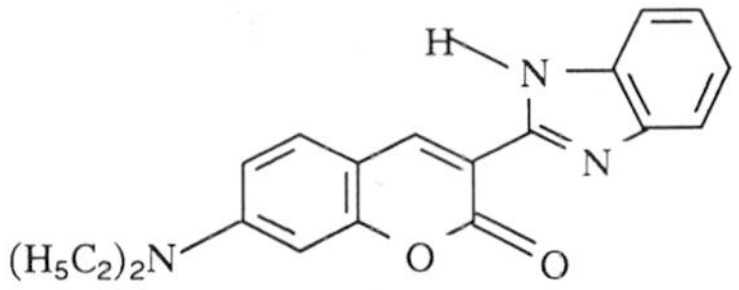
Coumarin 6 (Coumarin 540)	350.44	458	497	523 (Nd:YAG)[1]	methanol, ethanol	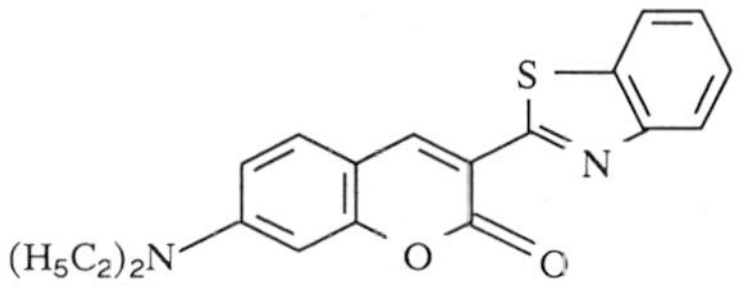
Fluorescein (Fluorescein 548; Fluorescein 27)	332.31	498	518	545 (Nd:YAG)[1]	methanol + 2% base	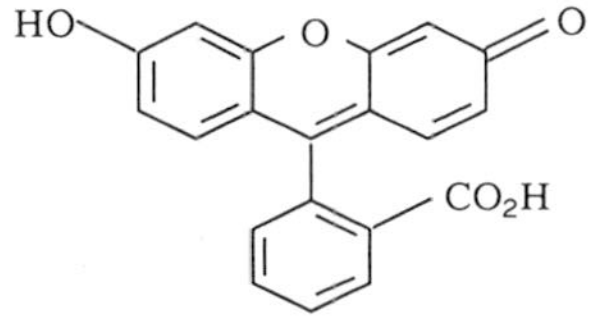

(continues)

Appendix of Laser Dyes

Name	Molecular Weight (au)	Maximum Absorption λ(nm)	Maximum Fluorescence λ(nm)	Maximum Lansing λ (nm) (Pump Laser)	Solvents	Molecular Structure
Rhodamine 110 (Rhodamine 560)	366.80	498	520	554 (N_2)	methanol, ethanol	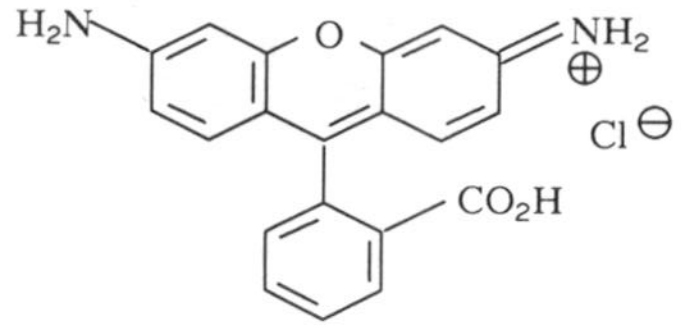
2',7'-Dichlorofluorescein	401.20	512	526	557 (Nd:YAG)[1]	methanol + 2% base	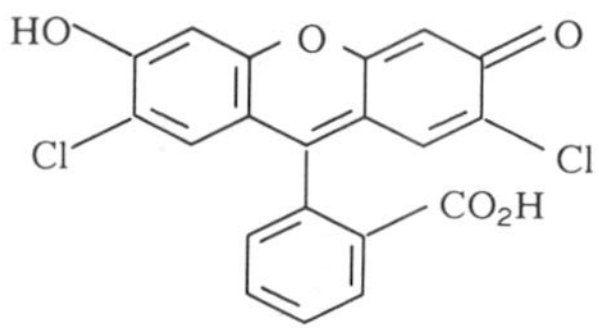
Rhodamine 6G Tetrafluoroborate (Rhodamine 590 Tetrafluoroborate)	530.37	528	547	581 (N_2) 563 (Nd:YAG)[2]	methanol, ethanol	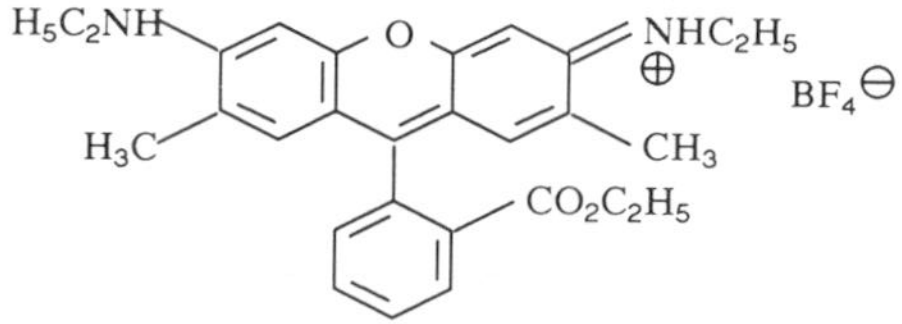

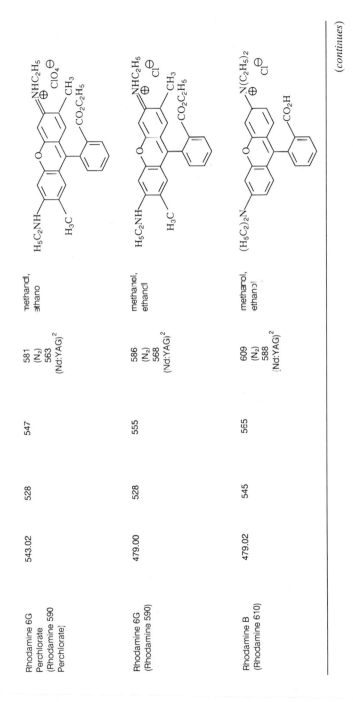

Rhodamine 6G Perchlorate (Rhodamine 590 Perchlorate)	543.02	528	547	581 (N_2) 563 (Nd:YAG)[2]	methanol, ethanol	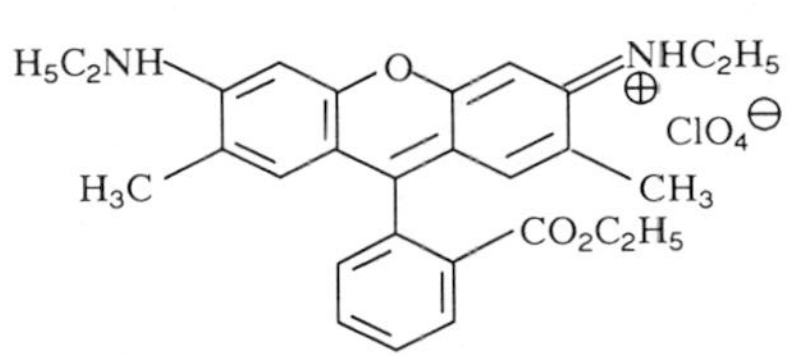
Rhodamine 6G (Rhodamine 590)	479.00	528	555	586 (N_2) 568 (Nd:YAG)[2]	methanol, ethanol	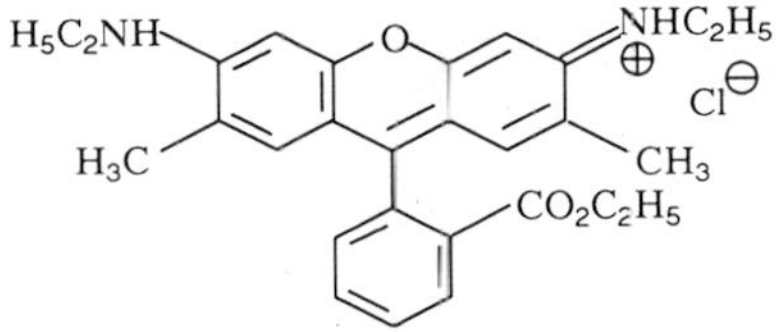
Rhodamine B (Rhodamine 610)	479.02	545	565	609 (N_2) 588 (Nd:YAG)[2]	methanol, ethanol	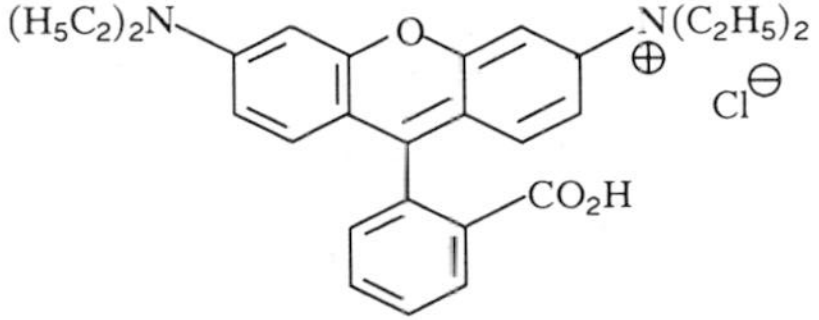

(*continues*)

Appendix of Laser Dyes

Name	Molecular Weight (au)	Maximum Absorption λ(nm)	Maximum Fluorescence λ(nm)	Maximum Lansing λ (nm) (Pump Laser)	Solvents	Molecular Structure
Sulforhodamine B (Kiton Red 620; Xylene Red B)	558.66	556	572	624 (N_2) 589 (Nd:YAG)[2]	methanol, ethanol	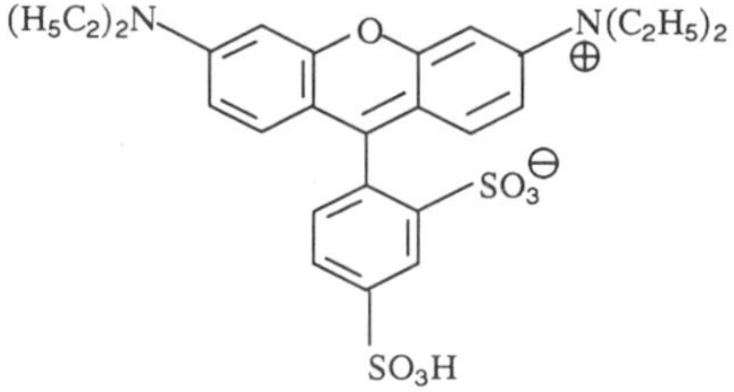
Sulforhodamine 101 (Sulforhodamine 640)	606.00	578	605	649 (N_2) 614 (Nd:YAG)[2]	methanol, ethanol	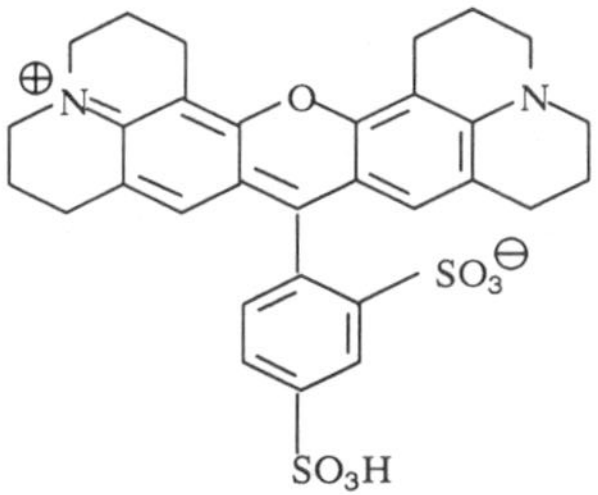
Cresyl Violet Perchlorate (Oxazine 9 Perchlorate; Cresyl Violet 670 Perchlorate)	361.74	593	615	702 (N_2) 641 (Nd:YAG)[2]	methanol	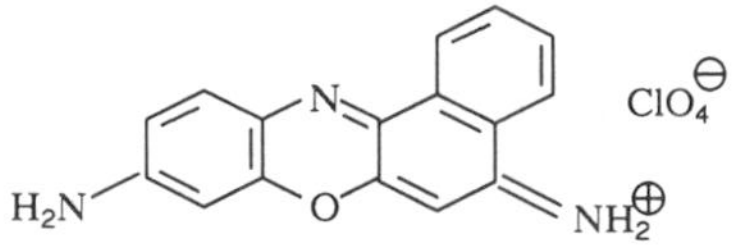
DODC Iodide (DODCI; 3,3'-Diethyloxadi carbocyanine Iodide)	486.35	578	605	647 (Nd:YAG)[2]	methanol	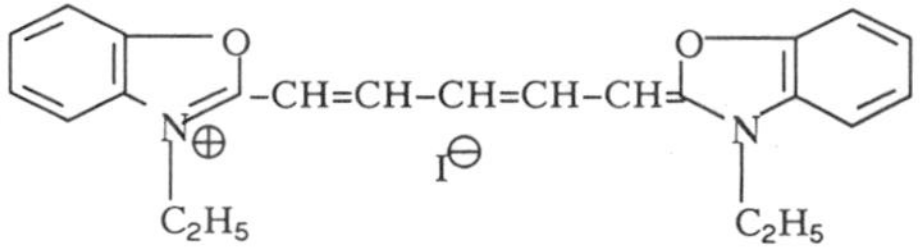

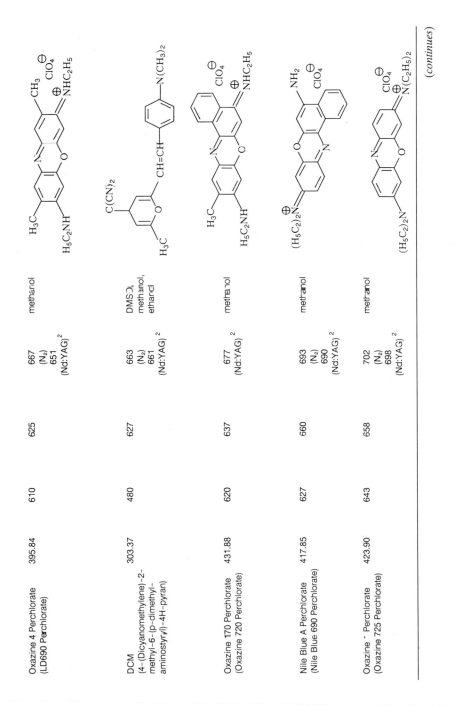

Oxazine 4 Perchlorate (LD690 Perchlorate)	395.84	610	625	667 (N_2) 651 (Nd:YAG)[2]	methanol	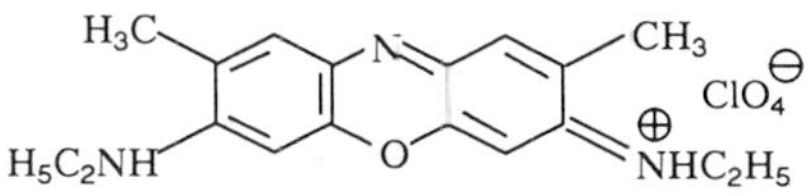
DCM (4-(Dicyanomethylene)-2-methyl-6-(p-dimethyl-aminostyryl)-4H-pyran)	303.37	480	627	663 (N_2) 661 (Nd:YAG)[2]	DMSO, methanol, ethanol	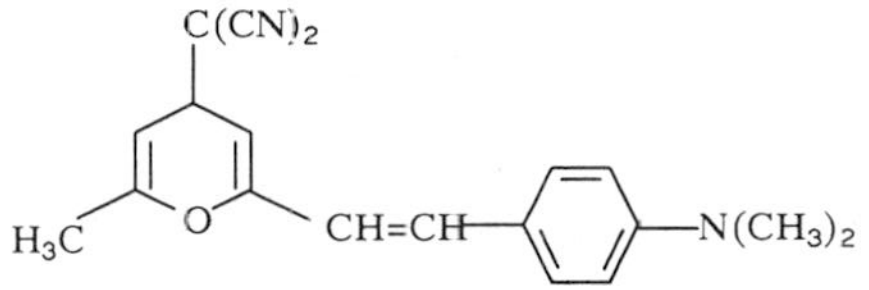
Oxazine 170 Perchlorate (Oxazine 720 Perchlorate)	431.88	620	637	677 (Nd:YAG)[2]	methanol	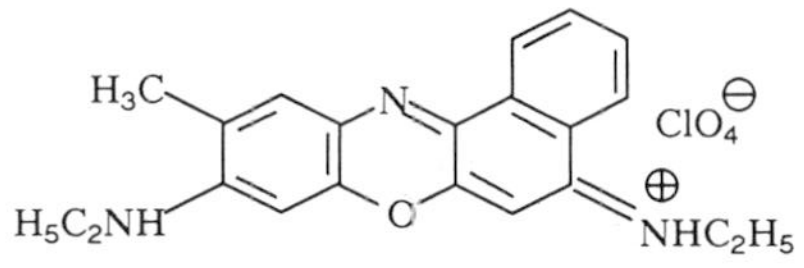
Nile Blue A Perchlorate (Nile Blue 690 Perchlorate)	417.85	627	660	693 (N_2) 690 (Nd:YAG)[2]	methanol	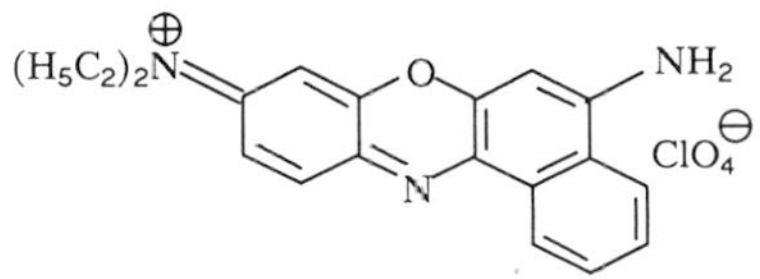
Oxazine 1 Perchlorate (Oxazine 725 Perchlorate)	423.90	643	658	702 (N_2) 698 (Nd:YAG)[2]	methanol	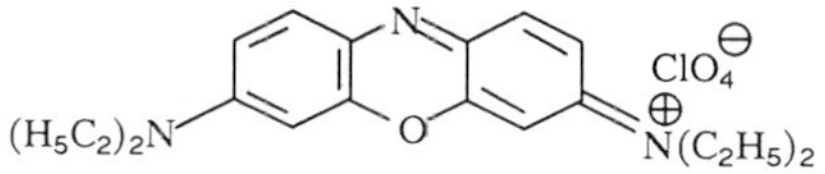

(*continues*)

Appendix of Laser Dyes

Name	Molecular Weight (au)	Maximum Absorption λ(nm)	Maximum Fluorescence λ(nm)	Maximum Lansing λ (nm) (Pump Laser)	Solvents	Molecular Structure
DTDC Iodide (DTDCI; 3,3'-Diethylthiadi carbocyanine Iodide)	518.48	662	679	743 (N_2) 698 (Nd:YAG)[2]	DMSO	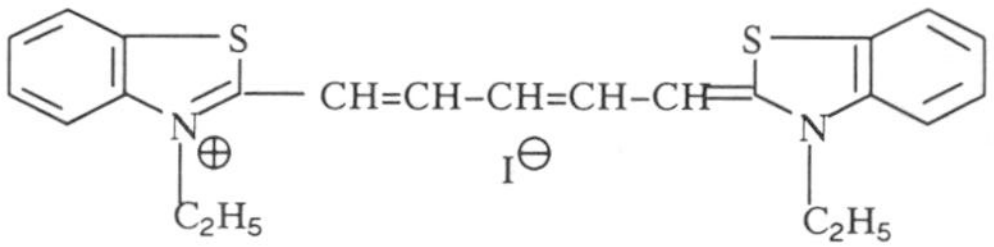
Styryl 7 (2-[4-(4-Dimethylamino-phenyl)-1,3-butadienyl]-3-ethylbenzothiazolium p-Toluenesulfonate)	506.6	560	704	715 (N_2)	methanol	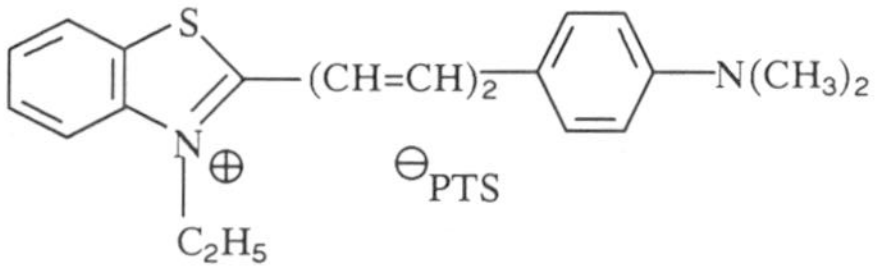
DOTC Iodide (DOTCI; 3,3'-Diethyloxatri-carbocyanine Iodide)	512.39	695	719	762 (Nd:YAG)[2]	DMSO	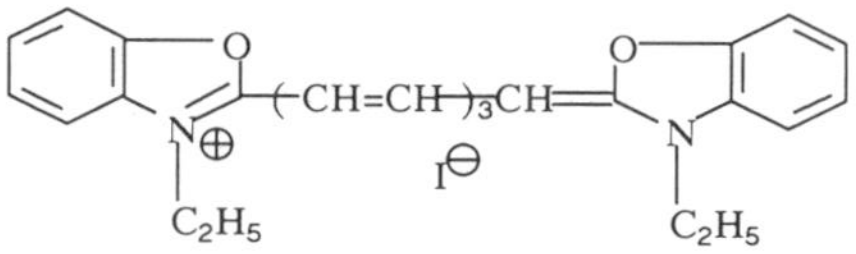
HITC Perchlorate (HITCP; 1,1',3,3,3',3'-Hexamethyl-indotricarbocyanine Perchlorate)	509.05	750	790	837 (N_2) 826 (Nd:YAG)[2]	DMSO	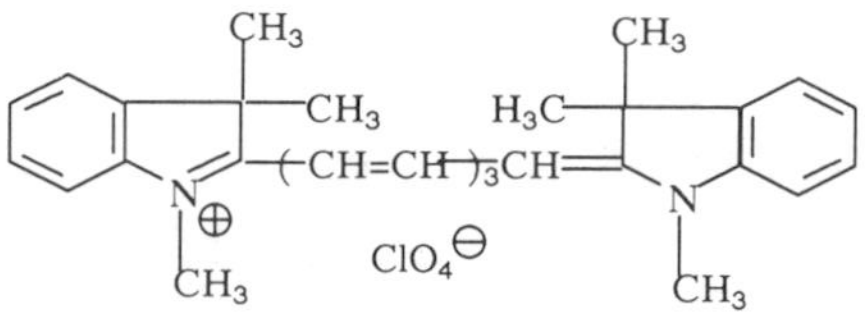

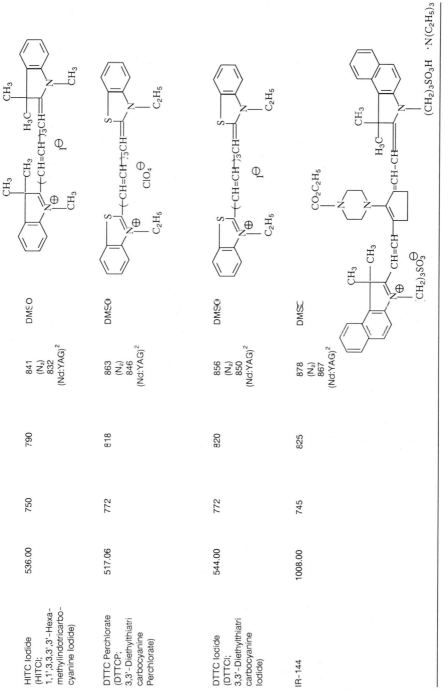

HITC Iodide (HITCI; 1,1',3,3,3',3'-Hexamethylindotricarbocyanine Iodide)	536.00	750	790	841 (N_2) 832 (Nd:YAG)[2]	DMSO	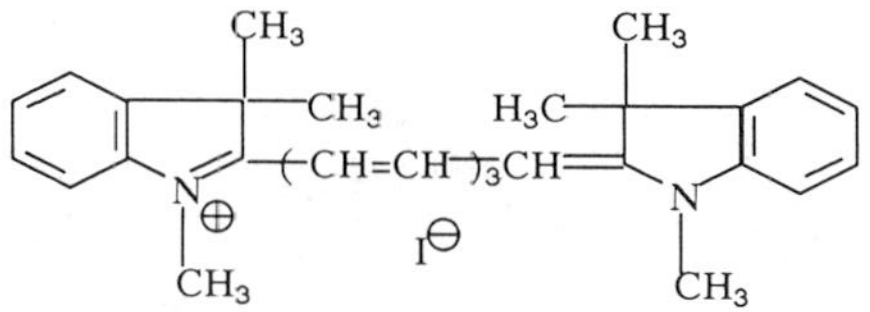
DTTC Perchlorate (DTTCP; 3,3'-Diethylthiatri carbocyanine Perchlorate)	517.06	772	818	863 (N_2) 846 (Nd:YAG)[2]	DMSO	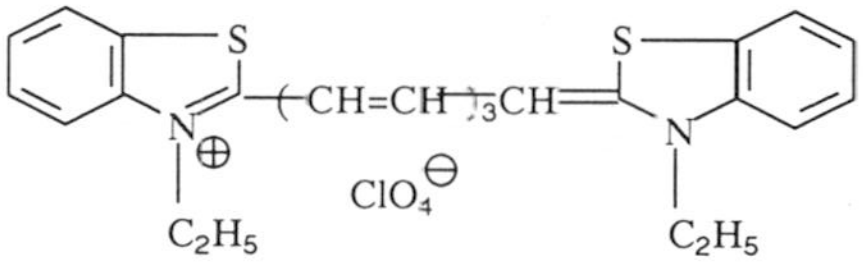
DTTC Iodide (DTTCI; 3,3'-Diethylthiatri carbocyanine Iodide)	544.00	772	820	856 (N_2) 850 (Nd:YAG)[2]	DMSO	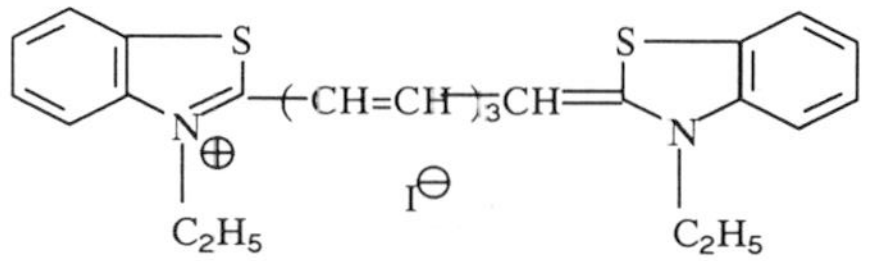
IR-144	1008.00	745	825	878 (N_2) 867 (Nd:YAG)[2]	DMSO	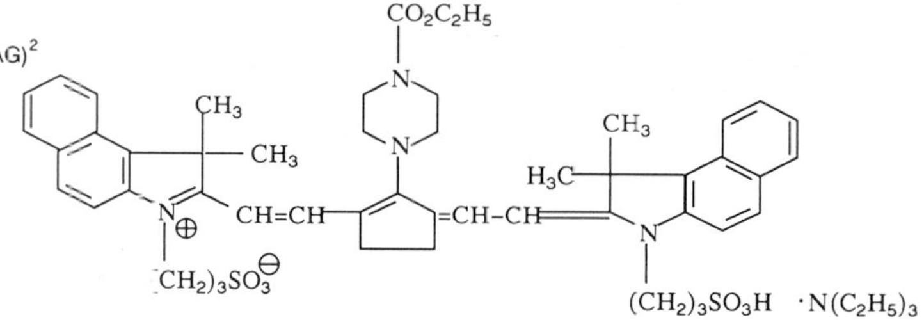

(continues)

Appendix of Laser Dyes

Name	Molecular Weight (au)	Maximum Absorption λ(nm)	Maximum Fluorescence λ(nm)	Maximum Lansing λ (nm) (Pump Laser)	Solvents	Molecular Structure
HDITC Perchlorate (HDITCP; 1,1',3,3,3',3'-Hexamethyl-4,4',5,5'-dibenzo-2,2'-indotricarbocyanine Perchlorate)	609.17	780	828	913 (N_2) 886 (Nd:YAG)[2]	DMSO	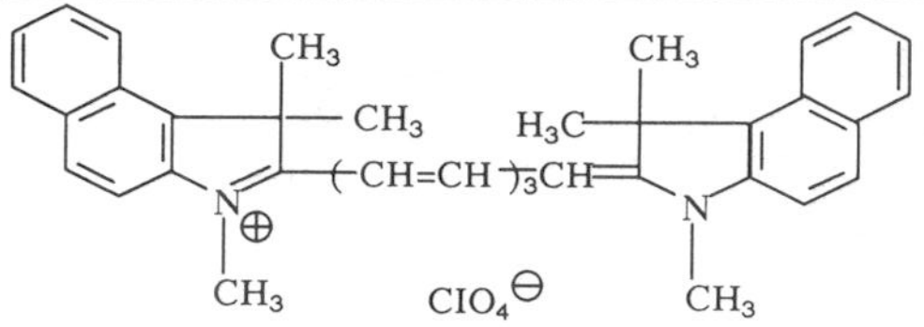
IR 140	779.00	826	882	943 (N_2) 897 (Nd:YAG)[2]	DMSO	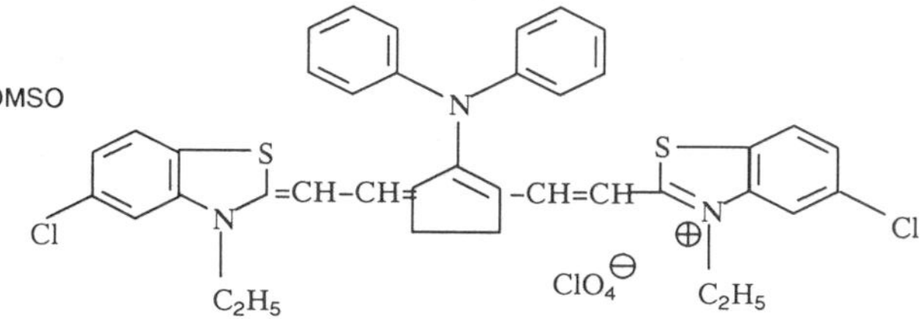
IR-132	954.55	832	905	918 (N_2) 914 (Nd:YAG)[2]	DMSO	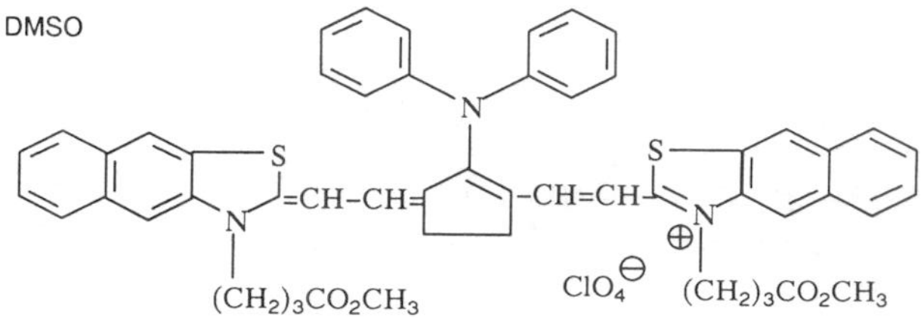

IR-125	774.00	795	833	914 (N_2) 915 (Nd:YAG) [2]	DMSO	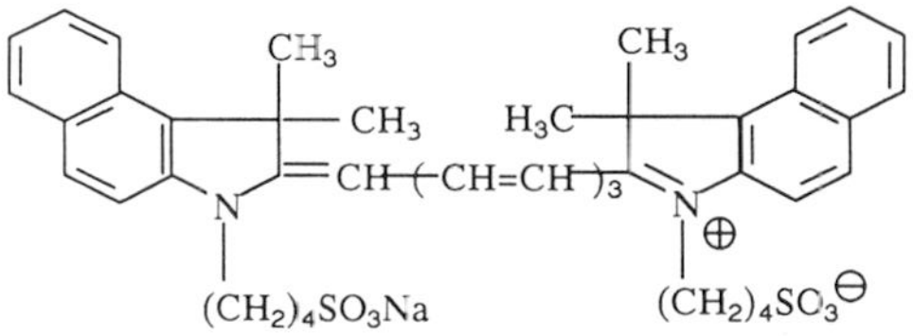

Notes: Absorption and emission data obtained with solvent listed in first place. For further alternative solvents, see Drexhage (1977) and Maeda (1984).
The information given in this table has been reproduced from KODAK Laser Dyes (Birge, 1987) courtesy of Eastman Kodak Company.
[1] Third Harmonic from Nd:YAG at 355 nm.
[2] Second Harmonic from Nd:YAG at 532 nm.

References:
Birge, R.R. (1987). KODAK Laser Dyes. KODAK Publication JJ-169. Eastman Kodak Company, Rochester, N.Y.
Drexhage, K.H. (1977). Structure and Properties of Laser Dyes. In Dye Lasers (Schäfer, F.P., ed.). Springer-Verlag, New York, pp. 144-193.
Maeda, M. (1984). Laser Dyes. Academic Press, New York.

Abbreviations used in Chapter 3:

DASBTI	2-(p-dimethylaminostyryl)-benzothiazo ylethyl iodide
DaQTeC	1,1′-diethyl-13-acetoxy-2,2′-quinotetracarbocyanine iodide
DCM	4-dicyanomethylene-2-methyl-6-p-dimethylaminostyryl-4H-pyran
DCCI	1,1′-diethyl-2,4′-carbocyanine iodide
= DCI	
DDCI*	1,1′-diethyl-2,2′-dicarbocyanine iodide
= DDI	
DOCI	3,3′-diethyl oxacarbocyanine iodide
DODCI	3,3′-diethyl oxadicarbocyanine iodide
DOTCI	3,3′-diethyl oxatricarbocyanine iodide
DQOCI	1,3′-diethyl-4,2′-quinolyoxacarbocyanine iodide
HICI	1,1′3,3,3′,3′-hexamethylindocarbocyanine iodide
HITCI	1,1′,3,3,3′,3′-hexamethylindotricarbocyanine iodide
MNA	2-methyl-4-nitroaniline

*Cryptocyanine	1,1′-Diethyl-4,4′-carbocyanine iodide
DTDCI	3,3′-Diethyl thiadicarbocyanine iodide
DQTCI	1,3′-Diethyl-4,2′-quinolthiacarbocyanine iodide

INDEX

D

Quantum Electronics—Principles and Applications

Edited by: Paul F. Liao, *Bell Communications Research, Red Bank, New Jersey*
Paul L. Kelley, *Lincoln Laboratory, Massachusetts Institute of Technology, Lexington, Massachusetts*

N.S. Kapany and J.J. Burke, *Optical Waveguides*
Dietrich Marcuse, *Theory of Dielectric Optical Waveguides*
Benjamin Chu, *Laser Light Scattering*
Bruno Crosignani, Paolo Di Porto, and Mario Bertolotti, *Statistical Properties of Scattered Light*
John D. Anderson, Jr., *Gasdynamic Lasers: An Introduction*
W.W. Duly, *CO_2 Lasers: Effects and Applications*
Henry Kressel and J.K. Butler, *Semiconductor Lasers and Heterojunction LEDs*
H.C. Casey and M.B. Panish, *Heterostructure Lasers: Part A, Fundamental Principles; Part B, Materials and Operating Characteristics*
Robert K. Erf, editor, *Speckle Metrology*
Marc D. Levenson, *Introduction to Nonlinear Laser Spectroscopy*
David S. Kliger, editor, *Ultrasensitive Laser Spectroscopy*
Robert A. Fisher, editor, *Optical Phase Conjugation*
John F. Reintjes, *Nonlinear Optical Parametric Processes in Liquids and Gases*
S.H. Lin, Y. Fujimura, H.J. Neusser, and E.W. Schlag, *Multiphoton Spectroscopy of Molecules*
Hyatt M. Gibbs, *Optical Bistability: Controlling Light with Light*
D.S. Chemla and J. Zyss, editors, *Nonlinear Optical Properties of Organic Molecules and Cyrstals, Volume 1, Volume 2*
Marc D. Levenson and Saturo Kano, *Introduction to Nonlinear Laser Spectroscopy, Revised Edition*
Govind P. Agrawal, *Nonlinear Fiber Optics*
F.J. Duarte and Lloyd W. Hillman, editors, *Dye Laser Principles: With Applications*

Yoh-Han Pao, *Case Western Reserve University, Cleveland, Ohio*, Founding Editor 1972–1979